Plasma Membrane Oxidoreductases in Control of Animal and Plant Growth

NATO ASI Series

Advanced Science Institutes Series

A series presenting the results of activities sponsored by the NATO Science Committee, which aims at the dissemination of advanced scientific and technological knowledge, with a view to strengthening links between scientific communities.

The series is published by an international board of publishers in conjunction with the NATO Scientific Affairs Division

A	**Life Sciences**	Plenum Publishing Corporation
B	**Physics**	New York and London
C	**Mathematical and Physical Sciences**	Kluwer Academic Publishers Dordrecht, Boston, and London
D	**Behavioral and Social Sciences**	
E	**Applied Sciences**	
F	**Computer and Systems Sciences**	Springer-Verlag
G	**Ecological Sciences**	Berlin, Heidelberg, New York, London,
H	**Cell Biology**	Paris, and Tokyo

Recent Volumes in this Series

Series A: Life Sciences

Plasma Membrane Oxidoreductases in Control of Animal and Plant Growth

Edited by

Frederick L. Crane and
D. James Morré

Purdue University
West Lafayette, Indiana

and

Hans Löw

Karolinska Institute
Stockholm, Sweden

Springer Science+Business Media, LLC

Proceedings of a NATO Advanced Research Workshop on
Plasma Membrane Oxidoreductases in Control of Animal and Plant Growth,
held March 21–25, 1988,
in Cordoba, Spain

Library of Congress Cataloging in Publication Data

NATO Advanced Research Workshop on Plasma Membrane Oxidoreductases in
Control of Animal and Plant Growth (1988: Corboda, Spain)
    Plasma membrane oxidoreductases in control of animal and plant growth /
edited by Frederick L. Crane and D. James Morré and Hans Löw.
        p.   cm.—(NATO ASI series. Series A, Life sciences; v. 157)
    "Proceedings of a NATO Advanced Research Workshop on Plasma Membrane
Oxidoreductases in Control of Animal and Plant Growth, held March 21–25, 1988
in Cordoba, Spain"—T.p. verso.
    "Published in cooperation with NATO Scientific Affairs Division."
    Includes bibliographies and index.
    ISBN 978-1-4684-8031-3   ISBN 978-1-4684-8029-0 (eBook)
    DOI 10.1007/978-1-4684-8029-0
    1. Cell membranes—Congresses. 2. Cellular control mechanisms—Congress-
es. 3. Growth—Congresses. 4. Oxidation-reduction reaction—Congresses. I.
Crane, Frederick L. II. Morré, D. James, 1935-   . III. Löw, Hans. IV. North Atlantic Treaty Organization. Scientific Affairs Division. V. Title. VI. Series.
QH601.N36   1988   88-39078
574.87′5—dc19   CIP

© 1988 Springer Science+Business Media New York
Originally published by Plenum Press New York in 1988
Softcover reprint of the hardcover 1st edition 1988

PREFACE

     The objective of this workshop was to examine the nature
of plasma membrane electron transport and how this electron
transport contributes to growth of cells. The workshop came at
a time when the study of the plasma membrane oxidoreductase
activity was beginning to attract more widespread attention
from researchers working with both plants and animals. The
rapid response of c fos and c myc Proto-oncogene to stimulation
of plasma membrane redox activity by external oxidants under-
scores a potential role of plasma membrane oxidoreductases in
growth control. Other experiments with isolated endosomes indicate emerging roles in endocytosis and lytic processes.
Primary attention was focused on transplasma membrane electron
transport which brings about the oxidation of cytosolic, NADH,
NADPH or other substrates by electron flow across the plasma
membrane to external oxidants including ferric iron, semidehydroascorbate or oxygen.

     A major theme in the workshop was the relation of this
electron flow to pH changes of the cytoplasm or the transfer
of protons to the external medium. The presence and role of
other oxidoreductases in the plasma membrane was documented,
especially in regard to peroxide production. In plant cells
this may contribute to cellular defense against invading para-
sites. A corresponding function in animals has been long known
and extensively discussed but was beyond the scope of this
workshop.

     The possible role of the electron transport systems in
growth and membrane transport provided significance to the
discussion. Basic observations to define these effects were
presented. Furthermore a beginning was made in defining the
mechanisms by which the redox system could contribute to control of cell function and transmembrane signalling. The possible role of the redox systems as a site of action for growth
factors, antitumor drugs and herbicides also was considered in
the context of growth control. The transplasmalemma redox enzyme also was discussed as essential for iron uptake from
transferrin in animals and for iron uptake by plants.

     This was the first meeting on this subject which brought
together researchers from both plant and animal fields from
many nations. The most important contribution may have been
the wide agreement on the presence of oxidoreductase enzymes
in the plasma membrane. This agreement, plus the widespread
range of physiological responses to stimulation or inhibition
of plasma membrane redox activities, in particular the trans-

membrane enzyme, provides an overture to these enzymes acting as cell surface constituents important for growth control and other vital cellular functions.

Imperatives for future investigation include focused studies to define redox chain constituents, analysis of how growth and the plasma membrane oxidoreductase are coupled, and the preparation of purified enzymes and antibodies.

F.L. Crane
D.J. Morré
H. Löw

ACKNOWLEDGEMENTS

The concept of significant oxidoreductase function in the plasma membrane has been slow to develop and has been controversial. We believe that the essential support of this workshop by the Scientific Affairs Division of the North Atlantic Treaty Organization was foresighted and courageous. We are grateful to The Secretary General of Scientific Affairs, Dr Henry Durant and the Advisory Panel for Advanced Research Workshops for their support and to Dr Craig Sinclair, Director of the NATO Advanced Research Workshop Program. Without this support the meeting would not have been possible. Important support was also provided by the University of Cordoba, Junta de Andalucia, Disputacion Provincial de Cordoba, Caja Provincial de Ahorros de Cordoba and Monte de Piedad y Caja de Ahorros de Cordoba and Farmitalia Carlo Erba.

A vital factor in the success of this meeting was the excellent planning and coordination of program and facilities by Prof. Placido Navas and the local arrangement committée from the University of Cordoba. The wholehearted cooperation of the administrator and staff of the Parador La Arruzafa provided an outstanding environment for scientific interaction by the participants. We also wish to acknowledge the support of the other members of the organizing committée, Dr F. Arcamone, Prof. M. Böttinger, Prof. W. Lucas and Prof. G. Garcia Herdugo.

CONTENTS

REDOX AND TRANSPORT

CONTROL OF CELL FUNCTION AND
DEVELOPMENT

MECHANISMS FOR REDOX CONTROL

POSTER ABSTRACTS

ELECTRON TRANSPORT

INTRODUCTORY REMARKS

F.L. Crane

Department of Biological Science
Purdue University
W. Lafayette IN 47907 USA

The concept of plasma membrane electron transport is not new. Various facets have been studied over the years especially in relation to ion-transport, fatty acid desaturation, hormonal control, and peroxidative defense mechanisms[1]. There has been evidence for some time that blue light control of plant growth can be expressed in the plasma membrane but conclusive evidence for this location has developed slowly[2]. Evidence that impermeable oxidants could stimulate cell division presented over 40 years ago was apparently never followed up[3]. Consequently the general idea that plasma membrane electron transport can control growth and development has developed only recently[4].

The diversity and location of oxidoreductase enzymes in plasma membranes is slowly being defined and progress has been more rapid with development of methods for preparation of highly purified plasma membranes[5,6]. The idea of loosely bound peroxidase on plant cells is of long standing[7]. Induction of peroxide generating dehydrogenase in neutrophils is now well established and under intensive investigation[8,9]. A system for ferric ion reduction at the cell surface is induced in some plant roots in response to iron deficiency[10]. NADH cytochrome b_5 reductase and cytochrome b_5 have been clearly demonstrated in erythrocyte membranes, but their presence in other plasma membranes has been subject to considerable controversy[11]. In addition to cytochrome b_5, cytochromes P450 and P420 have been observed in some plant and animal membranes but they have not been associated with any oxidoreduction reactions[12]. Only in the case of the blue light induced reactions of some plant plasma membranes has there been evidence for reduction and oxidation of flavin and cytochromes in the plasma membrane. Cyanide insensitive NADH oxidases have been observed in both plant and animal membranes but their electron carriers are unknown[13,14]. Furthermore NADH diferric transferrin reductase has been demonstrated in liver plasma membranes[15].

The orientation of oxidoreductases in the plasma membrane has been defined in some cases. They can be oriented exclusively on the inner surface, on the outer surface or across the membrane. Kant and Steck[16] showed that the NADH cytochrome b_5 reductase and cytochrome b_5 are exclusively on the cytosolic side of erythrocyte membranes. Peroxidase is an extrinsic protein on the outer surface of plant cells. The presence of a transplasma membrane electron transport system was first proposed by E.G. Baron and L. Hoffman[17] in 1929 on the basis that impermeable redox dyes could be reduced by starfish eggs and that the redox potential of the active dyes was much above the potential of dyes which could be reduced internally. Later study of ferricyanide reduction by erythrocytes and erythrocyte membranes defined the transmembrane enzyme as an NAD(H) ferricyanide reductase[18,19]. This type of transplasma membrane electron transport enzyme has been found in all cells which have been examined. The activity has been studied by following the reduction of ferricyanide and other impermeable electron acceptors at the cell surface and in isolated plasma membrane. It has also been demonstrated by histochemical staining using the Hatchet Brown procedure[20]. Two other transplasma membrane electron transport systems are known. One is the ferric reductase produced in roots of certain plants under iron deficiency[10]. The other is the NADPH oxidase induced in neutrophil plasma membrane when bacteria are captured and engulfed[8].

Early studies of the plasma membrane dehydrogenases showed that they were sensitive to low concentrations of certain hormones which suggested a relation to control of cell function! The discovery by Ellem and Kay[21] that ferricyanide stimulates the growth of melanoma cells opened up a new consideration of the plasma membrane electron transport as a participant in control of cell growth. It has since been shown that ferricyanide and other external oxidants which can act as electron acceptors can stimulate the growth of several different animal cell lines in the absence of fetal calf serum. These include HeLa, SV40 infected fetal rat liver, pineal, swiss 3T3, Ehrlich ascites and Hep G2 cells[22]. Diferric transferrin and other iron complexes have long been known to stimulate serum free cell growth and the effect has been related to iron supply to the cells. The stimulation of growth by oxidants which do not contain iron, eg indigosulfonates, indicated that the growth stimulation is not based exclusively on iron supply to the cells, but that the electron transport is also involved[23].

Auxin stimulated respiration in plants is inhibited by ferricyanide[24] which suggests that an auxin sensitive oxidase in the plasma membrane may be involved in growth control[14]. Ferricyanide and other impermeable high potential oxidants inhibit the growth of plant cells but are not cytoxic[25]. Thus, in contrast to animal cells the plasma membrane electron transport to artificial external oxidants inhibits growth of plant cells.

The basis for the growth effects of external electron acceptors for the plasma membrane electron transport remains to be found. Electron transport has been shown to activate proton release from cells[26,27], increase internal pH, increase cytosolic NAD^+[28] and change membrane potential[29]. Evidence for increase in cytosolic calcium has also been presented[30]. The response of other known messenger systems, such as phosphotidylinositol turnover, remains to be tested. There is also evidence that the plasma membrane electron transport can control transport of ions or amino acids into the cell[31].

The dramatic development in the understanding of receptor signalling systems on the plasma membrane introduces a new framework in which the plasma membrane oxidoreductase systems can contribute to the interplay of vital messenger and transport systems within the membrane.

REFERENCES

1. H. Löw and F.L. Crane, Redox functions in plasma membranes. Biochim Biophys Acta 515:141 (1978).

2. H. Senger, Blue Light Effects in Biological Systems, Springer-Verlag Berlin 1984.

3. M.M. Brooks, Activation of eggs by oxidation reduction indicators. Science 106:320 (1947).

4. F.L. Crane, I.L. Sun, M.G. Clark, C. Grebing and H. Löw, Transplasma membrane redox systems in growth and development. Biochim Biophys Acta 811:233 (1985).

5. F.L. Crane, H. Löw and M.G. Clark, Plasma membrane redox enxymes in The Enzymes of Biological Membranes Vol 4, ed. A.N. Martonosi, Plenum, N.Y. (1985) p.465.

6. D.J. Morré, P. Navas and F.L. Crane, Isolation, Purification, Evaluation and Quantitation of Plasma Membranes from Tissues and Cultured Cells: Methods and Applications to Cell Free Systems for Analysis of Growth Processes, in Redox Functions of the Eukaryotic Plasma Membranes. J. Ramirez ed. Consejo Superior de Investigaciones Cientificas, Madrid 1987, p. 93.

7. I.M. Møller and W. Lin, Membrane bound dehydrogenases in higher plant cells. Ann Rev Plant Physiol 37:309 (1986).

8. R. Mitchell, The lethal oxidase of leucocytes. Trends Biochem Sci 8:117 (1983).

9. L.M. Henderson, J.B. Chappell and O.T.G. Jones, Internal pH changes associated with the activity of NADPH oxidase of human neutrophils. Biochem J 251:563 (1988).

10. H.F. Bienfait, Regulated redox processes at the plasma-
    lemma of plant root cells and their function in iron
    uptake. J Bioenerg Biomemb 17:73 (1985).

11. H. Goldenberg, Plasma membrane redox activities.
    Biochim Biophys Acta 694:203 (1982).

12. G. Bruder, A. Bretscher, W.W. Franks and E.-D. Jarasch,
    Plasma membranes from intestinal microvilli and ery-
    throcytes contain cytochromes b_5 and P420. Biochim
    Biophys Acta 600:739 (1980).

13. D.P. Gayda, F.L. Crane, D.J. Morré and H. Löw, Hormone
    effects on NADH oxidizing enzymes of plasma membranes
    of rat liver. Proc. Indiana Acad Sci 86:385 (1977).

14. D.J. Morré, P. Navas, C. Penel and F. Castillo, Auxin
    stimulated NADH oxidase (semidehydroascorbate reduc-
    tase) of soybenn plasma membrane: Role in acidification
    of the cytoplasm. Protoplasma 133:195 (1986).

15. I.L. Sun, P. Navas, F.L. Crane, D.J. Morré and H. Löw,
    NADH diferric transferrin reductase activity in liver
    plasma membranes. J Biol Chem 262:15915 (1987).

16. J.A. Kant and T.L. Steck, Cation impermeable inside out
    and right side out vesicles from human erythrocyte
    membranes. Nature 240:26 (1972).

17. E.S. Guzman Barron and L.A. Hoffman, The catalytic
    effect of dyes on the oxygen consumption of living
    cells. J Gen Physiol 13:483 (1929).

18. R.K. Mishra and H. Passow, Induction of intracellular
    ATP synthesis by extra cellular ferricyanide in human
    blood cells. J Memb Biol 1:214 (1969).

19. C. Grebing, F.L. Crane, H. Löw and K. Hall, A trans-
    membrane NADH dehydrogenase in human erythrocyte memb-
    ranes. J Bioenerg Biomemb 16:517 (1984).

20. D.J. Morré, E.L. Vigil, C. Frantz, H. Goldenberg and
    F.L. Crane, Cytochemical demonstration of glutaralde-
    hyde resistant NADH ferricyanide oxidoreductase in rat
    liver plasma membranes and Golgi apparatus. Cytobiolo-
    gie 18:213 (1978).

21. K.A.O. Ellem and G.F. Kay, Ferricyanide can replace
    pyruvate to stimulate growth and attachment of serum
    restricted human melanoma cells. Biochem Biophys Res
    Communs 112:183 (1983).

22. F.L. Crane, H. Löw, I.L. Sun, P. Navas and D.J. Morré,
    Redox control of cell growth in Redox Functions of the
    Eukaryotic Plasma Membrane, ed. J. Ramirez, Consejo
    Superior de Investigaciones Cientificas, Madrid 1987,
    p.3.

23. I.L. Sun, F.L. Crane, H. Löw and C. Grebing, Trans-plasma membrane redox stimulates HeLa cell growth. Biochem Biophys Res Communs 125:649 (1984).

24. V.V. Polevoy and T. Salamatova, Auxin, proton pump and cell trophics, in E. Marré and O. Ciferri eds. Regulation of Cell Membrane Activities in Plants, Elsevier, Amsterdam 1977, p. 209.

25. F.L. Crane, R. Barr, T.A. Craig and D.J. Morré, Trans-plasma membrane electron transport in relation to cell growth and iron uptake. J Plant Nutrition in press 1988.

26. T.L. Dormandy and Z. Zardy, The mechanism of insulin action: the immediate electrochemical effects of insulin on red cell systems. J Physical 180:684 (1965).

27. I.L. Sun, F.L. Crane, C. Grebing and H. Löw, Properties of a transplasma membrane electron transport system in HeLa cells. J Bioenerg Biomemb 16:583 (1984).

28. P. Navas, I.L. Sun, D.J. Morré and F.L. Crane, Decrease of NADH in HeLa cells in the presence of transferrin or ferricyanide. Biochem Biophys Res Communs 135:110 (1986).

29. V.A. Novak and N.G. Ivankina, Influence of nitroblue tetrazolium on membrane potential and ion transport in water thymes. Soviet Plant Physiol 30:845 (1983).

30. H. Löw, F.L. Crane, E.J. Partick and M.G. Clark, Adrenergic stimulation of trans-sarcolemma electron efflux in perfused heart: Possible regulation of Ca^{++} channels by a sarcolemma redox system. Biochim Biophys Acta 844:142 (1985).

31. H.N. Christensen, Amino acid transport in Redox Functions of the Eukaryotic Plasma Membrane, ed. J. Ramirez, Consejo Superior de Investigaciones Cientificas, Madrid 1987, p. 117.

MODULATION OF REDOX REACTIONS INVOLVED IN DNA SYNTHESIS

BY OXYGEN AND ARTIFICIAL ELECTRON ACCEPTORS

Charles E. Wenner, Anthony Cutry, Alan
Kinniburgh, L.D. Tomei, and Kirk J. Leister

Roswell Park Memorial Institute
Buffalo, NY 14263

INTRODUCTION

The study of the oxygen requirements for DNA synthesis
in C3H 10T1/2 mouse embryonic fibroblasts offers several
advantages for the understanding of mitogen-induced cell
proliferation. Firstly, cell cycle kinetics have been well
defined, and these cells are capable of being staged in Go/Gl
so that the role of oxygen in different phases of the cell
cycle can be evaluated. Further, these cells are capable of
withstanding oxygen deprivation conditions necessary for
removal of trace levels of dissolved oxygen which would mask
the correct assessment of needs for oxygen.

The oxygen dependencies of mitogen-induced DNA synthesis
by post confluent C3H 10T1/2 fibroblasts and of factors
critical to DNA synthesis were initially examined. The
requirement of oxygen for early events in cell proliferation
such as enhanced (Na+/K+)ATPase activity and increases in
amino acid incorporation were factors under consideration
since Mueller reported an oxygen requirement for amino acid
transport (1). The ability of postconfluent quiescent cells to
incorporate tritiated thymidine into the acid insoluble
fraction when cells were transferred to a gas phase of 95%N2-
5%CO2 was also compared with that of logarithmic growing
fibroblasts. Comparison of log phase cells with cells staged
in early Gl enabled us to evaluate the dependence of oxygen
in cells at the Gl stage of the cell cycle. It might be
anticipated that cell cycle progression may be limited by
oxygen-dependent biosynthesis of pyrimidine or purine
nucleotides which occurs in the Gl phase of the cycle. It
should be noted that de novo synthesis of uridine
monophosphate which requires dihydroorotate dehydrogenase ,
and deoxyribonucleotide biosynthesis which requires
ribonucleotide diphosphate reductase, would be favored by
aerobic conditions (2,3).

The contribution of oxidative energy production to
proliferative events was also studied by comparisons of DNA

synthesis in the presence or absence of oxygen or with respiratory inhibitors.

Since Crane et al (4) had reported that impermeant oxidants such as ferricyanide can act as growth stimulants, the effect of an external electron acceptor on growth related proto-oncogene expression was examined to identify the components of the signal transducing pathways responsible for the activation of these genes.

RESULTS AND DISCUSSION

Oxygen Dependency of DNA Synthesis and of Related Biochemical Processes . Initially, the effects of oxygen deprivation on Dihydroteleocidin B (DHTB) -induced early responses required for DNA synthesis in postconfluent C3H 10T1/2 cells were examined. In view of the requirement of oxygen for amino acid transport (1) and since mitogens enhance amino acid incorporation and protein synthesis, it was of interest to compare DHTB-induced H3-leucine incorporation in the presence and absence of oxygen. Since amino acid transport has been closely associated with (Na+/K+)ATPase activity in mouse fibroblasts, comparisons of Rb+ (a K+ analog) uptake with leucine incorporation were also made.

In Fig. 1, an experiment is described where DHTB- induced 3H-dThd incorporation , 3H-leucine incorporation, and ouabain-sensitive 86Rb+ uptake by postconfluent, quiescent cells were compared under aerobic and anaerobic conditions. Cumulative

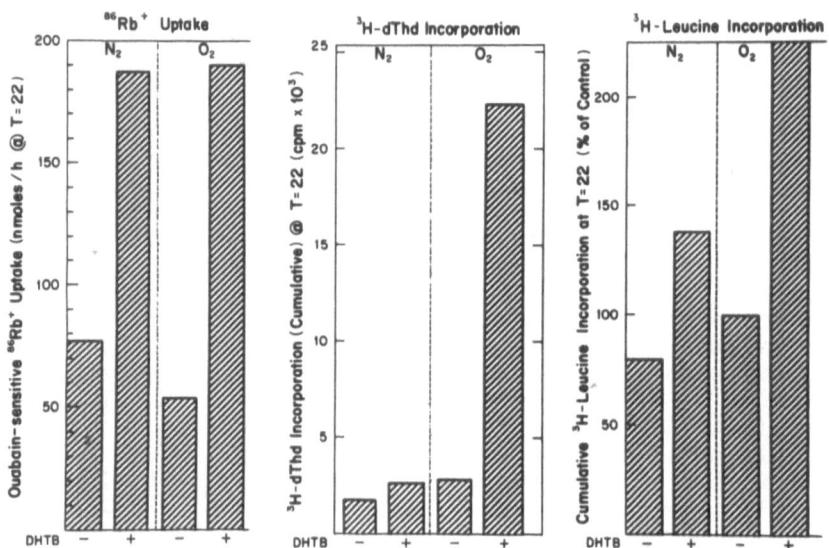

Figure 1. Postconfluent, quiescent C3H 10T ½ cells ± DHTB (10^{-8}M) were incubated with a gas phase of 95% air-5% CO_2 or 95% N_2-5% CO_2 for a period of 22 h. Cumulative ^{3}H-dThd and H-leucine incorporation were measured, and at T=22, ^{86}Rb+ uptake was measured in the presence or absence of ouabain following a one-hour incubation period to determine ouabain-sensitive (Na+/K+)-ATPase activity.

Table 1

SERUM-STIMULATED CELLS IN S-PHASE
ARE CAPABLE OF DNA SYNTHESIS UNDER ANAEROBIC CONDITIONS

Cumulative ^{3}H-Thymidine Incorporation @ T = 26 (cpm)

Gas phase	Control	% of aerobic phase	*TPA	% of aerobic phase
95% Air–5% CO_2	36,887 ±3,011		319,533 ±11,939	
95% N_2–5% CO_2	116,356 ±8,256	315%	295,622 ± 8,798	92.5%

Post confluent cells were subjected to a media change with 5% fetal calf serum. The cells were incubated for 24 h to allow cells to enter S and G_2 periods. At that time, cells subjected to anaerobiosis were pre-equilibrated in 95% N_2–5 CO_2. Then, ^{3}H-thymidine ± TPA (10^{-7}M) was introduced to cells in oxygen free environment as well as those in 95% air–5% CO_2. After 26 h incubation at 37°C, the reaction was stopped and incorporation of ^{3}H-thymidine into the acid-insoluble fraction was measured.

^{3}H-dThd and ^{3}H-leucine incorporations were measured following 22h of incubation. At T=22, 86Rb+ uptake was measured following a one-hour pulse in the presence and absence of ouabain. The marked stimulation of 3H-dThd incorporation induced by DHTB was completely eliminated when oxygen was removed. Similarly, DHTB induced 3H-leucine incorporation into the acid insoluble fraction was appreciably less under anaerobic conditions than under aerobic conditions. In contrast, ouabain-sensitive 86Rb+ uptake was similar under aerobic conditions and anaerobic conditions. This was also the case when rubidium uptake was measured at T=0.

It should be pointed out that glucose utilization was accelerated under anaerobic conditions. The values of glucose utilized in mg per cent during the 22 h incubation period were as follows: N2/CO2 control 17.1; + DHTB 45.3; air/CO2 control 9.6; +DHTB 33.3. Thus, a significant Pasteur effect was observed which suggests that energy deficiency was not the likely cause of the failure of DHTB to stimulate DNA synthesis under anaerobic conditions.

Oxygen Dependence of Gl Phase of Cell Cycle but Independence of Later Stages   The use of mitogen-stimulated postconfluent cells was examined next, and as seen in an experiment described in Table 1, the incorporation of 3H-dThd into the acid insoluble fraction was examined following transfer of these fibroblasts to an anaerobic gas phase. Cumulative thymidine incorporation was higher in control cells under anaerobic conditions suggestive that cells remained in S phase for a longer period in the absence of oxygen. Further, the phorbol ester tumor promoter, 12-O-tetradecanoylphorbol-13-acetate (TPA) stimulated thymidine incorporation under anaerobic conditions as well as under aerobic conditions. This experiment suggests that if cells are actively synthesizing DNA, deprivation of oxygen does not prevent completion of the cycle. In view of these findings and

Table 2

EFFECT OF ROTENONE ON DHTB INDUCED CELL RESPONSES

Additions @ T=0	[86]Rb+ Uptake @ T=25 nmoles/hr/10[6] cells		^{3}H-Leucine @ T=20 cpm	^{3}H-dThd @ T=25 cpm	Glucose in the Medium	
	Ouabain (10[-3] M)				@ T=0 mg/100 ml	@ T=25 mg/100 ml
	−	+				
Control	203 ±44	169 ± 3	4,639 ± 283	930 ± 82	104.1 ± 2.3	89.3 ± 2.6
+ DHTB (10[-8] M)	395 ± 5	226 ± 7	13,996 ± 393	24,320 ±1,502		68.6 ± 0.2
+ Rotenone (2 μM)	253 ± 1	172 ± 9	4,584 ± 144	887 ± 130		62.8 ± 3.4
+ DHTB + Rotenone	407 ±10	201 ± 2	11,061 ± 593	32,315* ±4,279		35.5 ± 2.4
+ DHTB + Rotenone + Succinate	391 ± 2	181 ± 1	7,417 ± 785	31,954 ±7,258		37.3 ± 2.4

Postconfluent C3H 10T1/2 cells were treated 72 h
after a medium change with glucose supplement and
DHTB (10-8M), rotenone (2uM), and/or succinate(1mM).
At the indicated times, one h pulse incorporation
measurements of 3H-leucine (1 uCi/plate, 3H-dThd
(1uCi/plate) were taken on replicate cultures
(n=3). 86Rb+-uptake measurements were taken as
described previously (5).

those described in the previous experiment, it is concluded
that there is an oxygen-dependent step in G1 required for S-
phase entry but when cells progress in the cell cycle past the
oxygen-dependent step these cells can continue to undergo DNA
synthesis in the absence of oxygen.

Effect of Respiratory Chain Inhibitors on Mitogen-Induced [3]H-
dThd Incorporation.   A further test of the requirement of
oxidative energy for S-phase entry was carried out with the
use of inhibitors of the respiratory chain. The effects of
rotenone, which blocks the respiratory chain at the NADH-Q
reductase, on 3H-dThd, 3H-leucine incorporation and ouabain
sensitive 86Rb+ uptake, and glucose utilization was examined
as described in Table 2. In contrast to what had been observed
with quiescent cells by the inhibition of respiration
following oxygen depletion, the introduction of rotenone did
not inhibit DHTB-induced 3H-dThd incorporation into the acid
insoluble fraction. In fact, rotenone slightly increased
DHTB-induced 3H-dThd incorporation at T=25h. Further, the
addition of succinate, which bypasses the rotenone block, did
not significantly alter this response. Nevertheless, rotenone
did exhibit the ability to alter other parameters such as
a marked increase in glucose utilization (indicative of a
significant Pasteur Effect). It would appear that a
respiratory deficiency is not responsible for the failure of
cells to enter S-phase under anaerobic conditions. It should
also be noted that rotenone did not inhibit (Na+/K+)-ATPase
activity, and had only a slight inhibitory effect on DHTB-
stimulated 3H-leucine incorporation.

Table 3

THE EFFECT OF ANTIMYCIN A ON
DHTB-INDUCED ^{3}H-dThd, or ^{3}H-LEUCINE INCORPORATION
AND GLUCOSE UTILIZATION AND LACTATE PRODUCED

	Control	DHTB (10^{-8}M)	DHTB + Antimycin (1 μM)
Cumulative ^{3}H-dThd incorporation into DNA.			
O$_2$	20,400 ±1,700	104,400 ±6,800	82,400 ±13,500
N$_2$	11,500 ±1,200	17,400 ±3,600	24,800 ±500
Cumulative ^{3}H-leucine incorporation into acid-insoluble fraction			
O$_2$	53,400 ±4,900	114,100 ±2,100	105,600 ±1,200
N$_2$	39,100 ±1,000	65,900 ±5,800	– –
Glucose Remaining in the Medium at T = 24 hr. @ T=0 106.0±1.8 mg/100ml			
O$_2$	102.4 (3.6)* ±2.8	71.9 (34.1) ±7.7	52.7 (53.3) ±1.6
N$_2$	96.2 (9.8) ±1.2	85.5 (20.2) ±4.0	– –
Lactate Produced measured at T = 24, at T = 0 29.2 ±1.4 umoles (umoles lactate produced)			
O$_2$	41.1 (11.9) ±0.1	74.7 (45.6) ±7.2	96.1 (66.9) ±7.5
N$_2$	43.9 (14.7) ±2.5	59.4 (30.3) ±3.3	– –

*Values in parenthesis are number mg glucose utilized/100ml. The value for antimycin by itself was 26.1.

The effect of antimycin, which blocks the respiratory chain at cytochrome b, on DHTB-induced changes associated with S-phase entry was also tested as described in an experiment reported in Table 3. In this experiment, DHTB induced a significant stimulation of 3H-dThd incorporation for the 24 h period of incubation. Antimycin produced an inhibitory effect (approximately 20%) of the stimulated thymidine incorporation and somewhat lesser inhibition of leucine incorporation. Glucose utilization was markedly enhanced by DHTB and further stimulated by antimycin. These changes were in parallel with increases in lactate production.

The question arises whether antimycin induces inhibition of DNA synthesis or whether it merely delays S phase entry. Therefore, an experiment was carried out as described in Fig. 2 in which the kinetics of TPA-induced 3H-dThd incorporation (1 h pulse) in the presence and absence of antimycin A were examined. When antimycin was added at T=0, the incorporation of 3-H-thymidine at T=24 was decreased. However, antimycin did not inhibit DNA synthesis as judged by measurement of thymidine incorporation over an extended time frame but its addition merely delayed the time at which cells entered S-phase. Thus, the inhibition of the respiratory chain, which may lower ATP production as reported by Loffler (6), is compensated for by an increased glucose utilization and lactate production

EFFECT OF A RESPIRATORY CHAIN INHIBITOR (ANTIMYCIN A) ON DHTB-INDUCED ^{3}H-Thd INCORPORATION

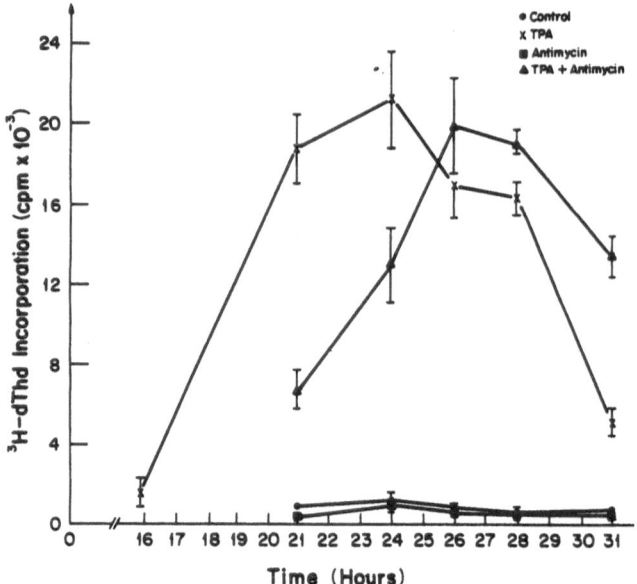

Figure 2. In this experiment, postconfluent C3H 10T ½ cells were treated with TPA (10^{-7}M) ± antimycin A (1 µM) at T=0. At designated time intervals, 1 hour pulse ^{3}H-Thymidine incorporation measurements were made.

as seen in Table 3. This indicates that glycolytic capacity enables these cells to tolerate conditions which otherwise would lead to energy deprivation and lack of DNA biosynthesis.

These differences in abilities to undergo DNA synthesis under anaerobic conditions vs inhibition by blockage of the cytochrome chain are attributed to the absolute requirement of oxygen for synthesis of pyrimidine nucleotide substrates, since Thelander et al found that ribonucleotide reductase involves an oxygen-dependent formation of tyrosyl radicals (6). Reichard has proposed that the balance between tyrosine radical formation and removal of this radical might serve as a regulatory effect on ribonucleotide reductase and since oxygen influences this balance ,it may have a regulatory effect on DNA synthesis by way of its influence on reductase activity (7). Loffler reported for Ehrlich ascites tumor cells that when oxygen is deprived, deoxycytidylate and thymidylate pools are depleted, which may account for the accumulation of cells at the end of G1 (5). Since supplementation of antimycin-treated ascites cells with exogenous pyrimidine nucleosides stimulated reprogression of G1 cells (5), respiratory insufficiency by antimycin-treatment may also result in a delay of formation of precursors for RNA and DNA synthesis.

In view of the oxygen requirements for the synthesis of nucleic acid pyrimidine and deoxy- nucleotides, our results are in accord with cell cycle arrest occurring in G1,the point of nucleotide biosynthesis.

Effect of ferricyanide on `immediate early gene´ expression and DNA synthesis. One of the earliest consequences of mitogenic stimulation of cells is the activation of a set of `immediate early genes´, two of which are c-fos and c-myc (8). In preliminary studies, we noted that an external electron acceptor , potassium ferricyanide, elicited a small mitogenic response in C3H 10T1/2 fibroblasts. Consequently, it was of interest to learn if this stimulatory effect was accompanied by inductions of c-fos and c-myc mRNAs. To investigate this question, we stimulated density arrested, serum depleted cells with either 100 or 500 uM ferricyanide for 30, 60 and 90 minutes. Whole cell RNA was isolated, electrophoresed on a 1.1% agarose gel, transferred to a nylon membrane and hybridized to radiolabelled c-fos or c-myc cDNA probes. The results of this experiment are shown in Figure 3.

Ferricyanide causes a rapid rise in c-fos mRNA levels which reaches a maximum at 30 minutes post-stimulation (Figure 3, top panel). The maximal level of c-fos mRNA is about 3.5-fold over control levels, as determined by computer aided densitometric analysis (not shown). c-fos mRNA levels fall rapidly after 30 minutes and return to basal levels by 90 minutes post-stimulation. These kinetics of c-fos induction are similar to those reported with peptide hormone growth factors such as nerve growth factor (9), transforming growth factor alpha (10) and serum (11,12). Similar results were obtained with 500 uM ferricyanide, i.e., at these concentrations we did not detect a dose dependent effect.

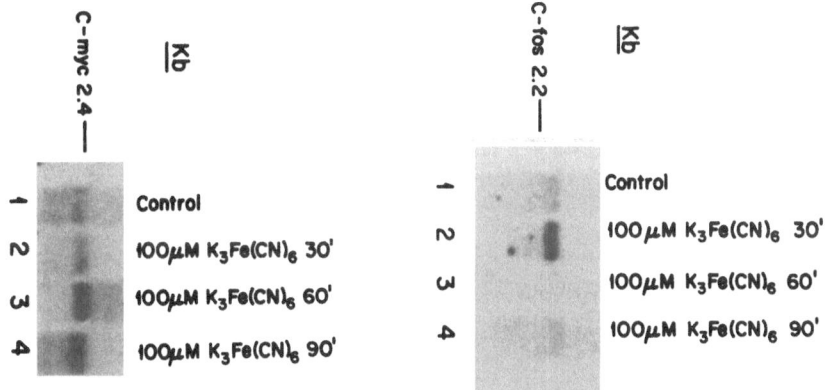

Figure 3. Induction of c-fos and c-myc by ferricyanide.
Quiescent 10T ½ cells were treated with 100 μM ferricyanide for the indicated times, at which whole cell RNA was isolated using 4M guanidium isothio-cyanate. RNA was purified by density gradient centrifugation in 5.8M CsCl. 10μg of RNA was denatured with glyoxal applied to each lane and electrophoresced through a 1.1% agarose gel. RNA was then transferred to a nylon-66 membrane (Hybond-N) by the method of Southern. cDNA probes were prepared using the random primer method with the 2.8 kb Xba I/BamHI fragment of c-fos DNA or the 1.5 kb Cla/EcoR1 fragment of c-myc DNA. Filters were hybridized overnight at 68°C in 50 ml hybridization buffer containing 2x10(8) cpm/ml of the appropriate probe. Filters were then washed and exposed to autoradio-graphic film (Kodak XAR) for seven days.

A smaller increase was observed with c-myc. 100 uM ferricyanide caused a 2.5-fold increase in c-myc mRNA levels 60 minutes post-stimulation. Similar mRNA levels persisted at 90 minutes, and these results are in accord with the temporal induction of c-myc mRNA by PDGF (13) and serum (14). As was the case with c-fos, 500 uM ferricyanide did not differ with respect to c-myc mRNA induction. These data indicate that the presence of an impermeable electron acceptor such as ferricyanide is capable of generating a response, albeit small, as measured by increases in c-myc and c-fos mRNAs.

We then studied the effect of ferricyanide on C3H 10T1/2 cells when it was used in conjunction with TPA or epidermal growth factor (EGF). We were interested to see how ferricyanide may modulate the effect of these mitogens on DNA synthesis. The results of a representative experiment with TPA, in which one hour pulses with 3H-dThd were conducted at the indicated times, are shown in Figure 4. As can be seen, ferricyanide can potentiate the effect of TPA on DNA synthesis. We used ferrocyanide as a control in this experiment. However, at the later time points, ferrocyanide also appears to potentiate the effect of TPA. We believe that this is a non-specific effect caused by the oxidation of ferrocyanide to ferricyanide over the time in which the experiment was done. Similar results were obtained with EGF (not shown). Ferricyanide, therefore, can stimulate c-fos and c-myc mRNA synthesis on its own, as well as potentiate the DNA synthetic response observed with either TPA or EGF.

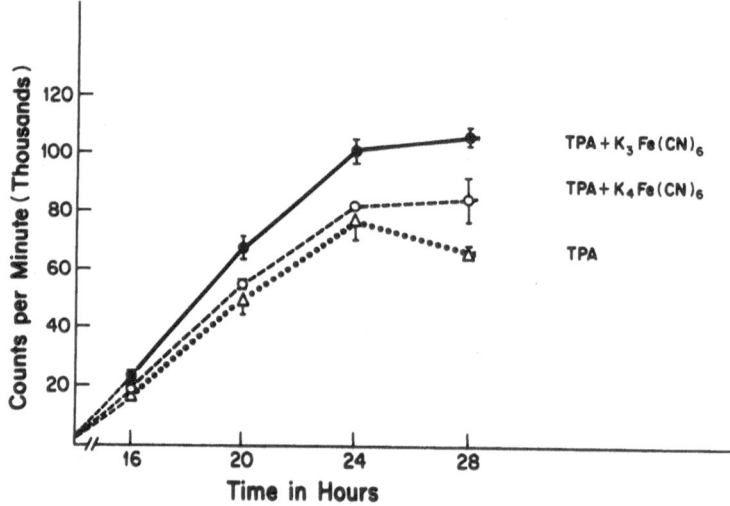

Figure 4. Ferricyanide potentiation of TPA-induced DNA synthesis. Quiescent cells were treated with TPA (100 nM) and 10 µM ferricyanide or ferrocyanide at T=0. At the indicated times cells were pulsed with tritiated thymidine. The reaction was stopped by aspirating the media and rinsing the plates twice with cold PBS. Cells were then treated with 5% trichloroacetic acid for 30 min. The acid was aspirated and the acid insoluble fractions were solubilized in 0.5 N NaOH for 2 h. These solutions were quantitatively transferred to scintillation vials, neutralized, and counted.

We believe that ferricyanide may be eliciting these effects through a membrane charge separation which results in alkalinization of the cytosol. Other labs have indicated that the effect of ferricyanide is dependent on amiloride sensitive sodium movements (15). This report also suggested that ferricyanide was causing a migration of protein kinase C to the membrane, implying that this kinase is activated. In light of some results obtained in our laboratory, showing that kinase C has a negative regulatory role in EGF signalling, we do not think kinase C is involved in ferricyanide mediated signal generation, since ferricyanide can potentiate the effects of EGF on DNA synthesis. Rather, we are developing a model which coincides with observations made in the study of ras oncogene function, that intracellular pH may be a key regulator of cell growth.

Several investigators have made use of the technique of microinjection of proteins into cells. A recent study (16) using this method to microinject p21 ras into NIH 3T3 cells showed that the oncogenic ras protein caused a rapid rise in intracellular pH while the normal p21 ras did not elicit this effect. Furthermore, it was demonstrated that the rise in intracellular pH was amiloride sensitive, implicating the sodium/proton antiport system in this effect. Evidence was also presented which suggests that the change in intracellular pH is involved in the progression of these cells through S phase. Others have demonstrated that microinjection of oncogenic ras induces c-fos expression (17). Since intracellular pH changes very rapidly upon microinjection of ras, it is tempting to speculate that the pH change is involved in increasing c-fos expression. Finally, several groups have postulated that rises in intracellular pH, caused by serum (18,19) or purified growth factors (20,21), mediated by the sodium/proton exchange system are involved in the control of cell cycle progression. We believe that our data with ferricyanide, particularly with c-fos and c-myc support this hypothesis. Experiments are currently in progress to delineate the role of the sodium/proton exchanger and intracellular pH in ferricyanide stimulated cells. Finally, ferricyanide stimulation of proto-oncogene expression, particularly since it appears to be independent of a receptor mediated event, can be a very useful tool for investigating the role of intracellular pH and the sodium/proton antiport system in control of early gene activation by serum and growth factors.

REFERENCES

1.Mueller, G.C., Wertz, P.W., Kwong, C. H., Anderson, K. and Wrighton, S.A., Dissection of the early events in the activation of lymphocytes by 12-O-Tetradecanoylphorbol-13-acetate in "Carcinogenesis:Fundamental Mechanisms and Environmental Effects". B. Pullman, P.O. Tso and H. Gelboin,eds. D.Reidel Publishing Co. Dordrecht, Netherlands(1980).
2. Chen, J. and Jones, M. E., The cellular location of dihydro-orotate dehydrogenase: Relation to de Novo biosynthesis of pyrimidines. Arch. Biochem and Biophys. 176: 82(1976).
3. Thelander, L., Graslund, A. and Thelander, M. Continual presence of oxygen and iron required for mammalian ribonucleotide reduction : Possible regulation mechanism.

Biochem. Biophys. Res. Commun. 110: 859 (1983).

4. Crane, F. L., Sun, I. L., Clark, M.G., Grebing, C. and Low, H., 1985, Transplasma-membrane redox systems in growth and development. Biochim. Biophys. Acta 811:233 (1985).

5. Leister, K.J., Wenner, C. E., and Tomei. L.D. Correlation of ouabain-sensitive ion movements with cell-cycle activation. Proc. Natl. Acad. Sci.82: 1599 (1985).

6. Loffler, M. Towards a further understanding of the growth-inhibiting action of oxygen deficiency. Expt´l Cell Research 157:195 (1985).

7. Reichard, P. Regulation of deoxyribotide synthesis. Biochemistry 26: 3245 (1987).

8. Lau, L.F. and Nathans, D. Expression of a set of growth-related immediate early genes in BALB/C 3T3 cells: Coordinate regulation with c-fos or c-myc. Proc. Natl. Acad. Sci. 84: 1182 (1987).

9. Kruijer, W., Schubert, D. and Verma, I.M. Induction of the proto-oncogene c-fos by nerve growth factor. Proc.Natl.Acad.Sci. 82: 7330 (1985).

10. Cutry, A.F., Kinniburgh, A.J., Twardzik, D.R. and Wenner, C.E. Transforming growth factor alpha (TGFa) induction of c-fos and c-myc expression in C3H 10T1/2 cells. Biochem. Biophys. Res. Comm. (in press, 1988).

11. Greenberg, M.E. and Ziff, E.B. Stimulation of 3T3 cells induces transcription of the c-fos proto-oncogene. Nature 311: 433 (1984).

12. Muller, R., Bravo, R., Burckhardt, J. and Curran, T. Induction of c-fos gene and protein by growth factors precedes activation of c-myc. Nature 312: 716 (1984).

13. Kelly, K., Cochran, B.H., Stiles, C.D. and Leder, P. Cell-specific regulation of the c-myc gene by lymphocyte mitogens and platelet-derived growth factor. Cell 35: 603 (1983).

14. Dean, M., Levine, R.A., Ran, W., Kindy, M.S., Sonenshein, G.E. and Campisi, J. Regulation of c-myc transcription and mRNA abundance by serum growth factors and cell contact. J.Biol.Chem. 261: 9161 (1986).

15. Malviya, A.N. and Anglard, P. Modulation of cytosolic protein kinase C activity by ferricyanide. FEBS Lett. 200: 265 (1986).

16. Hagag, N., Lacal, J.C., Graber, M., Aaronson, S. and Viola, M.V. Microinjection of ras p21 induces a rapid rise in intracellular pH. Mol.Cell.Biol. 7: 1984 (1987).

17. Stacey, D.W., Watson, T., Kung, H-F. and Curran, T. Microinjection of transforming ras proteins induces c-fos expression. Mol.Cell.Biol. 7: 523 (1987).

18. Moolenaar, W.H., Mummery, C.L., van der Saag, P.T. and de Laat, S. Rapid ionic events and the initiation of growth in serum-stimulated neuroblastoma cells. Cell 23: 789 (1981).

19. Pouyssegur, J., Chambard, J.J.C., Franchi, A., Paris, S. and van Obberghen-Schilling, E. Growth factor activation of an amiloride-sensitive Na+/H+ exchange system in quiescent fibroblasts: coupling to ribosomal protein S6 phosphorylation. Proc.Natl.Acad.Sci. 79: 3935 (1982).

20. L´Allemain, G., Franchi, A., Cragoe, Jr., E. and Pouyssegur, J. Blockade of the Na+/H+ antiport abolishes growth factor-induced DNA synthesis in fibroblasts. J.Biol.Chem. 259: 4313 (1984).

21. Pouyssegur, J., Chambard, J.J.C., Franchi, A., L´Allemain, G., Paris, S. and van Obberghen-Schilling, E. Growth-factor activation of the Na+/H+ antiporter controls growth of fibroblasts by regulating intracellular pH. Cancer Cells (Cold Spring Harbor) 3: 409 (1985).

REDOX AGENTS WHICH MODULATE THE GROWTH OF MELANOMA CELLS:

EVALUATION OF MECHANISMS

K.A.O. Ellem, G.F. Kay, A.J. Dunstan and D.J. Stenzel

Queensland Institute of Medical Research
Bramston Terrace
Brisbane, Qld. Australia

The growth of many types of cell under the artificial conditions of in vitro culture is dependent on the attachment of the cells to a surface. The chemistry of the surface attachment can determine the response of the cells by modulating surface receptors for growth factors, by altering the strength of adhesion and thus overall cell surface area[1] and in other ways which remain to be elucidated. An array of functional and structural changes in the surface of cells has been documented in cells undergoing the phenotypic expression of malignancy. The very basis of the commitment of cells to multiplication is the binding of growth factors to surface receptors as initial events in the cascade which leads to DNA synthesis and mitosis[2]. These initial events include phosphorylation of the growth factor receptors, and other proteins whose role in the cascade is yet to be determined, and very early changes in ion fluxes[3], including egress of protons. It is clear that the plasma membrane is very actively involved in the events which determine whether a cell divides rapidly or slowly, differentiates, or languishes in a quiescent state. This study is directed towards an evaluation of some of the difficulties with attempts to define this association, and to evaluate some of the evidence that such electron flow may have a significant role in modulating cell attachment and replication.

We became involved in relating cell growth responses to membrane mediated redox events from a study of the role of pyruvate as a growth requirement for some conditions of culture of mammalian cells[4] We had found that conditioning of culture medium by previous exposure to certain cell lines could enhance the growth of melanoma cells plated at low density ($<10^4$ cell cm^{-2}), in a commonly used culture medium (RPMI 1640) containing low serum concentrations i.e. less than 2.5% foetal bovine serum ($<2.5\%$ fbs). The conditioning factor was identified as a 2-oxocarboxylate, probably pyruvate[5], and the conditioning effect shown to be due to the 2-oxo function[4] rather than to the pivotal position that pyruvate occupies in intermediary metabolism. There was a strong correlation between the attachment of cells and their growth response to added 2-oxocarboxylates or serum, which suggested that the effects may be mediated by influencing activities of the cell membrane.

In seeking alternative roles for the 2-oxocarboxylates we noted that pyruvate and glyoxylate were both substrates for the plasma membrane dehydrogenases for which Crane and Low had proposed an active role in local membrane energetics in the late 1970's[6]. The ready permeability of the cell membrane to both these anions left open the issue of whether transmembrane electron flow, e.g. from NADH to an appropriate electron sink, might be the driving energy source. Ferricyanide was, therefore, chosen as another excellent, but impermeable substrate[6] to try for a growth stimulatory effect, despite the de novo improbability of the outcome. To our surprise and delight we found an unequivocal growth promoting effect in the concentration range 10-100μM[7]. The ferricyanide-ferrocyanide (FeIIICN-FeIICN) redox pair had previously been shown to act across the cell membrane in coupling electron flow and ATP levels to and from intraerythrocytic NAD[+]/NADH[8]. Another four ferricoordination complexes, chosen to explore a variety of geometries of the FeIII coordination shell were all found to be effective stimulants of growth and attachment[9]. We interpreted the effectiveness of these compounds, at first, as electron acceptors for intracellular NADH via a transmembrane dehydrogenase.

A corollary of this hypothesis would be that such activity might be demonstrable in the plasma membrane of the cells under study. In figure 1, using the procedure of Karnofsky modified by Morre et al[10] to detect extramitochrondrial NADH dehydrogenase activity, it can be seen that MM96 cells do have fine deposits of Hatchett's brown precipitate well localized to the plasma membrane. The plasma membranes were stained in cultures growing rapidly in 10% fbs or 1% fbs+1mM pyruvate (fig 1A, 1B) and even in those cultures showing no growth in 1%fbs without added pyruvate (fig.1C). In the nongrowing cultures the surface membrane had extensive rounded evaginations instead of the microvilli present on the growing cells. Typically, no staining was found when NADH was omitted from the staining mixture[10] (fig.1D). Two puzzling observations which need further investigation should be noted. In the positively stained cultures, a significant proportion of the cells have no reaction in their plasma membranes e.g. figure 1B shows 3 cells, two of which are reactive but these are separated by a cell which is non reactive. This heterogeneity is unexplained, but could be cell cycle related. Secondly, intranuclear staining was prominent and tended to be associated with more condensed areas of chromatin in both of the growing cultures (fig.1A & B). Such staining was absent from the majority of cells in the non growing, low serum cultures (e.g.fig.1C). Whether this has implications for NADH dehydrogenase activity in association with actively replicating chromatin, or not, remains to be elucidated.

An initial test of this hypothesis was by evaluating the effectiveness of other FeIII compounds, in particular comparing the easily reducible ferrihaem and its protein complexes such as cytochrome c and haemoglobin with the very difficultly reducible catalase. All were found to be active, and catalase was the most effective at a concentration 3 orders of magnitude less than the other haem proteins[9]. This changed the interpretation of the previous findings and suggested that hydrogen peroxide may have been present in the medium causing inhibition of cell growth and attachment and was being removed by catalatic or peroxidatic activity of the FeIII complexes, thereby releasing cell growth from its peroxide inhibited state.

Assay of the medium RPMI 1640 for H_2O_2 showed that it was in fact present following exposure to fluorescent light. The H_2O_2 accumulated at a rate of approximately 5μM/hour in 10ml lots of serum-free medium exposed in standard biohazard hood lighting (0.8W m[2] flux), and was a little less in larger volumes, presumably due to absorption of exciting wave lengths by the phenol red containing medium. Photodynamic production of H_2O_2 in culture media has previously been described and

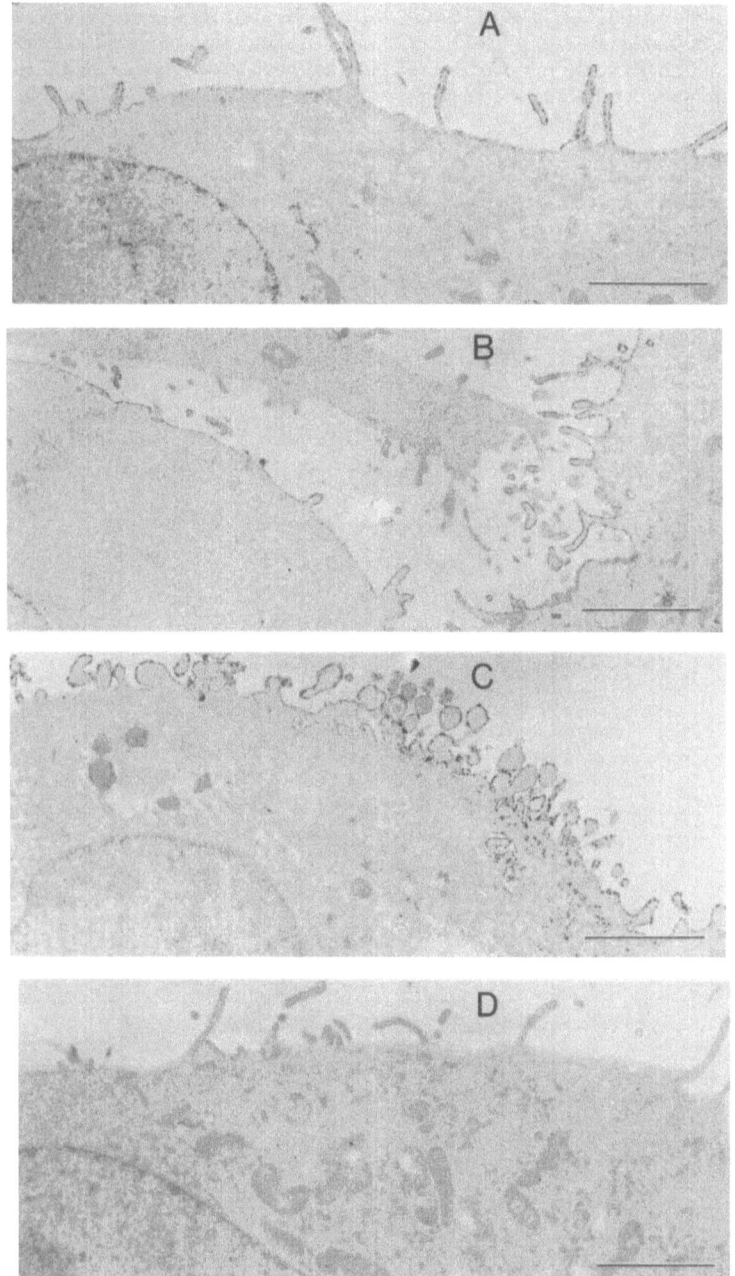

Figure 1. Electron micrographs of Hatchett's brown staining due to ferricyanide - NADH dehydrogenase activity in MM96 cells in RPMI 1640 medium with different additives. A:+10% fbs, plasma membrane and intranuclear staining. B:+1% fbs +1mM pyruvate, plasma membrane and intranuclear staining. C:+1% fbs no additives, plasma membrane stain alone. D:+10% fbs, NADH omitted from staining solution, no stain. Bars indicate 2μm.

shown to be due principally to riboflavin[11] and tryptophan[12]. It is noteworthy that four commonly used media formulations MEM, RPMI 1640, McCoy's 5a and Dulbecco's MEM all have a high riboflavin content (100,200,200 and 400μg/L) and that the latter has been found to require the inclusion of 1mM pyruvate. Pyruvate reacts as quickly with H_2O_2 as does catalase, undergoing oxidative decarboxylation as it consumes the H_2O_2 [13].

With the exception of the ferricoordination complexes, the stimulation of growth and attachment correlated well with the ability of serum, 2-oxocarboxylates and the haem proteins to remove H_2O_2 from the medium[9]. Extracellular superoxide had little to do with the peroxide effects since superoxide dismutase had a barely significant stimulatory effect by itself but this little was additive to that of catalase[9]. However, when the 6 ferricoordination complexes were tested the FeIII complexes with citrate (CT), nitrilotriacetate (NTA) and ethylenediamine tetra-acetate (EDTA) all halved the concentration of H_2O_2 (added to 20μM) in one hour whereas FeIII (CN), diethylene triamine pentaacetate (DTPA) and nitroprusside (NP) had no effect on the preadded peroxide. The status of the FeIII complexes will be discussed later. First it will

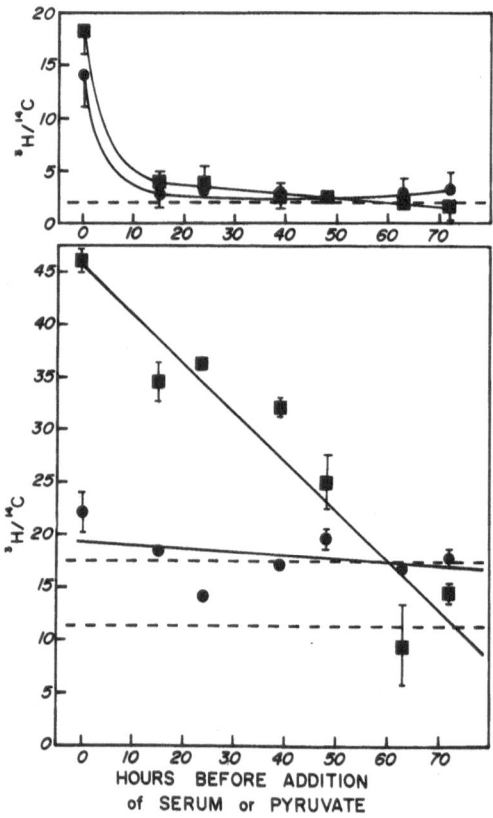

Figure 2. Decay of reversibility of loss of MM 96 cell growth due to cell seeding in "light exposed" medium RPMI 1640+1% fbs. Growth after addition of 1mM pyruvate (●) or 10% fbs (■) at indicated times. Top panel: Data using cells grown in RPMI 1640+1% fbs+ 1mM pyruvate for 3 days prior to experimental seeding. Bottom panel: Data using cells grown in RPMI 1640+18% fbs for 3 days prior to experiment. Absence of errorbars indicates error smaller than symbol size. Dashed lines indicate range of H/C values of unstimulated cultures.

be useful to look at certain details of the system being used to evaluate the effects of these compounds on cell growth to appreciate some of the confounding phenomena which can occur.

Cultures of MM96 used for testing were grown in RPMI 1640 medium with low serum levels (1% fbs) and containing 1mM pyruvate, since such cells usually were very sensitive to the medium inhibition and to the stimulatory effects of the various additives. The kinetics of the rescue of the growth and attachment of cells in low serum medium point towards the uniqueness of the first few hours of the stimulatory/protective activity. Figure 2(upper panel) shows, using cells which were grown in low serum and pyruvate prior to experimental use, that if the addition of pyruvate or serum is delayed until 16 hours after seeding MM96 cells into low serum (1% fbs) medium, the stimulatory effect is lost. In other experiments addition at shorter times after seeding the cells shows that this period of reversibility lasts only 2-4 hours with pyruvate and up to 8-10 hours with fbs. However, it is noteworthy that growth of the same cells in medium containing 10% serum (rather than low serum + pyruvate) for 3 days prior to the experiment changed the responsiveness of the cells to the stimulating effect of serum, but not that of pyruvate (figure 2 lower panel). Thus even 50 hours after seeding into low serum medium significant stimulation of growth could be achieved by adding fbs to a concentration of 10%. Pyruvate was still ineffective after the first few hours.

Clearly growth of cells in high serum concentrations leaves them in a more robust condition to withstand toxic medium, than does growth in low serum plus pyruvate. Also, either the added serum counteracts toxic products in addition to removing H_2O_2, or the serum aids repair of the H_2O_2 damage (e.g.membrane lipid peroxidation) in a fashion which pyruvate does not. This difference is also apparent in the interaction between pyruvate and serum described elsewhere[4], where, at low serum levels, cell growth and attachment are tightly correlated with added pyruvate levels, but have a maximum level of stimulation well short of that achieved with higher concentrations of dialyzed serum in the absence of pyruvate. At high levels of dialyzed serum (10%), pyruvate has a concentration dependent activity on growth, but not on attachment (since this already is maximal in dialyzed (10%) serum[4]). This shows that pyruvate can stimulate cell growth independently of its ability to encourage cell attachment in low serum, presumably due to its removal of H_2O_2 from the medium.

We are concerned with the possible influence of plasma membrane redox activity on cell growth. Systems for monitoring the target activity are perforce dependent on the demonstration of changes in the rate of cell division or more conveniently on the amount of DNA synthesis in unit time as a measure of S phase activity, with all the usual caveats concerning precursor pool artefacts[4]. Much of this workshop will be devoted to elegant biochemical studies using purified preparations of membranes or enzymes, but the cell culture systems used to study growth are complex and "dirty" in the chemical sense. The assumption that an excluded anion such as ferricyanide exerts its effects only as an electron sink for transmembrane dehydrogenases needs to be taken with caution. It is capable for instance of oxidizing sulphydryl group to disulphides,(e.g. cysteine to cystine), without producing H_2O_2 as a by product[14], such as occurs during their autoxidation when catalyzed by ubiquitous traces of Cu^{++} [15]. Could FeIII CN thus simulate the activity of the red kidney bean protein which has been reported to stimulate RNA synthesis in various organisms and which acts by chelating Cu^{++} [15]? FeIII CN would thereby eliminate cell inhibition caused by production of H_2O_2 from thiol oxidation by Cu^{++}.

FeIII complexes could directly oxidize components of the cell surface such as membrane protein thiols [14] which can have profound effects on e.g. anion [16] and cation transport, which in turn may induce the expression of the protooncogene c-fos, one of the earliest transcriptional responses to mitogenic stimulation[17]. Apart from chemical modification of the cell surface, FeIII complexes are well known for a variety of catalytic activities which could modify the chemical composition of the medium. We recently came across an example of this which may be just the tip of an iceberg. When the ultraviolet spectrum of RPMI 1640 medium which had been exposed to fluorescent light was compared with that of unexposed medium, a peak of absorptivity appeared with a maximum at 325nm, which was thus presumably either a direct or indirect end product of photodynamic activation. Addition of H_2O_2 to unexposed medium led to the development of the A_{325} peak without any light exposure, so that it is presumably the result of H_2O_2 oxidation of some component of the medium. To identify the constituent involved, solutions were made of the various classes of compounds used to make up the culture medium and the appearance of a peak of absorption at A_{325} was checked following the addition of H_2O_2 (to $20\mu M$). Streptomycin sulphate was found to be the compound undergoing oxidation. Several of the ferricoordination compounds effective in stimulating cell growth were tested (at $10\mu M$) for an effect on the streptomycin oxidation reaction. Table 1 shows that haemin and ferri nitrilotriacetate (FeIIINTA) increased the rate of streptomycin oxidation 22 fold, whereas FeIIICN and nitroprusside (FeIIINP) did not. This selectivity may be explained by the availability of aqua coordination sites in these different complexes determining the reactivity of the compound with H_2O_2 and thus substrate (e.g.[18]).

In the present context loss of streptomycin from the medium overall is not expected to have any physiological sequelae for the growth experiments (unless the products are toxic) other than by contributing to the removal of H_2O_2 from the medium. However, FeIII chelate catalysis of other changes involving essential nutrients (e.g. [19]) could alter the growth characteristics of the medium. Decreased concentration of a medium component from a particular medium formulation which is optimal for one cell type can lead to increased efficacy for the growth of another, as has been amply demonstrated by the meticulous studies of Ham's group[20]. There is certainly a significant amount of substrate for horseradish peroxidase (HRP) and H_2O_2 in the standard RPMI 1640 culture medium, since 4 moles of H_2O_2 were required for complete oxidation of 1 mole of scopoletin by HRP when reacted in the medium, whereas the reaction was equimolar (1:1) when reacted in phosphate buffered saline[9].

Table 1. Rate of oxidation of streptomycin in tissue culture medium by chelated iron compounds and H_2O_2 [a].

RPMI 1640 plus:	ΔA_{325} per hour
No additions	0.06
+ 0.5% FBS	0.08
+ Haemin $10\mu M$	1.12
+ Fe(III) CN	0.05
+ Fe(III) NP	0.14
+ Fe(III) NTA	1.30

[a] In these experiments the change in absorbance at 325nm was measured after 30min reaction of the various additives with tissue culture medium and H_2O_2 at room temperature. The samples were measured in a dual beam spectrophotometer with the control sample containing the medium and additive but no H_2O_2. The tissue culture medium contained streptomycin at the usual concentration ($100\mu g/ml$).

There are undoubtedly many compounds available in the culture media for catalytic oxidation by ferricoordination complexes.

Examples of ferricomplex catalysis of organic compounds abound but only a few of possible relevance can be mentioned. Haem is well known for a variety of catalytic functions both catalatic and peroxidatic, including oxidation of ascorbate by H_2O_2 [21]. Ascorbate is essential, among other things for the hydroxylation of proline and lysine residues during collagen synthesis, and its lack can lead to deficiencies in the extracellular matrix which can in turn influence the growth responses of various cell types[1]. It is of interest that FeIII NTA has been shown to have pyrocatechase activity[22] as well as peroxidatic and catalatic reactions[19].

Since modulation of cell adhesion appears to be integral to a significant proportion of the stimulating activities which we have observed, and active oxygen species are involved in at least part of the phenomena, we asked whether lipid soluble antioxidants might not affect cell growth or attachment. Table 2 shows that none of the tested antioxidants increased growth (each one tested over a wide range of concentrations) and only a small effect was seen on cell attachment. The slight increase in cell attachment observed occurred with the retinoids so that even this effect cannot be ascribed to their antioxidant chemistry because they have extensive biological effects in addition to this function, including enhancement of cell-substrate adhesion[23].

The question arises whether melanoma cells are unduly susceptible to redox active materials, and whether such a sensitivity may be masking underlying redox modulated growth effects mediated via the plasma membrane. It must be recalled that melanizing cells convert tyrosine by the polyphenol oxidase tyrosinase to melanin. Intermediate compounds include the highly redox reactive phenol quinoids 3,4 dihydroxyphenylalamine (dopa), 5,6 dihydroxyindole and their quinones. Exploitation of the unique biochemistry of the melanogenic pathway has been a goal of chemotherapy in melanoma because of the dispensibility of normal epidermal melanocytes (e.g.[24]).Parsons and colleagues at QIMR have made extensive studies of the sensitivity of melanoma cell lines to a series of redox active catechols. They found that MM96, for example, was from 3.6 to 7.7 times more sensitive to noradrenaline, dopa, 3,4

Table 2. Effect of lipid soluble antioxidants on MM96 cultures in low serum medium.

Additive	$^3H/^{14}C$	(+S.D.)	^{14}C	(+S.D.)
10%FBS	13.28[a]	(0.57)	12975[a]	(1098)
0.5%FBS	1.12	(0.21)	721	(145)
0.5%FBS+BHT 1mM	2.71	(1.17)	1870	(129)
0.5%FBS+BHA 1mM	0.24	(0.20)	2667	(356)
0.5%FBS+α-tocopherol 10μM	1.03	(0.43)	2631	(831)
0.5%FBS+retinol 1mM	0.12	(0.13)	3302	(1014)
0.5%FBS+catalase 1000U/ml	7.60[a]	(0.22)	6715[a]	(156)

The growth ($^3H/^{14}C$) and attachment (^{14}C) parameters for each medium additive were drawn from an experiment where an extensive dose response series was tested. The values shown in the table represent the maximum growth and its corresponding attachment response for a particular medium additive. The lipid soluble antioxidants butylated hydroxytoluene (BHT), butylated hydroxyanisol (BHA), α-tocopherol, and retinol were added to 0.5%FBS/RPM 1640. Catalase was added as a control to show the protection from H_2O_2 in the growth system.

dihydroxybenzylamine (DHBA), dopamine and 3,4 dihydroxyphenylacetate than HeLa cells, a difference which could not be attributed to any deficiencies in intracellular glutathione levels[25] or to any measurable differences in the levels of oxygen metabolizing enzymes, DNA repair efficiency or, in the case of dopa, to catechol uptake[26]. Also, amongst 6 different melanoma cell lines there was no correlation between their tyrosinase levels, or their melanin synthesizing capacity and their sensitivity to these agents. The toxic effects of the catechols were mediated by active oxygen species generated extracellularly, probably by autooxidative reactions, since the cells were protected by catalase, superoxide dismutase or peroxidase when added to the medium, and the toxicity was potentiated by the inhibitors of these enzymes (3-amino,1,2,4,triazole, diethyldithiocarbamate and D or L-penicillamine[26]). Since high cell density was also protective, endogenously produced pyruvate may be responsible by eliminating the H_2O_2 generated by catechol autooxidation.

Interestingly when HeLa cells and MM96 cells were tested in a survival/growth assay by seeding into medium containing H_2O_2, they were found to be equally sensitive (90% inhibition by 10-12μM)[26]. However, MM96 cultures were 3 times more sensitive than HeLa cells to H_2O_2 generated over a significant length of time by the addition of glucose oxidase to the medium, and were 6 times more sensitive to the combination of superoxide and H_2O_2 produced by the addition of xanthine oxidase [27]. A human retinoblastoma line, of neural crest origin like melanomas, although lacking the melanin series of oxidative hazards, proved to be as sensitive as MM96 to dopa and DHBA. As with melanoma cells, and HeLa cells exposed to the higher doses necessary for inhibition, the effects of the catechols were shown to be cell cycle specific, causing a block at the G_1-S interface[27,28].

Further data which highlight the peculiar sensitivity of many melanoma cell lines to agents with direct or indirect redox effects were found during studies to manipulate intracellular glutathione levels. Buthionine sulfoximine, an inhibitor of the rate limiting enzyme in glutathione synthesis (γ-glutamyl cysteine synthetase), although only producing a modest 20% fall in the glutathione pool of both MM96 and HeLa cells, proved to be a very selective inhibitor of melanoma cells (e.g. MM96 were 27 times more sensitive than HeLa cells[25]). Again the cationic dye rhodamine 123 which is concentrated by, and inhibitory to oxidative phosphorylation in, mitochondria, also proved to be 6-7 times more toxic to MM96 than HeLa cells[27]. The conclusion seems inescapable that, as a class, melanoma cell lines (and perhaps other neural crest derivatives) show marked sensitivity to diverse agents with known effects on different redox systems for which at present the common property seems to be an associated generation of active oxygen species. Demonstration of growth stimulation by putative electron sinks for plasma membrane dehydrogenases appears to be confounded in these cultures by virtue of this unique sensitivity to hydrogen peroxide, etc.

While these studies have been conducted in the essentially artificial but controllable microenvironment of tissue culture, the H_2O_2 sensitivity and resistance of tumour cells has been correlated with important biological phenomena. Unselected, spontaneously transformed, hamster embryo cells, having a low propensity to metastasis, are susceptible to the killing effects of H_2O_2. Selection of cell lines from lung metastases resulting from subcutaneous injection of high cell numbers leads to selection of variant lines which have an enhanced ability to form metastases (10-1000 fold) and a higher resistance to H_2O_2 (17 fold). In different lines there was a strong correlation between metastasizing activity and H_2O_2 resistance[29]. The in vivo selection pressure is presumed to be the result of cell mediated tumour rejection via the superoxide/H_2O_2 generated by cells of the macrophage/monocyte lineage.

The final and crucial question involves the ability of FeIII complexes to stimulate cell growth and/or attachment without H_2O_2 as a confounding variable. We found previously [9] that FeIIICN, FeIIINP and FeIIIDTPA at the optimal stimulating doses (30,10 and 10μM, respectively) did not remove preformed H_2O_2 from the medium. However, these 3 FeIII complexes were no less effective in stimulating the growth and attachment of MM96cells seeded in "light exposed" medium than were the complexes which did eliminate H_2O_2 (FeIIINTA, FeIIICT and FeIIIEDTA). Can we conclude that FeIII chelates such as FeIIIDTPA which lack coordination water and will not catalyse hydroxyl radical formation from active oxygen species in a Fenton reaction[30] do not effect the cellular changes observed by circumventing H_2O_2 cell damage? It should be noted that even FeIIIDTPA reduced the amount of H_2O_2 accumulating in the medium during fluorescent light exposure[9], although not removing preformed H_2O_2 under those conditions. As in the case of the even more efficient complexes FeIIICN and FeIIINP direct reaction of the FeIII chelates with the photo activated species must occur to deactivate the photo excited state without H_2O_2 production. Perhaps this opens up other possible modes of action of these compounds with culture medium or cell surface constituents which are alternate to the initial hypothesis that they are acting as electron sinks for plasma membrane dehydrogenases to promote cell growth and the surface activity which includes greater adhesion to surfaces.

References

1. D. Gospodarowicz, G. Greenberg and C.R. Birdwell, Determination of cellular shape by the extracellular matrix and its correlation with the control of cellular growth. Cancer Res. 38:4155 (1978).
2. R. James and R.A. Bradshaw, Polypeptive growth factors. Ann.Rev. Biochem. 53:259 (1984)
3. S. Schuldiner and E. Rosengurt, Na$^+$/H$^+$ antiport in Swiss 3T3 cells: Mitogenic stimulation leads to cytoplasmic alkalinization. Proc. Natl. Acad. Sci.,USA. 79:7778 (1982)
4. G.F. Kay and K.A.O. Ellem, Use of growth-attachment corelation plots to analyse human melanoma cell dependency on 2-oxocarboxylates and serum, Cell Tissue Kinet. 18:355 (1985).
5. K.A.O. Ellem and G.F. Kay, The nature of conditioning nutrients for human malignant melanoma cultures. J.Cell. Sci. 62:249 (1982).
6. F.L. Crane, H. Goldenburg, D.J. Morre and H. Low, Dehydrogenases of the plasma membrane, Subcell. Biochem. 6:345 (1979).
7. K.A.O. Ellem and G.F. Kay, Ferricyanide can replace pyruvate to stimulate growth and attachment of serum restricted human melanoma cells, Biochem. Biophys. Res. Comm. 112:183 (1983).
8. R.K. Mishra and H. Passow, Induction of intracellular ATP synthesis by extracellular ferricyanide in human red blood cells, J. Membr. Biol. 1:214 (1969).
9. G.F. Kay and K.A.O. Ellem, Nonhaem complexes of FeIII stimulate cell attachment and growth by a mechanism different from that of serum, 2-oxocarboxylates and haemproteins. J.Cell. Physiol. 126:275 (1986).
10. D.J. Morre, W.N. Yunghans, E.L. Vigil and T.W. Keenan, Liver organelles and endomembrane components, in "Methodological Developments in Biochemistry" E.Reid, ed., Longman, London. vol.4:195 (1974).
11. R.J. Wang and B.T. Nixon, Identification of hydrogen peroxide as a photoproduct toxic to human cells in tissue-culture medium

irradiated with "Daylight" fluorescent light In Vitro 14:715 (1978).

12. J.P. McCormick, J.R. Fisher, J.P. Pachlatko and A. Eisenstark, Characterization of a cell-lethal product from the photooxidation of tryptophan: Hydrogen peroxide. Science 191:468 (1976).

13. M.G. Sevag, Uber den Atmungsmechanisms der Pneumokokkon I. Justus Liebig's Ann. Chem. 507:92 (1933).

14. D.K. Kidby, Direct spectrophotometric estimation of ferrocyanide and its possible uses in sulfhydril oxidation states. Anal. Biochem. 28:230 (1969).

15. I. Fedorcsak, M. Harms-Ringdahl and L. Ehrenberg, Prevention of sulphydryl autoidation by a polypeptide from red kidney beans, described to be a stimulator of RNA synthesis. Exp. Cell. Res. 108:331 (1977).

16. N. Takeguchi, R. Joshima, Y. Inona, T. Kashiwagura and M. Morii, Effects of Cu^{2+}-o-Phenanthroline on gastric $(H^+ + K^+)$-ATpase. Evidence for opening of a closed anion conductance by S-S cross linkings. J. Biol. Chem. 258:3094 (1983).

17. J.I. Morgan and T. Curran, Role of ion flux in the control of c-fos expression. Nature 322:552 (1986).

18. J.H. Baxendale, Decomposition of hydrogen peroxide by catalysts in homogeneous aqueous solution, Adv. Catalysis, 4:31 (1952).

19. C. Walling, M. Kurz and H.J. Sugar, The iron (III)-ethylenediamine tetraaceticacid-peroxide system, Inorg. Chem. 9:931 (1970).

20. R.G. Ham and W.L. McKeehan, Media and growth requirements, in "Methods in Enzymology", W.B. Jakoby and I.H. Pastan, eds, Academic Press, New York, vol. LVIII:44 (1979).

21. M.L. Kremer, Haemin-catalyzed oxidation of ascorbic acid by H_2O_2, Trans. Faraday Soc. 63:1208 (1967).

22. M.G. Weller, Ferric nitrilotriacetate: An active centre analogue of pyrocatechase in "The coordination Chemisry of Metalloenzymes" I. Bertim, R.S. Drago and C. Luchinal, eds. D. Reidal Publishing Co., Dordrecht (1983).

23. S. Kato and L.M. DeLuca, Retinoic acid modulates attachment of mouse fibroblasts to laminin substrates, Exp. Cell. Res. 173:450 (1987).

24. M.M. Wick, The chemotherapy of malignant melanoma, J. Invest. Dermatol. 80:61s (1983).

25. E.P.W. Kable and P.G. Parsons, Sensitivity of human melanoma cells to dopa and buthionine sulfoximine. Cancer Res. Submitted.

26. P.G. Parsons, Modification of dopa toxicity in human tumour cells. Biochem. Pharmacol. 34:1801 (1985).

27. E.P.W. Kable and P.G. Parsons, Potency, selectivity and cell cycle dependence of catechols in human tumour cells in vitro. Biochem. Pharmacol. (in press).

28. E.P.W. Kable and P.G. Parsons, Melanin synthesis and the action of L-dopa and 3,4 dihydroxybenzylamine in human melanoma cells, Cancer Chemotherapy Pharmacol. Submitted.

29. G.I. Deichman and E.L. Vendrov, Characteristics of in vitro transformed cells essential for their in viro survival, selection and metastasizing activity. Int. J. Cancer 37:401 (1986).

30. E. Graf, J.R. Mahoney, R.G. Bryant, and J.W. Eaton, Iron catalysed hydroxyl radical formation: Stringent requirement for free iron coordination site. J. Biol. Chem. 259:3620 (1984).

POSSIBLE ROLE OF TRANSPLASMA MEMBRANE FERRICYANIDE REDUCTASE

IN MITOGENIC ACTIVATION OF RAT LIVER CELLS

R. García-Cañero

Servicio de Bioquímica Experimental
Clínica Puerta de Hierro
28035 Madrid, Spain

SUMMARY

Transplasma ferricyanide reductase of hepatocytes is activated by the specific mitogen, r-LGF, immediately upon addition, in a sodium-dependent, amiloride-sensitive way. The above phenomenon seems to be transient, and is abolished when the sodium gradient is reversed. Ferricyanide activates H^+ efflux and ^{22}Na uptake in parallel fashion. Ferricyanide by itself increases ^{22}Na uptake and H^+ efflux, and the natural substrate for the redox enzyme, diferric transferrin, increases DNA synthesis when added, together with submaximal doses of r-LGF, to cultured hepatocytes.

Phorbol esters such as TPA fail to activate ferricyanide reductase, and r-LGF is able to further activate ^{22}Na uptake in hepatocytes previously treated with TPA for 24 hours. Neither r-LGF nor ferricyanide promotes any change in the levels of soluble protein kinase C, nor do trifluoperazine, isobutyl-methylxantine, A23187 or H-7 decrease significantly the basal redox activity.

The above data point out substantial differences between the natural mitogen, r-LGF, and that hypothesized for phorbol esters via protein kinase C with regard to the early activation of Na uptake in hepatocytes.

INTRODUCTION

Transplasma membrane ferricyanide reductase of hepatocytes is activated by mitogens, depends on external Na^+ or Li^+

and pH, and is sensitive to amiloride[1]. Ferricyanide increases Na influx in a concentration-dependent way and promotes H^+ efflux[2].

The activity is linked to the growth state of cultured HeLa cells and to the Na/H antiport operation[2]. Thus, it seems to play a role in early mitogenic activation in eukaryotes[3], apparently in connection with the transferrin-receptor expression, linked to the reductive removal of iron from transferrin[4].

Considerable importance is ascribed to protein kinase C in signal transduction in many cell types[5], and to its role in activating Na uptake[6], although several discrepancies have emerged from the work of different authors[7].

In looking into the mitogenic activation mechanisms of the hepatocytes, we studied the effects of r-LGF, the hepatocyte's specific mitogen[8], on DNA synthesis, ^{22}Na uptake and ferricyanide reductase activation, and the possible link between the redox activity and protein phosphorylation in activating Na influx at short times, when the physiological mitogen is present.

MATERIAL AND METHODS

All the reagents came from Sigma. Hepatocyte culture medium (Waymouth M75/B2), fetal calf serum and antibiotics were from Flow. Culture plasticware was from Costar. ^{22}Na chloride (specific activity 18 Ci/mmol), ^{3}H-U-thymidine (specific activity 20 Ci/mmol) and γ-^{22}P-ATP (specific activity 2,000 Ci/mmol) were from New England Nuclear.

Diferric transferrin was prepared by the method of Karin and Mintz[9]. r-LGF was purified from plasma of partially hepatectomized rats[8].

Hepatocytes were isolated and cultured as described[8]. When used for NADH determination[2], protein phosphorylation[6], Na uptake[8] or reductase measurements[1], cells were resuspended in measuring buffer (TD) composed of 145 mM NaCl, 5 mM KCl, 1.2 mM Na_2HPO_4 and 25 mM Tris-ClH, pH=7.4. Reductase measurements were as previously described[1]. Thymidine incorporation into DNA after 24 hours of incubation was done according to Díaz-Gil et al.[8]. H^+ efflux was as described[1], at 22°C, using TD buffer in which Tris-ClH was lowered to 1.2 mM. Protein was measured by a modification of the Lowry method[10].

28

RESULTS AND DISCUSSION

100 µM ferricyanide and 5 µg/ml diferric transferrin, synergistically increase DNA synthesis in cultured hepatocytes when present for 24 hours, together with submaximal doses (5 ng/ml) of r-LGF, approaching the value obtained at maximum dose (10 ng/ml) as shown in Figure 1, although each substrate alone seems to increase thymidine incorporation to a lesser extent (not shown).

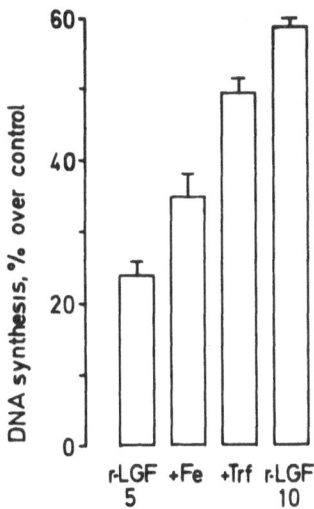

Fig. 1. Effect of 100 µM ferricyanide, 5 µg/ml diferric
transferrin or 5 or 10 ng/
ml r-LGF on DNA synthesis
of cultured hepatocytes.

The synergism observed points out the possible participation of the reductase activity in the signaling pathway for mitogenesis in hepatocytes[2], especially due to the increase in expression of the transferrin receptor when acted upon by growth factors[15] and to the fact that the reductase seems to form part of the transferrin receptor signaling pathway in other cell types[4].

At short times after addition, ferricyanide by itself decreases the NADH content of hepatocytes to a greater extent than does r-LGF alone, the effect depending upon the presence of Na in the external medium. The two together decrease this level further, as shown in Table I, in agreement with the proposed role of reductase in cell activation[4].

Likewise, ferricyanide increases medium acidification in

the presence of hepatocytes and sodium, as shown in Figure 2 trace b, although ferricyanide in absence of cells does not promote appreciable changes in medium acidification, trace a.

Table 1. NADH content of isolated rat hepatocytes in presence of r-LGF and ferricyanide.

Condition	NADH plus Na nmol/mg prot/5' ± SD	NADH minus Na nmol/mg prot/5' ± SD
Control	0.52 ± 0.04	0.55 ± 0.06
100 µM ferricyanide	0.35 ± 0.03	0.50 ± 0.02
10 ng/ml r-LGF plus ferricyanide	0.43 ± 0.02	—
plus r-LGF	0.20 ± 0.01	0.54 ± 0.10

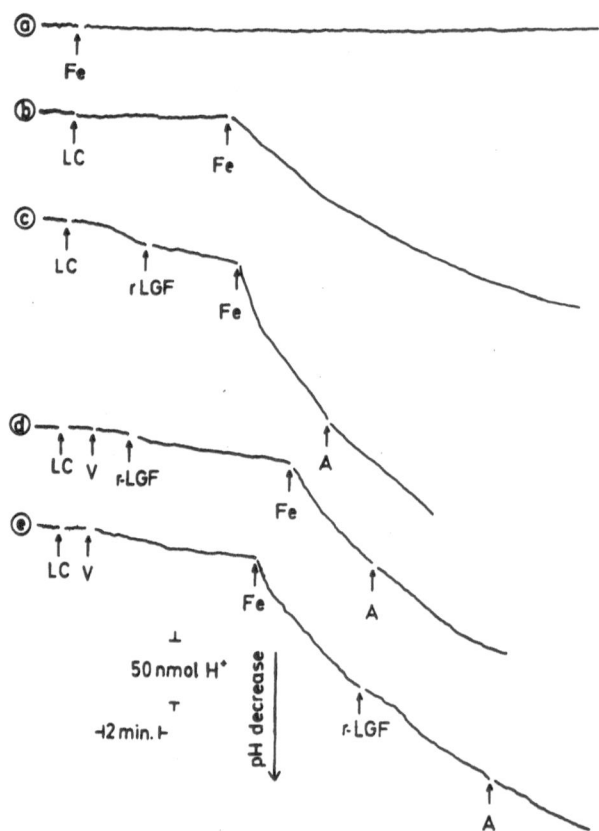

Fig. 2. H^+ efflux from isolated hepatocytes in presence of several effectors. Fe, 100 µM ferricyanide; LC, 1 µg/ml hepatocytes; A, 0.4 mM amiloride; and V, 5 µg/ml vinblastine sulfate.

The acidification produced by ferricyanide almost doubles when 10 ng/ml r-LGF is present, trace c, and shows partial sensitivity to 0.4 mM amiloride, indicating the relationship between ferricyanide reduction and H efflux in hepatocytes. The latter is probably governed in part by the Na/H antiport since it is responsive to mitogens in hepatocytes. More support comes from the effects of 5 μm/ml vinblastine on acidification in presence of ferricyanide and r-LGF, traces d and e, in which both the activation and the sensitivity to amiloride are abolished, indicating a link between the binding of the mitogen to its putative receptor and the activation of both the reductase and the antiport system. Such a link is consistent with the effects of cytoskeleton disruptors in transmembrane control of receptors[11,16].

The existence of said interaction is reinforced by the fact, shown in Figure 3, that the Na-loading of hepatocytes abolishes the r-LGF-promoted activation of the reductase activity, as well as its amiloride sensitivity, despite the lack of effect on the basal redox activity. Thus, it seems that the normal operation of the reversible Na/H antiporter is required for signal transduction via reductase activation in hepatocytes.

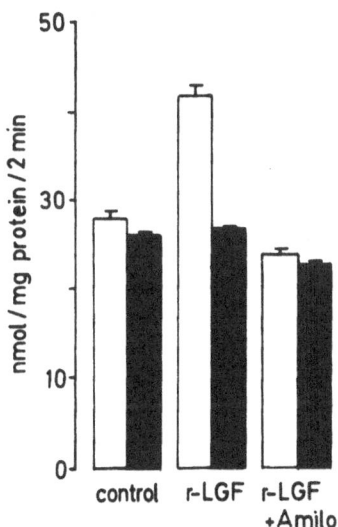

Fig. 3. Internal sodium dependence of ferricyanide reductase. Control unloaded cells, open bars; Na-loaded, closed bars.

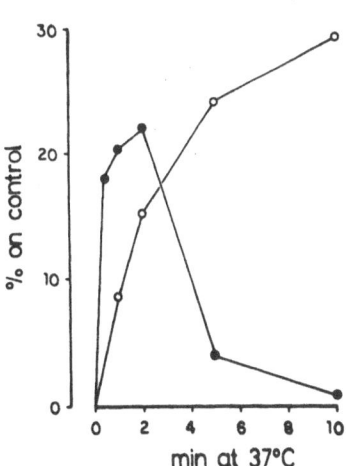

Fig. 4. Time dependence of 100 μm ferricyanide reduction (●) and Na uptake (o) by isolated hepatocytes.

The results suggest that the activation of the transmembranal enzyme by the mitogen is reversible and hence a transient phenomenon. As shown in Figure 4, r-LGF-activated ferricyanide reduction starts at least as early as Na uptake, reaching a maximum about 2 minutes after addition and decreasing rapidly to basal levels at 10 min, whereas [22]Na uptake continues, in agreement with the slow, time-dependent decrease in medium acidification shown in Figure 2, traces b and c.

Because a role in signal transduction for hormones and growth factors has been proposed for phosphoinositides and calcium[12], we looked into the effects of inhibitors of signal transfer through the above on the reductase activity, as shown in Table II.

Table II. Effect of several inhibitors on ferricyanide reductase of hepatocytes in presence of 1.2 mM external calcium.

Inhibitor	nmol/mg prot./2 min ± SD	%
None	51.2 ± 4.4	100
Amiloride, 0.4 mM	25.3 ± 4.5	49
H-7, 0.05 mM	40.9 ± 7.6	80
Trifluoperazine, 0.1 mM	50.9 ± 7.3	99
A-23187, 0.021 mM	55.3 ± 6.3	117
Isobuthylmethylxantine, 0.1 mM	53.4 ± 4.1	109

At concentrations active in other systems, the calcium ionophore A-23187, the calmodulin antagonist trifluoperazine and the phosphodiesterase inhibitor isobuthylmethylxanthine have no effect on basal, sodium-dependent reductase activity. However, H-7 (1,5-isoquinolinyl-sulfonyl-2-methylpiperazine) a protein kinase C inhibitor, slightly inhibits the reductase activity, while amiloride inhibits 50% of the activity.

To test the possible participation of protein phosphorylation in the activation of the reductase and the Na/H antiport, we studied the effects of tetradecanoyl-phorbol dimyristate (TPA) and diacylglycerol (OAG) on these activities. Figure 5 shows the lack of effect of TPA (50 nM to 0.5 µM) on ferricyanide reductase activity, either in the presence or absence of Na or Li in the external medium, indicating lack of TPA action in promoting the activation of electron transport.

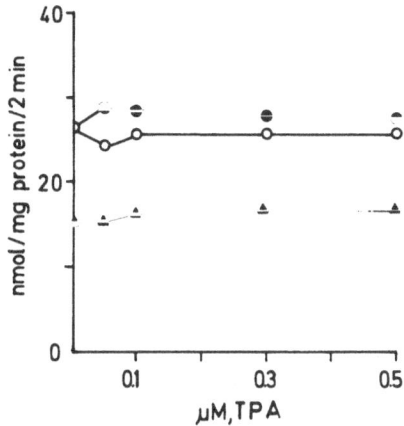

Fig. 5. Effect of TPA on 100 μM ferricyanide
reduction in hepatocytes, in presence of
Na (o), in presence of Li (•) or in
their absence (▲).

In contrast, similar to ferricyanide and r-LGF, TPA activates the amiloride-sensitive Na uptake in cultured hepatocytes, as Figure 6 shows, which affords the possibility that
protein phosphorylation is implicated in antiporter activation[13]. However, when cells are treated for 24 hours with 1
μM TPA, which depresses TPA receptors in other cells[14], ferricyanide and r-LGF activate further Na uptake, indicating the
existence of alternative and separate pathways for the Na/H
antiporter activation.

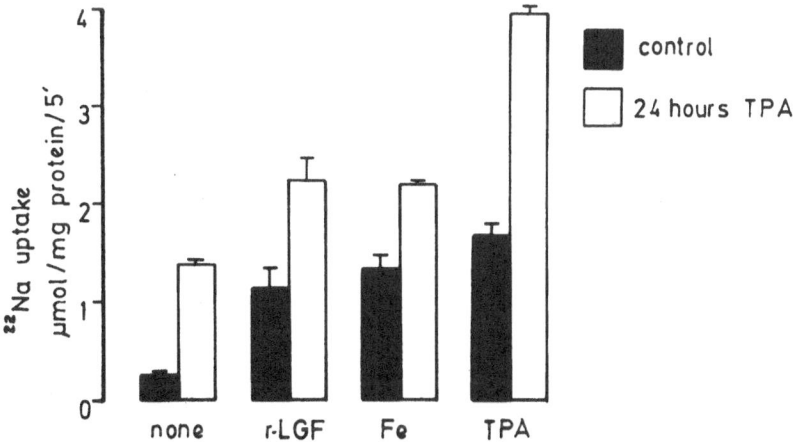

Fig. 6. Effect of 1 μM TPA preincubation on Na uptake activated by r-LGF and ferricyanide in cultured hepatocytes.

The above assertion is reinforced by the results shown in Table III, in which protein phosphorylation promoted by cytosolic, Ca- and phospholipid-dependent protein kinase C is decreased at short times, either by TPA or OAG, but remains unaltered when r-LGF or ferricyanide are present, in agreement with the effects observed for insulin and EGF in hepatocytes[7].

Table III. Effect of r-LGF, TPA, OAG and ferricyanide on soluble protein kinase C in hepatocytes (values are expressed as pmol ^{32}Pi incorporated/mg cell protein ± SD of three separate determinations.

Condition	Phosphate incorporation	
	Experiment 1	Experiment 2
Basal	264.0 ± 56.0	220.0 ± 22.0
then, 1 min at 37°C	309.1 ± 15.7	205.0 ± 10.0
+ TPA, 1 μM	171.4 ± 0.9	92.3 ± 30.0
+ OAG, 0.1 mg/ml	——	98.0 ± 5.0
+ r-LGF 10 ng/ml	271.3 ± 30.0	250.0 ± 7.0
+ ferricyanide 100 μM	355.8 ± 76.7	230.0 ± 23.0

The data support the idea that the early mitogenic activation of hepatocytes can occur through at least two different mechanisms. One, involves activation of the reductase prior to the activation of the Na/H exchanger, which is promoted by the presence and binding of the physiological hepatic mitogen to its receptor. This mechanism could require or be potentiated by reducible substrates. The other involved agents that are capable of activating the antiporter by membrane protein phosphorylation.

ACKNOWLEDGMENTS

This work has been supported partially by grants nos. 86/798 and 87/877 from the Fondo de Investigaciones Sanitarias, Spain. The author thanks M. Messman for her correction and typing of the text.

REFERENCES

1. R. García Cañero, J.J. Díaz Gil and M.A. Guerra, Is transmembrane ferricyanide reductase connected with early prereplicative signals in hepatocytes?, in: "Redox Functions of the Eukaryotic Plasma Membrane," J.M. Ramírez, ed., Madrid (1987).

2. I.L. Sun, R. García Cañero, W. Liu, W. Toole-Sims, F.L. Crane, D.J. Morré, and H. Löw, Diferric transferrin reduction stimulates the Na/H antiport of HeLa cells, Biochem. Biophys. Res. Comm. 145:467 (1987).

3. R. García Cañero, and M.A. Guerra, External sodium dependence and variation in activity of transmembrane ferricyanide reductase during tumor cell growth, Oncología, in press (1988).

4. H. Löw, I.L. Sun, P. Navas, C. Grebing, F.L. Crane, and D.J. Morré, Transplasmalemma electron transport from cells is part of a diferric transferrin reductase system, Biochem. Biophys. Res. Comm. 139:1117 (1986).

5. Y. Nishizuka, The role of protein kinase C in cell surface signal transduction and tumour promotion, Nature 308:693 (1984).

6. A.N. Malviya, and P. Anglard, Modulation of cytosolic protein kinase C activity by ferricyanide: priming event seems transmembrane redox signaling. FEBS 200:265 (1986).

7. F. Vara, and E. Rozengurt, Stimulation of Na/H antiport activity in EGF and insulin occurs without activation of protein kinase C, Biochem. Biophys. Res. Comm. 130:646 (1985).

8. J.J. Díaz Gil, P. Escartín, R. García Cañero, C. Trilla, J.J. Veloso, G. Sánchez, A. Moreno Caparrós, C. Enrique de Salamanca, R. Lozano, J. Gavilanes, and M. García Segura, Purification of a liver DNA-synthesis promoter from plasma of partially hepatectomized rats, Biochem. J. 235:49 (1986).

9. M. Karin, and B. Mintz, Receptor-mediated endocytosis of transferrin in developmentally totipotent mouse tecatocarcinoma stem cells, J. Biol. Chem. 256:3245 (1981).

10. M.A.K. Markwell, S.M. Haas, L.H. Bieber, and N.E. Tolbert, A modification of the Lowry method to determine protein in lipoprotein samples, Anal. Biochem. 87:206 (1978).

11. G.L. Nicholson, Transmembrane control of the receptors in normal and tumor cells, Biochem. Biophys. Acta 457:57 (1976).

12. M.J. Berridge, Inositol trisphosphate and diacilglycerol as second messengers, Biochem. J. 220:345 (1984).

13. W. Kruijer, H. Skelley, F. Bottieri, H. van der Putten, J.R. Barber, I.M. Verma, and H. Leffert, Protooncogene

expression in regenerating liver is stimulated in cultures of primary adult rat hepatocytes, J. Biol. Chem. 261:7927 (1986).

14. A. Rodríguez-Pena, and E. Rozengurt, Disappearance of Ca-sensitive, phospholipid-dependent protein kinase C activity in phorbol ester-treated 3T3 cells, Biochem. Biophys. Res. Comm. 120:1053 (1984).

15. R.J. Davis, and M.P. Czeh, Regulation of transferrin receptor expression at the cell surface by insulin-like growth factors, EGF and PDGF, EMBO J 5:653 (1986).

16. M. Prentki, H. Crettaz, and B.J. Jeanrenaud, Role of microtubules in insulin and glucagon stimulation of amino acid transport in isolated rat hepatocytes, J. Biol. Chem. 256:4336 (1981).

THE REQUIREMENT FOR TRANSMEMBRANAL ELECTRON FLOW FOR

CELL PROLIFERATION

F.L. Crane, H. Löw, I.L. Sun and M. Isaksson

Department of Biological Sciences
Purdue University, W. Lafayette, IN 47907 USA
and Department of Endocrinology, Karolinska
Institute, Stockholm, Sweden

INTRODUCTION

Diferric transferrin or other iron complexes have long been recognized as essential for cellular growth[1-5]. This requirement is primarily based on the essential role of iron in heme and DNA synthesis[6,7]. Removal of iron from cells inhibits growth[8]. The iron monement into the cell may depend on endocytosis of ferric transferrin attached to transferrin receptor in the plasma membrane[9] or ferric iron uptake directly through the membrane by undefined transport systems[5,10]. Ferric chelates have also been used in place of transferrin to stimulate growth[4,11]. The discovery of an electron transport system in the plasma membrane which could reduce ferric chelates as well as other impermeable oxidants[12,13] introduced a mechanism for ferric iron reduction outside the membrane before transport as ferrous iron. The discovery that impermeable ferricyanide stimulated cell growth under serum deficient conditions introduced a possible new role for ferric compounds as external oxidants to stimulate cell proliferation[13,14]. Since it has been shown that ferricyanide cannot stimulate growth of L1210 cells after iron depletion by desferrioxamine and cannot act as an iron supply for these cells[5] we examined its effect on growth of cells which had not been depleted of iron.

METHODS

Culture of L1210 cells without serum on RPMI 1640 was according to Basset et al[3]. 3T3 cells were cultured without serum on DMEM for 2 days with indicated ferricyanide concentration. DNA synthesis was measured by pulsed incorporation of H^3 thymidine at 0.1 μC/ml for 24 hr.

Reduction of ferric ammonium sulfate at 37° was measured by formation of ferrous bathophenanthroline disulfonate

(BPS) at 535-600 nm with the dual beam spectrophotometer. Reaction mixture contained 130 mM NaCl, 5 mM KCl, 1 mM $MgCl_2$ and 1 mM $CaCl_2$ in a total volume of 2.5 ml with 10 mM Hepes buffer pH 7.4, 0.5 mM bicarbonate, 15 µM BPS, 7.5 µM ferric ammonium citrate (Fe equivalent) and 2 to 15 x 10^6 cells[12]. Reduction of loosely-bound iron associated with diferric transferrin was measured in the same way using 17 µM diferric transferrin containing one free Fe^{+++} per mole. Ferricyanide reduction was measured at 420-500 nm with 0.2 mM potassium ferricyanide as described[15]. Protein was determined according by Bio Rad procedure.

RESULTS

Growth of L1210 cells in serum deficient media is stimulated by low concentrations of ferricyanide as measured by cell count (fig. 1), total protein (fig. 2) and H^3 thymidine incorporation (fig. 3). The stimulation reaches a maximum at 0.003 to 0.03 mM ferricyanide with a decline at higher concentrations. Diferric transferrin at a much lower concentration causes growth stimulation also. It should be noted that this diferric transferrin contained loosely bound iron at a one to two mole ratio to the tightly bound iron so the effects may be based on readily available ferric iron. Stimulation is seen by cell count (fig. 4), protein increase (fig. 5) and H^3 thymidine incorporation (fig. 3). An impermeable organic oxidant, indigo-disulfonate (indigo-carmine), can also be used to stimulate growth (fig. 6, fig. 7) but the growth stimulation is not as good as the growth with ferricyanide. The indigo-disulfonate, however, is only one third as active as ferricyanide as an external electron acceptor (not shown) with these cells.

Insulin has very little effect on growth of the L1210 cells and does not stimulate the transmembrane electron transport measured with ferric ammonium citrate, diferric transferrin or ferricyanide (Table 1).

Table 1. Reduction of external electron acceptor by L1210 cells.

Electron acceptor	Reduction rate
	nmole Fe min^{-1} $cells^{-6}$
Ferric ammonium citrate 7.5 µM	0.10
Fe^{+++} diferric transferrin 17 µM	0.08
Ferricyanide 0.2 mM	0.31

Assays as described in[12].

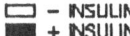

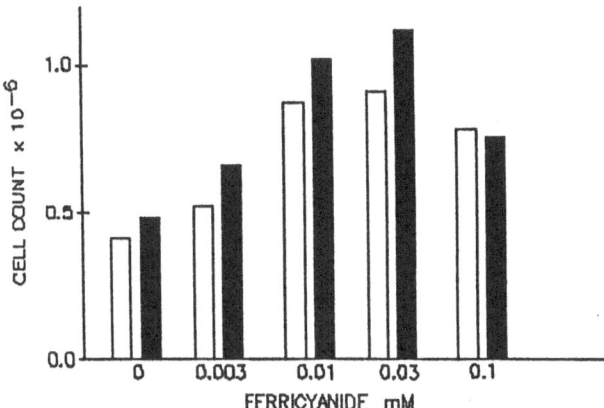

Fig. 1.   Stimulation of L1210 cell proliferation
by ferricyanide measured by cell count.
Growth in serum free RPMI medium for 24
hours with and without 1 μg/ml insulin.

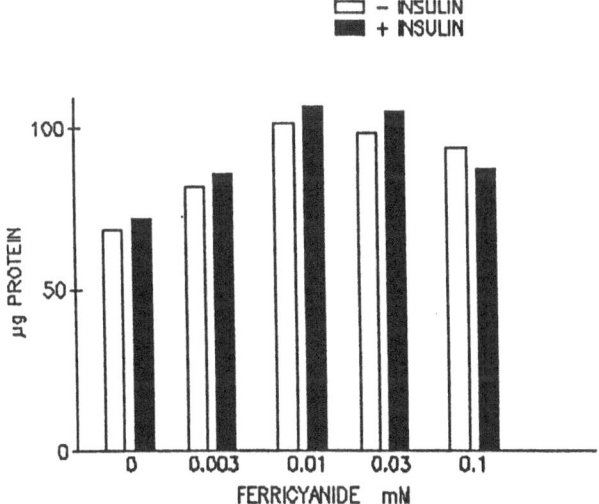

Fig. 2.   Stimulation of growth of L1210 cells
with ferricyanide measured by protein
content. Cells grown in serum free
media with and without 1 μg/ml insulin
for 24 hours.

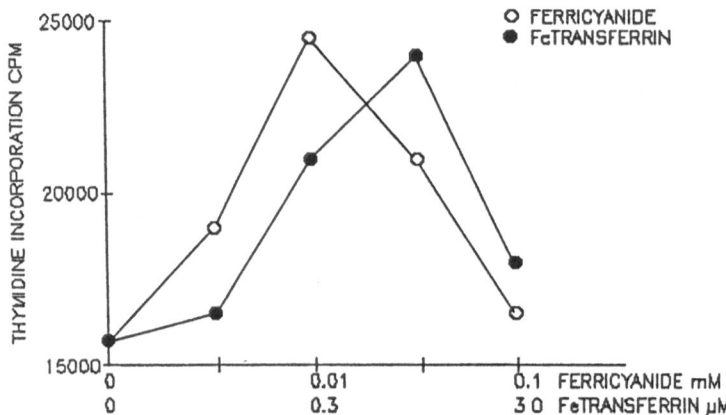

Fig. 3.   Stimulation of DNA synthesis in L1210
          cells by ferricyanide and ferric iron
          together with diferric transferrin.
          Measured by 24 hr incorporation of H^3
          thymidine.

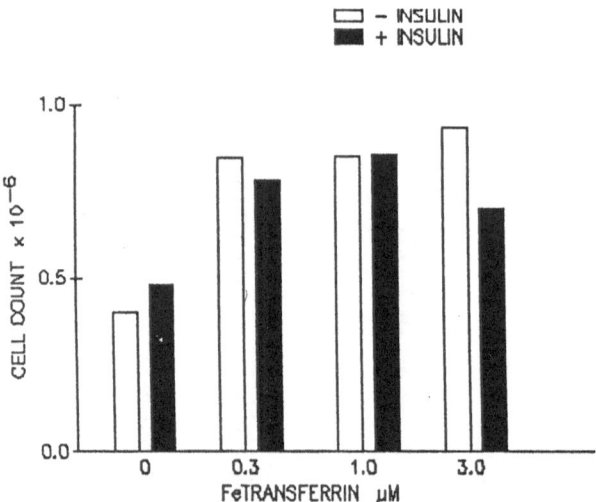

Fig. 4.   Stimulation of L1210 cell proliferation
          by diferric transferrin containing 1 mole
          per mole loosely bound iron with and with-
          out 1 μg/ml insulin. Same cell culture as
          used for fig. 1.

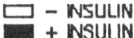

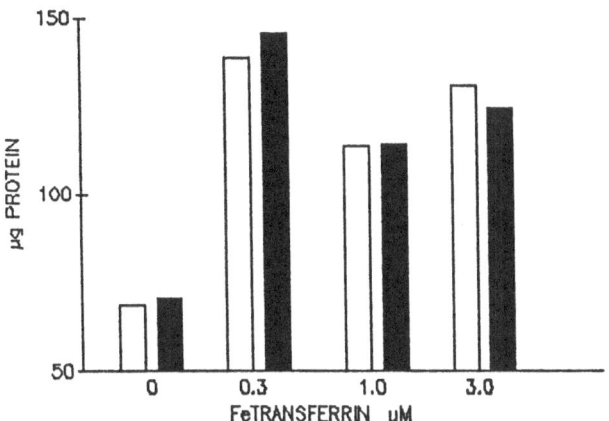

Fig. 5.   Effect of diferric transferrin plus
equimolar loosely bound iron on protein
increase of L1210 cells. Culture 24
hours with and without insulin as for
fig. 2.

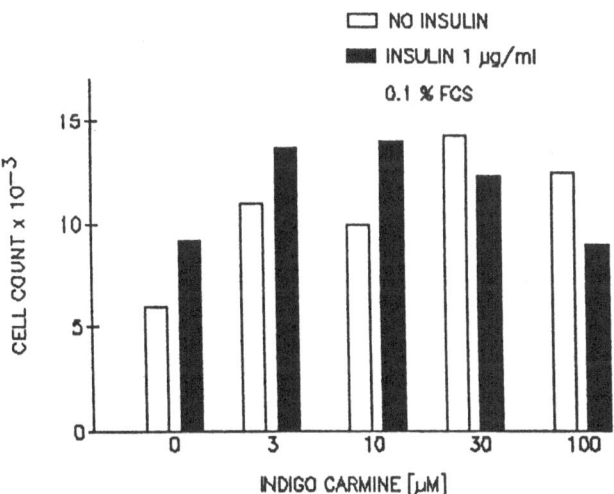

Fig. 6.   Stimulation of L1210 cells by indigo-
-disulfonate with and without 1 μg/ml
insulin in media with 0.1% fetal calf
serum growth for 24 hr.

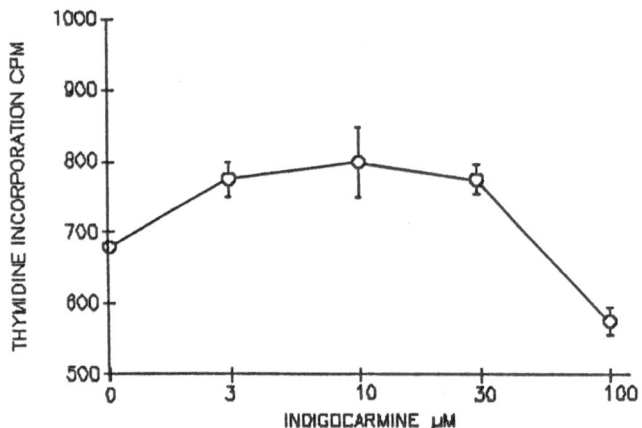

Fig. 7. Stimulation of DNA synthesis in L1210
cells by indigo-disulfonate with 0.1%
fetal calf serum as measured by incorporation of H^3 thymidine 24 hr.

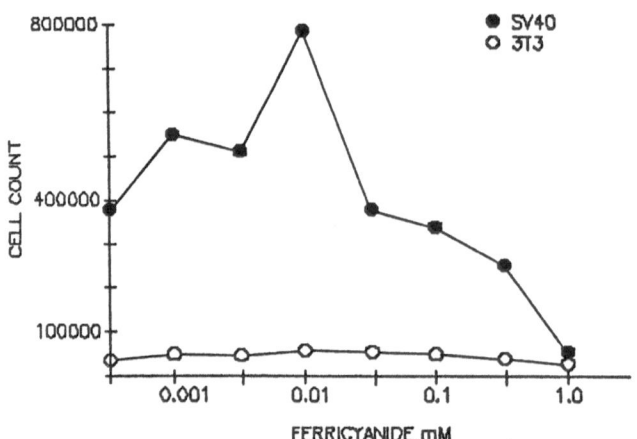

Fig. 8. Effect of ferricyanide on the growth of
swiss 3T3 and SV40 transformed 3T3 in
serum free media with no addition for
48 hours. Growth determined by cell count.

In contrast to L1210 cells, swiss 3T3 cells do not grow with simple addition of ferricyanide to minimal media. SV 40 transformed 3T3 cells do respond to added ferricyanide (fig. 8).

DISCUSSION

Basset et al[5] have shown that ferric citrate, but not ferricyanide, can reinitiate DNA synthesis in L1210 cells after depletion of iron by desferrioxamine. They find only a 100% increase in H^3 thymidine incorporation over controls in cells exposed to 0.1 mM potassium ferricyanide four hours after desferrioxamine treatment whereas ferric citrate gives an 800% increase. By considering the effect of heme synthesis inhibitors they conclude that the iron is required in ribonucleotide reductase for DNA synthesis[7].

These experiments indicate that ferricyanide is not a source of iron for L1210 cells. It has previously been shown that ferricyanide is impermeable to liver cells or erythrocytes[13,16]. Basset et al[5] indicate that it is unlikely that iron can stimulate growth by stimulation of transplasma membrane electron flow. This is clearly true when cells have been depleted of iron to inhibit DNA synthesis since transplasma membrane electron flow would not restore ribonucleotide reductase or other essential internal iron functions. However, if internal iron is undisturbed by chelator we show external electron acceptors such as ferricyanide, and even indigo-disulfonate[17] can stimulate L1210 cell proliferation in serum free media.

With HeLa cells the growth stimulation by ferricyanide and the rate of ferricyanide reduction were markedly increased by high concentrations of insulin[18]. It is clear that insulin has very little effect on oxidant stimulated growth or transmembrane electron transport in L1210 cells.

SV 40 transformed swiss 3T3 cells also show increased growth with ferricyanide but swiss 3T3 cells do not respond to ferricyanide alone. Since growth of untransformed 3T3 cells requires growth factors such as PDGF to initiate growth it appears that transmembrane electron transport cannot initiate the G_0 to G_1 transition but can be important at a later stage of the cell cycle[19].

In conclusion iron is essential for cell growth and cannot be replaced by other factors. External oxidants can stimulate growth if iron is in the cells (or media) but cannot stimulate in depleted cells. Ferric salts can however supply iron and act as external oxidants for transmembrane electron transport. In addition, in cells dependent on other growth factors for initiation of growth the transmembrane electron transport cannot act as a substitute for the other factors but it is not clear if the growth factors can eliminate the need for electron transport.

Supported in part by grants from the National Institutes of Health PO1CA36761 and GMK6-21839.

References

1. P.S. Rudland, H. Durbin, D. Chingan and L. Jimenez de Asua, Iron salts and transferrin are specifically required for cell division of 3T6 cells, Biochem Biophys Res Communs 75: 556 (1977).
2. Barnes and G. Sato, Serum free cell culture; a unifying approach. Cell 22: 649 (1980).
3. P. Basset, J. Zwiller, M.O. Revel and G. Vincendon, Growth promotion of transformed cells by iron in serum free culture, Carcinogenesis 6: 355 (1985).
4. W. Landschulz, I. Thelseff and P. Ekblom, A lipophilic iron chelator can replace transferrin as a stimulator of cell proliferation and differentiation. J Cell Biol 98: 596 (1984).
5. P. Basset, Y. Quesneau and J. Zwiller, Iron-induced L1210 cell growth: Evidence of a transferrin-independent iron transport. Cancer Res 46: 1644 (1986).
6. P. Aisen and I. Listowsky, Iron transport and storage proteins. Ann Rev Biochem 49: 357 (1980).
7. P. Reichard and A. Ehrenberg, Ribonucleotide reductase - a radical enzyme. Science 221: 514 (1983).
8. H.M. Lederman, A. Cohn, J.W.W. Lee, M.H. Freedman and E.W. Zelfand, Desferrioxamine: A reversible S-phase inhibitor of human lymphocyte proliferation. Blood 64: 748 (1984).
9. J-H. Octave, Y-J. Schneider, A. Trouet and R.R. Crichton, Iron uptake and utilization by mammalian cells. Trend Biochem Sci 8: 217 (1983).
10. C. van den Heul, A. Veldman, M.S. Kroos and H.G. van Eijk, Two mechanisms are involved in the process of iron uptake by rat reticulocytes. Int J Biochem 16: 383 (1984).
11. P. Ponka, R.W. Grady, B. Wilczynska and H.M. Schulman, Iron chelates supply iron for cell growth. Biochim Biophys Acta 802: 477 (1984).
12. H. Löw, C. Grebing, A. Lindgren, M. Tally, I.L. Sun and F.L. Crane, Involvement of transferrin in the reduction of iron by the transplasma membrane electron transport system. J Bioenerg Biomemb 19: 535 (1987).
13. F.L. Crane, I.L. Sun, M.G. Clark, C. Grebing and H. Löw, Transplasma-membrane redox systems in growth and development. Biochim Biophys Acta 811: 233 (1985).
14. G.F. Kay and K.A.O. Ellem, Nonhaem complexes of Fe III stimulate cell attachment and growth by a mechanism different from that of serum; 2-oxycarboxylates and hemoproteins. J Cellular Physiol 126: 275 (1986).
15. I.L. Sun, F.L. Crane, C. Grebing and H. Löw, Properties of a transplasma membrane electron transport system in HeLa cells. J Bioenerg Biomemb 16: 514 (1983).
16. M.G. Clark, E.J. Partick, G.S. Patten, F.L. Crane, H. Löw and C. Grebing, Evidence for the extracellular reduction of ferricyanide by rat liver. Biochem J 200: 565 (1981).
17. M.M. Brooks, Activation of eggs by oxidation-reduction indicators. Science 106: 320 (1947).
18. I.L. Sun, F.L. Crane, C. Grebing and H. Löw, Transmembrane redox in control of cell growth. Exp Cell Res 156: 528 (1985).
19. I. Zachary, P.J. Woll and E. Rozengurt, A role for neuropeptides in the control of cell proliferation. Develop Biol 124: 295 (1987).

ROLES FOR PLASMA MEMBRANE REDOX SYSTEMS IN CELL GROWTH

D. James Morré, Andrew Brightman, Juan Wang, Rita Barr and
Frederick L. Crane

Department of Medicinal Chemistry and Department of
Biological Sciences, Purdue University, West Lafayette
IN 47907 USA

INTRODUCTION

Correlative evidence from cultured mammalian cells suggests that
growth and plasma membrane redox activities are somehow related. This
comes in part from studies with hormones, growth factors and anticancer
drugs where substances that stimulate or inhibit growth respectively
stimulate or inhibit plasma membrane redox activities to corresponding
degrees and vice versa (Crane et al., 1985). One possibility is that
plasma membrane redox activities are somehow involved in cell cycle
control but the mechanisms by which that regulation might be achieved
remain obscure. In this report, we have extended the observations from
mammalian cells to elongating plant tissues and regulation by auxin
hormones where growth is due primarily, if not exclusively, to an
increase in cell size.

GROWTH FACTOR STIMULATION OF A TRANSPLASMAMEMBRANE REDOX SYSTEM IN HELA
CELLS

A transplasma membrane redox system that transfer electrons from
reducing agents in the cytoplasm to external impermeable oxidants is
ubiquitous among eukaryotic cells (Crane et al., 1985). Many growth
factors and hormones that stimulate growth increase the rate of oxidant
reduction (Goldenberg et al., 1979; Sun et al., 1985; Crane et al. 1985).
For example, insulin (30 μg/ml) enhances the growth and increases
simultaneously the rate of oxidant reduction of HeLa cells in a serum
free medium (Fig. 1). These growth studies were carried out with cells
harvested during the exponential growth phase. Additionally, the
coupling of proton release to this electron transport indicates that
local membrane energization may be affected by the transmembrane electron
flow or that intracellular pH may change (Sun et al., 1987). Activation
of a Na^+/H^+ antiport and an elevation of cytoplasmic pH is a common
response to growth factors (Moolenaar, 1986).

IMPERMEANT ELECTRON ACCEPTORS FOR THE TRANSPLASMA MEMBRANE REDUCTASE
STIMULATE GROWTH OF HELA CELLS

With the discovery that impermeable ferricyanide could stimulate
growth of serum-deficient melanoma cells it was considered that a redox
function of iron at the cell surface could be important to trigger or
energize cell growth (Ellem and Kay, 1983). Subsequently, a variety of

impermeable electron acceptors that stimulate plasma membrane redox activities were shown to stimulate cell growth such as ferric chloride (Basset et al., 1986), ferric isopyridoxylnicotinamide (Lanschulz et al., 1984; Ponca et al., 1984) or diferric transferrin (Barnes and Sato, 1980). Stimulation of growth by oxidants is not limited to iron compounds as the growth of HeLa cells also is stimulated by hexamine-ruthenium III chloride, a trivalent cation, and by indigo-tetrasulfonate (Sun et al., 1985). Impermeable oxidants which do not interact with the electron transport system, do not stimulate growth. The use of a series of impermeable indigosulfonates with different redox potentials showed that extra-cellular oxidants with a redox potential $E'_{7.0}$ above -175 mV could stimulate growth (Sun et al., (1984a).

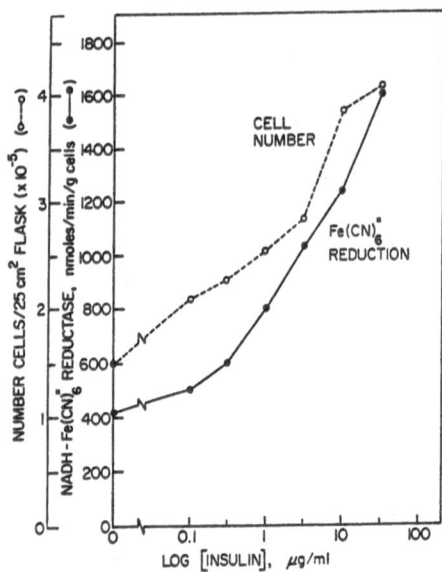

Fig. 1. Insulin promotion of cell growth (0--0) and rate of ferri-cyanide reduction (●—●) for HeLa cells. Redrawn from Sun et al. (1985).

INHIBITION OF PLASMA MEMBRANE NADH DEHYDROGENASES BY ADRIAMYCIN AND OTHER ANTITUMOR DRUGS

Certain clinically useful antitumor drugs inhibit transplasma membrane electron transport. These drugs, adriamycin, cis diammine dichloroplatinum II (cis-platin) and bleomycin, inhibit ferricyanide reduction by HeLa cells at the same concentrations of the drugs that inhibit growth or are cytotoxic (Sun and Crane, 1981; 1984a,b; 1985; Sun et al., 1986; Fig. 2). Others have shown that impermeable adriamycin derivatives can inhibit growth (Tritton and Yee, 1982; Rogers and Tokes, 1984) and, on this basis, have proposed a site of action at the plasma membrane.

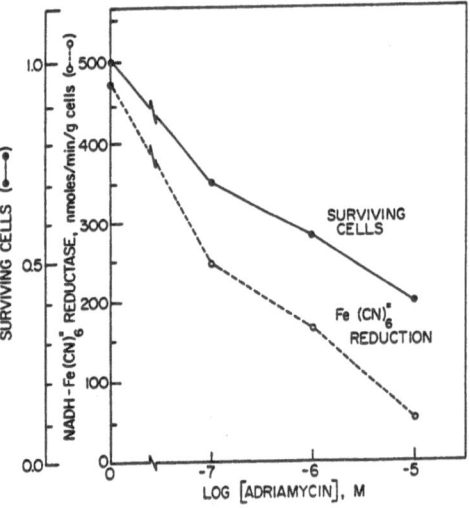

Fig. 2. Parallel effects of adriamycin on ferricyanide reductase and survival of HeLa cells treated with adriamycin. From Sun and Crane (1984b). Cell survival was measured after 1 h of drug treatment using the eosin Y exclusion method.

That the inhibition of NADH-ferricyanide reductase by adriamycin and other antitumor agents is through a direct effect of the drug on the enzyme is shown by studies where these drugs show inhibitions of activities of highly purified plasma membrane preparations comparable to those observed with cells (Fig. 3). With adriamycin, the NADH-ferricyanide reductase activity of mouse liver plasma membrane begins to be inhibited at adriamycin concentration of 5×10^{-8} M, with maximum inhibition at 10^{-6} M. Half maximum inhibition occurs at about 10^{-7} M (Sun and Crane, 1984b).

Finally, transformation modified the transplasma membrane electron transport system. The changes are seen both as a lower activity in transformed cells (Fig. 4) and by a greater sensitivity to antitumor drugs (Crane et al., 1985; Sun et al., 1986; Fig. 4). Studies with SV-40 transformed embryonic liver cells and with temperature sensitive A209-SV-40 pineal cells grown at permissive (33°) and restrictive (40°) temperatures to switch from transformed to untransformed phenotype, show the cells with the transformed phenotype to have a 10-fold or greater sensitivity to the antitumor agents than the same cell line when expressing the more normal phenotype at the restrictive temperature. The findings, taken together, suggest a fundamental modification of plasma membrane redox as one of the alterations of the trans-membrane signal system correlated with loss of growth control during cellular transformation. Its greater sensitivity to antitumor drugs makes it a likely target for regulatory intervention in therapy (See Faulk, this volume).

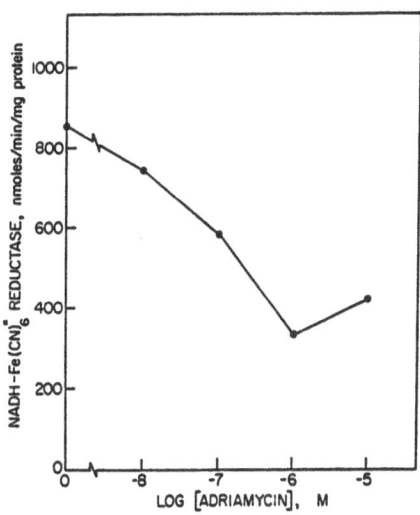

Fig. 3. Inhibition by adriamycin of NADH-ferricyanide reductase of mouse liver plasma membranes isolated according to Goldenberg et al. (1979). The dose response with the isolated plasma membranes is similar to that with cells. From Sun et al. (1984b).

PLASMA MEMBRANE REDOX AND EXPANSION GROWTH OF PLANT CELLS

Growth in a population of cells is normally a combination of an increase in cell number and an increase in cell size. The two processes of cell surface enlargement and increases in cell number (=cell division) have been studied most extensively as isolated cellular events in plant stems or stem-like structures where cell division is restricted to an apical zone and where growth by pure expansion is in a subapical zone and occurs over a period of many hours independently of cell division. A role for plasma membrane redox systems in the enlargement process of growth was evaluated by excising these zones of nearly pure cell elongation and observing their elongation in vitro.

Plasma membranes of plants contain redox systems similar to

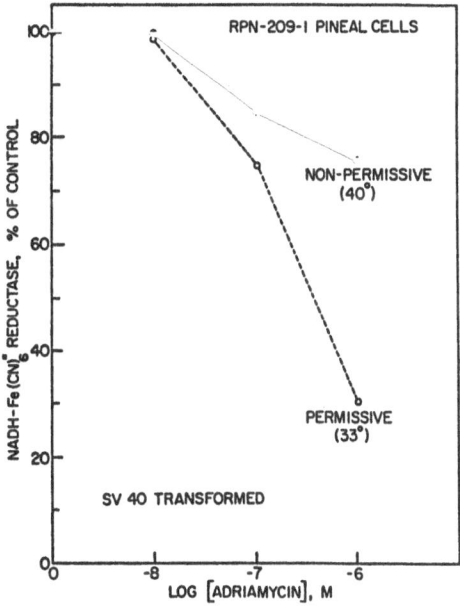

Fig. 4. Inhibition by adriamycin of ferricyanide reduction by SV-40 transformed pineal cells comparing the non-permissive (40°) (uninhibited rate of 214 nmoles min^{-1} g fr wt^{-1}) and permissive (33°) (uninhibited rate of 86 nmoles min^{-1} g fr wt^{-1}) phenotype. From Sun et al. (1986).

47

those of mammalian cells. Reduction of ferricyanide on plasma membranes of dark-grown (etiolated) soybean stems (hypocotyls) has been demonstrated by cytochemistry (Morré et al., 1987a) and biochemical assay using plasma membrane preparations highly purified either by free-flow electrophoresis (Barr et al., 1986) or by aqueous two-phase partition (Sandelius et al., 1986; Buckhout and Hrubec, 1986). The redox system of plant plasma membranes has been implicated in several functions including H^+ efflux, K^+ uptake, iron reduction and membrane polarization (Crane et al., 1985; Møller and Lin, 1986). More recently, a role for plasma membrane redox systems has been indicated in the energization of membrane displacements to form vesicles and in the migration of vesicles once formed (Sun et al., 1984c; Morré et al., 1987b). The findings that plasma membrane redox constituents may respond to hormones in mammalian cells (e.g. Löw and Crane, 1976; Gadya et al., 1977; Goldenberg et al., 1979) prompted us to design a series of experiments to determine if plasma membrane redox activities might also be involved in the expansion phase of growth in plants.

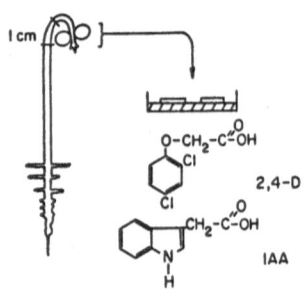

Fig. 5. Diagram of the ca 1 cm long subapical zone of plant stems where growth is by cell expansion without division. These regions, when excised and floated on solutions, continue to elongate for many hours. Elongation is promoted by growth hormones of the auxin type, both synthetic (2,4-D) and natural (IAA). See Text.

In initial studies, the same series of inhibitors and anti-proliferative agents as used to correlate growth and inhibition of plasma membrane redox activities in mammalian cells (Sun and Crane, 1981; 1984; 1985) were applied to the etiolated hypocotyls of soybean (Fig. 6). The results showed both an inhibition of elongation growth (Fig. 6) and, in purified plasma membrane vesicles, of NADH-ferricyanide reductase (Fig. 7) suggestive of a connection between plasma membrane redox activities and the elongation growth process. With both the reductase and growth, a concentration-dependent inhibition was observed for cis-platin, adriamycin and p-nitrophenylacetate, the latter an inhibitor of plasma membrane redox activities reported by Barr et al. (1986). Inhibition was much the same as for mammalian cells (Figs. 2 and 4) and membranes (Fig. 3). The growth active cis-platin was inhibitory whereas the growth inactive trans-platin was largely without effect except at very high concentrations.

As is evident in the data of Figure 6, soybean stem segments incubated in the presence of the synthetic auxin hormone, 2,4-D, elongated approximately two times faster than stem segments in its absence. Interestingly, this auxin-dependent growth was inhibited preferentially by antiproliferative ·drugs compared to control growth (Fig. 6). Nearly identical responses were observed at both 1 μM (Morré et al., 1988a) and 10 μM (Fig. 6) 2,4-D after 4, 16 or 20 h of incubation. Generally, the auxin-stimulated elongation was completely inhibited with only a 10 to 20% inhibition of control growth and approximately 50% of the auxin-stimulated elongation was inhibited at inhibitor concentrations that have no effect on control elongation. These findings suggested that the hormone might exert its growth controlling effects through some component of the plasma membrane redox system.

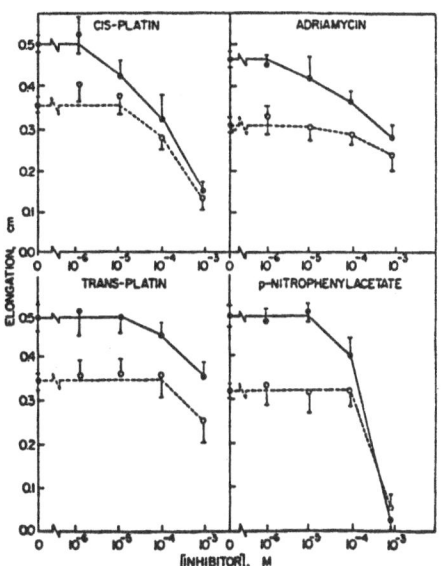

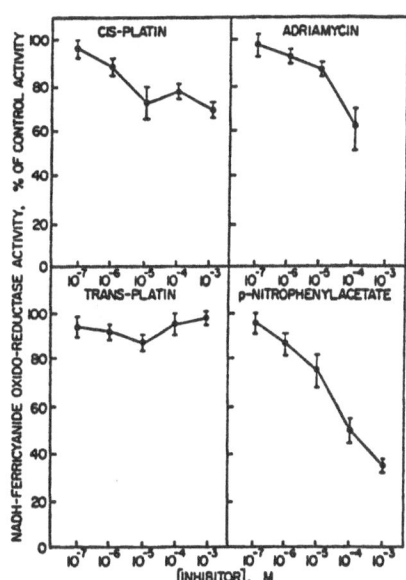

Fig. 6. Elongation of excised 1 cm segments of etiolated soybean seedlings floated for 16 h in the presence (solid symbols) or absence (open symbols) of aqueous solutions of 10 μM 2,4-D with or without the various inhibitors of plasma membrane redox activities. Values are averages of determinations from 3 different experiments ± standard deviations. Nearly identical results were reported by Morré et al. (1988a) for incubations with 1 μM 2,4-D for 20 h.

Fig. 7. NADH-ferricyanide reductase activities of soybean plasma membranes as a function of inhibitor concentration. Control rate (100%) in the absence of inhibitor was 190 ± 15 mmol (mg protein)$^{-1}$ min^{-1}. Values were from six different plasma membrane preparations ± standard deviation. Preparations were preincubated for 3 min with inhibitor. Rate was measured over 5 min after addition of ferricyanide and NADH. From Morré et al. (1988).

With external oxidants, however, the situation with plants differed from that with animals. Growth of carrot cells in culture was inhibited rather than stimlulated by external oxidants (Crane et al., 1984). With the excised soybean hypocotyl segments, ferricyanide had little or no effect over the concentration range 0.01 to 500 mM either with or without 10 μM 2,4-D for 16 h. Similarly, an equimolar mixture of ascorbate and dehydroascorbate (to generate ascorbate free radical as an impermeant external electron acceptor) was without significant effect or was slightly growth inhibitory (6% increase in elongation at 0.1 mM).

AUXIN HORMONE RESPONSIVENESS OF NADH OXIDATION BY PLANT PLASMA MEMBRANES

With plasma membrane vesicles from soybean hypocotyls isolated either by aqueous two-phase separations (Morré et al., 1988b) or by preparative free-flow electrophoresis (Morré et al., 1987), the synthetic plant growth hormone, 2,4-dichlorophenoxyacetic acid (2,4-D) stimulated the oxidation of NADH 2-fold or more (Fig. 8). The activity was stimulated both in the presence or absence of a detergent (Triton X-100) whereas auxin exerted little or no effect on the NADH-ferricyanide reductase of the same vesicle preparations (Brightman et al., 1988). The specificity of the response is illustrated by data of Table 1 where

49

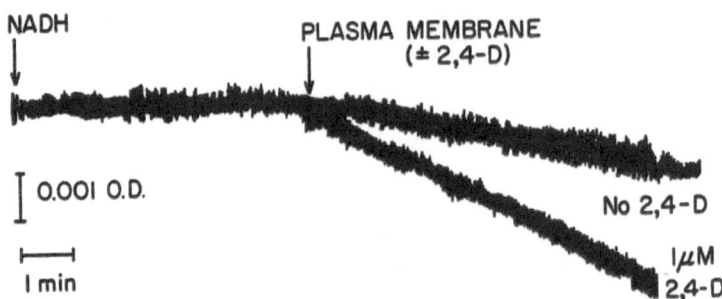

Fig. 8. Spectrophotometric tracing of NADH oxidase activity of plasma membrane vesicles from soybean isolated by aqueous two phase partition. Plasma membrane was added to the reaction to determine the control rate. Subsequently, 1 μM 2,4-D was added and the stimulated rate was measured. Control activity (no 2,4-D) was 0.4 nmol min^{-1} mg protein^{-1}.

Table 1. Specificity of the response to auxins by NADH oxidase of isolated plasma membrane vesicles from dark-grown soybean hypocotyls.

Compound	Concen- tration	Growth Active	Specific Activity
None			18 ± 2
2,4-dichlorophenoxyacetic acid (2,4-D)	1 μM	+	45 ± 15
2,3-dichlorophenoxyacetic acid (2,3-D)	1 μM	-	20 ± 2
α-Naphthyleneacetic acid (α-NAA)	0.1 μM	+	36 ± 3
ß-Naphthyleneacetic acid (ß-NAA)	0.1 μM	-	21 ± 3
Indole-3-acetic acid (IAA)	1 μM	+	32 ± 7
Benzoic acid	1 μM	-	15 ± 5

*nmoles min^{-1} mg protein^{-1}. Adapted from Brightman et al. (1988).

several acids and auxin analogs are compared. The synthetic auxin analog, 2-4-dichlorophenoxyacetic acid (2,4-D) and α-naphthaleneacetic acid (α-NAA) were active as was the natural auxin, indole-3-acetic acid (IAA). The growth inactive analogs, 2,3-dichlorophenoxyacetic acid (2,3-D) and ß-napthyleneacetic acid (ß-NAA) were inactive in stimulating the NADH oxidase as was benzoic acid when tested at the same concentrations as the active materials. The dose response curves for both 2,4-D and IAA in increasing the rate of cell elongation and in affecting the NADH oxidase of isolated membrane vesicles were similar (Morré et al., 1988b; Brightman et al., 1988) except that optimum stimulations of the NADH oxidase occurred at about 10-fold lower concentrations of auxin than did stimulations of growth. As the optimum concentrations were exceeded, both growth and NADH oxidase activities became less.

The auxin stimulated oxidation activity was subsequently solubilized with the zwitterionic detergent CHAPS and purified by ion exchange chromatography and gel filtration approximately 2000-fold over that of the total homogenate (Brightman et al., 1988). Both the partially purified fraction and an active band from nondenaturing gel electrophoresis revealed a complex of three peptide bands when analyzed by denaturing gel electrophoresis. When obtained from plasma membrane vesicles from the region of rapid elongation of soybean hypocotyls, the NADH oxidase complex retained auxin responsiveness throughout purification (3- to 5-fold stimulation by 1 micromolar 2,4-D). With CHAPS (but not Triton X-100), auxin responsiveness was lost at high detergent concentrations but was restored by the removal of the excess CHAPS suggesting that some degree of association between proteins of the complex was necessary for stimulation by the auxin hormones.

TRIIODOTHYRONINE STIMULATION OF NADH OXIDASE AND GROWTH OF LIVER CELLS

3,5,3'-Triiodothyronine (T_3) (Fig. 9) at 0.2 μM stimulates the growth of liver cells and of hepatoma cells to an even greater extent (Solomon and Mishkin, 1982). At 1 μM, T_3 stimulates the NADH oxidase of mouse liver plasma membrane by 250% (Fig. 10).

A dose-response curve for T_3 stimlulation of NADH oxidase of highly purified rat liver plasma membranes shows a degree of stimulation similar to that observed with plasma membranes of mouse liver with an optimum stimulation of about 350% at 2 μM. With fat globule membranes prepared from bovine milk, a much smaller stimulation was observed than with the liver plasma membranes (Fig. 11).

The significance of a growth hormone stimulated NADH oxidase activity for both plant and animal cells is not yet clear. However, the enzyme system does offer the opportunity for transfer of electrons from NADH to molecular oxygen. Among other things, the involvement of a transplasma membrane oxidase in growth or growth control would provide an explanation for oxygen control of cell division (Ghani and Hollenberg, 1978).

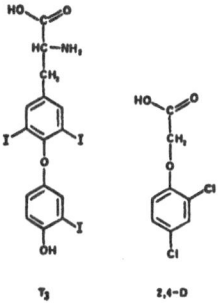

Fig. 9. Structures of triiodothyronine (T_3) and 2,4-dichlorophenoxyacetic acid.

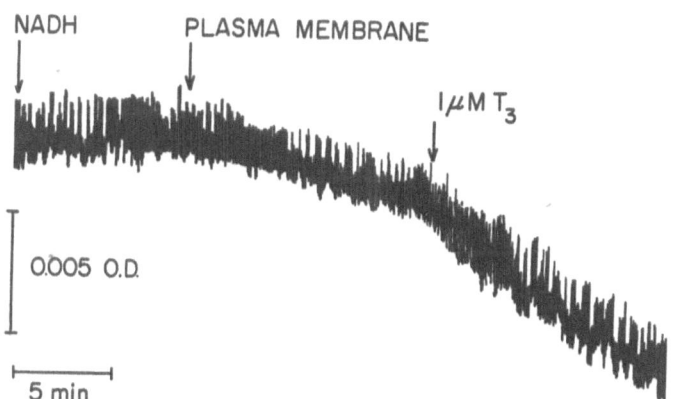

Fig. 10. Spectrophotometric tracing of NADH oxidase activity of plasma membrane vesicles from rat liver isolated by aqueous two-phase partition. Stimulation by 1 μM T_3 was determined by adding hormone to the control reaction. Control activity (no T_3) was 0.7 nmoles min^{-1} mg protein^{-1}.

PLASMA MEMBRANE REDOX ACTIVITY AND
CYTOPLASMIC pH CONTROL

Stimulation of transplasma
membrane reduction of NADH by growth
factors and hormones could have
important implications in the growth
control mechanism. One such
implication would be to help account
for cytoplasmic pH changes
associated with the initiation of
cell division (Busa and Nucchitelli,
1984).

In animal cells, a
sodium/proton antiport has been
described which so far is activated
by serum and all known growth
factors and mitogenic peptides and
is implicated in growth control
through regulation of cytosolic pH
(Moolenaar, 1986). A major
determinant in the regulation of the
Na^+/H^+ exchanger is cytoplasmic H^+.
When pH_i falls below a certain
threshold value, the activity of the
exchanger is stimulated
increasingly. Thus cytoplasmic pH
acts as an allosteric activator of
the Na^+/H^+ exchanger by binding to
an inward-facing regulator site.
Protonation at the regulatory site
sets the exchanger in motion

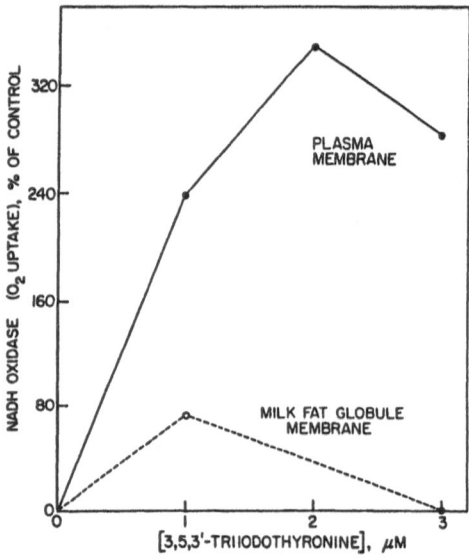

Fig. 11. The stimulation of NADH
oxidase activity of rat liver
plasma membrane vesicles when
treated with different
concentrations of 3, 5, 3'-
triiodothyronine compared to fat
globule membranes from bovine
milk. With the plasma membranes,
NADH oxidase activity is
stimulated by the T3 several fold.
Redrawn from Gayda et al. (1977).

(Aronson et al., 1982). The H^+ extrusion once the exchanger is activated
accounts for mitogen-induced rises in pH now thought to be one of the
permissive events which may be necessary but perhaps not sufficient for
progression through S-phase of the cell cycle (Moolenaar et al., 1986).

In HeLa cell, activation of plasma membrane redox activity by
external electron acceptors does lead to activation of a H^+/Na^+ antiport
(Sun et al., 1987). One possible mechanism would be through allosteric
activation from cytoplasmic acidification as a result of NADH oxidation.

In plants, there is evidence for cytoplasmic acidification from the
work of Felle where both an increase in membrane potential and a decrease
in cytoplasmic pH were measured in response to auxin treatment using a
double electrode system (Felle, 1987). The cytoplasmic acidification in
plants is followed by proton extrusion and subsequent cytoplasmic
alkalinization which oscillates for a time. A lag of 5 to 7 min between
auxin addition and the onset of external proton extrusion is well
established for plant stems (Cleland and Rayle, 1978). With plants, the
mechanism for proton release while paralleling the animal cell response,
is not the result of a Na^+/H^+ antiport (auxin growth is insensitive to
amelioride and amelioride derivatives and proton extrusion is sodium
independent).

While there is increasing evidence for a role for the plasma
membrane redox system in the growth mechanism, the observation of
hormone-sensitive NADH oxidases of both plant and animals cells capable
of electron transfer from NADH to molecular oxygen now add a new
dimension to the phenomenon. Stimulation of animal cell growth by

ferricyanide and parallel inhibition of growth and ferricyanide reduction by antitumor drugs in both plants and animals indicates that transmembrane electron transport can stimulate growth. Stimulation of electron transfer from NADH to oxygen by growth hormones, both plant and animal, further indicates some aspect of NADH oxidation as a rate-limiting step in the growth control process. To identify the mechanism whereby NADH oxidation (Navas et al., 1986) and growth control are coupled represents an important challenge for future research.

A role of plasma membrane redox activities in expansion growth is supported by two types of findings. First, NADH-ferricyanide reductase activity but not NADH oxidase activity is inhibited by growth-inhibitory drugs in parallel to their inhibition of hormone-induced growth. Second, NADH oxidase activity but not NADH-ferricyanide reductase activity is stimulated strongly by growth active auxins and T_3. These observations are reconciled diagramatically by a transmembrane flow of electrons from cytoplasmic NADH outward to the cell surface (Fig. 12). In the absence of external electron acceptors, NADH oxidase may serve as a rate limiting terminal oxidase to facilitate transfer of electrons directly to oxygen.

Fig. 12. Diagram illustrating the hypothetical arrangement of constituents leading to signal transduction in hormonal regulation of expansion growth. To reconcile the various findings, electron flow from cytoplasmic NADH is regarded as coupled to molecular oxygen via NADH oxidase. Impermeant electron acceptors such as ferricyanide or ascorbate radical (AA·) are reduced by the transmembrane activity but in plants, in contrast to animal cells, this has little or no effect on growth. Also indicated in the diagram is the ability of both NADH oxidase and the external oxidant reductase to oxidize external NADH directly as in the normal bio-chemical assays of these activities.

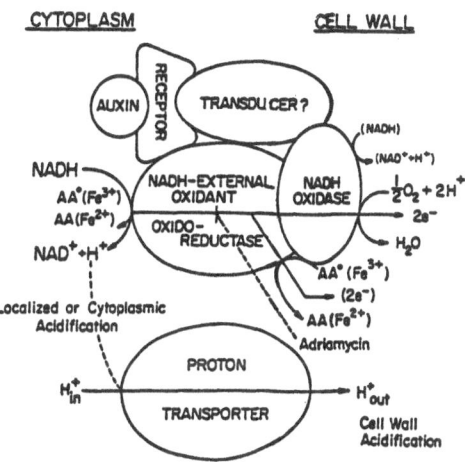

ACKNOWLEDGEMENTS

Work supported in part by grants from the National Institutes of Health CA 138801 and P01 CA36761. We thank Betty Leak and L.-Y. Wu for excellent technical assistance and Dr. Iris Sun and Wiletta Toole-Simms for helpful discussions and for contributed data.

References

Aronson, P. S., Nee, J., and Suhm, M. A. (1982) Modifier role of internal H^+ in activating the Na-H exchanger in renal microvillus membrane vesicles. Nature 299, 161-163.

Barnes, D. and Sato, G. (1980) Serum-free cell culture: a unifying approach. Cell 22, 649-655.

Barr, R., Craig, T. A., and Crane, F. L. (1985) Transmembrane ferricyanide reduction in carrot cells. Biochim. Biophys. Acta 812, 49-54.

Barr, R., Sandelius, A. S., Crane, F. L. and Morré, D. J. (1986) Redox reactions of tonoplast and plasma membranes isolated from soybean hypocotyls by free-flow electrophoresis. Biochim. Biophys. Acta 582, 254-261.

Basset, P., Quesneau, Y., and Zwiller, J. (1986) Iron induced L1210 growth: Evidence of a transferrin independent iron transport. Cancer Res. 46, 1644-1647.

Brightman, A. O., Barr, R., Crane, F. L., and Morré, D. J. (1988) Auxin-stimulated NADH oxidase purified from plasma membrane of soybean. Plant Physiol. (in press).

Buckhout, T. J., and Hrubec, T. C. (1986) Pyrimidine nucleotide-dependent ferricyanide reduction associated with isolated plasma membranes of maize (Zea mays. L.) roots. Protoplasma 135, 144-145.

Busa, W. B., and Nuccitelli, R. (1984) Metabolic regulation via intracellular pH. Am. J. Physiol. 246, R409-R438.

Cleland, R. E., and Rayle, D. L. (1978) Auxin, H^+-excretion and cell elongation. Bot. Mag. (Tokyo) Spec. Issue 1, 125-139.

Crane, F. L., Barr, R., Craig, T. A., and Misra, P. C. (1984) Growth control by proton pumping plasma membrane redox. Proc. Plant Growth Regul. Soc. Am. 11, 87-95.

Crane, F. L., Sun, I., Clark, M. G., Grebing, G., and Low, H. (1985) Transplasma membrane redox systems in growth and development. Biochim. Biophys. Acta 811, 233-264.

Ellem, K. A. O., and Kay, G. F. (1983) Ferricyanide can replace pyruvate to stimulate growth and attachment of serum restricted human melanoma cells. Biochem. Biophys. Res. Commun. 112, 183-190.

Felle, H. (1987) Proton transport and pH control in Sinapis alba root hairs: A study carried out with double-barrelled microelectrodes. J. Exp. Bot. 38, 340-354.

Gayda, D. P., Crane, F. L., Morré, D. J. and Low, H. (1977) Hormone effects on NADH-oxidizing enzymes of plasma membranes of rat liver. Proc. Indiana Acad. Sci. for 1976 86, 385-390.

Ghani, Q. P., and Hollenberg, M. (1978) Poly(adenosine diphosphate ribose) metabolism and regulation of myocardial cell growth by oxygen. Biochem. J. 170, 387-394.

Goldenberg, H., Crane, F. L., and Morré, D. J. (1979) NADH oxido-reductase of mouse liver plasma membranes. J. Biol. Chem. 254, 2491-2498.

Landschulz, W., Theslaff, I., and Ekblom, P. (1984) A lipophilic iron chelator can replace transferrin as a stimulator of cell proliferation and differentiation. J. Cell Biol. 98, 596-601.

Löw, H., and Crane, F. L. (1976) Hormone regulated redox function in plasma membranes. FEBS Lett. 68, 157-159.

Löw, H., Sun, I. L., Navas, P., Grebing, C., Crane, F. L., and Morré, D. J. (1986) Transplasmalemma electron transport from cells is part of a diferric transferrin reductase system. Biochem. Biophys. Res. Commun. 139, 1117-1123.

Møller, I. M., and Lin, W. (1986) Membrane-bound NAD(P)H dehydrogenases in higher plant cells. Ann. Rev. Plant Physiol. 37, 309-334.

Moolenaar, W. H. (1986) Cytoplasmic pH and free Ca^{2+} in the action of growth factors. In: Oncogenes and Growth Control', Kahn, P., and Graf, T., eds. Springer-Verlag, Berlin/Heidelberg/New York, pp. 170-176.

Morré, D. J., Navas, P., Penel, C., and Castillo, F. J. (1986) Auxin-stimulated NADH oxidase (semidehydroascorbate reductase) of soybean plasma membrane: Role in acidification of cytoplasm? Protoplasma 133, 195-197.

Morré, D. J., Auderset, G., Penel, C., and Canut, H. (1987a) Cytochemical localization of NADH-ferricyanide oxido-reductase in hypocotyl segments and isolated membrane vesicles of soybean. Protoplasma 140, 133-140.

Morré, D. J., Crane, F. L., Sun, I. L., and Navas, P. (1987b) The role of ascorbate in biomembrane energetics. Ann. N. Y. Acad. Sci. 498, 153-171.

Morré, D. J., Crane, F. L., Barr, R., Penel, C., and Wu, L.-Y. (1988a) Inhibition of plsma membrane redox activities and elongation growth of soybean. Physiol. Plant. 72, 236-240.

Morré, D. J., Brightman, A. O., Wu, L.-Y., Barr, R., Leak, B., and Crane, F. L. (1988b) Role of plsma membrane redox activities in elongation growth in plants. Physiol. Plant. (in press).

Navas, P., Sun, I. L., Morré, D.J., and Crane, F.L. (1986) Decrease of NADH in HeLa cells in the presence of transferrin or ferricyanide. Biochem. Biophys. Res. Commun. 135, 110-115.

Ponca, R., Brady, R. W., Wilozykska, A., and Schulman, H. M. (1984) The effect of various chelating agents on the mobilization of iron from reticulocytes in the presence and absence of pyridoxal isonicotinoyl hydrazone. Biochim. Biophys. Acta 802, 477-489.

Rogers, K. E., and Tokes, Z. A. (1984) Novel mode of cytotoxicity obtained by coupling inactive anthracycline to a polymer. Biochem. Pharmacol. 33, 605-608.

Sandelius, A. S., Barr, R., Crane, F. L., and Morré, D. J. (1987) Redox reactions of plasma membranes isolated from soybean hypocotyls by phase partition. Plant Sci. 48, 1-10.

Solomon, D., and Mishkin, S. (1982) Tri-iodothyronine (T_3) stimulates the growth of Morris hepatoma cells grown in culture. Biochem. Biophys. Res. Commun. 105, 1611-1617.

Sun, I. L., and Crane, F. L. (1981) Transplasmalemma NADH dehydrogenase is inhibited by actinomycin D. Biochem. Biophys. Res. Commun. 101, 68-75.

Sun, I. L., and Crane, F. L. (1984a) The antitumor drug, cis diaminedichloro-platinum, inhibits trans plsma membrane electron transport in HeLa cells. Biochem. Intern. 9, 299-306.

Sun, I. L., and Crane, F. L. (1984b) Effects of anthracycline compounds on transmembrane redox function of cultured HeLa cells. Proc. Indiana Acad. Sci. for 1983 93, 267-274.

Sun, I. L., and Crane, F. L. (1985) Bleomycin control of transplasma membrane redox activity and proton movement in HeLa cells. Biochem. Pharmacol. 34, 617-623.

Sun, I.L., Crane, F. L., Löw, H., and Grebing, C. (1984a) Transplasma membrane redox stimulates HeLa cell growth. Biochem. Biophys. Res. Commun. 125, 649-654.

Sun, I. L., Crane, F. L., Löw, H., and Grebing, C. (1984b) Inhibition of plasma membrane NADH dehydrogenase by adriamycin and related anthracycline antibiotics. J. Bioenergetics Biomemb. 16, 209-211.

Sun, I. L., Morré, D. J., Crane, F. L., Safranski, K., and Croze, E. M. (1984c) Monodehydroascorbate as an electron acceptor for NADH reduction by coated vesicle and Golgi apparatus fractions of rat liver. Biochim. Biophys. Acta 797, 266-275.

Sun, I. L., Putnam, J. E., and Crane, F. L. Control of cell growth by transplasmalemma redox: Stimulation of HeLa cell growth by impermeable oxidants. (1985) Proc. Indiana Acad. Sci. for 1984 94, 407-416.

Sun, I. L., Crane, F. L., and Chou, J. Y. (1986) Modification of transmembrane electron transport activity in plasma membranes of Simian virus 40 transformed pineal cells. Biochim. Biophys. Acta 886, 327-336.

Sun, I. L., Garcia-Canero, R., Liu, W., Toole-Simms, W., Crane, F. L., Morré, D. J., and Löw, H. (1987) Diferric transferrin reduction stimulates the Na^+/H^+ antiport of HeLa cells. Biochem. Biophys. Res. Commun. 145, 465-473.

Tritton, R. R., and Yee, G. (1982) The anticancer agent adriamycin can be actively cytotoxic without entering cells. Science 217, 248-250.

REDOX COMPONENTS IN THE PLANT PLASMA MEMBRANE

Ian M. Møller, Per Askerlund, Christer Larsson,
Alajos Bérczi* and Susanne Widell

Department of Plant Physiology, University of Lund
Box 7007, S-220 07 Lund, Sweden; *Institute of
Biophysics, Biological Research Center, Hungarian
Academy of Sciences, Box 521, H-6701 Szeged
Hungary

INTRODUCTION

As early as 1945, Lundegårdh proposed that redox processes took place in the plasma membrane (PM) of plants. He further envisaged that ion uptake, specifically anion uptake, across the PM of root cells was directly coupled to the flow of electrons to oxygen. This anion respiration was thought to be catalyzed by the respiratory redox components known at that time, mainly the cytochromes. When these around 1950 were discovered to reside in an organelle, the mitochondrion, and not in the PM, the theory of Lundegårdh became less tenable (see Lundegårdh, 1955) and it gradually lost prominence. However, in hindsight we can now see that his main idea was correct - there is at present plenty of evidence that the PM of both animals and plants contains redox components which can participate in a number of redox processes (see Crane et al., 1985a,b; Lüttge and Clarkson, 1985; Møller and Lin, 1986; for comprehensive reviews). In the present review we will critically evaluate the evidence for each redox component proposed to be present in the PM of plants. We will then briefly discuss the possible location of these redox components as well as the various processes they might be involved in.

In this context it is important to have in mind that a given tissue contains several cell types, the PM of which might differ quite considerably due to different functions of the cells. The PM of a given cell type may also be heterogeneous, i.e. have a lateral heterogeneity so that certain components are concentrated in some parts. Thus, PM preparations may contain several different populations of PM vesicles and the results cited are mere averages and may not reflect precisely the properties of the PM of any given cell.

THE REDOX COMPONENTS

Purification of the Plasma Membrane

Intact roots catalyze ferricyanide reduction, NADH-ferri-cyanide activity and NADH oxidation (e.g. Sijmons et al., 1984 a,b; Bienfait, 1985; Qui et al., 1985). The same is true for cultured cells and protoplasts (e.g. Craig and Crane, 1981; Lin, 1982; Novak and Miklashevich, 1986). These responses indicate that several different redox components may be present in the PM.

However, not all activities are due to true PM components. Therefore, it is necessary to isolate pure PM vesicles to identify the individual redox components and characterize the redox activities in more detail. As markers for PM vesicles, naphthylphthalamic acid binding, silicotungstic acid staining, glucan synthase II and vanadate-inhibited K^+, Mg^{2+}-ATPase acti-vities have been used (Larsson, 1985; Sandelius et al., 1986 a, b; Larsson et al., 1987). The latter activity should, however, not be used as an absolute marker since it appears also to be present in another membrane fraction at least in maize and wheat roots (Bérczi et al., 1988).

Many reports on redox components in the literature are based on experiments with microsomal fractions or PM-enriched fractions obtained by sucrose gradient centrifugation. The latter contain only 50-60% PM (Bérczi and Møller, 1986; Hodges and Mills, 1986; Larsson et al., 1987) but should clearly still be preferred to unpurified microsomal fractions. However, puri-fication by aqueous polymer two-phase partitioning is a rapid and simple method for obtaining >90% pure PM vesicles (Widell and Larsson, 1981; Larsson, 1985; Larsson et al., 1987) and fractions obtained by this method, or possibly by free-flow electrophoresis (Sandelius et al., 1986 a)- though this method is not as convenient - should be used for the definitive locali-zation and purification of specific PM-associated redox compo-nents. Using these methods only cyt b (including cyt P-450/420) and flavins have been localized and quantified with certainty in the plant PM.

Flavins and b-Type Cytochromes

The first cytochrome reported in plant PM was a blue light-reducible b-type cyt (Brain et al., 1977; Jesaitis et al., 1977). Their observations were made on PM-enriched fractions from corn coleoptiles, and light-reducible b-cyt has later been reported in PM from cauliflower inflorescences (Widell and Larsson, 1981; 1983), spinach leaf (Kjellbom and Larsson, 1984) and several other species, and seems to be a general constituent of plant PM (for review, see Widell 1987). This blue light-reducible b-cyt is measured as a light-induced absorbance change (LIAC) under low oxygen tension in the presence of a reductant like EDTA. LIAC is thought to be caused by the transfer of electrons from the reductant via a flavin to a b-type cyt (see Widell, 1987; and references therein). About half the amount of LIAC is located in the PM (Widell and Larsson, 1983). Low temperature spectra of the light-reducible cyt show only one component in corn coleoptile PM (Leong et al., 1981), cauliflower PM (Widell et al., 1983) and spinach leaf PM (Kjellbom and Larsson, 1984), with the peak of the α-band at 555-558 nm, which coincides with

Table 1. Redox components identified with certainty in the plant PM.

Component	Source	Concentration nmol (mg prot.)$^{-1}$	Reference
Cytochrome b	Corn coleoptiles	0.12[a]	Jesaitis et al. 1977
	Oat roots	0.18	Ramirez et al. 1984
	Spinach leaves	0.51	Askerlund et al. 1988
	Sugar beet leaves	0.13	Askerlund et al. 1988
	Soybean hypocotyls	0.5	Barr et al. 1986
	Cauliflower inflorescences	1.9[b]	Caubergs et al. 1986
		0.23	Askerlund et al. 1988
Cytochrome b (light-reducible)	Corn coleoptiles	0.03[c]	Leong et al. 1981
	Cauliflower inflorescences	0.08[c]	Widell et al. 1983
	Spinach leaf	0.33[c]	Kjellbom and Larsson, 1984
Cytochrome P-450	Cauliflower inflorescences	0.02[c]	Kjellbom et al. 1985
Cytochrome P-420	Cauliflower inflorescences	0.10[c]	Kjellbom et al. 1985
Flavins	Oat roots	0.21	Ramirez et al. 1984
	Soybean hypocotyls	0.5	Barr et al. 1986

a. Calculated from Table III using an extinction coefficient for the Soret band of 171 mM^{-1}cm^{-1}
b. Calculated from Fig. 8 using an extinction coefficient for the α-band of 20 mM^{-1}cm^{-1}
c. Calculated from data given in results

that of the dithionite-reducible b-cyt, suggesting that the light-reducible b-cyt is identical to the major b-cyt of the PM.

Extraction and quantification of non-covalently bound flavin from purified PM vesicles has yielded values of 0.2-0.5 nmol (mg protein)$^{-1}$ (Table 1). It is possible that there are also covalently bound flavins in the PM, but there is no experimental evidence for that.

The amount of dithionite-reducible cyt b in purified PM as measured by difference spectrophotometry is generally about the same as for flavins (Table 1). Figure 1 shows the results of a more detailed study of the b-type cyt in PM from spinach leaves. A small part of the cyt was oxidized in the sample when ferricyanide was added to the reference cuvette (Fig. 1B). Apparently this cyt was partly reduced even in the absence of added reductants. NADH reduced about 30% of the b-cyt, but only a single peak at 557 nm was observed (Fig. 1C). Cyt b_5 has a split α-band and would normally be reduced by NADH (Bruder et al., 1978; Madyastha and Krishnamachary, 1986). Ascorbate caused a greater reduction than NADH (Fig. 1D), but considerably less than dithionite (Fig. 1E). Similar results were obtained with cauliflower PM (Caubergs et al., 1986). The difference spectrum between dithionite-reduced and ascorbate-reduced showed a split α-band (Fig. 1F) which is probably due to cyt b_5 since ascorbate does not necessarily reduce cyt b_5 (Madyastha and Krishnamachary, 1986). The absence of a split α-band in the presence of NADH (Fig. 1C) can therefore be interpreted to be due to the absence of cyt b_5 reductase in the PM. The presence of cyt b_5 could be

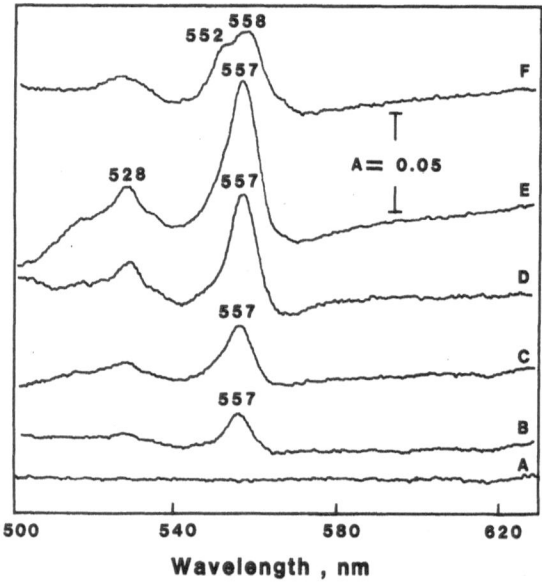

Wavelength , nm

Fig. 1. Difference spectra (77 K) of PM
from spinach leaves purified by
two-phase partitioning.

A. Base-line
B. Air ox minus FeCN ox
C. NADH red minus FeCN ox
D. Ascorbate red minus FeCN ox
E. Dithionite red minus FeCN ox
F. Dithionite red minus ascor-
   bate red
Note the complete abscence of
cyt c oxidase, which has a peak
at 600 nm. (Data from Askerlund
et al. 1988.)

due to membrane flow to the PM, and the cyt b_5 may be inactive because its reductase is missing.

Difference spectra of PM from cauliflower inflorescences are almost identical to those shown in Fig. 1, whereas difference spectra of PM from barley roots and leaves are slightly different. In the latter cases the spectrum for NADH minus ferricyanide shows a shoulder at 552 nm consistent with the presence of a small amount of cyt b_5 reduced by cyt b_5 reductase. However, the ascorbate minus ferricyanide spectra also show this shoulder (Askerlund et al., 1988).

At least some of the b-type cyt in the PM of cauliflower (Kjellbom et al., 1985) and spinach leaves (Askerlund et al., 1988) is cyt P-450/420. The amounts reported are probably underestimates since not all cyt P-420 binds CO (15-60% binds, Bruder et al., 1978). Heme-staining of gels showed a band at 93 kDa with PM from cauliflower inflorescences (Kjellbom et al., 1985), however, this band disappears after washing the PM at pH 9.7 indicating that it is due to a contaminating peroxidase. The non-covalently bound heme in the b-type cyt thus appears to be lost upon solubilization of the PM for gel electrophoresis (Askerlund et al., 1988).

Redox titrations of the b-type cyt in purified PM vesicles from plants have been done only in a few cases. Leong et al. (1981) reported the presence of one component in the PM from corn coleoptiles with a midpoint potential of -65 mV. Caubergs et al. (1986) found two b-type cyt in PM from cauliflower with midpoint potentials of -23 mV (28% of total cyt b) and 146 mV (72%). Redox titrations of PM from both spinach leaves and cauliflower inflorescences gave results very similar to those of Caubergs et al. (1986); the major cyt had a midpoint potential around 150 mV whereas a minor component had a lower midpoint potential (Askerlund et al., 1988).

Other Redox Components

Quinones have been reported to be present in certain animal PM (Crane et al., 1985 a,b), but very little information is available from plants. Lin (1984) reported an NADH minus oxidized difference spectrum of a pentane extract of trypsin-released components from corn root protoplasts. It had minima at 280 and 345 nm and a maximum at 315 nm. This was interpreted to be due to ubiquinone (Lin, 1984) in spite of the fact that the spectrum of ubiquinone looks quite different (Crane and Barr, 1971). Furthermore, Lin (1982) did not exclude contamination by mitochondria or chloroplasts released by the trypsin treatment.

By using methanol/light petroleum-extraction of pure PM vesicles, a method which extracts all the ubiquinone from mitochondria (Kröger, 1978), quinones are indeed extracted from the PM. The extracts all show a reduced-oxidized difference spectrum similar to that of ubiquinone-10 and the concentration in the PM is around 1.5 nmol (mg protein)$^{-1}$ (P. Askerlund, unpublished results) or about 5 times higher than for flavins and cytochromes (Table 1) This is about the same ratio as in the respiratory chain of mitochondria where ubiquinone has a central role. Thus, it is quite possible that a quinone has a function in the plant PM.

It has been suggested that one of the functions of the PM redox chain is to keep SH-groups reduced (Crane et al., 1985 a). Animal PM contain about 80 nmol SH-groups (mg protein)$^{-1}$ (Crane et al., 1985 a); however, so far there are no studies on plant PM available.

Iron-sulfur proteins are important redox carriers in all the energy-transducing membranes, the inner mitochondrial membrane, the thylakoid membrane and the bacterial plasma membrane. PM from animals contain 4-8 nmol iron (mg protein)$^{-1}$ (Crane et al., 1985 b) but there are no reports on iron-sulfur proteins in plant PM. It appears worthwhile to conduct an EPR study and to measure acid-extractable iron and hydrogen sulfide of purified PM vesicles.

REDOX ACTIVITIES AND POSSIBLE FUNCTIONS

Redox activities

Although antimycin A-insensitive NAD(P)H-cyt c reductase activity is commonly referred to as a marker for the endoplasmic reticulum it is also present in most other plant cell membranes including the PM (Lundborg et al., 1981; Møller and Lin, 1986). Purified PM vesicles often contain a relatively low activity compared to the microsomal fraction (e.g. Widell and Larsson, 1983; Buckhout and Hrubec, 1986; Hodges and Mills, 1986); however, in some tissues the PM contains up to 5 times higher specific activities of NAD(P)H-cyt c reductase than the micro-somal fraction (Caubergs et al., 1986; Larsson et al., 1987). This variation probably depends on the relative amounts of endo-plasmic reticulum and PM in the microsomal fraction. NAD(P)H-cyt c reductase activity is due to electron transfer from a flavoprotein (a cyt b reductase), via a b-type cyt to cyt c (Crane et al., 1985 a). Unfortunately, this gives no clue as to the function since cyt c can not be the natural electron acceptor in the PM.

NAD(P)H-ferricyanide reductase activity is catalyzed by most flavoprotein dehydrogenases and the PM contains high activities (Barr et al., 1986; Buckhout and Hrubec, 1986; Sandelius et al., 1986 b; Bérczi et al., 1988; Luster and Buckhout, 1988). A partial purification from PM vesicles has been reported recently by Luster and Buckhout (1988). PM vesicles can also use duroquinone, ascorbate free radical (Luster and Buckhout, 1988) and ferric-EDTA (Barr et al., 1986) as acceptors of electrons from NAD(P)H.

Protoplasts and right side-out PM vesicles also oxidize NAD(P)H in the absence of an added electron acceptor. The final electron acceptor is oxygen and the final product appears to be hydrogen peroxide. The oxidase activity is strongly enhanced by phenols and phenol-like substances (Lin, 1982; Barr et al., 1985; Møller and Bérczi, 1985, 1986; Askerlund et al., 1987). Several properties of this system are reminiscent of perioxi-dases (Askerlund et al., 1987), although only part of the activity was removed by washing with 0.7 M NaCl indicating that the activity was due to a firmly bound or integral protein. However, in later experiments with cauliflower PM a wash at pH 9.7 removed most of the oxidase activity as well as the heme band at 93 kDa (Askerlund et al., 1988). At the moment it is not possible to draw any general conclusion as to whether all oxidase activity in PM preparations is due to peroxidase(s) and whether the peroxidase(s) is an artifact caused by adsorption during isolation.

Duroquinone-stimulated NAD(P)H-dependent oxygen uptake has also been reported (Pupillo et al., 1986; Askerlund et al., 1987).

Kinetics of NAD(P)H Dehydrogenases

Several studies have dealt with the kinetics of NAD(P)H-cyt c and NAD(P)H-ferricyanide reductase activities in purified PM fractions (Barr et al., 1986; Buckhout and Hrubec, 1986; Sandelius et al., 1986 b). It is outside the scope of this review to discuss the details of these findings. However, when planning kinetic measurements on membrane-bound (or cell wall-bound) enzymes with a charged substrate (NAD(P)H and ferri-cyanide are negatively charged and cyt c is positively charged) it is essential to be aware of the problem of electrostatic interactions since all membranes, including the plant PM (Møller et al., 1984), have a net negative charge at neutral pH. This charge will affect the concentration of the substrate near the active site and thus the apparent Km as has been shown for the NADH dehydrogenase on the outer surface of the inner mito-chondrial membrane of plant mitochondria (Edman et al., 1985). To minimize such interactions and to ensure that kinetic results are comparable, experiments should be performed in the presence of a high concentration of cations (100 mM KCl or perhaps better 5-10 mM $MgCl_2$) which screen the negative charges of membrane surfaces and reduces the size of the surface potential.

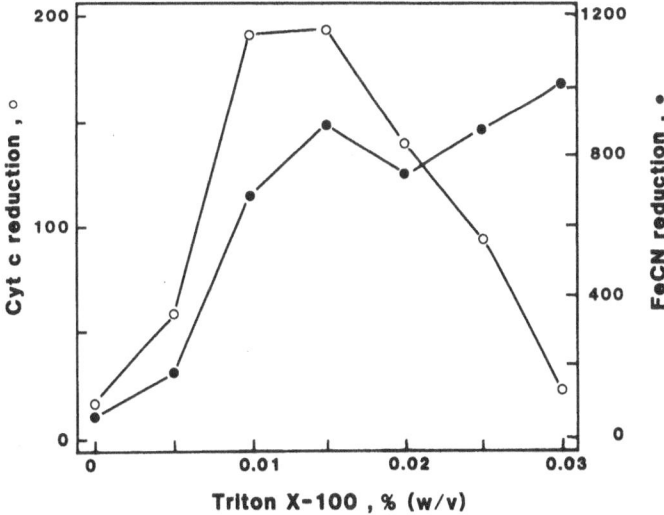

Fig. 2. The effect of Triton X-100 on NADH-ferri-cyanide and NADH-cyt c reductase activities in PM vesicles purified from cauliflower in-florescences by two-phase partitioning. (Data from Askerlund et al., 1988.)

Orientation of Redox Enzymes

Until recently all PM preparations purified by two-phase partitioning were reported to be predominantly right-side out as judged by the high latency of ATPase and glucan synthase II activity (Larsson et al., 1984; Bérczi and Møller, 1986; Hodges and Mills, 1986). NAD(P)H-ferricyanide and NAD(P)H-cyt c reductase activities of these vesicles also often show high latencies (Buckhout and Hrubec, 1986; Bérczi et al., 1988; Luster and Buckhout, 1988). One problem in these experiments is the type and concentration of detergent used to determine the enzyme latency. Different detergents have very different effects (Buckhout and Hrubec, 1986; Bérczi and Møller, 1987) and while NADH-ferricyanide reductase activity is not particularly sensitive to the detergent concentration, NADH-cyt c reductase activity is quite sensitive and probably never reaches its maximum before inhibition becomes severe (Fig. 2).

Taken together the above results indicate: 1) that the binding sites for NAD(P)H and/or ferricyanide and cyt c are on the inner, cytoplasmic surface of the PM and 2) that several components (viz flavoprotein and b-type cyt) are involved in NADH-cyt c reductase and fewer components in NADH-ferricyanide reductase activity (Crane et al., 1985 a).

Very recently Larsson et al. (1988) succeeded in purifying inside-out PM vesicles, and results with such preparations show that the inner cytoplasmic surface of PM vesicles from cauliflower inflorescences contain active sites for NADH, NADPH, cyt c and ferricyanide (Askerlund et al., 1988). The outer surface may contain active sites for ferricyanide (see below) and possibly sites for NAD(P)H. However, in the latter case it is difficult to be certain, since the problem of bound peroxidase/ integral peroxidase/NADH oxidase is still unresolved (see above). The binding site for oxygen is unknown for the same reason.

Natural Electron Donors and Acceptors and Possible Functions of the PM Redox Activities

According to Bienfait (1985) there are two redox systems shuttling electrons across the PM. The "Standard" system is constitutive in all plants, it reduces ferricyanide but not other ferric chelates, its natural electron acceptor is oxygen, it uses NAD(P)H as the electron donor and it functions in membrane polarization. The second system, the "Turbo" system, is inducible by iron deficiency, it is found in non-graminaceous plants possibly only in the epidermis of young roots, it is active with both ferricyanide and ferric chelate, it may use NADPH as the electron donor, and its function is iron reduction (Bienfait, 1985). Most of the experimental evidence on which this hypothesis is based comes from experiments on intact tissues or whole cells. There is clearly a need to confirm these findings with purified PM preparations as suggested by Lüttge and Clarkson in 1985.

PM redox processes are most likely coupled to H^+-efflux and membrane polarization. E g addition of ferricyanide to intact roots causes a depolarization of the membrane potential (Sijmons et al., 1984 b), and ferricyanide reduction by carrot cells (Craig and Crane, 1981) or microalgae (Novak and

Miklashevich, 1986) is accompanied by an increased efflux of H^+. However, there is a wide variability in the ratio of electrons transported by the redox system to protons extruded (Bienfait, 1985; Crane et al., 1985 a,b). If the redox processes work in parallel with the H^+-pumping ATPase in generating a proton gradient across the PM the relative ratio of these activities could be an important regulatory factor for active uptake across the PM.

The hydrogen peroxide produced by the external peroxidase could be used in cell wall synthesis although the source of extracellular NADH is still obscure. Furthermore, since by far most of the peroxidase activity is located in the cell wall and not in the PM, it is doubtful whether this can be called a function of the PM.

The action spectrum of LIAC (for definition, see above) is very similar to that of blue-light photomorphogenesis (Widell et al., 1983). Thus, the flavoprotein-cyt P-450/420 complex could be the receptor for the light signal (Kjellbom et al., 1985; see Widell 1987 for references). How the light signal is converted to an intracellular message is completely unknown.

Finally, anticancer drugs, which inhibit both growth and PM redox activities in mammalian cells, have been observed to inhibit growth of soybean hypocotyls as well as NADH-ferricyanide reductase activity in PM vesicles isolated from the hypocotyls. This led to the suggestion that there is a connection between PM redox activities and control of elongation growth (Morré et al., 1988).

We have attempted to summarize the present knowledge of the PM redox components and their possible functions into a model (Fig. 3): A flavoprotein spanning the membrane is

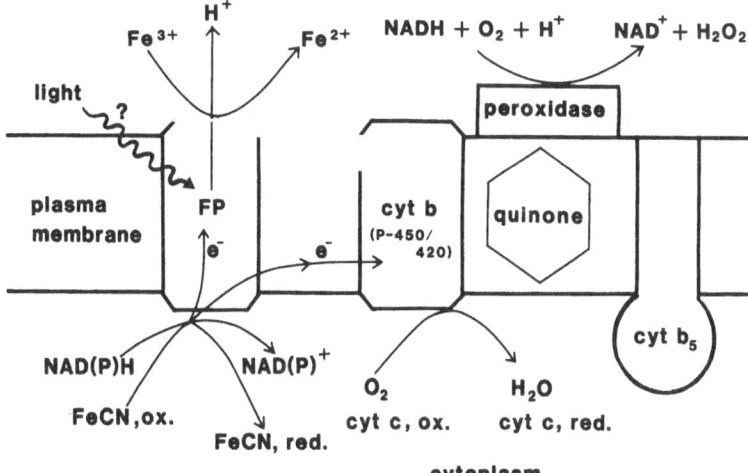

Fig. 3. Model of the arrangement of redox components in the plant plasma membrane. FeCN, ferricyanide. For details, see text.

suggested to be responsible for NAD(P)H oxidation on the inner
surface of the PM. The electrons are either passed on to a cyt
b or shuttled across the membrane to reduce Fe^{3+}-chelates on
the outer surface. Protons may be pumped out of the cell as part
of this process. There is no evidence for the presence of
several flavoproteins, e.g. one for the "standard system", one
for the "Turbo system" (see above) and one for the blue-light
receptor. The cyt b receiving electrons from the flavoprotein
is suggested to be identical to the cyt P-450/420 and it is
responsible for cyt c reduction on the inner surface in assays
in vitro. In vivo the natural electron acceptor may be oxygen.
We have not attempted to speculate on the localization of the
quinone. However, pentane extraction of the quinone (Ernster
et al., 1978) did not affect NAD(P)H-cyt c reductase activity
(P. Askerlund, unpublished results), so it is unlikely that the
quinone is involved in the transfer of electrons between the
flavoprotein and the cyt P-450/420. The cyt b_5 has been placed
on the inner surface since it may arrive by membrane flow from
the endoplasmic reticulum where it is located on the outer
surface. We do not suggest a function for cyt b_5. Finally, a
peroxidase is placed on the outer surface (though some of it
could be integral; Askerlund et al., 1987) where it is responsible for some of (or all of) the oxygen consumption induced by
NADH or NADPH addition. Note that the reaction mechanism is more
complex than shown in Fig. 3 (see Askerlund et al. 1987).

REFERENCES

Askerlund, P., Larsson, C., Widell, S., and Møller, I.M., 1987,
    NAD(P)H oxidase and peroxidase activities in purified
    plasma membranes from cauliflower inflorescences, Physiol.
    Plant.,71:9.
Askerlund, P., Larsson, C., and Widell, S., 1988, manuscript in
    preparation.
Barr, R., Sandelius, A.S., Crane, F.L., and Morré, D.J., 1985,
    Oxidation of reduced nucleotides by plasma membranes of
    soybean hypocotyl, Biochem. Biophys. Res. Commun., 131:943.
Barr, R., Sandelius, A.S., Crane, F.L., and Morré, D.J., 1986,
    Redox reactions of tonoplast and plasma membranes isolated
    from soybean hypocotyls by free-flow electrophoresis,
    Biochim. Biophys. Acta, 852:254.
Bérczi, A., and Møller, I.M., 1986, Comparison of the properties
    of plasmalemma vesicles purified from wheat roots by phase
    partitioning and by discontinuous sucrose gradient centri-
    fugation, Physiol. Plant., 68:59.
Bérczi, A., and Møller, I.M., 1987, Mg^{2+}-ATPase activity in
    wheat root plasmalemma vesicles: Time-dependence and effect
    of sucrose and detergents, Physiol. Plant., 70:583.
Bérczi, A., Larsson, C., Widell, S., and Møller, I.M., 1988,
    Separation of wheat root microsomal membranes by counter-
    current distribution. An evaluation of plasma membrane
    markers, in "Proc. of the 3rd Int. Symp. "Structure and
    Function of Roots", Nitra, Czechoslovakia (In press).
Bienfait, H.F., 1985, Regulated redox processes at the plasma-
    lemma of plant root cells and their function in iron
    uptake, J. Bioenerg. Biomembr., 17:73.
Brain, R.D., Freeberg, J.F., Weiss, C.V., and Briggs, W.R.,1977,
    Blue light-induced absorbance changes in membrane fractions
    from corn and Neurospora, Plant Physiol., 57:948.
Bruder, G., Fink, A., and Jarasch, E.-D., 1978, The b-type cyto-
    chrome in endoplasmic reticulum of mammary gland epithelium

and milk fat globule membranes consists of two components, cytochrome b5 and cytochrome P-420, Exp. Cell Res., 117:207.

Buckhout, T.J., and Hrubec, T.C., 1986, Pyridine nucleotide-dependent ferricyanide reduction associated with isolated plasma membranes of maize (Zea mays L.) roots, Protoplasma, 135:144.

Caubergs, R.J., Asard, H.H., De Greef, J.A., Leeuwerik, F.J., and Oltmann, F.L., 1986, Light-inducible absorbance changes and vanadate-sensitive ATPase activity associated with the presumptive plasma membrane fraction from cauliflower inflorescences, Photochem. Photobiol., 44:641.

Craig, T.A., and Crane, F.L., 1981, Evidence for a trans-plasma membrane electron transport system in plant cells, Proc. Indiana Acad. Sci., 90:150.

Crane, F.L., and Barr, R., 1971, Determination of ubiquinones, Methods Enzymol.,18:137.

Crane, F.L., Löw, H., Clark, M.G., 1985a, Plasma membrane redox enzymes, in: The Enzymes of Biological Membranes, A.N. Martonosi, ed., Plenum Press, New York.

Crane, F.L., Sun, I.L., Clark, M.G., Grebing, C., and Löw, H., 1985b, Transplasma-membrane redox systems in growth and development, Biochim. Biophys. Acta, 811:233.

Edman, K., Ericson, I., and Møller, I.M., 1985, The regulation of exogenous NAD(P)H oxidation in spinach leaf mitochondria by pH and cations, Biochem. J., 232:471.

Ernster, L., Glaser, E., and Norling, B., 1978, Extraction and reincorporation of ubiquinone in submitochondrial particles, Methods Enzymol. 53:573.

Hodges, T.K., and Mills, D., 1986, Isolation of the plasma membrane, Methods Enzymol., 118:41.

Jesaitis, A.J., Heners, P.R., Hertel, R., and Briggs, W.R., 1977, Characterization of a membrane fraction containing a b-type cytochrome, Plant Physiol., 59:941.

Kjellbom, P., and Larsson, C., 1984, Preparation and polypeptide composition of chlorophyll-free plasma membranes from leaves of light-grown spinach and barley, Physiol. Plant., 62:501.

Kjellbom, P., Larsson, C., Askerlund, P., Schelin, C., and Widell, S., 1985, Cytochrome P-450/420 in plant plasma membranes: A possible component of the blue-light-reducible flavoprotein-cytochrome complex, Photochem. Photobiol., 42:779.

Kröger, A., 1978, Determination of contents and redox states of ubiquinone and menaquinone, Methods Enzymol., 53:579.

Larsson, C., 1985, Plasma membranes, in: Cell Components. Modern Methods of Plant Analysis, New Series, Vol. 1, pp. 87-104, H.F. Linskens, J.F. Jackson, eds, Springer-Verlag, Berlin.

Larsson, C., Kjellbom, P., Widell, S., and Lundborg, T., 1984, Sidedness of plant plasma membrane vesicles purified by partition in aqueous polymer two-phase systems, FEBS Lett., 171:271.

Larsson, C., Widell, S., and Kjellbom, P., 1987, Preparation of high-purity plasma membranes, Methods Enzymol., 148:558.

Larsson, C., Widell, S., and Sommarin, M., 1988, Inside-out plant plasma membrane vesicles of high purity obtained by aqueous two-phase partitioning, FEBS Lett., in press.

Leong, T.-Y., Viestra, R.D., and Briggs, W.R., 1981, A blue light-sensitive cytochrome-flavin complex from corn coleoptiles. Further characterization, Photochem. Photobiol. 34:697.

Lin, W., 1982, Isolation of NADH oxidation system from the plasmalemma of corn root protoplasts. Plant Physiol., 70:326.

Lin, W., 1984, Further characterization on the transport property of plasmalemma NADH oxidation system in isolated corn root protoplasts, Plant Physiol., 74:219.

Lundegårdh, H., 1945, Absorption, transport and exudation of inorganic ions by the roots, Arkiv Bot., 32A(12):1

Lundegårdh, H., 1955, Mechanisms of absorption, transport, accumulation, and secretion of ions, Annu. Rev. Plant Physiol., 6:1.

Lundborg, T., Widell, S., and Larsson, C., 1981, Distribution of ATPases in wheat root membranes separated by phase partition, Physiol. Plant., 52:89.

Luster, D.G., and Buckhout, T.J., 1988, Multiple electron transport activities in plasma membranes from maize (Zea mays) roots, Physiol. Plant., in press.

Lüttge, U., and Clarkson, D.T., 1985, Mineral nutrition: Plasmalemma and tonoplast redox activities, Progr. Bot., 47:73.

Madyastha, K.M., and Krishnamachary, N., 1986, Purification and partial characterization of microsomal cytochrome b_{555} from the higher plant Catharanthus roseus, Biochem. Biophys. Res. Commun., 136:570.

Møller, I.M., and Bérczi, A., 1985, Oxygen consumption by purified plasmalemma vesicles from wheat roots. Stimulation by NADH and salicylhydroxamic acid, FEBS Lett., 193:180.

Møller, I.M., and Bérczi, A., 1986, Salicylhydroxamic acid-stimulated NADH oxidation by purified plasmalemma vesicles from wheat roots, Physiol. Plant., 68:67.

Møller, I.M., and Lin, W., 1986, Membrane-bound NAD(P)H dehydrogenases in higher plant cells, Annu. Rev. Plant Physiol. 37:309.

Møller, I.M., Lundborg, T., and Bérczi, A., 1984, The negative surface charge density of plasmalemma vesicles from wheat and oat roots, FEBS Lett., 167:181.

Morré, D.J., Crane, F.L., Barr, R., Penel, C., and Wu, L.-Y. 1988, Inhibition of plasma membrane redox activities and elongation growth of soybean, Physiol. Plant., 72:236.

Novak, V.A., and Miklashevich, A.I., 1986, Ferricyanide reductase and ferrocyanide oxidase activities of the microalga Scenesdesmus acuminatus, Sov. Plant Physiol., 32:694.

Pupillo, P., Valenti, V., De Luca, L., and Hertel, R., 1986, Kinetic characterization of reduced pyridine nucleotide dehydrogenases (duroquinone-dependent) in Cucurbita microsomes, Plant Physiol., 80:384.

Qui, Z.-S., Rubinstein, B., and Stern, A.I., 1985, Evidence for electron transport across the plasma membrane of Zea mays root cells, Planta, 165:383.

Ramirez, J.M., Gallego, G.G., and Serrano, R., 1984, Electron transfer constituents in plasma membrane fractions of Avena sativa and Saccharomyces cerevisiae, Plant Sci. Lett., 34:103.

Rubinstein, B., and Stern A.I., 1986, Relationship of trans-plasmalemma redox activity to proton and solute transport by roots of Zea mays., Plant Physiol., 80:805.

Sandelius, A.S., Penel, C., Auderset, G., Brightman, A., Millard, M., and Morré, D.J., 1986a, Isolation of highly purified fractions of plasma membrane and tonoplast from the same homogenate of soybean hypocotyls by free-flow electrophoresis, Plant Physiol., 81:177.

Sandelius, A.S., Barr, R., Crane, F.L., and Morré, D.J., 1986b,
Redox reactions of plasma membranes isolated from soybean
hypocotyls by phase partition, Plant Sci., 48:1

Sijmons, P.C., van den Briel, W., and Bienfait, H.F., 1984a,
Cytosolic NADPH is the electron donor for extracellular
Fe^{III} reduction in iron-deficient bean roots, Plant
Physiol., 75:219.

Sijmons, P.C., Lanfermeijer, F.C., de Boer, A.H., Prins, H.B.A.,
and Bienfait, H.F., 1984b, Depolarization of cell membrane
potential during trans-plasma membrane electron transfer
to extracellular electron acceptors in iron-deficient
roots of Phaseolus vulgaris L., Plant Physiol., 76:943.

Widell, S., 1987, Membrane-bound blue light receptors - Possible
connection to blue light photomorphogenesis, pp. 89-98,
in: Blue Light Responses: Phenomena and Occurrence in
Plants and Microorganisms, Vol. II, H. Senger, ed., CRC
Press, Boca Raton, FL.

Widell, S., and Larsson, C., 1981, Separation of presumptive
plasma membranes from mitochondria by partition in an
aqueous polymer two-phase system, Physiol. Plant., 51:368.

Widell, S., and Larsson, C., 1983, Distribution of cytochrome
b photoreductions mediated by endogenous photosensitizers
or methylene blue in fractions from corn and cauliflower,
Physiol. Plant., 57:196.

Widell S., Caubergs, R.J. and Larsson, C., 1983, Spectral
characterization of light-reducible cytochrome in a plasma
membrane-enriched fraction and in other membranes from
cauliflower inflorescences, Photochem. Photobiol., 38:95.

REDUCTION-OXIDATION ACTIVITIES OF PLANT PLASMAMEMBRANES

J.O.D. Coleman and J.D.C. Chalmers

Department of Plant Sciences
University of Oxford
Oxford, U.K.

INTRODUCTION

Plant cells catalyse a number of reduction-oxidation (redox) reactions at the cell surface. There is good evidence that the plasmamembrane contains a redox charge transfer system which conducts electrons from internal reductants to external oxidants such as non-permeant anionic ferricyanide. This ferricyanide reductase activity which is associated with a net acidification of the external medium has been observed in cultured cells (Craig and Crane, 1981; Barr, Crane and Craig, 1984; Chalmers, Coleman and Walton, 1984) in roots (Federico and Giartosio, 1983; Rubinstein and Stern, 1986; Sijmons, Lanfermeijer, DeBoer, Prins and Bienfait, 1984) and in leaf cells (Ivankina, Novak and Miclashevich, 1984; Neufeld and Bown, 1987).

In addition to the transplasmamembrane ferricyanide reductase activity, a plasmamembrane associated system for the oxidation of exogenous NAD(P)H has been identified in plant protoplasts. (Lin, 1984; Thom and Maretski, 1985; Chalmers, Coleman and Hawes, 1986).

We have studied the ferricyanide reductase activity and the external NAD(P)H oxidase activity in carrot cells grown in suspension culture and in protoplasts derived from these cells.

MATERIAL AND METHODS

Cell Cultures

Cell suspension cultures of carrot (Daucus carota L.) were grown and harvested as described by Chalmers and Coleman (1983).

Protoplasts

Protoplasts were prepared from cultured carrot cells as described by Slabas, Powel and Lloyd (1980).

Viability of Cells and Protoplasts

Estimates of cell and protoplast viability were made using
fluorescein diacetate as described by Widholm (1972). Cells and
protoplasts suspensions with greater than 95% viability were used.

Measurement of Ferricyanide Reduction and Ferrocyanide Oxidation

Measurement of ferricyanide reduction and ferrocyanide oxidation were
made using the amperometric system described by Chalmers, Coleman and
Walton (1984). In short, this electrochemical device consisted of a
4.5 ml glass incubation vessel, fitted with three electrodes. A gold
(working) electrode, a platinum gauze (counter) electrode and a
silver-silver chloride (reference) electrode (E˙ = -56 mV vs. standard
calomel electrode at 20°C) were connected to a three electrode
potentiostat. The potentiostat was used to maintain the gold electrode at
a preset potential difference relative to the reference electrode. When
the gold electrode was maintained at an oxidising potential of +360 mV,
ferrocyanide was oxidised at the gold electrode and an oxidation current
was recorded. When cells or protoplasts reduced ferricyanide to
ferrocyanide the resultant oxidation current was a measure of the rate of
ferricyanide reduction. Similarly, the rate of ferrocyanide oxidation by
cells was measured by maintaining the gold electrode at a reducing
potential of 0 mV.

Except where otherwise stated the reduction medium for these
measurements consisted of 10 mM KCl and 0.5 mM $CaCl_2$ adjusted to pH 6.2
with KOH. 300 mg fresh weight of carrot cells (i.e. 7×10^5 ml^{-1}
cells) were added to give a final volume of 4.2 ml. The reaction was
started by the addition of ferrocyanide or ferricyanide to give a final
concentration of 1 mM. The flow of current was monitored continuously
with a recorder.

Simultaneous measurements of pH were made with a fast response pH
electrode inserted into the reaction medium and the pH was recorded with a
potentiometric recorder.

RESULTS

Reduction of Ferricyanide by Carrot Cells

Use of the amperometric technique had the advantage of a sensitive
method which maintained a constant concentration of ferricyanide in the
reaction medium and therefore permitted measurements for long experimental
periods.

Actively growing and dividing carrot cells catalysed a rapid
reduction of ferricyanide. Figure 1A shows a representative current-time
profile for cells 8 days after sub-culture. The data demonstrate that the
rate of ferricyanide reduction was not constant but increased steadily
with time. It is apparent from the current-time profile that the
increasing rate of ferricyanide reduction took place in two phases. An
initial phase (Phase I) which lasted for 30-40 minutes followed by a
second phase (Phase II) which is characterised by a rapid rise in the rate
of reduction reaching a maximum at 80 minutes. Coinciding with the
ferricyanide reduction was a progressive acidification of the medium.
Acidification was slow in the initial phase, the greatest change of pH
occurred during the second phase of ferricyanide reduction.

When cells were supplied with ferrocyanide and the working electrode maintained at 0 mV a reduction current was observed indicating the oxidation of ferrocyanide by cells Figure 1B shows that the rate of ferrocyanide increased with time reaching a maximum rate after 30 minutes, which was maintained for 20 minutes followed by a slow fall off to a steady state rate after 80 minutes. During the oxidation of ferrocyanide the pH of the medium increased slightly and remained at a level which was higher than the control.

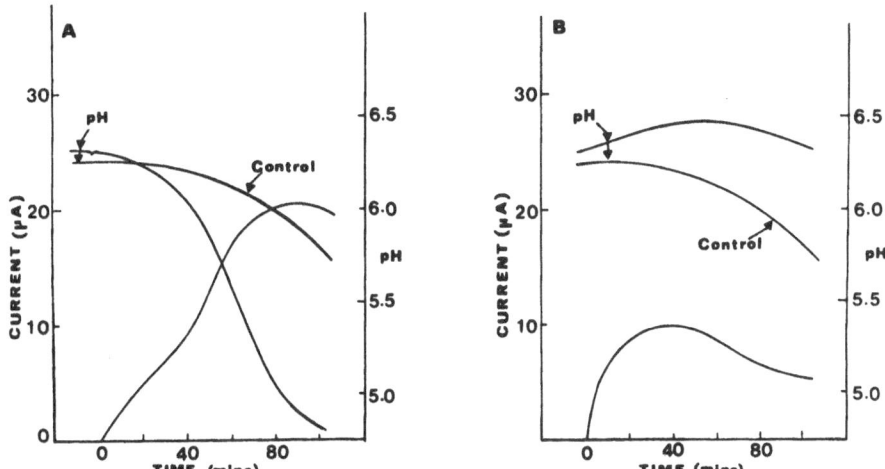

Fig. 1. Amperometric measurement of ferricyanide reduction A., and ferrocyanide oxidation B. by carrot cells. pH changes occurring in the presence and absence of ferricyanide are also shown.

Changes of Ferricyanide Reductase and Ferrocyanide Oxidase Activities during Culture Growth

The capacity of cells to reduce ferricyanide and also to oxidise ferrocyanide was dependent upon the time from sub-culture. Measurements of this capacity were made by calculating the charge transferred (Coulombs) for the first hour after addition of the mediator.

Figure 2 shows that the ferricyanide reductase capacity rose during the first half of the growth cycle reaching a peak around 8 days. The rate then fell off as the culture approached the stationary phase, 12-14 days.

The pattern for ferrocyanide oxidase capacity was similar to that of ferricyanide reductase except that the time of maximum charge transfer from ferrocyanide occurred a day earlier.

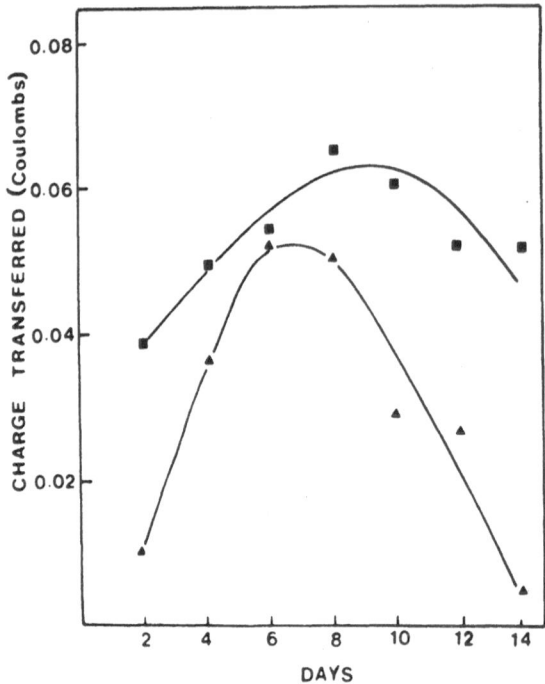

Fig. 2. The effect of cell culture age (days) on ferricyanide
reduction (■) and ferrocyanide oxidation (▲) , measured as
charge transferred in 60 minutes.

Factors affecting Ferricyanide Reduction

We have examined the effects of a number of factors on ferricyanide
reductase in 8 day old cells (period of highest activity). These effects
were related to the two phases of ferricyanide reduction.

Preincubation

As shown in Figure 1A ferricyanide reduction was biphasic for freshly
harvested cells. However preincubation of these cells in the reaction
medium for 90 minutes before the addition of ferricyanide resulted in
ferricyanide reduction which increased rapidly with time reaching a
maximum rate after 25 minutes. Figure 3 shows the current time profiles
and pH changes which occurred when cells were pre-incubated for 10 minutes
and for 90 minutes. In both cases the pH of the medium was readjusted to
6.2 immediately before the addition of ferricyanide. Preincubation
increased the rate of acidification of the medium by cells which was
further increased during ferricyanide reduction.

Oxygen Effects

When cells were incubated with oxygen free air (anaerobic conditions)
the charge transferred in the first 30 minutes of ferricyanide reduction
increased by 2.5 times of that observed under aerobic conditions.
However, after 60 minutes the rate of ferricyanide reduction was the same
in both cases. This indicates that anaerobiosis had a pronounced effect
on Phase I of ferricyanide reduction.

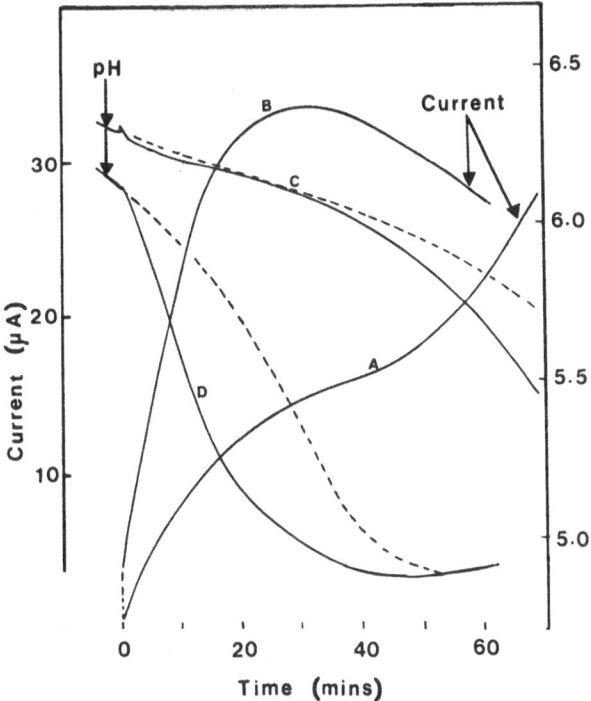

Fig. 3. The effects of preincubation on ferricyanide reduction and
medium acidification. A = current 10 minutes preincubation.
B = current 90 minutes preincubation. C = pH, 10 minutes
preincubation. D = 90 minutes preincubation. The dotted
line represents the respective pH changes in the absence of
ferricyanide.

Effects of Ionophores

The effects of the protonophore FCCP and the K^+ ionophore
valinomycin were tested. Addition of these compounds were made 10 minutes
after ferricyanide reduction had commenced. Valinomycin (10^{-7}M) had no
effect on the subsequent reduction of ferricyanide whereas FCCP (10^{-6} M)
caused a dramatic decrease in the rate of ferricyanide reduction to a
value which was 10% of the maximum value.

Effects of Inhibitors

A number of inhibitors which are known to inhibit either redox
reactions or ion-motive ATPases were tested for their ability to affect
ferricyanide reduction. Inhibitors were added during a preincubation
period of 10 minutes before the addition of ferricyanide. The degree of
inhibition was determined from the quantity of charge transferred in the
first 60 minutes, and the phase of ferricyanide reduction in which these
compounds had their affects assessed. The results obtained are shown in
Table 1.

Table 1.   The Effects of Electron Transfer and ATPase Inhibitors on Ferricyanide Reduction by Carrot Cells.

Inhibitor	Concn (M)	Ferricyanide Reduced Stimulation or Inhibition (%)	Phase Affected
DES	10^{-4}	-84	I
VANADATE	10^{-4}	+34	I
DCCD	10^{-4}	-65	II
HOQNO	5×10^{-5}	-67	II
ROTENONE	5×10^{-7}	+35	
AZIDE	10^{-3}	-53	II
NEM	10-3	-60	II
QUERCETIN	2×10^{-6}	+88	I
RUTIN	10^{-4}	+116	I

(+ indicates stimulation, - indicates inhibition relative to control)

Ferricyanide Reduction by Protoplasts and the effects of Osmotic Potential

Protoplasts derived from carrot cells reduced ferricyanide at a rate which was significantly greater than the corresponding intact cells.   In the first 10 minutes the activity was 3 fold greater and the total charge transferred in 60 minutes was 5 fold greater than the intact cells.

One of the ways in which the protoplasts differed from the whole cells is that they were suspended in the incubation solution which contained 0.4 M sorbitol as an osmoticum.   When intact cells were incubated in the presence of increasing sorbitol concentrations, there was an increase in the ferricyanide reductase activity with sorbitol concentrations above 0.1 M (122 mOsM) reaching a peak at 0.3 M (334 mOsM) (Chalmers and Coleman, 1986).   Concentrations above 0.3 M progressively induced plasmolysis.   This increase in electron transfer was accompanied by an increase in the rate of proton extrusion.   Replacement of sorbitol with manitol or sucrose produced similar results.   Control experiments showed that the increased ferricyanide reduction did not result from an increase extrusion of soluble reductants.

Oxidation of Exogenous NAD(P)H by Isolated Carrot Protoplasts

Addition of NAD(P)H to isolated carrot protoplasts caused a marked increase in oxygen consumption.   The media and methods used are described by Chalmers, Coleman and Hawes (1986).

The NAD(P)H oxidation was absent or negligible in freshly prepared protoplasts but increased to a maximal value (75 nmoles $O_2/10^6$ protoplasts per min) 1.5-2h after preparation.

The NAD(P)H oxidase activity had the following characteristics:

(i) A pH optimum of 6.5.
(ii) Caused a marked alkalisation of the medium.

(iii) Was stimulated by Mn^{2+} (10^{-3}M optimal) and by monophenols
       (p-coumarate 3 x 10^{-5}M optimal). Stimulations by Mn^{2+} and
       p-coumarate were additive.
 (iv) The activity was inhibited by cyanide (1.0mM), superoxide dismutase
       and by catalase.
  (v) The activity resulted in the formation of H_2O_2.
 (vi) The activity could be released from the protoplasts by washing in
       the suspending medium. Full activity was recovered in the cell
       free supernatant.

These observations show that the NAD(P)H oxidation was catalysed by
peroxidase. This was confirmed by the presence of very high peroxidase
activity in the supernatant.

The supernatant activity was low in freshly prepared protoplasts but
increased after 1.5-2h. During this period there was no significant
rupturing of the plasmamembrane. The presence of peroxidase was confirmed
by cytochemical localisation at the electron microscopy level. The
reactions involved in the oxidation of NAD(P)H by peroxidase are described
by Chalmers, Coleman and Hawes (1986).

DISCUSSION

By employing an electrochemical measuring technique which recycled
oxidant and reductant, we were able to measure ferricyanide reduction and
ferrocyanide oxidation for long experimental periods. It is clear from
the data obtained that the rate of ferricyanide reduction was influenced
by the metabolic state of the cell suspension. In freshly harvested cells
the rate of ferricyanide reduction increased for 80 minutes before a
steady state rate was reached. It is also clear that the increase in rate
took place in two well defined phases (Phase I and Phase II). An increase
in the rate of ferricyanide reduction with time has also been reported by
Craig and Crane (1982). We visualise the sequence of events which caused
the increasing rate of ferricyanide reduction as follows: Initially the
rate of ferricyanide reduction was limited by the supply of NAD(P)H in the
cytosol. The transfer of cells from growth medium to the incubation
medium triggered a metabolic re-adjustment which resulted in an increasing
availability of NAD(P)H from primary oxidative metabolism (glycolysis
and/or the pentose phosphate pathway). The biphasic nature of the
increasing rate and the differential effects of a number of factors and
inhibitors on these two phases suggests that the cytosolic reductant may
be supplied by two different metabolic pathways.

These carrot cells had a very high rate of aerobic glycolysis and
accumulated fermentation products in the medium (Chalmers and Coleman, in
preparation). For this to occur a ready supply of Pi and ADP would be
required. A possible source could be the plasmamembrane H^+
translocating ATPase. As described earlier preincubation stimulated an
acidification of the medium probably by the operation of the H^+-ATPase.
Preincubation also increased dramatically the rate of ferricyanide
reduction. It is tempting to suggest that increased turnover of
H^+-ATPase increased the rate of glycolysis and the supply of reductant
for ferricyanide reduction. Increasing the osmolality of the
extracellular medium to a level which caused incipient plasmolysis greatly
enhanced the rate of ferricyanide reduction, and increased the rate of
H^+ extrusion. This correlates with the observations that reduced turgor
enhances H^+ extrusion by plant cells, (Reinhold, Seiden and Volokita,
1984; Kinraide and Wyse, 1986).

In all cases ferricyanide reduction was accompanied by acidification of the medium. However, in the initial stages of Phase I the acidification was no greater than the control. This raises the question of the mechanism of ferricyanide reduction and H^+ extrusion. Figure 4 attempts to summarise some of the possibilities in two models. Model A shows the ferricyanide reductase as a H^+ extruding redox system. Model B shows the ferricyanide reductase system purely as an redox system with no H^+ pumping activity. In the latter H^+ extrusion occurs as a result of the stimulation of the H^+-ATPase by the increased cytosolic $[H^+]$ (Rubinstein and Stern, 1986). An additional factor must be considered.

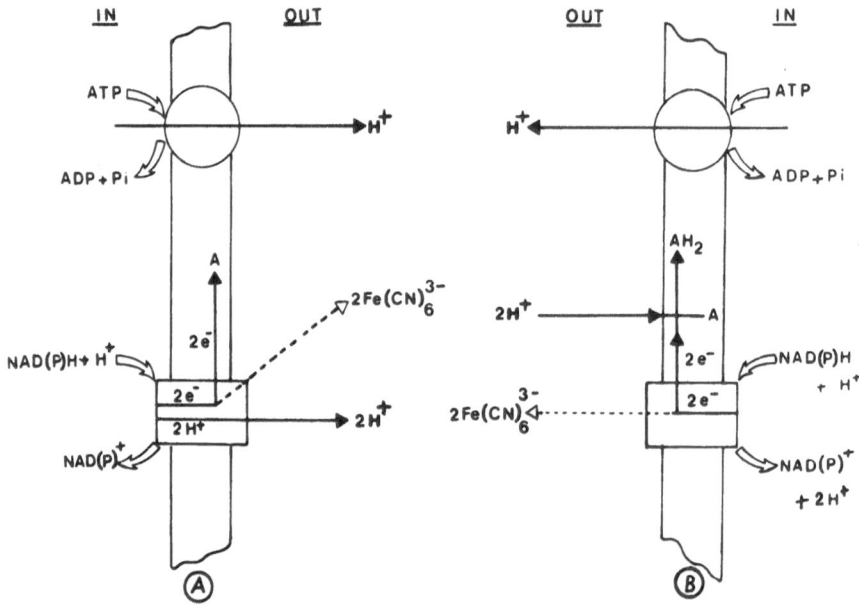

Fig. 4. Models for the mechanism of H^+ excretion and transplasmamembrane redox activity. A = physiological acceptor of reducing equivalents from the redox system.

The complete redox system is not necessarily involved in the ferricyanide reductase and the physiological acceptor is still unknown. If the reduction of this acceptor involves the uptake of an H^+ from the external medium, then under conditions of high flux in this system, diversion of electrons to ferricyanide would spare the uptake of H^+ and result in a net acidification of the medium. This possibility is to some extent supported by the finding that ferrocyanide oxidation is accompanied by an H^+ uptake, although we cannot be certain that the site of ferrocyanide reduction is the physiological acceptor of the ferricyanide reductase system.

It is not an easy matter distinguishing which of the two models more closely fit the redox system as clearly there is crosstalk between the H^+-ATPase and the redox system, and the level of the crosstalk is a reflection of the metabolic state of the cell.

The precise function of the redox system is still unknown, and this may only come with the resolution of the complete system. However from what is known it would appear to provide the cell with a range of options to meet the needs of growth and development. For example together with a suitable external, physiological mediator it may be a means of synchronising the redox state of adjacent cells and so influence growth and development.

External NAD(P)H Oxidase

Our evidence show that the external NAD(P)H oxidase associated with the plasmamembrane is peroxidase which is actively exocytosed by the cells (Chalmers, Coleman and Hawes, 1986). These conclusions have recently been supported by the findings of Askerlund, Larsson, Widell and Møller (1987). The high level of peroxidase in plant tissues and its association with membrane fractions must be taken into account when attempts are made to identify the NAD(P)H linked redox systems of the plasmamembrane.

ACKNOWLEDGEMENTS

We would like to thank Drs N.J. Walton and H.A.O. Hill of the I.C.L. Oxford for their assistance with the electrochemistry and for the use of facilities.

REFERENCES

Askerlund, C., Larsson, C., Widell, S., and Møller, I. M., 1987, NAD(P)H oxidase and peroxidase activities in purified plasmamembranes from cauliflower inflorescences, Physiol. Plant., 71:9.

Barr, R., Crane, F. L., and Craig, T. A., 1984, Transmembrane ferricyanide reduction in tobacco callus cells, Plant Growth Regul., 2:243.

Chalmers, J. D. C., and Coleman, J. O. D., 1983, Ethanol stimulated proton extrusion by cells of carrot (Daucus carota L.) grown in suspension culture, Biochem. Int., 7:785.

Chalmers, J. D. C., Coleman, J. O. D., and Walton, N. J., 1984, Use of an electrochemical technique to study plasmamembrane redox reactions in cultured cells of Daucus carota L., Plant Cell Reports, 3:243.

Chalmers, J. D. C., and Coleman, J. O. D., 1986, The effect of osmotic stress on trans-plasma membrane electron transport in plants, Biochem. Soc. Trans., 14:108.

Chalmers, J. D. C., Coleman, J. O. D., and Hawes, C. R., 1986, Oxidation of exogenous NAD(P)H by isolated carrot protoplasts is catalysed by isoperoxidase, Biochem. Soc. Trans., 14:734.

Craig, T. A., and Crane, F. L., 1981, Evidence for a transplasmamembrane electron transfer system in plant cells, Proc. Indiana Acad. Science, 90:150.

Craig, T. A., and Crane, F. L., 1982, Hormonal control of a transplasmamembrane electron transport system in plant cells, Proc. Indiana Acad. Science, 91:150.

Federico, R., and Giartosio, C. E., 1983, A transplasmamembrane electron transport system in maize roots, Plant Physiol., 73:182.

Ivankina, N. G., Novak, V. A., and Miclashevich, A. I., 1984, Redox reactions and active H^+ transport in the plasmalemma of Elodea leaf cells, in: "Membrane transport in Plants," W. J. Cram, K. Janacek, R. Rybova, and S. Sigler, eds., J. Wiley and Sons, England.

Kinraide, T. B., and Wyse, R. E., 1986, Electrical evidence for enhanced proton extrusion by reduced turgor in sugar beet tap root, Plant Physiol. Suppl., 80:57.

Lin, W., 1984, Further characterization on the transport properties of plasmalemma NADH oxidation system in isolated corn root protoplasts, Plant. Physiol., 74:219.

Neufeld, E., and Bown, A. W., 1987, A plasmamembrane redox system and proton transport in isolated mesophyll cells, Plant Physiol., 83:895.

Rubinstein, B., and Stern, A. I., 1986, Relationship of transplasmalemma redox activity to proton and solute transport by roots of Zea mays, Plant Physiol., 80:805.

Reinhold, L., Seiden, A., and Volokita, M., 1984, Is modulation of the rate of proton pumping a key event in osmoregulation?, Plant Physiol., 75:846.

Sijmons, P. C., Lanfermeijer, F. C., De Boer, A. H., Prins, H. B. A., and Bienfait, H. F., 1984, Depolarisation of cell membrane potential during transplasma membrane electron transfer to extracellular electron acceptors in iron deficient roots of Phaseolus vulgaris L., Plant Physiol., 76:943.

Slabas, A. R., Powel, A. J., and Lloyd, C. W., 1980, An improved procedure for the isolation and purification of protoplasts from carrot suspension cultures, Planta, 147:283.

Thom, M., and Maretski, A., 1985, Evidence for a plasmalemma redox system in sugar cane, Plant Physiol. 77:873.

Widholm, J. M. 1972, The use of fluorescein diacetate and phenosafranine for determining viability of cultured plant cells. Stain Technol., 47:184.

PURIFICATION OF NADH-FERRICYANIDE AND NADH-DUROQUINONE

REDUCTASES FROM MAIZE (*Zea mays* L.) ROOT PLASMA MEMBRANES

Thomas J. Buckhout and Douglas G. Luster

Plant Photobiology Laboratory, Beltsville Agricultural
Research Center, Agricultural Research Service, Beltsville, MD
20705, USA

The plasma membrane of eukaryotic cells contains endogenous, integral electron transport activities. These activities include NAD(P)H: ferricyanide reductase (FCR) and NAD(P)H:duroquinone reductase (DQR). DQR and FCR exhibit unique properties with regard to specificity for pyridine nucleotides, degree of stimulation of activity by detergents, solubilization characteristics and degree of purification following affinity chromatography . Both activities have been effectively solubilized using Triton X-100 and partially resolved and purified using dye-ligand affinity chromatography on Cibacron blue 3G-A-agarose. Elution of Triton-solubilized activities bound to Cibacron blue 3G-A-agarose with 25 µM NADH resulted in a purified protein with FCR and DQR activities. Evidence suggests that DQR and FCR eluted from Cibacron blue 3G-A-agarose with 25 µM NADH represent dual activities on the same protein, while a second enzyme with only FCR activity is eluted at higher concentrations of NADH.

INTRODUCTION

Advances in purification techniques for plant plasma membranes (PM) have facilitated the discovery of a number of PM electron transport activities (Crane et al. 1985; Møller and Lin 1986). Enzyme(s) in PM vesicles obtained by aqueous two-phase partitioning of microsomal membranes isolated from monocotyledenous and dicotyledenous species will transfer electrons from NAD(P)H to electron acceptors. The question thus arises: do these activities, as defined by their electron acceptors, exist as a single protein with broad acceptor specificity, as a set of enzymes with separate activities and corresponding physiological functions or as a series of components in an electron transfer chain, as found in mitochondrial inner membranes and prokaryotic plasma membranes? We are attempting to answer this question through characterization of the activities found in maize root PM. We have found significant differences in selected characteristics for 2 enzyme activities located in the maize PM; NAD(P)H: ferricyanide reductase (FCR) and NAD(P)H:duroquinone reductase (DQR). The activities were partially resolved upon solubilization at selected detergent-to-protein ratios and were further separable using dye-ligand affinity chromatography. In addition, we have raised monospecific antibodies to identify, characterize and purify these respective activities.

MATERIALS AND METHODS

Plant material and membrane isolation. Maize (*Zea mays* L. cv Golden Cross Bantam) roots were grown, harvested and microsomal membranes prepared as previously described (Buckhout and Hrubec 1986; Luster and Buckhout 1988). PM was isolated from the microsomal fraction using the 6.2% (w/w) polymer aqueous 2-phase system, and the upper phases were diluted with 5 volumes of 15 mM tris(hydroxymethyl)aminomethane (Tris) - 2-(N-morpholino)-ethanesulfonic acid (MES), pH 7.5, containing 0.25 M sucrose (assay medium) and centrifuged at 113 000g_{max} (SW 28 rotor, Beckman Instruments, Inc., Palo Alto, CA, USA). The resulting PM pellets were resuspended in assay medium containing 0.5 mM PMSF, 25 µg ml^{-1} chymostatin (Gallagher et al. 1986) and 1 mM EDTA for storage at -80°C. Stored PM preparations were diluted in and washed with 10 volumes of 0.5 M KCl in assay medium using 10 strokes of a Teflon™ pestle homogenizer and pelleted at 113 000g_{max} for 45 min. Pellets were then washed in 5 volumes of assay medium and repelleted before use in enzyme assays or detergent treatments.

Enzyme assays. Assays were conducted spectrophotometrically at 30°C in assay medium with 0.01% (w/v) Triton X-100 included where indicated. This Triton X-100 concentration resulted in a detergent-to-protein ratio shown to give maximum stimulation of NAD(P)H-ferricyanide reductase (Buckhout and Hrubec 1986). Reagents were prepared fresh daily in assay medium as 100-fold concentrated stock solutions, excepting duroquinone, which was prepared in absolute ethanol. FCR was assayed at 420 nm as described (Buckhout and Hrubec 1986) subtracting the sum of the rates obtained without enzyme and without NADH. An extinction coefficient of 1.02 mM^{-1} was used in calculations of ferricyanide reduction. NAD(P)H:duroquinone reductase (EC 1.6.5.1) was assayed at 340 nm using 0.16 mM NADH and 0.1 mM duroquinone (tetramethylquinone; Sigma Chemical Co., St Louis, MO, USA). The rate obtained prior to addition of duroquinone (i.e. NADH oxidase) was less than 5-10% of the duroquinone-dependent rate and therefore was not subtracted in the calculations. An extinction coefficient of 6.22 mM^{-1} was used in calculations of NAD(P)H:DQR.

Protein in PM fractions was measured using the Bradford (1976) colorimetric reagent (Bio-Rad, Rockville Centre, NY, USA).

Detergent solubilization. PM fractions from several isolations were pooled, washed with KCL and pelleted at 113 000g_{max} for 45 min. The pellet was resuspended in assay medium plus 0.5 mM PMSF and 25 µg ml^{-1} chymostatin and brought to selected ratios of detergent-to-protein (w/w) with the addition of Triton X-100. Stock solutions of 10% (w/v) Triton (Calbiochem, San Diego, CA, USA) were made fresh weekly from detergent, which had been stored protected from light under vacuum to reduce peroxide formation (Lever 1977). The membrane plus detergent suspension was mixed by 10 strokes of a Teflon™ pestle homogenizer, incubated on ice for 30 min and centrifuged for 60 min at 192 000g_{max} (SW 50.1 rotor). Where measurement of pelleted activities was desired, pellets were resuspended in 1-2 ml of assay medium.

Dye-ligand affinity chromatography. PM redox activities that had been solubilized with Triton X-100 at a detergent-to-protein ratio of 1.0 were applied to a 1.5 x 20 cm column of Cibacron blue 3G-A agarose (Affi-Gel blue, Bio-Rad) previously equilibrated with 10 bed volumes of 15 mM Tris-MES, pH 7.0, 0.25 M sucrose, 1 mM KCl, 1 mM CaCl$_2$, 5 mg ml^{-1} chymostatin and 0.06% (w/v) Triton X-100 (column buffer) at a flow rate of 0.2 ml min^{-1}. The column was washed with 2-3 bed volumes of column buffer at a flow rate of 0.5 ml min^{-1} to remove unbound protein. Reductase activities were eluted with 0.025 or 10 mM NADH (measured by absorbence at 340 nm) in column buffer at a flow rate of 0.1 ml min^{-1}. All procedures were conducted at 5°C. Protein was measured in column fractions by the method of Markwell et al. (1981) following precipitation with 6% (w/v) TCA plus 200 mg ml^{-1} deoxycholate (Bensadoun and Weinstein 1976).

SDS-PAGE. One ml from 10 mM NADH-eluted Cibacron blue 3G-A agarose chromatography column fractions was precipitated with TCA and deoxycholate as described above for protein assays. Precipitates were washed in ice-cold 100% ethanol and reprecipitated. The pellets resulting from TCA precipitation were resuspended in 5% (w/v) SDS sample buffer (Laemmli 1970) and heated at 90°C for 5 min. Samples were applied to a 1.5 mm thick mini-vertical polyacrylamide slab (6x8 cm; Hoefer Scientific Instruments, San

Francisco, CA, USA) composed of 10.5% total monomers with a 1.0 cm, 4% stacking gel and electrophoresed at 20 mA for 1 h. Silver staining was performed according to Merril et al. (1983). Standard proteins of known molecular weight (Bio-Rad, 14-97 kDa) were included in separate lanes for molecular weight estimations.

Immunization and fusion. Triton-solubilized PM proteins were diluted with 2 volumes of phosphate-buffered saline (PBS; 10 mM Na-phosphate, pH 7.2, and 0.15 M NaCl) and emulsified with an equal volume of Freund's complete adjuvant (Sigma Chemical Co.). BALB/c, female mice were injected subcutaneously with 200 µl (ca. 90 µg protein) of the emulsified antigen at multiple sites. Mice were boosted interperitoneally (I.P.) with approximately 20 µg of protein at 2-week intervals with protein fractions eluted with 0.1 mM NADH (Luster and Buckhout 1988) and emulsified in an equal volume of Freund's incomplete adjuvant (Sigma Chemical Co.). Production of antibodies to the immunogen was determined by ELISA, probing with goat antimouse antibodies conjugated to horseradish peroxidase (Kierkegaard and Perry, Gaithersburg, MD USA). Three days prior to fusion, mice were given a final I.P. injection with antigen prepared as described above. On the day of fusion, 3 mice were sacrificed by cervical dislocation and the spleens removed. Spleenocytes were fused with log-phase NS-1 myeloma cells (Kohler and Milstein 1975) at a ratio of spleenocytes to myeloma cells of 4:1 by the method of Galfré and Milstein (1981). Cells were plated into ten 96-well microtiter plates, and hybridomas were selected on hypoxanthine, aminopterin and thymidine (HAT) medium. After 12 days on HAT medium, cells were refed with hypoxanthine-thymidine (HT) medium. Medium was assayed for antibody production by ELISA when the majority of the wells had cells at 40 to 50% confluency. Positive cells were expanded into 24-well microtiter plates (miniclones) and the medium was retested for positive antibody production. Hybridomas were cloned by the soft agar method of Civin and Banquerigo (1983).

Dot-blot screening. Dot-blot analyses were conducted using a dot-blotting apparatus (Bio-Rad) and were performed according to the manufacturer's recommendations. Triton-solubilized PM proteins or enzyme fractions eluted from a Cibacron blue columns with 25 µM NADH were applied to nitrocellulose sheets, the nitrocellulose blocked with 3% (w/v) bovine serum albumin and 1:3 dilutions of culture supernatant was reacted with the antigen. Specific binding of antibodies was detected using goat antimouse antibodies conjugates to alkaline phosphatase. Alkaline phosphatase was assayed according to the manufacturer's suggestion (Sigma Chemical Co.).

RESULTS

The PM preparations used in this work were previously demonstrated to consist of ca. 95% PM with a predominantly right-side-out orientation as determined by Mg^{2+}-dependent ATPase latency (Larsson et al 1984; Buckhout and Hrubec 1986). The endogenous electron transport activities were clearly not due to contamination with endomembranes such as ER or mitochondria (Buckhout and Hrubec 1986) and thus represent PM-associated redox activities. Redox activities which have been identified in the PM include NAD(P)H-dependent reductions of a variety of electron acceptors including ascorbate free-radical (Morré et al. 1986; Luster and Buckhout 1988), cytochrome (cyt.) c (Lundborg et al. 1981; Buckhout and Hrubec 1986; Sandelius et al. 1986), dichloroindophenol (Buckhout and Hrubec 1986), ferricyanide (e.g. Sandelius et al. 1986) and indigotetrasulfonate (Buckhout and Hrubec 1986), as well as NAD(P)H oxidases (Møller and Lin 1986; Møller and Berczi 1986). It is unclear whether the various NAD(P)H-dependent activities found in PM represent a single enzyme with broad substrate specificities or multiple enzymes. We report here results on the characterization, purification and immunological identification of 2 PM redox activities, namely FCR and DQR.

Solubilization of PM redox proteins. When salt-washed PM were incubated in the presence of Triton X-100 at maximum detergent to protein ratios as described in the Materials and Methods, approximately 80% of the activity for NADH:FCR and NADH:DQR remained in the 192 000g_{max} supernatant (Fig. 1). The optimum ratio of detergent-to-protein for maximum solubilization of FCR with minimum solubilization of total protein was ca. 1.0. This ratio was therefore routinely used in solubilization procedures for affinity chromatography. While FCR was more effectively solubilized at somewhat lower detergent-to-protein ratios than DQR, the solubilization profiles were not greatly different. Significant purification was not obtained with detergent solubilization since the percent of the total protein solubilized paralleled the solubilization of the 2 redox activities (Fig. 1). Under these conditions, the vanadate-sensitive, Mg^{2+}-ATPase was not solubilized (data not shown).

Purification of PM redox proteins by affinity chromatography. As an initial purification

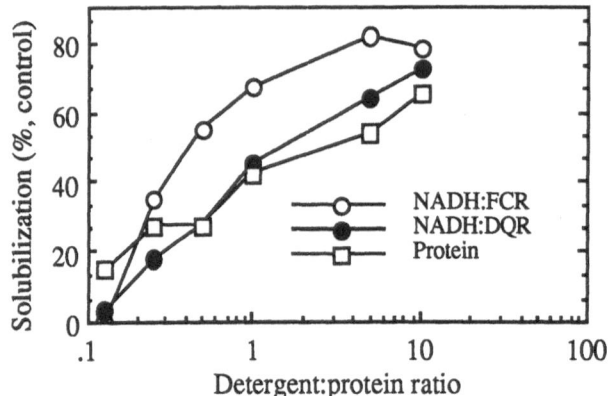

Figure 1. Solubilization of PM redox activities with Triton X-100. KCl-washed membranes (1.0 mg/ treatment) were homogenized with detergent using a Teflon™ pestle, incubated on ice for 30 min and centrifuged at $192\,000g_{max}$ for 60 min. Supernatants were assayed for NADH-dependent redox activities and expressed here as percentage of the control (no detergent) pellet activity. Results are from a typical experiment (n>10).

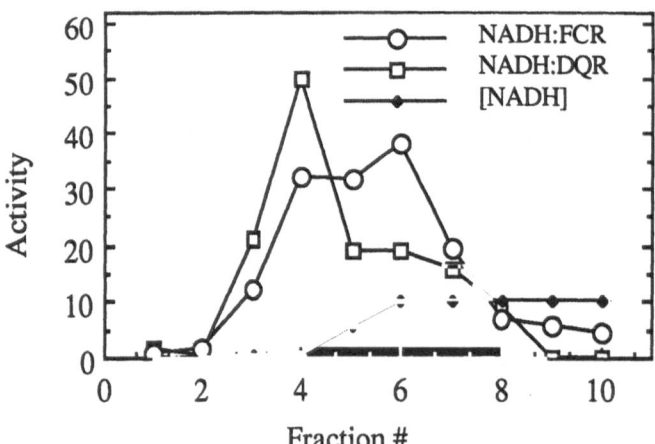

Figure 2. Partial resolution of NADH:FCR and NADH:DQR activities on Cibacron blue F3G-A agarose. PM proteins solubilized at a detergent-to-protein ration of 1 were applied to the Cibacron blue column. The column was washed extensively and was eluted with 10 mM NADH. Fractions eluting with NADH were analyzed for FCR and DQR. The corresponding protein profiles on SDS-PAGE for fractions 3 - 8 are shown in Fig. 3. The values on the y-axis correspond to mUnits (nmol min^{-1}) for NADH:DQR, mUnits x 0.1 for NADH:FCR and mM for NADH.

step for FCR and DQR, Cibacron blue 3G-A agarose columns loaded with Triton-solubilized PM proteins were eluted with column buffer containing 10 mM NADH (Fig. 2). Columns eluted under these conditions consistently demonstrated a partial resolution of the activity profiles of the two enzymes. Although all fractions eluting from the column contained both DQR and FCR, DQR consistently eluted before FCR. Interestingly, the bulk of the DQR activity and a substantial portion on the FCR activity eluted prior to any significant increase in 340 nm absorbence indicating that this portion of the DQR and FCR were eluting at a concentration significantly less than 0.1 mM NADH. Conversely, nearly one-half of the

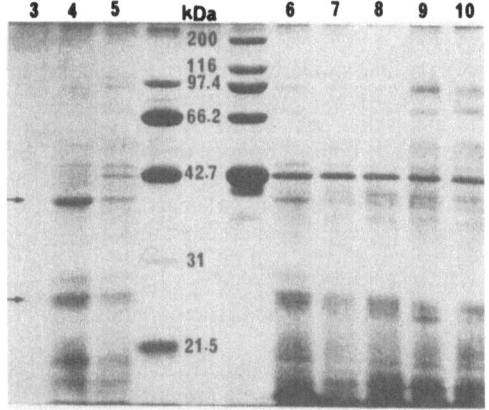

Figure 3. SDS-gel electrophoresis of fractions eluted with 10 mM NADH from the Cibacron blue 3G-A agarose column. Fractions corresponding to those in Fig. 2 were prepared as described in Materials and Methods. Arrows indicate the location of prominent bands of 28 and 40 kDa corresponding to the elution profile of the NADH:DQR activity.

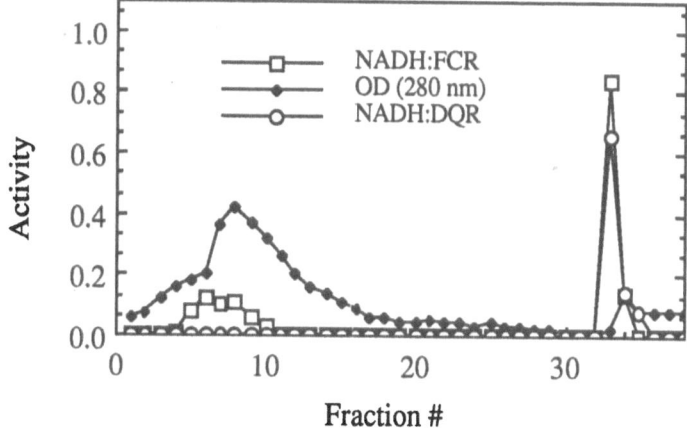

Figure 4. Activity profiles of DQR and FCR eluted from a Cibacron blue column with 25 µM NADH. Conditions for chromatography are described in Fig. 2 and in the Materials and Methods. The values on the y-axis correspond to units (mmol min^{-1}) for NADH:FCR, units x 10 for NADH:DQR.

FCR and a much smaller portion of DQR were eluted only after the NADH concentration had exceeded 5 mM. Thus, distinct FCR and DQR activities can be partially resolved based on their binding affinity to Cibacron blue.

Samples corresponding to the fractions illustrated in Fig. 2 were prepared and electrophoresed on SDS gels as described in the Materials and Methods. As illustrated in Fig. 3, numerous proteins eluted from the column with 10 mM NADH but, the protein patterns displayed in fractions eluting at < 1 mM were distinct from those eluting at ≥ 5 mM NADH. Polypeptide bands at 28 kDa and 40 kDa were found to correlated with the DQR activity profile shown in Fig. 2 (Fig. 3, arrows). Similar results were obtained by sequential elution of Cibacron blue agarose columns with 1.0 and 10.0 mM NADH (Luster and Buckhout 1988).

Because the majority of the DQR activity eluted at < 1 mM NADH, a series of experiments were conducted using decreasing concentrations of NADH in the column buffer. Linear gradients of 0 - 10 mM NADH were also employed but resulted in poor resolution of FCR activity (data not shown). Column buffer containing 25 μM NADH resulted in specific elution of a single, sharp peak of FCR and DQR activity (Fig. 4). The peaks of activity were coincident and the activity profiles similarly symmetrical. Neither the pH of the elution buffer nor the conductivity was measurably changed by the addition of NADH. Thus, elution of FCR and DQR at this concentration of NADH suggests a biospecific interaction of the dye molecule with the NADH binding domain on the protein. SDS-PAGE analysis of the TCA precipitated eluted peak FCR/DQR fraction revealed a single silver stained polypeptide band of molecular mass 28 kDa (data not shown) suggesting that FCR and DQR represent multiple activities of the same protein. The ratio of total FCR:DQR activity in the purified, 25μM-eluted fraction was approximately half the ratio found in either PM fractions or Triton-solubilized PM fractions. This suggests that a substantial proportion of FCR activity in the total PM fraction remains bound to the Cibacron blue affinity matrix and is not eluted at NADH concentrations below 5 mM.

Monoclonal antibodies. Five mice were initially immunized with Triton-solubilized proteins and subsequently boosted 4-times with proteins eluted by 0.1 mM NADH. Following fusion and selection on HAT medium, > 80% of wells (i.e. 800 wells) showed cell growth. These wells were assayed by ELISA using both total solubilized and NADH-eluted proteins as antigens. Twelve wells gave positive reactions against both antigens and were each expanded into a single well of a 24-well plate and cloned into soft-agar. The resulting clones were recovered from soft-agar, rescreened by ELISA as above and finally by dot-blot analysis against solubilized PM proteins and purified redox proteins eluted from the dye affinity column with 25 μM NADH. Results of a partial dot-blot screening are presented in Fig. 5. Clones P1A2, P1A3 and P1B1 showed strong

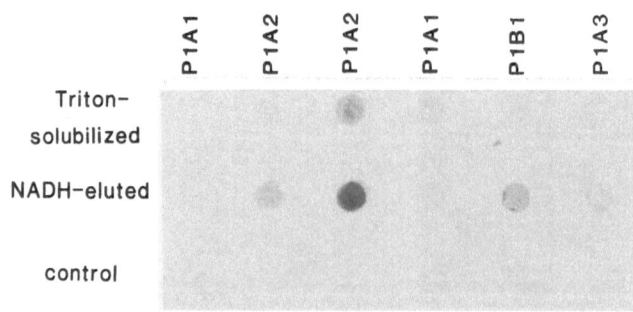

Figure 5. Dot-blot analysis of monoclonal antibodies against PM electron transport proteins. Dot-blot assays were conducted as described in the Materials and Methods. The control row contained no antigen, Triton-solubilized contained Triton X-100-solubilized PM proteins as described in the Materials and Methods and NADH-eluted contained the NADH:DQR/FCR protein eluted from Cibacron blue agarose with 25 μM NADH. Columns were treated from culture supernatant from 4 hybridoma clones.

positive reactions to both solubilized PM and purified redox proteins. Clone P1A1 did not react under these assay conditions and represents an internal, negative control. The preliminary finding that the fraction eluted with 25 µM NADH contains only a single, 28 kDa polypeptide suggests that the 3 positive clones shown here contain antibodies against the DQR/FCR redox protein of the PM.

DISCUSSION

Two classes of electron acceptors were analyzed in this study, namely: ferricyanide, a model acceptor for iron reduction at the root cell surface (Rubinstein et al. 1984) and duroquinone, which should mimic the proposed role of quinones as electron carriers in NADH oxidation at the outer surface of maize protoplasts (Lin 1982). We have been partially successful in resolving these 2 activities on a Cibacron blue affinity column (Fig. 2). Activities eluting at low NADH concentration contained the majority of the DQR activity as well as approximately one-half of the FCR activity while fractions eluting at higher NADH concentrations contained a greater portion of FCR activity. Eluting the affinity column with 25 µM NADH results in recovery of a single peak of activity containing both DQR and FCR activities (Fig. 4).

Experiments conducted by Luster and Buckhout (1988) reported on sequential elution of a Cibacron blue column first with 1 mM and subsequently with 10 mM NADH. Initial elution with 1 mM produced coincident peaks of FCR and DQR while subsequent elution with 10 mM produced a second peak of FCR activity which contained little DQR activity. Based on results presented here and elsewhere (Luster and Buckhout 1988), we propose that two distinct pyridine nucleotide dehydrogenases exist on the maize root PM, one capable of transferring electrons to both FCR and DQR and a second which transfers electrons only to FCR. The fact that specific immunological markers are now available to at least one of these dehydrogenases should greatly facilitate their characterization and cellular localization.

ACKNOWLEDGMENT

The mention of vendor or product does not imply that they are endorsed or recommended by U.S. Department of Agriculture over vendors of similar products not mentioned.

REFERENCES

Bensadoun, A., and Weinstein, D., 1976, Assay of proteins in the presence of interfering materials, Anal. Biochem., 70: 241-250.

Bradford, M.M., 1976, A rapid sensitive method for the quantitation of microgram quantities of protein utilizing the principle of protein dye-binding, Anal. Biochem., 72: 248-254.

Buckhout, T.J., and Hrubec, T.C., 1986, Pyridine nucleotide-dependent ferricyanide reductases associated with isolated plasma membranes of maize (*Zea mays* L.) roots, Protoplasma, 135: 144-154.

Civin, C.I., and Banquerigo, M.L., 1983, Rapid, efficient cloning of murine hybridoma cells in low gelation temperature agarose, J. Immunol. Meth., 61: 1-8.

Crane, F.L., Sun, I.L., Cark, M.G., Grebing, C., and Löw, H., 1985, Transplasma-membrane redox systems in growth and development, Biochim. Biophys. Acta, 811: 233-264.

Galfré, G., and Milstein, C., 1981, Preparation of monoclonal antibodies: strategies and procedures, Methods Enzymol., 73: 1-46.

Gallagher, S.R., Carroll, E.J., and Leonard, R.T., 1986, A sensitive diffusion plate assay for screening inhibitors of protease activity in plant cell fractions, Plant Physiol., 81: 869-874.

Köhler, G., and Milstein, C., 1975, Continuous cultures of fused cells secreting antibody of predefined specificity, Nature, 256: 495-497.

Laemmli, U.K., 1970, Cleavage of structural proteins during the assembly of the head of bacteriophage T4, Nature, 227: 680-685.

Larsson, C., Kjellbom, P., Widell, S., and Lundborg, T., 1984, Sidedness of plant plasma membrane vesicles purified by partitioning in aqueous two-phase systems, FEBS Lett., 171: 271-276.

Lever, M., 1977, Peroxides in detergents as interfering factors in biochemical analysis, Anal. Biochem., 83: 274-279.

Lin, W., 1982, Responses of corn root protoplasts to exogenous NADH: oxygen consumption, ion uptake, membrane potential, Proc. Natl. Acad. Sci. USA, 79: 3773-3776.

Lundborg, T., Widell, S., and Larsson, C., 1981, Distribution of ATPase in wheat root membranes separated by phase partition, Physiol. Plant., 52: 89-95.

Luster, D.G., and Buckhout, T.J., 1988, Characterization and partial purification of multiple electron transport activities in plasma membranes from maize (*Zea mays* L.) roots, Physiol. Plant., in press.

Markwell, M.A.K., Haas, S.M., Tolbert, N.E., and Bieber L.L., 1981, Protein determination in membrane and lipoprotein samples: manual and automated procedures, Methods Enzymol., 72: 296-303.

Merril, C.R., Goldman, D., and Van Keuren, M.L., 1983, Silver stain methods for polyacrylamide gel electrophoresis, Methods Enzymol, 96: 230-239.

Møller, I.M., and Berczi, A., 1986, Salicylhydroxamic acid-stimulated NADH oxidation by purified plasmalemma vesicles from wheat roots, Physiol. Plant., 68: 67-74.

Møller, I.M., and Lin W., 1986, Membrane-bound NAD(P)H dehydrogenases in higher plant cells, Ann. Rev. Plant Physiol., 37: 309-334.

Morré, D.J., Navas, P., Penel, C., and Castillo, F.J., 1986, Auxin-stimulated NADH oxidase (semidehydroascorbate reductase) of soybean plasma membrane: role in acidification of cytoplasm?, Protoplasma, 133: 195-197.

Rubinstein, B., Stern, A.I., and Stout, R.G., 1984, Redox activity at the surface of oat root cells, Plant Physiol., 76: 386-391.

Sandelius, A.S., Barr, R., Crane, F.L., and Morré, D.J., 1986, Redox reactions of plasma membranes isolated from soybean hypocotyls by phase partition, Plant Sci., 48: 1-10.

THE TURBO REDUCTASE IN PLANT PLASMA MEMBRANES

H.F. Bienfait

Oudegracht 285 bis, 3511 PA Utrecht, The Netherlands

Plants get hold of the necessary inorganic constituents by absorption from the soil. Ions like K^+, Mg^{2+}, NO_3^- are normally present in the soil solution. However, in normal soils, iron is only present at very low concentrations. Ferric is precipitated by OH^- and can attain levels above 1 μM only at pH values below 3. In the presence of oxygen and organic matter, ferrous is rapidly oxidized to ferric. In water culture, plants need soluble iron concentrations (realized by chelation with compounds like EDTA) around 1 μM (Lindsay and Schwab, 1982).

In order to realize a sufficient supply of iron, plants may develop different activities (strategies, Römheld and Marschner, 1986). Grasses excrete phytosiderophores which solubilize ferric ions from complexes in the soil (Takagi et al., 1984). Dicotyledons and non-grass monocots develop, when iron in the roots is getting low (Bienfait et al., 1987), specialized 'transfer' cells in the root epidermis (Kramer et al., 1980). These cells excrete protons, acidifying the soil immediately surrounding the roots, and their plasma membrane contains a redox system which can transfer electrons to extracellular ferric salts (Chaney et al., 1972). This redox system has a very low specificity, as it reduces different kinds of ferric chelates (Bienfait et al., 1983). The combined action of the proton excreting and ferric reducing activities results in mobilization of iron from ferric precipitates in the soil, and in production of ferrous ions at the root surface. Ferrous ions are easily taken up by roots.

The ferric reducing system which is induced in the roots by iron deficiency is different from the system which is constitutively present in grass roots and other plant tissues, reducing ferricyanide but no ferric chelates of the EDTA type. I called this latter system Standard, because of its apparently ubiquitous presence in plant cells, in contrast to the system induced by iron deficiency, 'Turbo' because of its high rates (Bienfait, 1985). Table I shows typical activities of both systems.

Table I. Standard and Turbo reductase activities in roots of bean and barley, grown with and without Fe. Rates are given as V_{max}. (From Bienfait, 1988.)

Substrate	Activity measured	Bean +Fe	-Fe	Barley +Fe	-Fe
		(μmol reduced/h.g fresh weight)			
Ferricyanide	Standard + Turbo	3.7	7.5	3.8	3.8
Fe(III)EDTA	Turbo	0.5	5.0	0.1	0.1

The Standard system is the subject of many contributions in this volume. A problem, not yet settled at this moment, is what the natural acceptor of this system is, and linked to this question, what its physiological function may be. A possibility is that it serves to keep proteins reduced in the plasma membrane, which are involved in ion uptake and ATPase activity (Lüttge and Clarkson, 1985, Bienfait and Lüttge, 1988).

The function of the Turbo system is clearly to promote iron uptake. Thus, a mutant of tomato which cannot develop this system is incapable of growth in normal soils or even in water culture with chelated iron in levels which are sufficient for the wild type; only when given large amounts of chelated iron it will grow healthily and produce tasty tomatoes (Brown et al., 1973).

The electron donor of the Turbo system is most probably NADPH (Sijmons et al., 1984). The activity of the system is extra stimulated when the roots have recently excreted high amounts of protons (Sijmons and Bienfait, 1986). Proton excretion by roots is accompanied by citrate formation in the same cells (Landsberg, 1986). We proposed that citrate, through the combined action of mitochondrial aconitase, leading to isocitrate, and cytosolic isocitrate dehydrogenase,exerts a strong reducing power on the NADP couple (Fig. 1, from Lubberding et al., 1988).

Electron acceptors of the system are, as mentioned above, ferric chelates and ferricyanide. However, Cakmak found that superoxide dismutase inhibited ferric-EDTA reduction, but only when oxygen was present. In the absence of oxygen, the rate was the same as with oxygen, but now superoxide had no effect (Cakmak et al., 1987). From his experiments we drew the conclusion that the final electron donor in the plasma membrane is a two-electron carrier such as a flavin or a quinone, in which the electrons in the wholly reduced form are at another redox potential than in the half-reduced form (Fig. 2).

Conceivably, Fe deficiency might turn a Standard redox system into a Turbo system through the loss of an Fe-containing component. But this is not probable, as the development of Turbo activity depends on the presence of the FER gene (Wann and Hills, 1973, Brown et al. 1973). Fig. 3 shows how this gene may control synthesis of the Turbo reductase.

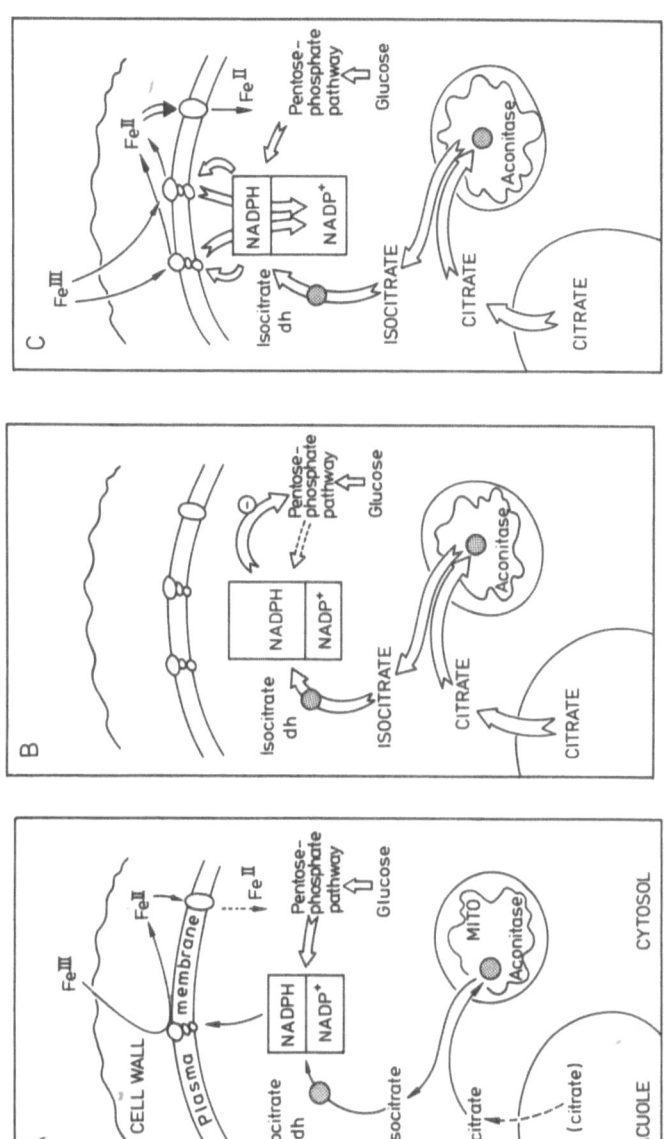

Figure 1. Mechanism for generation of reducing power in rhizodermal transfer cells of iron-deficient bean plants. A, control (= iron-sufficient) plant; B, iron-deficient plant. Citrate accumulation results in a strongly reduced state of NADP; C, iron-deficient plant after addition of a reducible ferric salt. MITO: mitochondrion. (From Lubberding et al., 1988.)

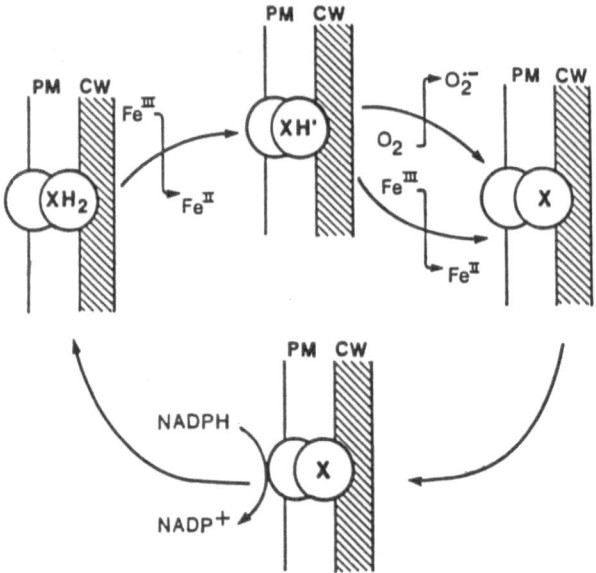

Figure 2. Electron donation to ferric chelates and oxygen at the plasma membrane of epidermal transfer cells in iron-deficient bean roots. X is a two-electron carrier. PM: plasma membrane; CW: cell wall. (From Cakmak et al., 1987.)

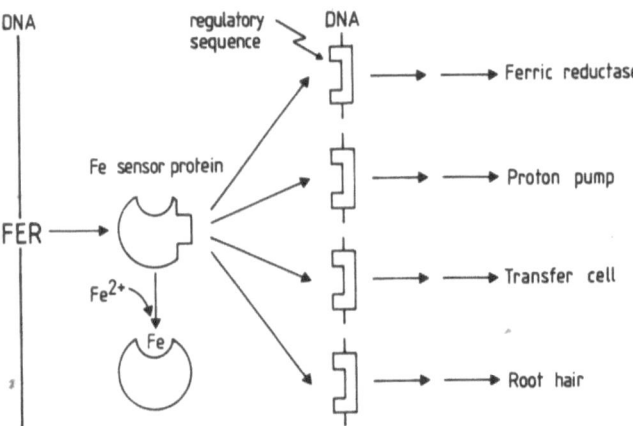

Figure 3. Hypothesis for the regulation of Fe efficiency reactions in tomato. The FER gene encodes a regulatory protein that can bind to common sequence elements which activate genes involved in Fe efficiency reactions, inducing transcription. The regulatory protein can bind ferrous ions and in doing so changes its conformation so that it can no longer bind to the genes' regulatory sequences. (From Bienfait, 1988.)

References

Bienfait, H.F., 1985, Regulated redox processes at the plasmalemma of plant cells and their function in iron uptake. J. Bioenerg. Biomembr., 17:73.

Bienfait, H.F., 1988, Proteins under the control of the FER gene for Fe efficiency. (Subm. to Plant Physiol.)

Bienfait, H.F., Bino, R.J., Van der Bliek, A.M., Duivenvoorden, J.F., and Fontaine, J.M., 1983, Characterization of ferric reducing activity in roots of Fe-deficient Phaseolus vulgaris. Physiol. Plantarum 59:196.

Bienfait, H.F., De Weger, L.A., and Kramer, D., 1987, Control of development of iron-efficiency reactions in potato as a response to iron deficiency is located in the roots. Plant Physiol., 83:244.

Bienfait, H.F., and Lüttge, U., 1988, On the function of two redox systems that can transfer electrons across the plasma membrane, Plant Physiol Biochem (in press).

Brown, J.C., Chaney, R.L., and Ambler, J.E., 1973, A new tomato mutant inefficient in the transport of iron. Physiol. Plantarum 25:48.

Cakmak, I., Van de Wetering, D.A.M., Marschner, H. and Bienfait, H.F., 1987, Involvement of superoxide radical in extracellular ferric reduction by iron-deficient bean roots. Plant Physiol. 85:310.

Chaney, R.L., Brown, J.C., and Tiffin, L.O., 1972, Obligatory reduction of ferric chelates in iron uptake by soybeans. Plant Physiol. 50:208.

Kramer, D., Römheld, V., Landsberg, E., and Marschner, H., 1980, Induction of transfer cell formation by iron deficiency in the root epidermis of Helianthus annuus L., Planta 147:335.

Landsberg, E.C., 1986, Function of rhizodermal transfer cells in the Fe stress response mechanism of Capsicum annuum L. Plant Physiol. 82:511.

Lindsay, W.L., and Schwab, A.P., 1982, The chemistry of iron in soils and its availability to plants. J. Plant Nutr. 5:821.

Lubberding, H.J., De Graaf, F.H.J.M., and Bienfait, H.F., 1988, Ferric reducing activity in roots of iron-deficient Phaseolus vulgaris: source of reducing equivalents. Biochem. Physiol. Pflanzen (in press).

Lüttge, U., and Clarkson, D.T., 1985, Mineral nutrition: plasmalemma and tonoplanst redox activities, in: 'Progress in Botany', Vol 47, Springer, Berlin, p 73.

Römheld, V., and Marschner, H., 1986, Mobilization of iron in the rhizosphere of different plant species. Adv.Plant Nutr., 2:155.

Sijmons, P.C., and Bienfait, H.F., 1986, Development of FeIII reduction activity and H^+ extrusion during growth of iron-deficient bean plants in a rhizostat. Biochem. Physiol. Pflanzen 181:283.

Sijmons, P.C., Van den Briel, W., and Bienfait, H.F., 1984, Cytosolic NADPH is the electron donor for extracellular FeIII reduction in iron-deficient bean roots. Plant Physiol. 74:219.

Takagi, S., Nomoto, K. and Takemoto T., 1984, Physiological aspect of mugineic acid, a possible phytosiderophore of gramineceous plants. J. Plant Nutr. 7:469.

Wann, E.V., and Hills, W.A., 1973, The genetics of boron and iron transport in the tomato. J.Heredity, 64:370.

SPECIFIC INTERACTION OF THE HERBICIDE SETHOXYDIM

WITH THE PLASMALEMMA REDOX SYSTEM OF PLANT CELLS

Elke Fischer, Angela Weber, Helmut Schipp von Branitz, and
Ulrich Lüttge

Institut für Botanik
Technische Hochschule Darmstadt
Darmstadt, FRG

INTRODUCTION

Sethoxydim, a cyclohexanedione derivative (Fig. 1), is a selective post-emergence herbicide, which shows a high toxic activity against gramineous plants and low activity against dicotyledonous and other mono-cotyledonous plants (Iwataki et al., 1983).

After foliar application sethoxydim is rapidly absorbed and translo-cated to meristematic zones which are thought to be the primary sites of action and where sethoxydim induces, in sensitive plants chlorosis, cessation of leaf and root growth, and finally necrosis (Ishikawa et al., 1980; Lichtenthaler and Meier, 1984). The biochemical mechanisms that lead to a differential sensitivity in plants are still the subject of in-vestigation. There is, however, increasing evidence that sethoxydim at low doses inhibits fatty acid biosynthesis in sensitive plants by blocking acetyl-CoA carboxylase (Lichtenthaler et al., 1987; Focke and Lichtenthaler, 1987).

The uptake of sethoxydim is a non-specific passive diffusion process which proceeds by equal rates in sensitive and tolerant plants (Struve et al., 1987). Sethoxydim absorption involves the equilibration of lipid compartments of the cell, which are mainly cell membranes, with the lipophilic non-dissociated weak acid molecule of sethoxydim.

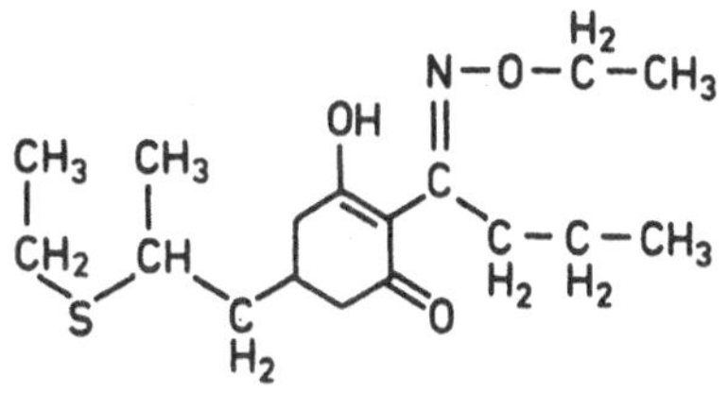

Fig. 1. Chemical structure of sethoxydim; 2-(1-(ethoxyimino)butyl)-5-(2-ethylthio-propyl)-3-hydroxy-2-cyclohexene-1-one; BAS 9052 OH; Poast[R].

The inhibition of lipid biosynthesis and a partitioning of the lipophilic herbicide into cell membranes imply that sethoxydim may directly or indirectly induce a perturbation of structure and function of the plasmalemma. A differential interference with transport systems located at the plasmalemma in sensitive and tolerant plants in the early stages of sethoxydim action could lead to more or less severe consequences for the maintenance of cell metabolism especially in metabolically highly active dividing and elongating meristematic cells.

For these reasons we studied the effect of sethoxydim on plasmalemma transport systems, i.e. the H^+-transporting ATPase and the electron transport system in vivo using leaf segments and in vitro using isolated plasmalemma vesicles. We chose two plant species within the genus Poa and Festuca which have been found to be tolerant to sethoxydim, i.e. Poa annua and Festuca rubra, and compared their responses to the herbicide to the responses of the sensitive species P. pratensis, F. ovina and Zea mays.

EFFECT OF SETHOXYDIM ON ΔE OF LEAF CELLS AND ON H^+-ALANINE COTRANSPORT

The first indication in favour of a direct interference of sethoxydim with plasmalemma transport processes arose from experiments testing the effect of the herbicide on the electrical membrane potential (ΔE) of grass leaf cells (Weber et al., 1988). The application of sethoxydim to leaf segments of sensitive and tolerant grass species at concentrations higher than 100 μM produced a reversible depolarization of the cell membrane with only a few minutes delay (Fig. 2 a). The degree of depolarization was linearly related to the herbicide concentration (Fig. 2 b). The reduction of ΔE appeared to originate in an interaction of the herbicide with active transport mechanisms at the plasmalemma since sethoxydim did not affect the passive diffusion component of ΔE (Fig. 2 c).

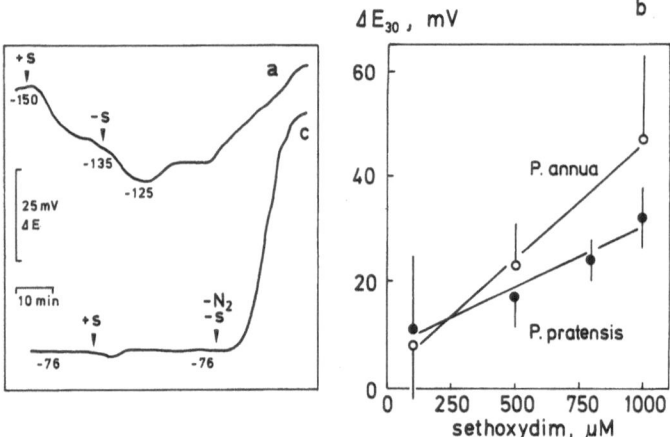

Fig. 2. (a) Effect of sethoxydim (800 μM) on ΔE of a Poa annua leaf cell. Sethoxydim (s) was applied for 20 min. The maximum depolarization of the plasmalemma occurred ca. 30 min after applying sethoxydim to the standard buffer solution, i.e. 10 min after removing sethoxydim from the medium. (b) Dependence of the maximum reduction of ΔE, i.e. after 30 min as shown in (a) of leaf cells of Poa pratensis (sensitive) and Poa annua (tolerant) on sethoxydim concentration. Sethoxydim was applied for 20 min at concentrations as indicated in the graph. Values are mean values ± SD ($n \geq 3$). (c) Effect of sethoxydim (800 μM) on the passive component of ΔE of a Poa pratensis leaf cell in the presence of N_2.

To characterize the effect of sethoxydim on the active component of ΔE in more detail we tested the influence of the herbicide on the H^+-amino acid cotransport system of Poa and Festuca leaf cells. After a pretreatment of leaf segments with sethoxydim for 20 minutes the alanine-induced transient depolarization of the plasmalemma which coincided with the onset of H^+-cotransport was larger than in control tissue (Table 1). The depolarization of the plasmalemma which is associated with amino acid uptake is caused by an increased H^+ influx during the H^+-amino acid cotransport. The depolarization of the plasmalemma and the acidification of the cytoplasm leads to an activation of the H^+-transporting ATPase at the plasmalemma. Hence, the maximum transient reduction of ΔE is determined by the amino acid uptake rate and the rate of activation of the H^+-ATPase (Fischer and Lüttge, 1980).

Since sethoxydim increased the alanine-induced transient maximum depolarization of the plasmalemma but simultaneously inhibited the uptake of alanine we concluded that sethoxydim directly interferred with the H^+-ATPase located at the plasmalemma.

EFFECT OF SETHOXYDIM ON THE H^+-ATPase OF ISOLATED PLASMALEMMA VESICLES

The results we had obtained in vivo using leaf segments prompted us to approach the problem of sethoxydim action in a direct way. We isolated plasmalemma vesicles by two different techniques: 1. by aqueous polymer two-phase partitioning from leaves of P. pratensis and P. annua and 2. by differential and sucrose gradient centrifugation from roots of Poa and Z. mays. The plasmalemma preparations were highly enriched in Mg-dependent ATPase activity, and according to the criteria of vanadate-, NO_3^-- and azide-sensitivity (Bowman et al., 1978) they were lacking significant contamination of mitochondrial and tonoplast origin (Table 2).

The application of sethoxydim had no significant effect on the ATP hydrolysis of plasmalemma preparations of sensitive and tolerant Poa leaves and roots, and roots of sensitive Z. mays between pH 6.5 and 5.0, independently of the technique of membrane isolation (Fig. 3 a). Sethoxydim did not affect the ATP-dependent H^+ transport into tightly sealed inside-out plasmalemma vesicles of maize roots (Fig. 3 b).

Table 1. Effect of Sethoxydim (800 μm) on the Alanine-Induced Transient Depolarization of the Plasmalemma of Leaf Cells and the Alanine Uptake into Leaf Segments of Sensitive and Tolerant Grasses. Leaf Segments were Pretreated with Sethoxydim for 20 min. ^{14}C Alanine Uptake was Subsequently Measured for 30 min. Experiments were Performed in the Dark at pH 5.7.

		Alanine-Induced Depolarization after Sethoxydim Treatment in % of Control	Inhibition of Alanine Uptake to % of Control	
		Alanine Concentration		
		1 mM	0.1 mM	1 mM
Poa pratensis	sensitive	111	52	71
Poa annua	tolerant	180	35	55
Festuca ovina	sensitive	117	-	-
Festuca rubra	tolerant	164	-	-

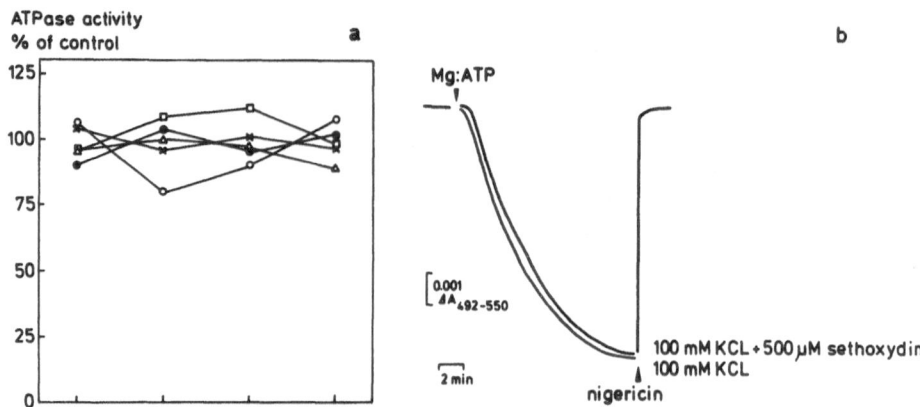

Fig. 3. (a) Effect of sethoxydim (500 µM) on the ATP hydrolysis of plasmalemma preparations of Poa pratensis leaves (Δ) and roots (●),
Poa annua leaves (□) and roots (×), and Zea mays roots (o). For
specific activities at pH 6.5 see Table 2. (b) Time course of
ATP-dependent H$^+$ transport into plasmalemma vesicles isolated
from maize roots in the presence and absence of sethoxydim (500
µM). The intravesicular acidification was measured as the decrease of absorbance of acridine orange. Measurements were conducted at pH 6.5.

Therefore the effect of sethoxydim on the active component of ΔE
and the H$^+$-amino acid cotransport system of leaf cells was apparently
not directly exerted via an inhibition of the plasmalemma-bound H$^+$-ATPase.

Table 2. ATPase Activity of Plasmalemma Preparations Isolated from
Poa pratensis and Poa annua Leaves (Two-Phase Partitioning
(Kjellbom and Larsson, 1984) and Roots, and Zea mays Roots
(Sucrose Gradient Centrifugation, 30 % - 40 % Interphase;
De Michelis and Spanswick, 1986). Effect of Inhibitors:
Vanadate (100 µM), NO$_3^-$ (100 mM), and Azide (1 mM). ATP
Hydrolysis was Measured in the Presence of 0.01 % Triton,
P$_i$ Liberated was Determined According to Ames (1966).

ATPase Activity
µmol (mg protein)$^{-1}$ h^{-1}

| | pH 6.5 | | pH 8.0 | | |
	Control	+ Vanadate	Control	+ NO$_3^-$	+ Azide
Leaves					
Poa pratensis	61.9	14.8	11.7	9.1	11.3
Poa annua	52.2	10.4	10.9	10.7	10.4
Roots					
Poa pratensis	41.2	10.8	17.4	16.0	12.7
Poa annua	36.9	12.2	15.4	12.9	12.0
Zea mays	16.3	3.8	5.2	4.5	4.8

EFFECTS OF SETHOXYDIM ON THE PLASMALEMMA REDOX SYSTEM AND THE ELECTRON
TRANSPORT SYSTEMS OF ISOLATED CHLOROPLASTS AND MITOCHONDRIA

Evidence for redox activity at the surface of Poa and Festuca leaf
cells was obtained by monitoring the reduction of ferricyanide used as an
impermeable external electron acceptor (Weber and Lüttge, 1988). The
linear time course of ferricyanide reduction for at least 4 h after the
beginning of the experiment excluded the participation of reducing agents
that may leak out of the cut edges of leaf segments (Fig. 4). Hence, we
concluded that ferricyanide reduction in Poa and Festuca leaves is the
result of an electron transfer across the plasmalemma from an endogenous
electron donor to extracellular ferricyanide.

Application of sethoxydim inhibited the reduction of ferricyanide by
leaf cells of the sensitive and tolerant Poa and Festuca species (Fig. 4;
Table 3). Sethoxydim reduced the electron transport activity in the
presence of ferricyanide more severely in sensitive than in tolerant
species. This difference in response may indicate that the interference
of sethoxydim with a redox system at the plasmalemma of leaf cells could
possibly not only contribute to a more general toxic effect of sethoxydim
but may play a role in a selective action of the herbicide.

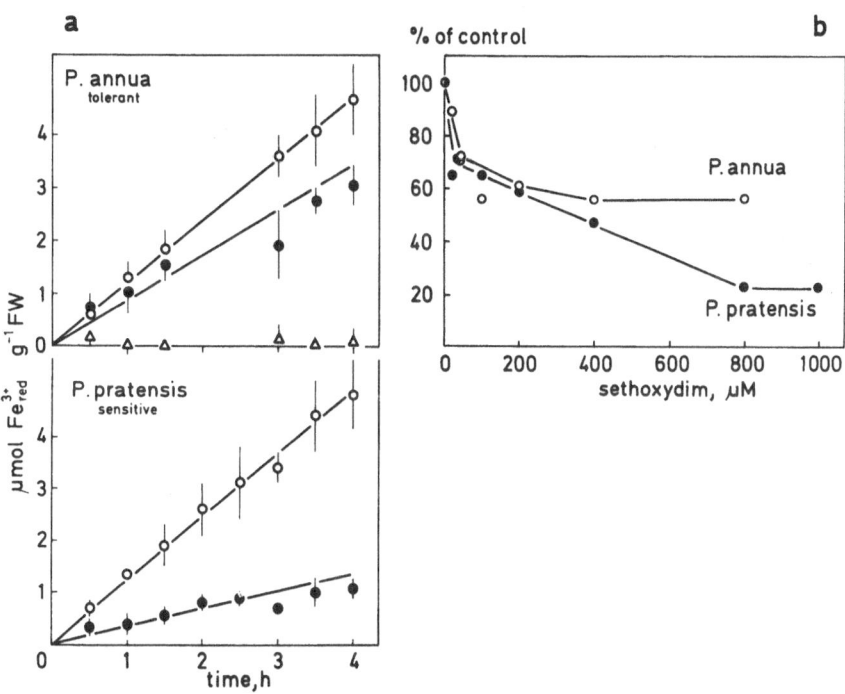

Fig. 4. (a) Time course of ferricyanide reduction in the absence and
presence of sethoxydim (800 μmol) of leaf segments of the sensitive grass species Poa pratensis and the tolerant grass species
Poa annua. Experiments were performed in the dark at pH 5.7.
Control (o), sethoxydim (●), ferricyanide plus sethoxydim without
plant material (Δ), ferricyanide concentration: 1 mM. (b) Relation of ferricyanide reduction of Poa leaf segments to the
sethoxydim concentration. Ferricyanide reduction was measured
over a time period of 2 h.

Table 3. Effect of Sethoxydim (800 µM) on the
Ferricyanide Reduction of Leaf Segments
of Sensitive and Tolerant Grasses. Ferri-
cyanide Reduction was Measured over a
Time Period of 4 h in the Dark at pH 5.7.
Mean Values $\pm$ SD (n $\geq$ 3).

		Control	Sethoxydim
		μmol Fe^{3+} (g FW)$^{-1}$ h^{-1}	
Poa pratensis	sensitive	1.16 $\pm$ 0.20	0.26 $\pm$ 0.01
Poa annua	tolerant	1.12 $\pm$ 0.25	0.75 $\pm$ 0.10
Festuca ovina	sensitive	1.30 $\pm$ 0.29	0.35 $\pm$ 0.04
Festuca rubra	tolerant	1.43 $\pm$ 0.36	0.64 $\pm$ 0.08

Although sethoxydim has not been found to be an inhibitor of respiration or photosynthesis in vivo (Hatzios, 1982; Asare-Boamah and
Fletcher, 1983), it reduced the electron transport activity of isolated
chloroplasts and mitochondria (Fig. 5). Electron transport activities
were measured as ferricyanide reduction of isolated chloroplasts and as
succinate-ferricyanide oxidoreductase activity of isolated mitochondria.
Chloroplasts had been osmotically shocked to break the outer membrane and
allow the transfer of electrons from photosystem II to ferricyanide. The
electron flow to the artificial electron acceptor ferricyanide of isolated mitochondria was completely inhibited by antimycin A, indicating
that ferricyanide did not penetrate the inner mitochondrial membrane and
thus only accepted electrons from the endogenous cytochrome c located on
the outer face of the inner membrane (Palmer and Kirk, 1974). It was
intriguing to find that the inhibition of electron transport systems of
chloroplasts and mitochondria by sethoxydim again appeared to be higher
in organelles isolated from herbicide-sensitive plants. The electron
transport activity of chloroplasts isolated from leaves of the sensitive
P. pratensis was reduced by 500 µM sethoxydim to 69 % of the control,
whereas the electron transport activity of chloroplasts of the tolerant
P. annua did not respond to sethoxydim at all. The electron transport
activity of mitochondria of sensitive Z. mays was severely inhibited by
500 µM sethoxydim to 25 % of the control activity. Ferricyanide reduction
rates of tolerant potato mitochondria at this concentration only decreased to 68 % of the control activity.

Since sethoxydim not only inhibited electron transport processes
associated with the plasmalemma of leaf cells but also interferred with
electron transport activities of isolated chloroplasts and mitochondria,
we concluded that sethoxydim may directly affect membrane-bound redox
systems in general.

EFFECT OF SETHOXYDIM ON THE NAD(P)H-DEPENDENT FERRICYANIDE REDUCTION OF
ISOLATED PLASMALEMMA PREPARATIONS

Plasmalemma preparations of maize and Poa roots obtained by discontinuous sucrose gradient centrifugation showed a NAD(P)H-dependent ferricyanide reduction activity at rates in the same order of magnitude as
described in the literature (Buckhout and Hrubec, 1986; Macri and
Vianello, 1986). This redox activity was insensitive to KCN, antimycin A,

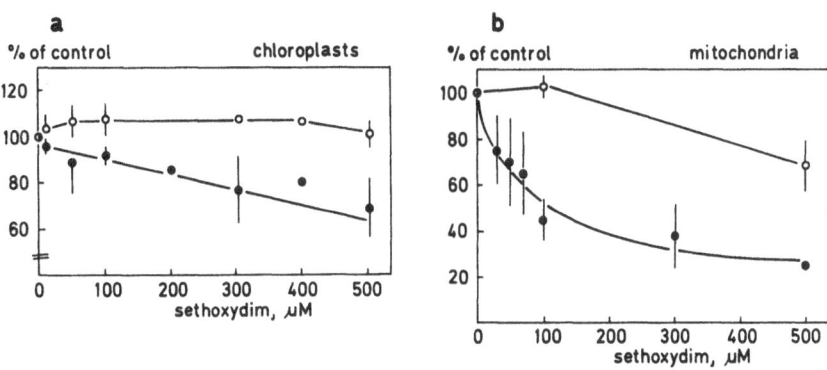

Fig. 5. (a) Effect of sethoxydim on ferricyanide reduction of chloroplasts isolated from leaves of the sensitive Poa pratensis (●)
and the tolerant Poa annua (o) at herbicide concentrations as
indicated. The ferricyanide reduction was measured over a time
period of 1 h at 450 μmol quanta m^{-2} s^{-1} at pH 7.6. Absolute
reduction rates (= 100 %): P. pratensis 47.2 $\pm$ 10.1 μmol Fe^{3+}
(mg Chl)$^{-1}$ h^{-1}, P. annua 46.4 $\pm$ 18.7 μmol Fe^{3+} (mg Chl)$^{-1}$ h^{-1}.
(b) Effect of sethoxydim on the succinate-ferricyanide oxido-
reductase activity of mitochondria isolated from sensitive Zea
mays coleoptiles (●) and from tolerant Solanum tuberosum
tubers (o) as described by Hampp (1979) and Friemert (1986).
The ferricyanide reduction was measured in the presence of
1 mM KCN to block the electron transfer to oxygen. Mitochondria were preincubated with sethoxydim for 3 min. Absolute
reduction takes (= 100 %): Z. mays: 197 $\pm$ 61 nmol Fe^{3+} (mg
protein^{-1}) min^{-1}, S. tuberosum: 2 000 nmol Fe^{3+} (mg protein^{-1})
min^{-1}. Mean values $\pm$ SD (n $\geq$ 3).

oligomycin and could not be initiated by succinate. Therefore, we believe
that the reduction rates we determined were devoid of any mitochondrial
activity. When we measured the NAD(P)H-dependent ferricyanide reduction
of plasmalemma preparations in the presence of sethoxydim we found an
increasing inhibition of the redox activity with decreasing pH values of
test-medium below pH 6.5 (Table 4). We observed an inhibition of the
plasmalemma redox activity of isolated membranes to a similar extent in
membrane preparations of roots of sensitive Z. mays and P. pratensis and
of roots of the tolerant P. annua. Thus the apparent differential sensitivity of the plasmalemma redox system to sethoxydim we had observed in
vivo in leaf segments of sensitive and tolerant grasses had disappeared
in vitro in isolated plasmalemma preparations.

CONCLUSIONS

     Sethoxydim reversibly reduced the active component of ΔE, increased
the alanine-induced depolarization at the onset of H$^+$-alanine cotransport
and simultaneously inhibited alanine uptake in sensitive and tolerant
grasses. These effects of sethoxydim pointed to an inhibition of the
plasmalemma H$^+$-translocating ATPase. However, the ATP hydrolysis of plasmalemma preparations isolated by two different techniques from leaves and

Table 4. Effect of Sethoxydim (500 μM) on the NADPH-Dependent
Ferricyanide Reduction of Plasmalemma Vesicles.

	Ferricyanide Reduction μmol Fe^{3+} (mg protein)$^{-1}$ min^{-1}			
	Control	Sethoxydim	Control	Sethoxydim
	pH 6.5		pH 5.0	
Poa pratensis	0.78	–	0.55	0.27
Poa annua	0.81	0.80	0.56	0.11
Zea mays	1.24	1.31	0.87	0.42

roots of sensitive and tolerant plants was not affected by sethoxydim.
Apparently the changes in transport characteristics caused by sethoxydim
in vivo in leaf segments originated in a specific interference of the
herbicide with the plasmalemma redox system. Sethoxydim inhibited the
electron transport activity that proceeds at the plasmalemma of grass
leaf cells in the presence of ferricyanide. Since the herbicide also
reduced the electron transport activities of isolated chloroplasts and
mitochondria and the NAD(P)H- dependent ferricyanide reduction of isolated plasmalemma preparations, we feel that sethoxydim interacts with
membrane-bound redox systems in a direct way. We can exclude that the
effect of the lipophilic herbicide is the result of an unspecific perturbation of membrane structure and function since sethoxydim inhibited
plasmalemma redox activity but not ATP hydrolysis and ATP-dependent H^+
transport in isolated plasmalemma vesicles. The differential sensitivity
of the electron transport system at the plasmalemma of leaf cells and the
inhibition of electron transport activities to a different degree in
chloroplasts and mitochondria of sensitive and tolerant plants indicate
that the interference of sethoxydim with the plasmalemma redox system may
contribute to the selective action of the herbicide in vivo. The NAD(P)H-
dependent ferricyanide reductase activity of plasmalemma preparations
isolated from roots of sensitive and tolerant plants, however, was equally inhibited by sethoxydim. Thus, it appears possible that sethoxydim
indeed specifically inhibited the plasmalemma-bound redox system, but
that factors decisive for a selective action of the herbicide observed in
intact cells were lost during the isolaton procedure, e.g. due to the
removal of the cell wall or cytoplasmic phase, or are not operating in
the in vitro membrane system.

It is, however, important to point out that sethoxydim seems to
specifically inhibit the NAD(P)H-dependent ferricyanide reduction in
isolated plasmalemma preparations, whereas the H^+-ATPase is not affected.
Thus, sethoxydim could be applied in future experiments as a specific
inhibitor of the plasmalemma electron transport to study the co-operation
of plasmalemma redox system and H^+-translocating ATPase.

ACKNOWLEDGEMENT

This work was supported by BASF-Aktiengesellschaft, D-6703 Limburgerhof, FRG, in a frame-work of a co-operation agreement with the
Technische Hochschule Darmstadt.

REFERENCES

Ames, B.N., 1966, Assay of inorganic phosphate, total phosphate and phosphatases, Methods in Enzymol. 8:115.

Asare-Boamah, N.K., and Fletcher, R.A., 1983, Physiological and cyto-logical effects of BAS 9052 OH on corn (Zea mays) seedlings, Weed Science, 31:49.

Bowman, B.J., Mainzer, S.E., Allen, K.E., and Slayman, C.W., 1978, Effects of inhibitors on the plasma membrane and mitochondrial adenosine triphosphatases of Neurospora crassa, Biochim. Biophys. Acta, 512:13.

Buckhout, T.J., and Hrubec, T.C., 1986, Pyridine nucleotide-dependent ferricyanide reduction associated with isolated plasma membranes of maize (Zea mays L.) roots, Protoplasma, 135:144.

De Michelis, M.I., and Spanswick, R.M., 1986, H$^+$ pumping driven by vana-date-sensitive ATPase in membrane vesicles from corn roots, Plant Physiol., '81:542.

Fischer, E., and Lüttge, U., 1980, Membrane potential changes related to active transport of glycine in Lemna gibba G1, Plant Physiol., 65:1004.

Focke, M., and Lichtenthaler, H.K., 1987, Inhibition of the acetyl-CoA carboxylase of barley chloroplasts by cycloxydim and sethoxydim, Z. Naturforsch., 42:1361.

Friemert, V., 1986, Untersuchungen über die CO_2-Abgabe einiger Pflanzen mit Crassulaceen-Säure-Stoffwechsel im Licht, Dissertation, TH Darmstadt.

Hampp, R., 1979, Kinetics of mitochondrial phosphate transport and rates of respiration and phosphorylation during greening of etiolated Avena leaves, Planta, 144:325.

Hatzios, K.K., 1982, Effects of sethoxydim on the metabolism of iso-lated leaf cells of soybean (Glycine max (L.) Merr.), Plant Cell Reports, 1:87.

Ishikawa, H., Okunuki, S., Kawana, T., and Hirono, Y., 1980, Histologi-cal investigation on the herbicidal effects of alloxydim-sodium in oat, J. Pestic. Sci., 5:547.

Iwataki, I., Shibuya, M., and Ishikawa, H., 1983; in: "Pesticide Che-mistry. Human Welfare and the Environment", I. Miyamoto and P.C. Kearney, eds., vol. 1, Pergamon Press, Oxford.

Kjellbom, P., and Larsson, C., 1984, Preparation and polypeptide-composi-tion of chlorophyll-free plasma membranes from leaves of light grown spinach and barley. Physiol. Plant., 62:501.

Lichtenthaler, H.K., Kobek, K., and Ishii, K., 1987, Inhibition by sethoxydim of pigment accumulation and fatty acid biosynthesis in chloroplasts of Avena seedlings, Z. Naturforsch., 42:1275.

Lichtenthaler, H.K., and Meier, D., 1984, Inhibition by sethoxydim of chloroplast biogenesis, development and replication in barely seedlings, Z. Naturforsch., 39:115.

Macri, F., and Vianello, A., 1986, Independence of trans-plasma membrane proton gradient from NAD(P)H-ferricyanide oxidoreduction in maize root microsomes, Plant Science, 43:25.

Palmer, J.M., and Kirk, B.I., 1974, The influence of osmolarity on the reduction of exogenous cytochrome c and permeability of the inner membrane of Jerusalem artichoke mitochondria, Biochemical Journal, 140:79.

Struve, I., Golle, B., and Lüttge, U., 1987, Sethoxydim-uptake by leaf slices of sethoxydim resistant and sensitive grasses, Z. Natur-forsch., 42:279.

Weber, A., Fischer, E., Schipp von Branitz, H., and Lüttge, U., 1988, The effects of the herbicide sethoxydim on transport processes in sensitive and tolerant grass species. I. Effects on the electrical membrane potential and alanine uptake, Z. Naturforsch., 43c: in press.

Weber, A., and Lüttge, U., 1988, The effects of the herbicide sethoxydim on transport processes in sensitive and tolerant grass species. II. Effects on membrane-bound redox systems in plant cells, Z. Naturforsch., 43c: in press.

NADH OR NADPH ?

Susanne Krüger and Michael Böttger

Institut für Allgemeine Botanik
Ohnhorststr. 18
D-2000 Hamburg 52, FRG

SUMMARY

Using the method of enzymic cycling, the dynamics of pyridine nucleotide concentrations in maize roots and cultured carrot cells after various treatments were studied. The changes were related to transplasmamembrane electron- and proton- transfer rates measured by a pH - redoxstat.

Ethanol increased H^+- efflux and, if plants were pretreated with IAA, electron transfer was enhanced too. It has been proposed that this effect is caused by a rise of NADH content due to the action of ADH. In fact, NADH level increased after ethanol treatment. These results would favour NADH as electron donor for the redox system.

In carrot cells, HCF III treatment decreased both NADH and NADPH levels within five minutes. In maize roots, however, the content of NADH was not affected, but a significant decline of NADPH concentration occured after 30 seconds. Similar results could be obtained with the new electron acceptor hexachloroiridate (IV) in maize roots.

In all these experiments, [NADPH + $NADP^+$] and [NADH + NAD^+] remained constant. This indicates that the changes in the contents of the reduced forms really reflect their oxidation rather than alterations of the nucleotide synthesis of the plant. Our results also indicate that NAD-kinase which could link the ATP level with the concentrations of the pyridine nucleotides, does not play a significant role in the nucleotide turnover and in the regulation of the redox pump.

Comparing NAD(P)H concentrations with the rates of electron transfer we conclude that there must be a rapid mechanism for back-regulation of the nucleotide content. Otherwise, the whole pool of reduced nucleotides would be completely oxidized within seconds.

Abbrevations:
ADH, alcoholdehydrogenase; HCF III, Hexacyanoferrate III;
HCl IV, Hexachloroiridate IV; NBT, Nitrobluetetrazolium.

INTRODUCTION

According to Bienfait (1985), the regnum plantarum can be
divided into two subunits: Plants, which are able to install
a so-called "Turbo system", and those which are not. "Turbo"
plants are specially equipped to reduce external Fe III
sources like chelates. It has been reported, that chelators
are excreted by roots (Römheld and Marschner, 1983). Iron deficiency increases the efficiency of the iron uptake system
by physiological (Sijmons et al., 1984) and morphological
adaptation (Kramer et al., 1980). This response enables roots
to reduce some artificial iron containing compounds as well.
The hypothesis claims that only dicots and "non-grass" monocots are able to install this "Turbo system" in the rhizodermis of young root parts if iron is lacking (reviewed by
Bienfait, 1985).

All important crops are only equipped with the so-called
"Standard system" that is also able to reduce non-iron-
containing compounds with a high redox potential like NBT
(Elzenga and Prins, 1987) and HCl IV (Lüthen and Böttger,
1988). It has been proposed that oxygen is the natural electron acceptor and that these artificial compounds compete for
oxygen (Böttger and Lüthen, 1986). Evidence is given that the
"Standard system" takes part in the maintainance of the electrochemical gradient between cytoplasm and apoplast represen-
ting the driving force for uptake mechanisms. The importance
of this system and a possible linkage between the redox chain
and the well established H^+- pumping ATPase is still under
discussion. The electron donor of the "Standard system" might
be NADH (Craig and Crane, 1981; Barr et al., 1985; Böttger
and Lüthen, 1986) or NADPH (Qiu et al., 1985), while there is
evidence that NADPH fuels the "Turbo system" (Sijmons et al.,
1984). Evidence for NADH participation in the electron transfer derives from experiments with ethanol, which increases
the NADH-level by the action of ADH. However, this situation
is abnormal for the plant.

The aim of the present investigation was to determine the
source of redox equivalents.

Information was achieved by increase in the level of
redox equivalents or by acceleration of the plasmamembrane
electron transfer owing to addition of artificial electron
acceptors. We used intact roots in order to minimize effects
of cut surfaces. On the other hand access of the e^-- acceptor
might be restricted to the cell layers next to the surface.
This has to be emphasized since the first subrhizodermal
layer might be equipped with a casparian band which blocks
the apoplastic transport (Peterson, 1987). Symplastic path is
unlikely because HCF III does not permeate into the cell
(Craig and Crane, 1981; Federico and Giartosio, 1983;
Rubinstein et al., 1984). We therefore used suspensions of
cultured carrot cells for comparison.

MATERIALS AND METHODS

Plant Material

Seeds of Zea mays L. c.v. Goldprinz were grown as described recently (Lüthen and Böttger, 1988). Carrot suspension
cultures were obtained from Mrs. Grieb, Justus Liebig University, Giessen.

Extraction and measurement of pyridine nucleotides

The extraction of pyridine nucleotides were performed
according to the description of Qiu et al. (1985) with the
following modifications:

Instead of root segments we used intact maize seedlings
for the assays.

After the incubation period the roots respectively the
carrot cells were washed in ice-cooled medium for 5 seconds.
Then the tissue was frozen in liquid N_2, disrupted in a pre-
cooled mortar and lyophilisized for four days.

The dried powder was extracted either with 0.1 N NaOH,
for assay of the reduced forms, or with 0.1 N HCl for assay
of the oxidized forms. The extracts were subsequently heated
for 10 (alkaline extract) or 20 (acid extract) minutes. After
addition of Tris/HCl pH 7.5 (final concentration = 0.075 M)
the extracts were adjusted to pH 7.5 with 1 N HCl or 1 N NaOH
and made up to equal volumes with water.

Pyridine nucleotide levels were estimated by the enzymic
cycling method, described by Monéger et al. (1977).

The redox-staining technique with NBT was described by
Sijmons and Bienfait (1983).

RESULTS AND DISCUSSION

It has already been shown that primary alcohols serving
as substrates for ADH increased H^+ - efflux in roots. The same
result was obtained with starved maize coleoptiles lacking
NADH (Böttger and Lüthen, 1986). In starved carrot cells, a
concomitant increase in HCF III reduction was observed (Craig
and Crane, 1981). This effect was described for maize roots,
too, but only after decrease of the initial rate by IAA
(Böttger and Hilgendorf, 1988).

As expected and shown in figure 1, application of ethanol
to carrot suspension cells increased NADH level without affecting the amount of NADPH. A peak of NADH- formation has
been observed within 30 seconds after addition of ethanol.
The concentration remained on a higher level for at least 10
minutes. A similar behaviour was observed using maize roots,
but lag phase, extent and duration of the response was different (fig. 2). This might be explained by different velo-
cities of NADH- producing and - consuming processes. Moreover, both plant materials investigated completely differed
in structure and diffusion conditions for ethanol.

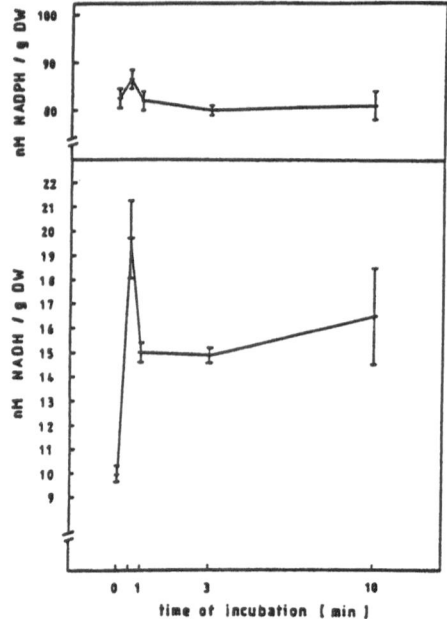

Fig. 1: Effect of ethanol (0.5 %)
on the NADH- and NADPH-
level of carrot cells

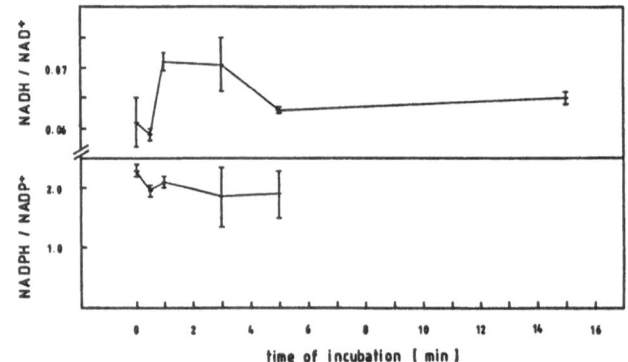

Fig. 2: Effect of ethanol on maize roots
Time course of changes in the
NADH/NAD$^+$- and NADPH/NADP$^+$- ratio
after incubation with 0.5 % ethanol

The effect may therefore be due to different uptake
rates. The subsequent decrease might be explained by use of
NADH for e$^-$- transport and redox-chain-linked proton transfer
or by a back-regulation of the internal NADH generating
system.

Another approach was to drain the system by addition of
electron consuming agents like HCF III or HCI IV instead of
feeding the possible source of redox equivalents. As figure 3
depicts, application of HCF III to carrot cells induced a decrease in NADH- as well as in NADPH- concentration. The effect of HCF III depended on its concentration as demonstrated
in figure 4.

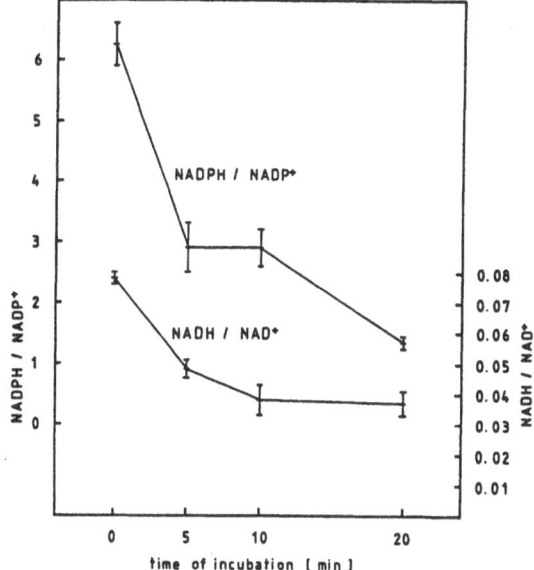

Fig. 3: Effect of HCF III on carrot
cells.
Time course of changes in the
NADPH/NADP+ - and NADH/NAD+ -
ratio after incubation with
5 mM HCF III

As shown in figure 3 and 4, NADH- and NADPH- contents
dramatically decreased after HCF III- treatment. Such a pronounced decrease should be based solely on effects on the
cytosolic NAD(P)H- level, because most of NADH and NADPH
seems to be located in the cytosol (Hampp et al., 1985). The
cytosol is the compartment concerned first, other compartments will be affected subsequently and may contribute to the
observed change.

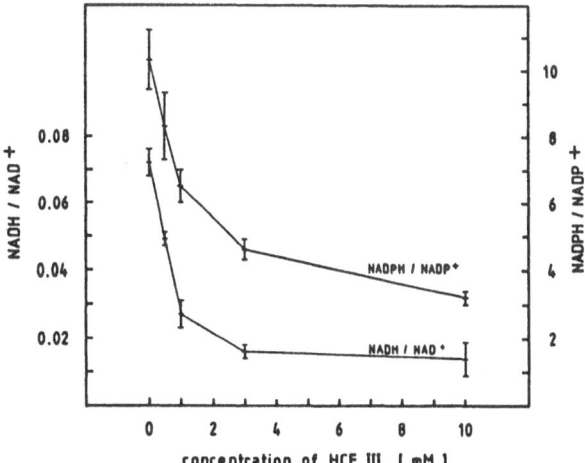

Fig. 4: Effect of HCF III- concentration on the NADPH/NADP+ -
respectively NADH/NAD+ -
ratio of carrot cells

The source of redox equivalents for transmembrane e⁻-
transport may be different in suspension cells supplied with
sucrose and intact roots with attached seedlings.

There were remarkable differences in the reaction of
maize roots after HCF III application. In contrast to carrot
cells, the amount of NADH remained unchanged even at high
HCF III concentrations while the ratio between NADPH and
NADP⁺ dropped significantly within 30 seconds after
application of HCF III (fig. 5). This finding is supported by
Qiu et al. (1985) and is in accordance with results reported
by Sijmons and Bienfait (1983) using iron-deficient bean
plants with "Turbo system". The lower level reached at 30
seconds remained unchanged during the time observed. Complete
response was obtained by 3 mM HCF III (fig. 6).

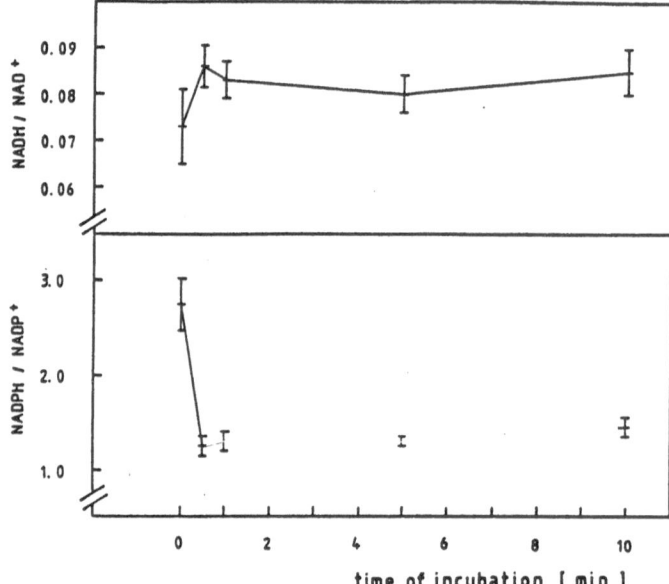

Fig. 5: Time course of changes in the
NADPH/NADP⁺ and NADH/NAD⁺ - ratio
in maize roots after incubation
with 5 mM HCF III

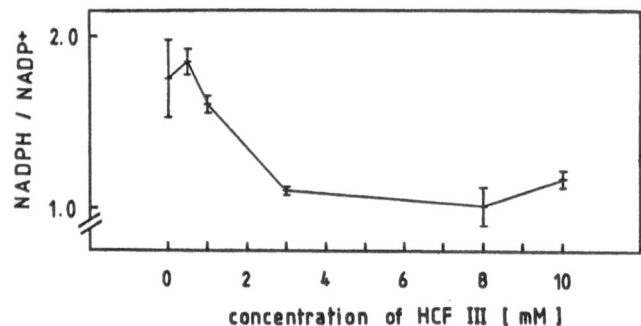

Fig. 6 : Effect of HCF III- concentration on
the NADPH/NADP⁺ - ratio in maize roots.
Ratios were estimated after 30 seconds
incubation at the HCF III- concentrations indicated

One has to keep in mind, however, that oxygen and HCF III considerably differ in redox potential. HCI IV, a new e⁻ - acceptor, much more powerful than HCF III was therefore applied. The redox potential of HCI IV is comparable to oxygen and like HCF III this complex does not permeate into the cells. Lüthen and Böttger (1988) compared the enhancement of proton secretion by HCF III and HCI IV and described the iridate complex to be an effective e⁻ - acceptor.

In accordance with these results HCI IV lowered the NADPH level in maize roots even at micromolar concentrations (fig. 7). A higher reduction rate of HCI IV in comparison to HCF III has been described by Lüthen and Böttger (1988).

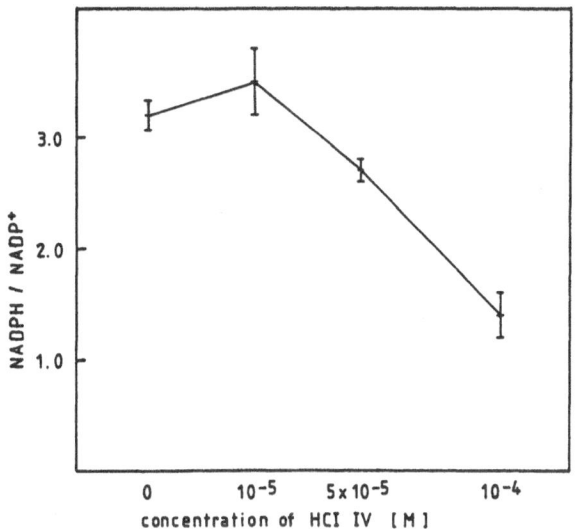

Fig. 7: Effect of HCI IV- concentration
of NADPH/NADP⁺- ratio in maize
roots

The use of intact roots for this kind of investigation has some disadvantages: The complete tissue was assayed for pyridine nucleotide level but due to a possible second casparian band (Peterson, 1987) in the subrhizodermal layer only the NADPH-level in the rhizodermis might be affected. This hypothesis is supported by the short duration of the lag phase. Furthermore the reaction is complete after 30 seconds. It is unlikely that the artificial e⁻ - acceptor HCF III diffuses to all cells within 30 seconds. On the other hand the main quantity of NADPH might be present only in the rhizodermal layer. The latter assumption is evidenced by redox staining with NBT showing most of NAD(P)H in the rhizodermal layer (fig. 8). If redox equivalents are located predominantly in the rhizodermis, the decrease appears less dramatic. The change is considerable if cells of the cortex contribute a high background.

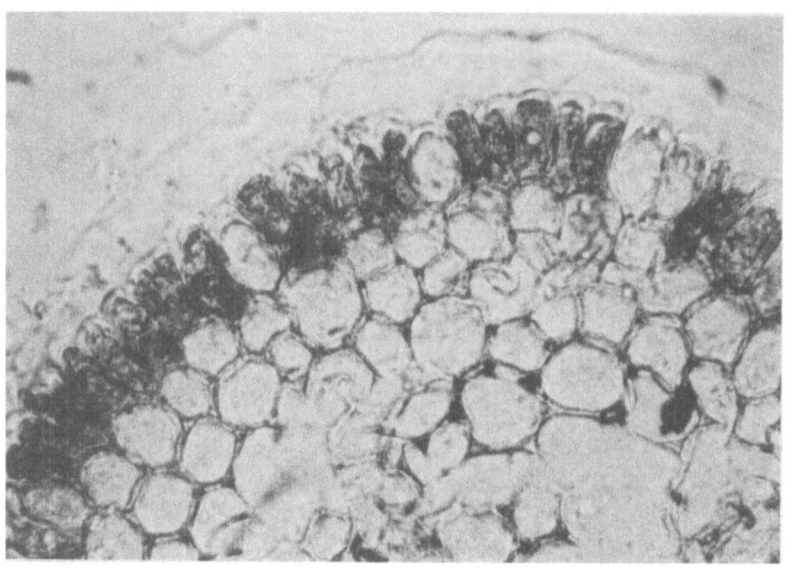

Fig. 8: Redox staining with NBT; only the rhizoder-
        mal layer is stained.
        It is impossible to distinguish between NADH
        and NADPH by this method.

The total amount of pyridine nucleotides
[NAD(P)H + NAD(P)$^+$] did not change during the experiment. The
induction of a de novo synthesis by application of HCF III
therefore may be excluded. An action of a kinase converting
NAD$^+$ into NADP$^+$ is unlikely because the amount of NAD$_{tot}$ to
NADP$_{tot}$ does not alter during the time obseved.

Nevertheless it is possible that NADH serves as source of
transmembrane electron flow in maize roots. In this case it
has to be assumed that the NADH- generating system is
efficiently compensating all changes in NADH- concentration
in the cytoplasm. It has to be kept in mind, however, that
the energy and electron transferring systems in the cell are
interacting. The NADH pool might be as well replenished by
NADPH via the activity of transhydrogenase. If the rate of
conversion was as high as the rate of consumption, a decrease
of NADH could not be seen.

The decrease in the NADPH level 30 seconds after application of HCF III is slight compared to the amount of HCF II
produced. A replenishment of the NADPH pool during that time
has to be postulated.

By measurement of the reduction rate of maize roots in a
pH - redoxstat (described by Böttger and Hilgendorf, 1988)
and the determination of the pool sizes we are able to estimate the turnover of NAD(P)H needed for transmembrane e$^-$-
flow. If NADPH was the source, the additional turnover would
be 10-12 times the pool size per minute. Due to the lower
amount of NADH its turnover would be one order of magnitude

higher. The change in contents might be used as a signal for subsequent processes like stimulation of the NAD(P)H generation.

The question arises what the reasons are for the different behaviour of carrot cells and maize roots. Carrots are able to induce the so called "Turbo system", but the cells are well supplied with iron (given as Fe^{III}-EDTA). Therefore the differences most nearly result from different back-regulation.

If NADPH is used as electron donor the replenishment probably is performed by the hexose monophosphate shunt. NADPH used in the "Turbo system" might be regenerated by action of the cytosolic isocitrate dehydrogenase (Bienfait, 1988). In carrot cells, NADH and NADPH seem to be donors for transmembrane electron transfer. Experiments with ethanol indicate a possible involvement of NADH parallel to NADPH in the electron transfer of maize roots and carrot cells. This idea is supported by experiments performed by Barr et al. (1985). Inhibitors of glycolysis are shown to inhibit HCF III reduction.

If glucose-6-phosphate is lacking, glycolysis and thus NADH generation will be decreased. Therefore it is possible that both, NADH and NADPH levels in carrot cells decrease, if glucose-6-phosphate is merely used for replenishment of NADPH

Acknowledgements:
Part of this work is supported by BMFT.
The help of S. Seidenberg and C. Rensch is gratefully acknowledged.

LITERATURE CITED

Barr, R.; Craig, T.A.; Crane, F.L., 1985: Transmembrane ferricyanide reduction in carrot cells, Biochim. Biophys. Acta 812: 49-54

Bienfait, H.F., 1985: Regulated redox processes at the plasmalemma of plant root cells and their function in iron uptake, J. Bioenerg. Biomem. 17 (2): 73-83

Bienfait, 1988:
    in: Nato ASI Series, this issue

Böttger, M.; Lüthen, H., 1986: Possible linkage between NADH-oxidation and proton secretion in Zea mays L. roots, J. Exp. Bot. 37: 666-675

Böttger, M.; Hilgendorf, F., 1988: Hormone influenced e⁻- and H^+- efflux, Plant Physiol. 86, in press

Craig, T.A.; Crane, F.L., 1981: Evidence for a transplasma membrane electron transport system in plant cells, Proc. Indiana Acad. Sci. 90: 150-155

Elzenga, J.T.M.; Prins, H.B.A., 1987: Light induced polarity of redox reactions in leaves of Elodea canadensis Michx, Plant Physiol. 85: 239-242

Federico, R.; Giartosio, C.E., 1983: A transplasmamembrane electron transport system in maize roots, Plant Physiol. 38: 642-648

Hampp, R.; Goller, M.; Füllgraf, H.; Eberle, I., 1985:
    Pyridine and adenine nucleotide status and pool sizes of
    a range of metabolites in choroplasts, mitochondria, and
    the cytosol / vacuole of Avena mesophyll protoplasts
    during dark / light transition: effect of pyridoxal
    phosphate, Plant Cell Physiol. 26 (1): 99-108
Kramer, D.; Römheld, V.; Landberg, E.; Marschner, H., 1980:
    Induction of transfer- cell formation by iron deficiency
    in the root epidermis of Helianthus annuus L., Planta
    147: 251-259
Lüthen H.; Böttger, M., 1988: Hexachloroiridate IV as an
    electron acceptor for a plasmalemma redox system in maize
    roots, Plant Physiol. 86, in press
Monéger, R,; Vermeersch, J.; Lechevallier, D.; Richard, D.,
    1977: Micro- analyse du NADP et du NAD réduits et oxydés
    dans les tissus foliaires et dans les plastes isolés de
    spirodèle et de blé.
    1. Problèmes posés par le dosage séparé du extraits
    végétaux, Physiol. Vég. 15: 29-62
Peterson, C.A., 1987: The significance of the exodermal
    casparian band for ion uptake and transport in roots
    In: XIV. Botanical Congress, Berlin (West), Germany,
    Ed.: Grenta, W.; Zimmer, B., Benke, H.-D.
Qiu Z.S.; Rubinstein,B.; Stern, A.I., 1985: Evidence for
    electron transport across the plasma membrane of Zea mays
    root cells, Planta 165: 383-391
Römheld, R.; Marschner, H., 1983: Mechanism of iron uptake by
    peanut plants. I. Fe^{III} reduction, chelate splitting, and
    release of phenolics, Plant Physiol. 71: 949-954 16.
Rubinstein B.; Stern, A.I.; Stout, R.G., 1984: Redox activity
    at the surface of oat root cells, Plant Physiol. 76:
    386-391
Sijmons, P.C.; Bienfait, H.F., 1983: Source of electrons for
    extracellular Fe (III) reduction in iron- deficient bean
    roots, Physiol. Plant. 59: 409-415
Sijmons, P.C.; van den Briel, W.; Bienfait, H.F., 1984:
    Cytosolic NADPH is the electron donor for extracellular
    Fe^{III} reduction in iron- deficient bean roots, Plant
    Physiol. 75: 219-221

MECHANISM AND CONTROL OF IRON UPTAKE BY ERYTHROID AND NON-ERYTHROID CELLS

P. Ponka and H.M. Schulman

Lady Davis Institute for Medical Research and
Departments of Physiology and Medicine, McGill
University, Montreal, Quebec, Canda

Iron is required by all eukaryotic cells for growth
and survival. For most cells, with the possible exception
of reticuloendothelial cells, transferrin is the only physiological iron donor. Iron uptake involves the binding of
iron-transferrin to specific cell surface receptors and
internalization of the transferrin receptor complex in
acidic endosomes where iron is released. The transferrin:
receptor complex is then recycled to the cell surface where
apotransferrin is released (1). Transferrin receptors are
highly expressed on immature erythroid cells which require
large amounts of iron for the formation of hemoglobin (2).
They are also highly expressed on proliferating cells such
as neoplastic cells, mitogen-stimulated lymphocytes and
many established cell lines (3).

We have shown that in erythroid cells heme inhibits
iron uptake from transferrin (4) and that the acquisition
of iron from transferrin limits the overall rate of heme
synthesis (2). Since heme controls the rate of globin mRNA
translation, the removal of iron from transferrin is probably rate limiting for the hemoglobinization of erythroid
cells. In contrast to erythroid cells the removal of iron
from transferrin by non-erythroid cells is not regulated by
the level of intracellular heme (5) leading us to postulate
that the mechanism of removal of iron from transferrin is
different in erythroid and non-erythroid cells.

The molecular mechanism and step at which heme inhibits the acquisition of iron in erythroid cells is unknown.
Heme may inhibit the release of iron from transferrin following its uptake by erythroid cells or, as suggested by
others, it may inhibit the endocytosis of transferrin. Our
recent data indicate that heme primarily inhibits iron
uptake subsequent to transferrin endocytosis and that the
inhibition of transferrin endocytosis caused by relatively
high concentrations of heme is a secondary effect (6).

Since heme does not appear to inhibit the acidification of endosomes we postulate that in erythroid cells heme may inhibit a reduction step which is involved in the release of iron from transferrin. This conclusion is supported by our observation that Zn(II)-protoporphyrin, which unlike Fe(III)-protoporphyrin is not reducible, is virtually without effect on the uptake of iron from transferrin by erythroid cells.

1. E.H. Morgan. Transferrin biochemistry, physiology and clinical significance. Mol. Ascepts Med. 4, 1 (1981)
2. J.D. Laskey, P. Ponka and H.M. Schulman. Control of heme synthesis during friend cell differentiation: Role of iron and transferrin. J. Cell. Physiol. 129, 185 (1986)
3. W.S. May Jr and P. Cuatrecasas. Transferrin receptor: Its biological significance. J. Membrane Biol. 88, 205 (1985)
4. P. Ponka and H.M. Schulman. Regulation of heme synthesis in erythyroid cells: Hemin inhibits transferrin iron utilization but not protophyrin synthesis. Blood 65, 850 (1985)
5. H.M. Schulman, A. Wilczynska and P. Ponka. Transferrin and iron uptake by human lymphoblastoid and K-562 cells. Biochem. Biophys. Res. Commun. 100, 1523 (1981)
6. P. Ponka, H.M. Schulman and J. Martinez-Medellin. Haem inhibits iron uptake subsequent to endocytosis of transferrin in reticulocytes. Biochem. J. 251, 105 (1988)

THE RELEASE OF IRON FROM A SUBFRACTION OF RAT LIVER HIGHLY ENRICHED IN

ENDOSOMAL ORGANELLES REQUIRES BOTH A FUNCTIONAL H^+-ATPase AND NADH

Torgeir Flatmark and Moududur R. Khan

Department of Biochemistry
University of Bergen
Årstadveien 19, N-5009 Bergen, Norway

INTRODUCTION

The liver parenchyma (hepatocytes) is second only to the erythroid bone marrow in its capacity for iron exchange with the plasma iron pool.[1] The iron uptake by reticulocytes and hepatocytes seems to occur by two receptor mechanisms,[1-4] i.e. (i) the well-known receptor-mediated endocytosis of iron-transferrin, and (ii) the less studied receptor-mediated endocytosis of ferritin (originally termed rhopheocytosis).[5,6] Thus, ferritin introduced into the plasma is rapidly cleared from the blood by a receptor-mediated uptake in hepatocytes.[4] The two iron proteins may share a common endocytic mechanism and may be processed within the cell in a similar way.[2,3] Thus, the reductive mobilization of iron is most effective at acidic pH values for both carrier proteins.[7,8]

In the present study it is shown that iron can be released from a pool of non-heme iron in a subcellular fraction of rat liver, highly enriched in endosomal organelles, containing a latent pool of non-heme Fe(III). The iron was slowly released to the medium (using bathophenanthroline sulfonate (BPS) as an external impermeable chelator) on acidification of the vesicles by MgATP (ATP regenerating system) and Cl^-, and it required reducing equivalents from NADH. The endosomal fraction contains not only an efficient proton pump, but also a NADH:acceptor oxidoreductase which seems to be similar to that reported for the plasma membrane of mammalian cells.[9] The proton pump activity was found to be partly inhibited by NADH and BPS.

MATERIALS AND METHODS

Chemicals

Lactate dehydrogenase, pyruvate kinase, phosphoenolpyruvate, NADH, ATP, acridine orange, dithiothreitol, and bathophenanthroline sulfonate

---

Abbreviations: BPS, bathophenanthroline sulfonate; H^+-ATPase, H^+-transporting adenosine triphosphatase (EC 3.6.1.34); ΔpH, pH difference across the membrane; Hepes, N -2-hydroxyethylpiperazine- N' -2-ethanesulfonic acid ; Pipes, piperazine- N , N -bis-2-ethanesulfonic acid; SDS, sodium dodecyl sulfate.

were obtained from the Sigma Chemical Company (MO, U.S.A.). Digitonin was purified by recrystallization. A standard solution of iron (0.01 mg Fe/mg) was obtained from G.F. Smith Chemical Co. (OH, U.S.A.).

Analytical differential centrifugation

Analytical differential centrifugation was carried out essentially as described.[10-12]

Animals and preparation of an endosomal fraction

Male albino rats (Wistar strain, purchased from Mollegård, Denmark) of 150 g body weight, fed an ordinary pellet diet, were fasted for 4 h, stunned and decapitated. The livers were homogenized (two strokes at 460 rpm) and the crude microsomal fraction prepared using 0.25 M sucrose, 5 mM Hepes buffer, 1 mM dithiothreitol and 0.2 mM EDTA as the homogenization medium; pH 7.2.[13] A centrifugal effect (i.e. a time integral of $(rpm)^2 dt$) of 4×10^9 min^{-1} was used to sediment the large granule fraction, including the three populations of lysosomes (lowest S-value of about 4500 S).[11,12] The microsomal fraction (pellet) was resuspended in the homogenization medium and further subfractionated by discontinuous sucrose gradient centrifugation.[13] The fraction highly enriched in endosomal organelles (at the interphase of sucrose density 1.08 and 1.10 g ml^{-1}) were used immediately or stored in liquid nitrogen.

Isolation of lysosomes

A subcellular fraction highly enriched in "heavy" lysosomes was prepared by a combination of differential and density gradient centrifugation on a Percoll gradient in isotonic sucrose.[12]

Assay of proton pump activity and internal acidification

The rate at which the pH gradient was generated across vesicular membranes, was measured with acridine orange as a probe.[13,14] The standard incubation medium (1 ml) contained 7.5 mM Pipes buffer (pH 7.0), 50 mM KCl, 0.5 mM dithiothreitol, 5 µg oligomycin, 3 µM acridine orange and the endosomal fraction (approx. 50 µg of protein). Kinetic constants were calculated as described.[15] In the studies on iron mobilization a maximal steady-state acidification of the endosomes was achieved by an ATP regenerating system.[14]

Assay of marker enzymes

NADH : glyoxylate oxidoreductase activity was assayed as described.[16] Galactosyltransferase was measured using $\underline{N}$-acetyl-glucosamine as acceptor.[17] NADPH : cytochrome $\underline{c}$ oxidoreductase activity was assayed as described.[18] β-$\underline{N}$-acetyl-D-hexosaminidase was assayed as described.[19] Protein was determined by the Bradford procedure[20] using bovine serum albumin as a standard.

Assay of iron

Total iron in the endosomal fraction was assayed spectrophoto-metrically as chelated with bathophenanthroline sulphonate (BPS) after reduction with sodium dithionite,[21] and in the presence of digitonin. The release of iron from endosomal fractions under native conditions was assayed by the BPS method, using the standard incubation medium found to be optimal for proton pump activity (see results).

RESULTS

Analytical differential centrifugation of lysosomes

Analytical differential centrifugations of liver homogenates revealed the presence of more than one population of lysosomes, when acid phosphatase and β- $\underline{N}$ -acetyl-D-hexosaminidase was used as marker enzymes. By statistical analysis of the sedimentation profiles, at least three populations of lysosomal particles were obtained with the average S-values of about 34 000, 13 500 and 4500 S, in agreement with previous analyses.[11,12]

Recovery, purity and functional properties of the endosomal fraction

The endosomal fraction was recovered at the interphase of sucrose density 1.08 and 1.10 g ml^{-1}, i.e. close to the median equilibrium density (1.11 g ml^{-1}) of the early labeling after receptor-mediated uptake of asialo-transferrin in isolated hepatocytes.[22] The fraction (accounting for 168 $\pm$ 37 µg of protein per g wet weight of liver (mean $\pm$ SD, $\underline{n}$ = 15)), represents endosomes free from organelles of S-value > 3000 S (i.e. lysosomes, mitochondria and peroxisomes) and free from contamination by Golgi elements, but it contained a significant amount of smooth microsomes.[13] The H$^+$-ATPase activity was rather selectively enriched in this fraction.[13] The functionally best preparations were obtained by using a weak shearing force during homogenization. Protease inhibitors were not required, but once removed, the liver had to be processed rapidly to obtain preparations with a high proton pump activity.

β- $\underline{N}$ -acetyl-D-hexosaminidase activity was found to be within the range of 1.5 - 5.2 nmol min^{-1} mg^{-1} protein, which was very low compared to the activity in lysosome-enriched fractions (53.3 - 118.6 nmol min^{-1} mg^{-1} protein). These numbers show that lysosomes were effectively separated from microsomal elements by differential centrifugation, in agreement with our previous findings.[11,12]

MgATP, but not NADH, support internal acidification of the endocytic vesicles

The overall Mg^{2+}-ATPase activity in the endosomal fraction was found to be in the range of 27 to 41 nmol min^{-1} mg^{-1} protein ($\underline{n}$ = 4) when measured at 25 $^{\circ}$C in the presence of 50 mM chloride (see below). The activity was inhibited 44 % by NEM (100 µM) and 21 % by DCCD (200 µM) and oligomycin (5 µg), but vanadate had no effect. Although absolute measurements of steady-state pH gradients are preferentially obtained by e.g. isotope distribution methods,[23] measurement of initial rates for the generation of the pH gradient (ΔpH/Δt) is preferred for kinetic studies.[13,14] As shown in Fig. 1, a stable base-line was obtained within 5 min when the endosomal fraction was incubated in the standard incubation medium containing acridine orange (Fig. 1). No proton pump activity (acidification) could, however, be measured on the addition of MgATP in the absence of a permeable anion, notably Cl$^-$ (Fig. 1, trace a). On the other hand, in the presence of 50 mM KCl (standard medium) a pH gradient (acid inside) was generated, reaching its half maximal value about 3 min after addition of MgATP (Fig. 1, trace b) and a maximum value after about 16 min (on an average). The steady-state level of the gradient varied from one preparation to the other. In general, preparations with high initial proton flux reached a higher steady-state level. The generated proton gradient was maintained at a high energized state for another 6-8 min, followed by a slow decay. When an ATP regenerating system was included in the medium, the pH gradient could be sustained at its maximum level for at least 1 h (Fig. 1, trace c). Thus,

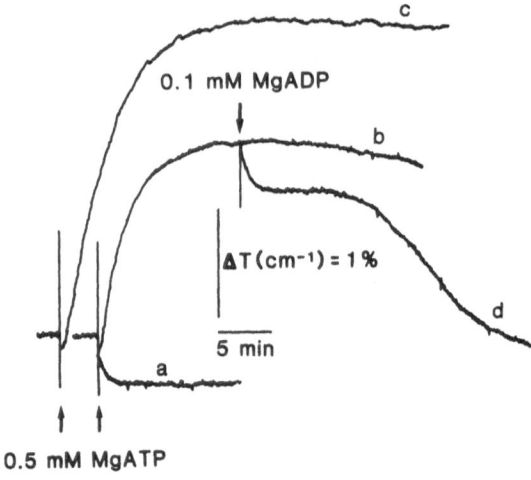

Fig. 1. (a) and (b). The time course for the change in absorbance
of acridine orange induced by 0.5 mM MgATP in the endosomal
fraction in the absence (trace a) and presence (trace b)
of 50 mM KCl. (c) The effect of an ATP regenerating system
(in the presence of 50 mM KCl). (d) The effect of MgADP on
the dissipation of the pH gradient. The spectral change of
acridine orange was followed in a dual-wavelength spectrophotometer using the wavelength pair 492 nm - 540 nm.

the decay (trace b) was mainly due to an accumulation of ADP, rather than
instability of the endosomes. If NH_4Cl or monensin was added, the pH
gradient was immediately dissipated.[13]

No pH gradient could be generated across the endosomal membrane in
the absence of an external permeant anion. Chloride and bromide were
effective whereas no proton gradient could be generated in the presence of
bicarbonate, phosphate or thiocyanate. A hyperbolic concentration-
response curve was obtained with an apparent K_m for chloride of 35 ± 5
mM (Fig. 2), indicating that an anion channel/transporter is involved.
Pyridoxal 5'-phosphate and 4-acetamide-4'-isothiocyanostilbene-2,2'-
disulphonic acid (DIDS) were found to completely inhibit the ATP-driven
proton translocation.[13] By blotting of endosomal proteins separated by
SDS gel electrophoresis, we have found that DIDS binds to three
polypeptides (data not shown). Proton translocation was insensitive to
oligomycin (5 µg ml^{-1}), vanadate (168 µM) and ouabain (1 mM). However,
it was effectively inhibited by N-ethylmaleimide (NEM) and N, N'
-dicyclohexylcarbodiimide (DCCD). Complete inhibition was obtained by 40
µM NEM and 200 µM DCCD, and half-maximal inhibition at 5.2 µM and 28 µM,
respectively. Dithiothreitol was always found to increase both the initial
rate of proton translocation and the steady-state level of the proton
gradient at maximal energization.

The K_m-value for MgATP was found to be 147 ± 8 µM. MgADP caused
a concentration dependent dissipation of the MgATP induced proton gradient
at its maximally energized state (Fig. 1, trace d). It was also found to
competitively inhibit the proton pump activity (data not shown). An ATP
regenerating system was therefore used to preserve a maximal pH-gradient
over an extended period of time.

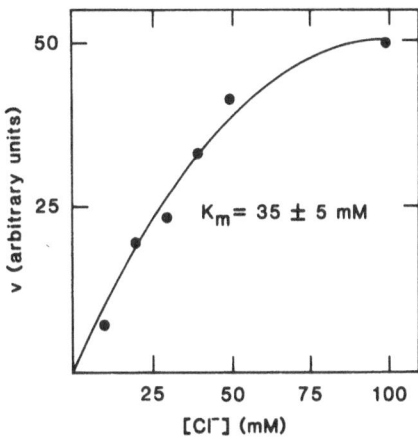

Fig. 2. The dependence of chloride concentration on the proton
pump activity of the endosomal fraction. Standard incubation medium.

Partial inhibition of the H^+-ATPase by NADH and bathophenanthroline sulfonate

No proton pump activity was observed on addition of NADH, but the
initial rate of proton pump activity was found to be inhibited
substantially when ATP was added shortly after the addition of NADH (data
not shown). Furthermore, bathophenanthroline sulfonate (BPS) was also
found to inhibit the H^+-ATPase at concentrations used in the following
iron mobilization studies, i.e. 50 % inhibition at 95 µM of BPS.

Proton pump activity in lysosomes

Proton translocation was also observed in the isolated lysosomal
fractions, but the specific activity was low as compared to endosomes, and
was short lived. Thus, it could be observed only if the liver was
processed rapidly and the fractions were assayed for activity immediately
after collection from the Percoll gradient. Unlike the endosomal fraction,
the lysosomal fraction also lost its proton pump activity on storage in
liquid nitrogen, although there was no loss in β- N -acetyl-D-
hexosaminidase activity. Percoll apparently had no deleterious effect on
the proton pump activity. Thus, the endosomal fraction was equally active
in the presence and absence of added Percoll.

Iron content of the endosomal fraction

The total amount of non-heme iron of the endosomal fraction was found
to be 13.4 ± 1.95 nmol mg^{-1} protein (n = 3), as determined by the
Fe(II)-BPS formation (Fig. 3, trace e) in the presence of digitonin and
dithionite.[21]

Effect of MgATP and NADH on the mobilization of iron from the endosomal fraction

In Fig. 3 is shown a typical series of experiments on the
mobilization of iron (as Fe(II)) from an endosomal fraction with high
proton pump activity. When the fraction was preincubated for 1 h at 25
°C in the medium found to be optimal for proton pump activity, a slow
formation of Fe(II)-BPS complex was seen even in the absence of any added
reducing agent (Fig. 3, trace a); 0.93 nmol iron mg^{-1} protein was

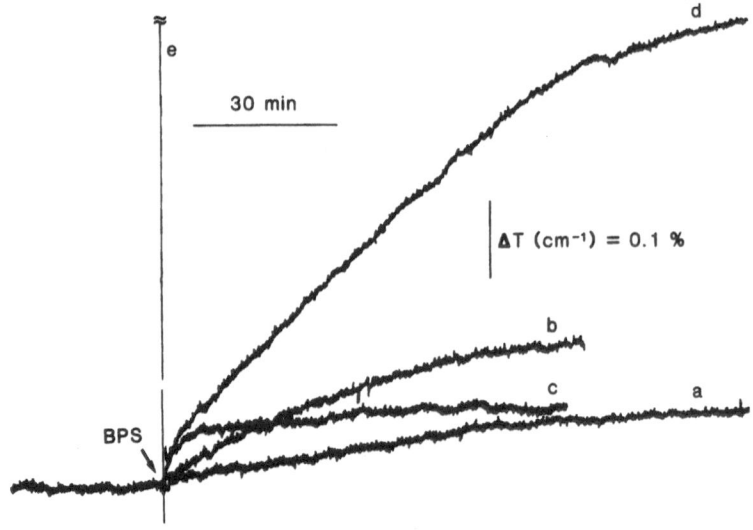

Fig. 3. The time course of the formation of the Fe(II)-batho-
phenanthroline sulfonate (BPS) chelate in the endosomal
fraction. (a) The endosomal fraction was preincubated 1 h
in the standard incubation medium before addition of BPS;
(b) 450 µM NADH included in the medium; (c) 250 µM MgATP
(plus an ATP regenerating system) included in the medium;
(d) 450 µM NADH and 250 µM MgATP included in the medium;
(e) The chelation of Fe(II) in the presence of 20 µM
digitonin and excess dithionite (amounts to about 1 nmol
of iron). The color development was followed in a dual-
wavelength spectrophotometer using the wavelength pair
540 nm - 575 nm.

chelated by BPS in 80 min after addition of the chelator. The presence of
either 250 µM MgATP (in an ATP regenerating system) or 450 µM NADH
increased the number to 3.04 and 1.25 nmol iron mg^{-1}, respectively
(Fig. 3, traces b and c). The most effective mobilization was, however,
obtained when both MgATP and NADH were present in the preincubation
mixture (Fig. 3, trace d). Under these conditions 8.59 nmol iron mg^{-1}
protein was chelated in about 180 min after addition of the chelator, i.e.
about 45 pmol min^{-1} mg^{-1} protein.

DISCUSSION

    Endosomes comprise a heterogeneous population of vacuoles and tubules
to which receptors, receptor-bound ligands, and extracellular solutes are
delivered after endocytosis, and are therefore difficult to isolate free
from other organelles. In the present study a population of endosomal
organelles has been isolated from the microsomal fraction of rat liver
homogenates. The procedure relies on a combination of differential and
density gradient centrifugation, which essentially eliminates intact
lysosomes, mitochondria and peroxisomes, but is contaminated by smooth
microsomes, which, however, have no proton pump activity.[13] The
oligomycin sensitive Mg^{2+}-ATPase activity (about 20 %) compares well
with the oligomycin sensitive proton pump activity and is assigned to
contaminating submitochondrial particles. Oligomycin was therefore
included in the standard incubation medium. Lysosomes were removed by
differential sentrifugation, by selecting a centrifugal effect which

sediments particles of S > 3000 S, including the lysosomal particles of lowest S-value (i.e. about 4500 S). Furthermore, considering the approx. 400-fold higher relative ratio (average value) between proton pump activity and β-N-acetyl-D-hexosaminidase in the endosomal than in the lysosomal fraction, as well as the rapid decay of proton pump activity in isolated lysosomes, the lysosomal contamination may account for only a negligible fraction of the H^+-ATPase activity of the endosomal fraction.

NADH did not support the generation of a proton gradient in the endosomal fraction. This distinguishes our endosomes from <u>trans</u> -Golgi elements, which have been reported to generate a proton gradient using a redox system coupled to NADH oxidation.[24]

The proton pump of the endosomal fraction revealed several properties in common with that of the secretory granules of chromaffin cells,[14] i.e. with respect to the K_m-value for MgATP, trinucleotide specificity, dependence on Cl^-, sensitivity to ADP and other inhibitors (NEM and DCCD) and insensitivity to vanadate. Thus, the H^+-ATPase of the endosomal fraction also contains one or more functional SH-group(s) involved in ATP-hydrolysis and/or proton translocation. The inhibition by ADP may provide the cell with a mechanism for controling the activity of the H^+-ATPase. The absolute dependence of endosomal acidification on Cl^- and the stimulation of the overall Mg^{2+}-ATPase activity by this anion, support a coupling between H^+ and Cl^- transport, but the mechanism of this coupling is not yet clear.

In the present study it is shown that the endosomal fraction contains a pool of non-heme iron (amounting to 13.4 $\pm$ 1.95 nmol mg^{-1}). Mobilization of this pool can be rapidly obtained in the presence of digitonin, a strong reducing agent (dithionite) and BPS. A fraction of this iron pool can be very slowly mobilized (about 45 pmol min^{-1} mg^{-1} protein) by acidification of the endosomes with MgATP (in an ATP-regenerating system), Cl^- and NADH, whereas acidification or NADH alone revealed only a small effect (Fig. 3).

Although the chemical nature of the endosomal non-heme iron pool is not known, it seems to be present largely as Fe(III), since its mobilization required reduction for chelation by BPS (used as external sink of Fe(II)). Only a negligible amount of the 13.4 nmol iron mg^{-1} may represent transferrin iron, since iron is rapidly released (to the cytosol) from transferrin taken up by hepatocytes by receptor-mediated endocytosis.[25] Part of the non-heme iron pool may represent ferritin iron or "iron-cores" derived from this iron protein. Thus, circulating ferritin is rather selectively taken up in hepatocytes by receptor-mediated endocytosis,[1,4] and the possibility of an intercellular iron transfer (transcytosis), as proposed for iron-transferrin,[26] should be considered also for ferritin and other forms of transport iron. Such a metabolic cooperation has been demonstrated for the intercellular transfer of ferritin from macrophages to erythroid cells by receptor-mediated endocytosis.[4,5] If present as ferritin, the endosomal iron would represent a minimum amount of about 3 pmol ferritin mg^{-1} protein.

The mechanism of iron release from both transferrin and ferritin is not known, but reduction of Fe(III) is a pre-requisite. Horse spleen ferritin "iron core" has a pH-dependent redox potential (-190 mV at pH 7.0), which becomes less negative at acidic pH.[27] In agreement with these redox measurements, the rate of the reductive mobilization of iron from ferritin increases by lowering the pH of the medium.[8] An even more pronounced pH dependence has been reported for the release of iron from diferric transferrin.[7] However, from such in vitro studies it is also known that a pH of about 5.5 alone is not sufficient for the rapid release of iron. Reduction of Fe(III) favours the release of iron from transferrin,[7] and the same is true also for the mobilization of iron from ferritin.[8] It was therefore not unexpected to find that the release of iron from the endosomal fraction also requires reducing equivalents as well as an external sink of Fe(II), in addition to the internal acidification (Fig. 3). The effect of NADH on the mobilization of iron from endosomes may indicate that the endosomal membrane contains a NADH:acceptor oxidoreductase, supporting a transmembrane electron transfer of the type described for the plasma membrane.[9] The transmembrane electron transport enzyme of this membrane can act as a diferric transferrin reductase in cells which have transferrin receptors.[28] The electron transport system reduces Fe(III) in transferrin bound to the cell surface, as measured by the formation of a Fe(II)-BPS complex outside the cell, and internal NADH is oxidized. Our findings support the conclusion that iron is mobilized from the endosomal preparation by a similar mechanism, i.e. by the combined action of a H^+-ATPase and a dehydrogenase linked to the oxidation of NADH. Further studies are, however, required to characterize the pool of non-heme iron present in the endosomal fraction, the mechanism by which the iron is released, including its reductive mobilization and transport across the vesicular membrane.

ACKNOWLEDGEMENTS

This study was supported by the Norwegian Research Council for Science and the Humanities, the Norwegian Cancer Society and the Norwegian Society for Fighting Cancer.

REFERENCES

1. E. H. Morgan and E. Baker, Iron uptake and metabolism by hepatocytes, Fed. Proc. 45:2810 (1986).
2. G. D. Blight and E. H. Morgan, Receptor-mediated endocytosis of transferrin and ferritin by guinea-pig reticulocytes. Uptake by a common endocytotic pathway, Eur. J. Cell Biol. 43:260 (1987).
3. G. D. Blight and E. H. Morgan, Transferrin and ferritin endocytosis and recycling in guinea-pig reticulocytes, Biochim. Biophys. Acta 929:18 (1987).
4. U. Mack, L. W. Powell, and J. W. Halliday, Detection and isolation of a hepatic membrane receptor for ferritin, J. Biol. Chem. 258:4672 (1983).
5. M. C. Bessis and J. Breton-Gorius, Iron particles in normal erythroblasts and normal and pathological erythrocytes, J. Biophys. Biochem. Cytol. 3:503 (1957).
6. A. Policard and M. Bessis, Sur un mode d'incorporation des macromolcules par la cellule, visible au microscope electronique: la rhopheocytose, Comp. Rend. Hebd. Acad. Sci. 246:3194 (1958).
7. A. Dautry-Varsat, A. Chiechanover, and H.F. Lodish, pH and the recycling of transferrin during receptor-mediated endocytosis, Proc. Natl. Acad. Sci. U.S.A. 80:2258 (1983).
8. F. Funk, J.-P. Lenders, R. R. Crichton, and W. Schneider, Reductive mobilization of ferritin iron, Eur. J. Biochem. 152:167 (1985).
9. F. L. Crane, I. L. Sun, M. G. Clark, C. Grebing, and H. Löw, Transplasma-membrane redox systems in growth and development,

Biochim. Biophys. Acta 811:233 (1985).

10. E. Slinde and T. Flatmark, Determination of sedimentation coefficients of subcellular particles of rat liver homogenates. A theoretical and experimental approach, Anal. Biochem. 56:324 (1973).

11. R. K. Berge, H. Osmundsen, and T. Flatmark, Enhancement of long-chain acyl-CoA hydrolase activity in peroxisomes and mitochondria of rat liver by peroxisomal proliferators, Eur. J. Biochem. 141:637 (1984).

12. A. Tangerås, Iron content and degree of lipid peroxidation in liver mitochondria isolated from iron-loaded rats, Biochim. Biophys. Acta 757:59 (1983).

13. T. Flatmark, J. Haavik, M. Grønberg, J. L. Kleiveland, S. Wahlstrøm Jacobsen, and S. Vik Berge, Isolation from the microsomal fraction of rat liver of a subfraction highly enriched in uncoated endocytic vesicles with high H^+-ATPase activity and a 50 kDa phosphoprotein, FEBS Lett. 188:273 (1985).

14. M. Grønberg and T. Flatmark, Studies on Mg^{2+}-dependent ATPase in bovine adrenal chromaffin granules. With special reference to the effect of inhibitors and energy coupling, Eur. J. Biochem. 164:1 (1987).

15. I. Knack and K. H. Röhm, Microcomputers in enzymology. A versatile BASIC computer program for analyzing kinetic data. Hoppe-Seyler's Z. Physiol. Chem. 362:1119 (1981).

16. F. L. Crane and H. Löw, NADH oxidation in liver and fat cell plasma membranes, FEBS Lett. 68:153 (1976).

17. A. J. Hymes and F. Mullinax, Assay of galactosyltransferase by high-performance liquid chromatography, Anal. Biochem. 139:68 (1984).

18. N. E. Tolbert, Isolation of subcellular organelles of metabolism on isopycnic sucrose gradients, Meth. Enzymol. 31:734 (1974).

19. D. Robinson and P. Willcox, 4-Methylumbelliferyl phosphate as a substrate for lysosomal acid phosphatase, Biochim. Biophys. Acta 191:183 (1969)

20. M.M. Bradford, A rapid and sensitive method for the quantitation of microgram quantities of protein utilizing the principle of protein-dye binding, Anal. Biochem. 72:248 (1976).

21. T. Flatmark and A. Tangerås, Mitochondrial "non-heme non-FeS iron" and its significance in the cellular metabolism of iron, in : "Proteins of Iron Metabolism," E.B. Brown, ed., Grune & Stratton, Inc., New York, pp. 349-358 (1977).

22. H. Tolleshaug, Intracellular segregation of asialo-transferrin and asialo-fetuin following uptake by the same receptor system in suspended hepatocytes, Biochim. Biophys. Acta 803:182 (1984).

23. D. Nicholls, "Bioenergetics," pp. 190, Academic Press, London (1982).

24. R. Barr, K. Safranski, I. L. Sun, F. L. Crane, and D. J. Morré, An electrogenic proton pump associated with the Golgi apparatus of mouse liver driven by NADH and ATP, J. Biol. Chem. 259:14064 (1984).

25. J.-C. Sibille, J.-N. Octave, Y.-J. Scheider, A. Trouet, and R. Crichton, Subcellular localization of transferrin protein and iron in the perfused rat liver. Effect of Triton WR 1339, digitonin and temperature, Eur. J. Biochem. 155:47 (1986).

26. T. Kishimoto and M. Tavassoli, Transendothelial transport (transcytosis) of iron-transferrin complex in the rat liver, Am. J. Anat. 178:241 (1987).

27. G. D. Watt, R. B. Frankel, and G. C. Papaefthymiou,
     Reduction of mammalian ferritin,
     Proc. Natl. Acad. Sci. U.S.A. 82:3640 (1985).
28. H. Löw, I. L. Sun, P. Navas, C. Grebing, F. L. Crane, and
     D. J. Morré, Transplasmalemma electron transport from
     cells is part of a diferric transferrin reductase system,
     Biochem. Biophys. Res. Commun. 139:1117 (1986).

UPTAKE OF IRON FROM FERRIC IRON COMPOUNDS BY ISOLATED HEPATOCYTES

Ketil Thorstensen,Liv Thommesen and Inge Romslo

Department of Clinical Chemistry,University of Trondheim
University Hospital
Trondheim,Norway

SUMMARY

Reticulocyte uptake of iron from transferrin occurs by receptor-mediated endocytosis.In contrast,in the hepatocyte uptake of iron from transferrin,and also from ferric nitrilotriacetate (Fe-NTA),depends on a reductive process at the cell surface.The uptake of iron from ferritin by isolated hepatocytes depends partly on endocytosis of ferritin to lysosomes with subsequent release of iron to cytosol,and partly on a reductive process at the cell surface.

There is no correlation between hepatocyte NAD(P)H levels and iron reduction and uptake.

INTRODUCTION

Cellular uptake of iron from transferrin is generally thought to occur by receptor-mediated endocytosis of transferrin with subsequent release of iron from transferrin within the acidic endosome (1).However,this view has been challenged (2),and in the hepatocyte evidence in favour of a reductive process at the cell surface has been presented (3).

In the present paper we report on experiments showing that hepatocyte uptake of iron from ferric iron compounds depends on reductive processes at the cell surface.Also,we report experiments showing that the mechanism of iron uptake from transferrin by isolated rat hepatocytes is different from that by reticulocytes.

MATERIALS AND METHODS

Hepatocytes were isolated from male Wistar-Möll rats by the two step collagenase perfusion method of Seglen (4) as previously described (5). Rat reticulocytes were obtained following intraperitoneal injections of phenylhydrazine according to Harding et al.(6) as previously described (7). Incubations were performed in Erlenmeyer flasks in a shaking water bath set at 37°C.In order to obtain high (300 uM) or low (20 uM) oxygen concentration the flasks were gently flushed with O_2 or N_2,respectively. At the time indicated duplicate aliquots were withdrawn and the cells were washed at 4°C.

^{59}Fe-labelled ferritin was isolated from hepatocytes which had been incubated with ^{59}Fe-labelled diferric transferrin for 30 min at 37°C.The isolation procedure included heating (10 min,76°C),precipitation with 40% ammonium sulphate,gel filtration (Sephacryl S-300) and centrifugation (100,000 x G,17 h).

^{59}Fe-labelled nitrilotriacetate (Fe-NTA) was prepared by mixing equal volumes of aqueous ^{59}FeCl$_3$ (10 mM) and nitrilotriacetate (20 mM).Following 1 h at room temperature the solution was diluted to 400 uM Fe in incubation buffer.

Determination of NAD(P)H was performed essentially as described by Kalhorn et al.(8).Transferrin was labelled with ^{59}Fe by the method of Bates and Schlabach (9).The concentration of diferric transferrin in the incubations was 1 uM unless otherwise stated.Human transferrin was used throughout.Agents to be tested were added to the cell suspension just prior to the addition of the ferric iron compound.

RESULTS AND DISCUSSION

An important difference between hepatocyte and reticulocyte uptake of iron from transferrin relates to oxygen (Fig. 1).Whereas iron uptake from transferrin by hepatocytes is markedly enhanced at low oxygen concentration (Fig. 1,see also refs. 3 and 5),the concentration of oxygen has no effect on reticulocyte iron uptake.From a physiological point of view such findings would not be unexpected (7),but as far as we are aware are not generally emphasized.Moreover,in the reticulocyte the classical evidence

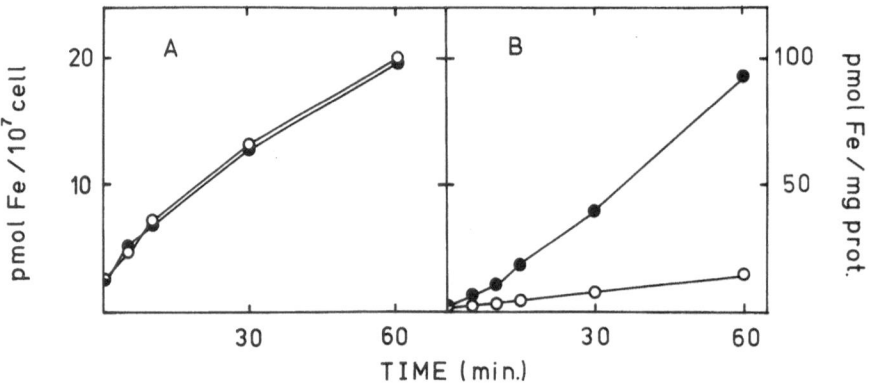

Figure 1. The effect of oxygen concentration on the uptake of iron from transferrin by reticulocytes and hepatocytes.

Reticulocytes (5·10^7 cells/ml) or hepatocytes (5·10^6 cells/ml) were incubated with 1 uM ^{59}Fe-labelled diferric transferrin at 37°C.At the time indicated duplicate aliquots were withdrawn and the amount of radio-activity in the cells was determined.

A: Reticulocytes. B: Hepatocytes. Open symbols: 300 uM oxygen. Closed symbols: 20 uM oxygen.

of receptor-mediated endocytosis are prominent,i.e. iron uptake is inhibited by weak bases,ionophores and metabolic inhibitors (TABLE I).By comparison, hepatocyte uptake of iron from transferrin apparently occurs by reduction at the cell surface (3).

In addition to transferrin,hepatocytes take up iron from other ferric compounds as well,e.g. Fe-NTA and ferritin (Fig. 2).Iron uptake from Fe-NTA is qualitatively similar to the uptake of iron from transferrin,i.e. low oxygen concentration enhances iron uptake (Fig. 2) whereas ferricyanide and bathophenanthroline disulphonate (BPS) inhibit iron uptake (TABLE II), as do divalent metal cations with ionic radii similar to or less than ferrous iron (Fig. 3).

TABLE I. Effect of weak bases,ionophores,metabolic inhibitors,iron chelators and ferricyanide on the uptake of iron from transferrin by hepatocytes and reticulocytes.

Hepatocytes or reticulocytes were incubated 30 min at 37°C with 1 uM ^{59}Fe-labelled transferrin.The oxygen concentration was approx. 20 uM. The results are given in per cent of controls without additions.The table shows mean ± S.D. (number of experiments).

IRON UPTAKE (%)

	Hepatocytes	Reticulocytes
CCCP (5 uM)	112 ± 9 (4)	53 ± 14 (4)
Methylamine (30 mM)	97 ± 17 (3)	53 ± 2 (4)
Antimycin A (10 uM)	99 ± 15 (3)	56 ± 21 (4)
BPS (50 uM)	14 ± 5 (3)	105 ± 4 (6)
Desferal (50 uM)	3 ± 1 (3)	98 ± 6 (4)
Ferricyanide (100 uM)	9 ± 5 (4)	97 ± 10 (4)

TABLE II. Uptake of iron from ferric iron compounds by isolated hepatocytes at low and high oxygen concentration.

Hepatocytes were incubated with ^{59}Fe-labelled ferric iron compounds in the presence of ferricyanide or BPS,as described in Fig. 1.The results are given in per cent relative to controls without any additions.The table shows mean ± S.D.,n=4.

IRON UPTAKE (%)

IRON COMPOUND	O_2 (uM)	Ferricyanide (100 uM)	BPS (50 uM)
Fe-NTA (5 uM Fe)	20	3 ± 1	15 ± 2
	300	16 ± 2	59 ± 22
TRANSFERRIN (2 uM Fe)	20	9 ± 5	14 ± 5
	300	49 ± 24	71 ± 26
FERRITIN (15 uM Fe)	20	65 ± 10	48 ± 7
	300	87 ± 6	96 ± 8

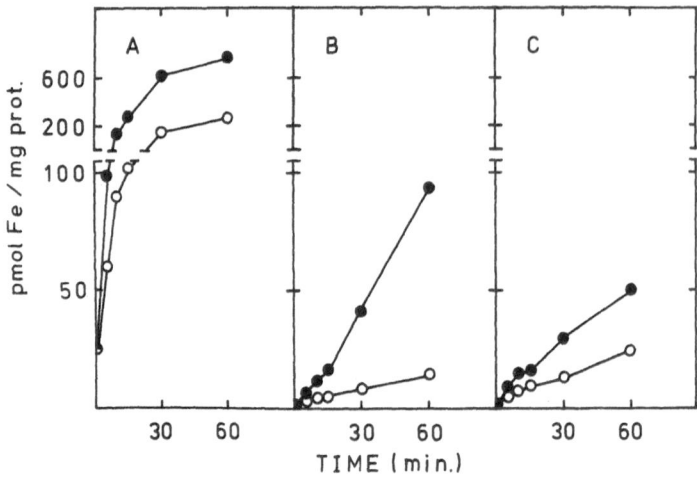

Figure 2. The effect of oxygen concentration on the uptake of iron from ferric iron compounds by isolated hepatocytes.

Hepatocytes were incubated with ferric iron compounds as described in Fig. 1. (A) Fe-NTA (5 uM Fe),(B) transferrin (2 uM Fe),(C) ferritin (15 uM Fe). Open symbols: 300 uM oxygen. Closed symbols: 20 uM oxygen.

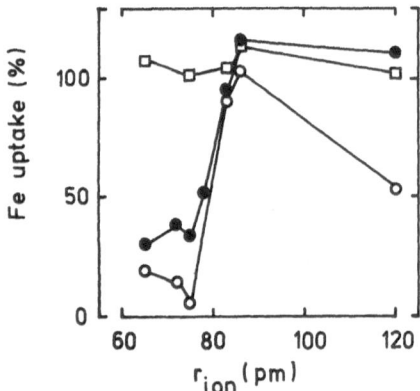

Figure 3. The effect of divalent metal cations on the uptake of iron from ferric iron compounds by isolated hepatocytes.

Hepatocytes were incubated with ferric iron compounds as described in Fig. 1. (O) Fe-NTA, (●) transferrin, (□) ferritin. The concentration of ferric iron compounds were as in Fig. 2. The metal ions (500 uM) were from left to right: Co,Cu,Zn,Fe (transferrin curve only),Mn,Pd and Pb.

Thus,iron uptake from Fe-NTA,which is independent of receptor-mediated endocytosis and the transferrin receptor,apparently occurs by a process similar to that of iron uptake from transferrin,i.e. by reduction of iron at the cell surface by the NADH:ferricyanide reductase system (3).

Iron uptake from ferritin is apparently more complex than uptake of iron from Fe-NTA and transferrin,in particular since we here are faced with the additional endocytosis and degradation of the ferric iron compound (10).Mobilization of iron from ferritin could thus be accomplished by reduction at the cell surface,by appropriate mediators within the endosome or from the iron core following degradation of ferritin within lysosomes.

As shown here (Fig. 2),hepatocyte uptake of iron from ferritin is enhanced at low oxygen concentration.Moreover,iron uptake is reduced in the presence of ferricyanide or BPS (TABLE II).However,compared to Fe-NTA or transferrin the effects are less with ferritin (Fig. 2 and TABLE II), in particular that of divalent metal cations (Fig. 3).These observations are interpreted to mean that iron uptake from ferritin partly occurs at the level of the plasma membrane and partly by endocytosis.

In line herewith are the results from subcellular fractionation of hepatocytes following incubation with ^{59}Fe-labelled ferritin.Density gradient centrifugation on 20% Percoll (11) showed that ^{59}Fe from ferritin accumulated in the lysosomes,i.e. following 5,30 and 60 min incubation at 37°C 7.5%, 13.7% and 20.6% of total radioactivity was found in the lysosomal fractions.The amount of radioactivity in the cytosol reached approx. 33% of total at 60 min.This is in contrast to iron uptake from transferrin or Fe-NTA which is almost exclusively localized to cytosol (data not shown,see also ref.11).Thus,uptake of iron from ferritin occurs also by receptor-mediated endocytosis of ferritin to lysosomes with sub-sequent degradation of the protein and release of iron to the cytosol.On the other hand,oxygen,BPS or ferricyanide do not affect endocytosis (3,5). Thus,the effects of these perturbants (Fig. 2C and TABLE II) are not consistent with iron uptake from ferritin by endocytosis only,but rather with a mechanism involving reductive mobilization of iron from ferritin at the cell surface (3).The presence of such process would also imply that ferritin which has released some of its iron must escape endocytosis and recycle back to the medium,otherwise the iron would still enter the cell by endocytosis of ferritin.Support to this interpretation are the results presented in Fig. 4.As shown here,following 30 min of incubation at 37°C with ^{59}Fe-labelled ferritin,washing of the cells at 4°C and reincubation at 37°C in the absence of ferritin,20-25% of cell associated ^{59}Fe was rapidly released from the cells.Of the ^{59}Fe released (87 $\pm$ 4)% could be pelleted at 100,000 x G,17 h.

Thus,with three different ferric iron compounds,of which two obviously not interact with the cell via the transferrin receptor,iron uptake could be explained,partly or entirely,by a reductive process at the cell surface.

Cytosolic NADH has been proposed as the intracellular electron donor to the ferricyanide reductase system (12).This conclusion is based on the findings that addition of ferricyanide or diferric transferrin to HeLa cells decreases the intracellular concentration of NADH but not NADPH (13), that iodoacetamide inhibit ferricyanide reduction (14) and that dihydroxy-acetone increases ferricyanide reduction by rat liver (15).However,NADPH is just as effective as NADH in promoting the release of iron from trans-ferrin by isolated rat hepatocyte plasma membranes (2),dihydroxyacetone acts as a reductant able to reduce ferricyanide in the absence of cells (data not shown),and in isolated hepatocytes the only perturbation found to affect intracellular NADH concentration is anaerobiosis (Fig. 5). Ferricyanide,diferric transferrin,iodoacetate,dihydroxyacetone,NaF or

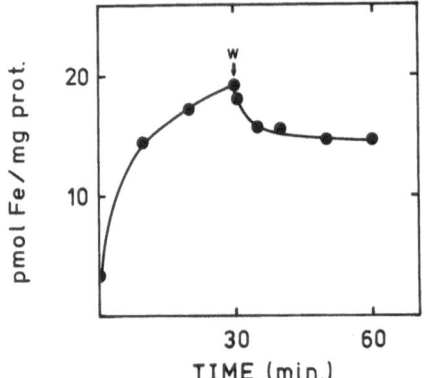

Figure 4. Uptake and release of ferritin by isolated hepatocytes.

Hepatocytes were incubated with ^{59}Fe-labelled ferritin (15 uM Fe) at 37°C.Following 30 min incubation the cells were washed at 4°C (arrow) and reincubated at 37°C in the absence of ferritin.At the time indicated duplicate aliquots were removed and the amount of radioactivity in the cells was determined.

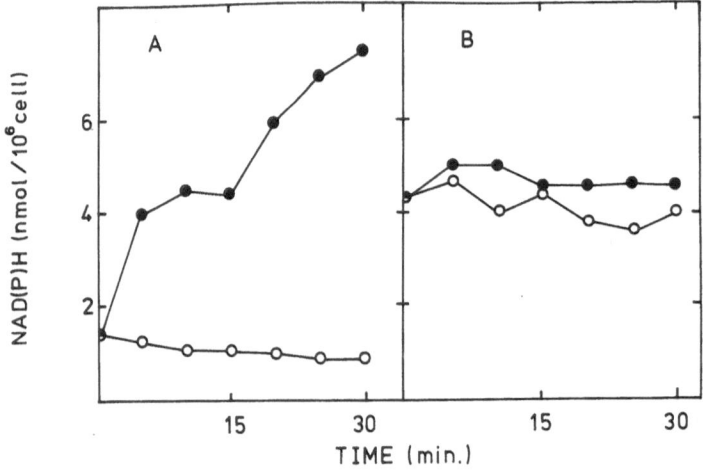

Figure 5. Nicotine adenine nucleotides in isolated hepatocytes.

The concentration of NADH (A) and NADPH (B) was determined in isolated hepatocytes during incubation at 37°C at low (closed symbols) or high (open symbols) oxygen concentration.The figure shows a typical experiment.

or chloroquine have no effect on intracellular NAD(P)H levels (data not shown).As iodoacetate,NaF and chloroquine inhibit both hepatocyte ferricyanide reduction and iron uptake from transferrin (3) no apparent correlation between intracellular NADH concentration,ferricyanide reductase activity and iron uptake exist in the hepatocyte.

In conclusion,hepatocytes take up iron from ferric iron compounds by a process which partly or entirely depends on reduction of iron at the cell surface.By comparison,reticulocytes take up iron,at least from transferrin,by receptor-mediated endocytosis.The reduction of iron by the hepatocyte occurs independently of the transferrin receptor,at least with ferritin and Fe-NTA,and the efficiency of the reduction process depends on the nature of the ferric iron compound.In hepatocytes no correlation between intracellular NAD(P)H concentration and ferricyanide reductase activity or iron uptake is evident.

REFERENCES

1. Dautry-Varsat A. (1986) Biochimie 68;375-381
2. Morley C.G.D. and Bezkorovainy A. (1985) Int. J. Biochem. 17;553-564
3. Thorstensen K. and Romslo I. (in press) J. Biol. Chem.
4. Seglen P.O. (1976) Methods in Cell Biology 13;29-83
5. Thorstensen K. and Romslo I. (1984) Biochim. Biophys. Acta 804;200-208
6. Harding C.,Heuser J. and Stahl P. (1983) J. Cell Biol. 97;329-339
7. Thorstensen K. (submitted) J. Biol. Chem.
8. Kalhorn T.F.,Thummel K.E.,Nelson S.D. and Slattery J.T. (1985) Anal. Biochem. 151;343-347
9. Bates G.W. and Schlabach M.R. (1973) J. Biol. Chem. 248;3228-3232
10. Sibille J-C.,Kondo H. and Aisen P. (in press) Hepatology
11. Thorstensen K. and Romslo I. (1987) Scand. J. Clin. Lab. Invest. 47;837-846
12. Löw H.,Sun I.L.,Navas P.,Grebing C.,Crane F.L. and Morré D.J. (1986) Biochem. Biophys. Res. Comm. 139;1117-1123
13. Navas P.,Sun I.L.,Morré D.J. and Crane F.L. (1986) Biochem. Biophys. Res. Comm. 135;110-115
14. Sun I.L.,Crane F.L.,Grebing C. and Löw H. (1984) J. Bioenerg. Biomemb. 16;583-595
15. Clark M.G.,Partick E.J. and Crane F.L. (1982) Biochem. J. 204;795-801

HEPATOCYTE IRON UPTAKE - A PLASMA MEMBRANE PROCESS OR AN ENDOSOMAL PROCESS ?

L. Thommesen, K. Thorstensen, I. Romslo, T. Nilsen and O. Gederaas

Department of Clinical Chemistry, University of Trondheim University Hospital, N-7000 Trondheim (Norway)

INTRODUCTION

The mechanisms by which eucaryotic cells accumulate iron from transferrin are still controversial, as visualized by the two major counterparts, the plasma-membrane model (1) and the endocytotic model (2). Recent studies suggest, however, that the two models rather than being contradictory are complementary, differing only in expression in different cells (3). Thus, in hepatocytes iron uptake from transferrin is largely linked to plasma membrane symport proton- and electron-fluxes (3), whereas in reticulocytes receptor-mediated endocytosis predominates (2,4). As to be expected, in hepatocytes iron uptake is extremely sensitive to oxygen. Similar sensitivity is not found in reticulocytes (5).

Based on these findings, and in an attempt to further substantiate these results, it was worthwhile to isolate endosomes following loading of the hepatocytes with transferrin, and to describe in quantitative and qualitative terms the processing of iron in these cell fractions.

MATERIALS AND METHODS

Animals, preparation of hepatocytes, loading with ^{59}Fe-transferrin, preparation of endosomes and other experimental procedures were as described (3). Following harvesting of the endosomes from the Percoll-gradient, fractions 7-12 from the top were pooled and centrifuged 100.000 x g for 30 min. The pellet thus obtained was gently resuspended in the suspension buffer and used in the subsequent studies.

RESULTS AND DISCUSSION

Quenching of endocytosed FITC-transferrin fluorescence upon addition of ATP and the reversal by protonophores is a well-known procedure to identify endosomes (6). As seen from Fig.1 centrifugation in Percoll gives fractions with relatively high FITC-quenching, but these are largely overlaid by plasma-membranes and Golgi. A further purification step was therefore included (see MATERIALS AND METHODS). The preparations

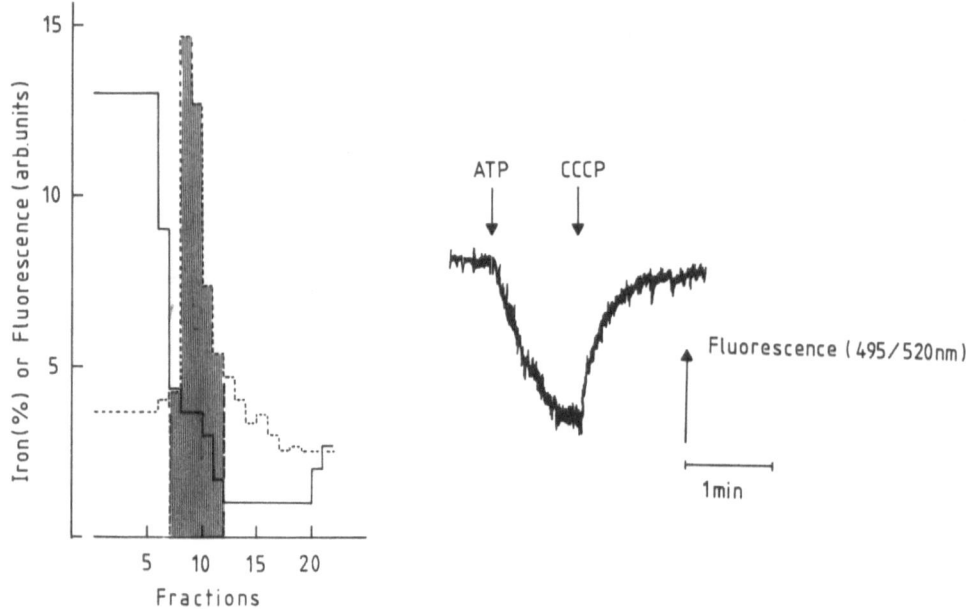

Fig.1 (left)

Distribution of [59]Fe (bold line) and FITC-transferrin quenching (stippled line) in hepatocytes as obtained by Percoll gradient centrifugation. Hatched area: fractions, pooled, recentrifuged and used in study.

Fig.2 (right)

Quenching of FITC-transferrin fluorescence by endosomal fractions used in study.

thus obtained retained significant FITC-quenching activity (Fig.2) The endosomal fractions were found to contain 15-25 pmol radioactive iron/ mg protein. The total amount of iron as determined by atomic absorption spectrophotometry was approx. 20 nmol/mg protein. Of the radioactive iron approx.80 per cent could be precipitated by TCA, 20 per cent was heat stable (76°C for 10 min) and only traces could be extracted into acidic cyclohexanone. Obviously the endosomal fractions contained iron species other than transferrin, ferritin and heme. As yet,however, the nature of these compounds remains unknown.

The mechanisms by which iron are removed from the vesicular membrane system to escape into cytosol and ferritin is unknown. Several years ago it was postulated that low molecular weight polyphosphates could bridge the gap between endosomes and the mitochondria or other intracellular destinations for iron (7). As yet, however, definitive proof for such functions of intraceelular polyphosphates is lacking, although a series of studies have shown that polyphosphates are potent modulators of iron metabolism (8).

The results reported in Fig.3 clearly show that iron in the endosomal fractions can be removed by ATP and pyrophosphate (additive effects)

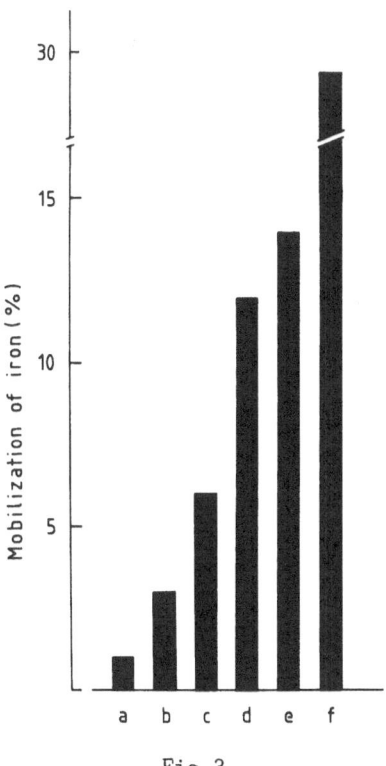

Fig.3

Mobilization of ^{59}Fe from endosomal fractions prepared from hepatocytes after loading with ^{59}Fe-transferrin.

a,control;b,with ATP;c,with pyrophosphate; d,with ATP plus pyrophosphate;e,with desferrioxamine;f.with cytosol.

but even better with desferrioxamine or soluble cytosol.NADH (or NADPH) and ferrous iron chelators are ineffective (date not shown). With inhibitors of plasma membrane ferricyanide oxidoreductase, protonophores or weak bases, insignificant effects on iron mobilization were observed.
Compared to the amount of iron to be processed by the liver in situ and the capacity of the hepatocyte to accumulate and process iron in vitro the results reported here are several orders of magnitude below that necessary to cope with these findings.
We therefore conclude that the results here are in line with the concept that the plasma membrane rather than the endosome is the important site for transferrin iron mobilization in hepatocytes.

REFERENCES

1. Morley,C.G.D. and Bezkorovainy,A.(1985) Int.J.Biochem.17,553-564
2. Morgan,E.M.,Smith,G.D. and Peters,T.J.(1986) Biochem.J. 237,163-173
3. Thorstensen,K.and Romslo,I.(1988) J.Biol.Chem.(in press)
4. Dautry-Varsat,A.(1986) Biochimie,68,375-381
5. Thorstensen,K.(1988) J.Biol.Chem.(submitted)
6. Galloway,C.J. et al.(1983) Proc.Natl.Acad.Sci USA,80,3334-3338
7. Jacobs,A.(1977) Blood,50,433-439
8. Nilsen,T. and Romslo,I.(1985) Biochim.Biophys.Acta, 842,162-169

REDUCTIVE RELEASE OF IRON FROM TRANSFERRIN AND RECEPTOR
MEDIATED RECYCLING

H. Goldenberg, M. Eder, R. Pumm, and B. Dodel

Department of Medical Chemistry
University of Vienna
Vienna, Austria

INTRODUCTION

Mammalian cells acquire iron mainly by receptor mediated endocytosis of the iron transport protein transferrin and release of iron from the protein. The bulk of the experimental evidence at present suggests that the critical step for iron release is the acidification of the intracellular endocytic compartment, because lowering the pH significantly lowers the affinity of transferrin for iron. Apotransferrin is then recycled to the cell surface undegraded and released from the plasma membrane because of its low receptor binding affinity at pH above 7 (see 1,2 for review).

In erythroid cells the endosomal pH ranges between 5.5 and 6.5 (3,4). In this range ferric ions are highly insoluble, so an intracellular chelator would be necessary to bind the iron freed from transferrin and to enable its passage through the endosomal membrane. Reduction of iron to the divalent state would greatly facilitate iron release from transferrin (5,6). Ferrous ions are released from transferrin much more quickly (7), and they are more soluble.

Indeed, iron uptake by erythroid cells is inhibited by ferrous ion chelators like bipyridine, but not ferric ion chelators like desferrioxamine (8-11). Intact reticulocytes release iron from transferrin to bipyridine (8,9,12), which may be caused by a transmembrane ferric transferrin reductase residing in the plasma membrane (13). This enzyme would oxidize reduced pyridine nucleotide on the cytosolic side and reduce transferrin on the extracellular side of the membrane, coupled to a transport of protons to enable iron reduction which would otherwise be thermodynamically unfavorable (14,15). In the absence of a chelator iron would not be released at the outside, but leave the protein in the endosomes, where low pH would facilitate its release. Ferrous ions could then pass the membrane to be incorporated into cytosolic ferritin or mitochondrial heme.

Another explanation would be that iron is reduced by an oxidoreductase or intracellular chemical reductants after release from transferrin (16). In both cases, bipyridine, which is lipophilic and partitions into the membrane, would bind ferrous ions, and leave the cell as a hydrophilic complex thereby inhibiting iron uptake (9,10).

This study aims to clarify this question by investigating the characteristics of chelator inhibition of iron uptake, release of iron and transferrin reduction by intact cells and influence of chelators on the subcellular distribution of transferrin and iron during endocytosis in K562 cells

METHODS

K562 cells (17) were grown in logarithmic phase in RPMI 1640 medium supplemented with 2mM glutamine, gentamycin and 10% fetal calf serum at a density between 10^5 and 10^6 cells per ml.

For each experiment described the cells were harvested at a density of approximately 7×10^5 cells per ml by centrifugation at 150g for 10min and washed twice in icecold RPMI. They were incubated for 20min at 37° in RPMI + 1% BSA to wash off endogenous transferrin (18,19). In experiments where effects of inhibitors were studied, they were included in this step at the indicated concentrations. The cells were again washed twice in icecold RPMI and resuspended in icecold RPMI + BSA at a concentration of 10^8 cells per ml.

Iron release from transferrin

5×10^6 cells were incubated in a total volume of 120ul with 2mM 2,2´-bipyridine, effectors and 0.5uM $^{55}Fe_2$-transferrin for 1h. In a parallel series 25uM unlabelled diferric transferrin was included to account for unspecific reactions. The cells were removed by centrifugation and 100ul of the supernatant were mixed with 1ml of icecold absolute ethanol to precipitate transferrin (12). In a control tube, the cells were added immediately before centrifugation. The precipitated proteins were removed by centrifugation at 3000rpm for 10min in a refrigerated clinical benchtop centrifuge. The supernatants were counted for radioactivity in a Packard Liquid Scintillation Counter with appropriate quench correction.

Uptake of transferrin and iron

Between 2 and 3 million cells were incubated in a total volume of 300ul with 100nM $^{55}Fe_2$-transferrin or ^{125}I-Fe_2-transferrin and inhibitors as listed in RESULTS for 30min at 0° on ice. A control for each experiment to correct for unspecific uptake containing 10uM unlabelled Fe_2-transferrin was run in parallel. For internalization of transferrin and uptake of iron the mixtures were warmed to 37° and incubated for the indicated times. The reaction was terminated by

addition of 1ml of icecold RPMI + albumin, and
the cells were washed twice in this medium.

To discriminate between total and surface bound
transferrin and iron the cells were suspended in 140mM
NaCl/25mM sodium acetate/2mM $CaCl_2$, pH 4.5 containing 50uM
DTPA. After immediate centrifugation, the cells were
resuspended in RPMI containing 50uM DTPA, incubated on ice for
10min and recentrifuged (20). After another washing,
radioactivity was determined by counting iodine in a LKB-
Wallac Mini Gamma Counter and iron in the liquid
scintillation counter after the cells had been dissolved in
Beckman Readysolv. The supernatant from the last wash was
always free of detectable radioactivity.

Short Pulse Uptake and Subcellular Fractionation

$2x10^8$ cells were incubated for binding of transferrin as
described above in a total volume of 5ml. After incubation at
0°, the cells were washed twice in icecold RPMI to remove
excess transferrin. To initiate endocytosis they were then
rapidly suspended in 5ml of warm (37°) RPMI + 1% BSA and
incubated at this temperature for 5 minutes (19,21-24). The
reaction was terminated by addition of a twentyfold surplus of
icecold RPMI and centrifugation.

Where indicated, surface bound transferrin and iron were
removed by treatment with DTPA as described above. After two
more wahings, the cells were suspended in 0.25M sucrose
buffered with 10mM Hepes/NaOH, pH 7.2, and homogenized with 25
strokes in a glass-teflon Potter homogenizer at approx.
2000rpm. After centrifugation at 150g for 10min, the
homogenization was repeated with the pellet. After another
centrifugation the supernatants were combined, sucrose was
added to a volume of 2ml and the homogenate centrifuged for
10min at 3000rpm (1000xg) to remove nuclei and debris. The
postnuclear supernatant (PNS) was mixed with 8ml of an
isotonic Percoll solution in sucrose to yield a Percoll
concentration of 10% and centrifuged in the Ty65 rotor of the
Beckman L8-55 ultracentrifuge at 12.500rpm for 45min. 50
fractions of 0.2ml each were collected by suction from the
bottom of the tube.

Photometric assay of transferrin iron reduction

Isotonic buffer (145mM NaCl, 10mM Hepes, pH 7.2),
containing inhibitors as indicated, was mixed with $2x10^6$ cells
and 25ul of 200mM bipyridine, 200ul of transferrin solution
(assay concentration 10uM) were added in a stirred cuvette
and the absorbance change at 524nm was followed in a Hitachi
150 dual beam recording spectrophotometer with buffer and
bipyridine as reference in a total volume of 3ml.

Appropriate blanks were run for cell independent
reactions, which were negligible under these conditions.

Assays with cellular homogenates were carried out in the
same manner. For pH 6 Hepes was replaced by MES as buffer.

An absorption coefficient of $10.5mM^{-1}$ was used for the
ferrous- bipyridine chelate complex.

Enzymatic reduction of transferrin iron

Membranes from 8 to 10 million cells (see below) were mixed with sodium phosphate buffer (50mM, pH as indicated), 20uM pyridine nucleotide (NADH or NADPH), 1mM KCN, 2mM bipyridine and inhibitors as indicated. After 10min preincubation, the reaction was started by addition of ^{55}Fe-labelled transferrin (0.6uM). The total volume was 100ul. After incubation for 60min at 37° serum albumin (1%) was added to enhance precipitation, and the protein precipitated with 1ml of icecold ethanol. The mixture was centrifuged and the supernatant counted for released radioactivity (see above). Appropriate blanks were run without membranes and without pyridine nucleotides.

Preparation of total cellular membranes

The cells were homogenized as described under Subcellular Fractionation, centrifuged at 600xg for 10min and then at 200000xg for 30min. The pellet was resuspended in 150mM NaCl containing 0.1% CHAPS and used as total membrane fraction.

Transferrin was loaded with unlabelled or radioactive iron ^{55}Fe according to (25).The iron loaded transferrin was iodinated with the iodogen method to a specific activity of 10^5 cpm/ug protein (26). Adriamycin (adriblastin) was a gift of Farmitalia. The B3/25 monoclonal antitransferrin receptor antibody was a gift of Dr.I.Trowbridge (The Salk Institute, LaJolla,Ca.)

Abbreviations: RPMI, Roswell Park Memorial Insitute. BSA, bovine serum albumin. Hepes, 2-[4-(2-hydroxyethyl)-1-piperazinyl] - ethanesulfonic acid. MES, (2[N-morpholino] - ethanesulfonic acid. DTPA, diethylenetriamine - pentaacetic acid. BPS, bathophenanthroline disulfonate. CHAPS, 3-[(3-cholamidopropyl-)dimethylammonio]-1-propane sulfonate.

RESULTS

The ferrous iron chelators 2,2´-bipyridine and BPS inhibited endocytosis of transferrin in K562 cells without affecting transferrin receptor binding (Fig.1). Bipyridine was more effective than BPS and completely inhibited iron uptake, which was depressed to about 50% by BPS (not shown).
Kinetic analysis of transferrin uptake into K562 cells showed the characteristics of recycling, while the iron content increased linearly with time (Fig.2) Surface bound label was removed by washing with DTPA(20), so only intracellular transferrin and iron are shown in the Figure. All data are corrected for unspecific uptake. Presence of 2mM bipyridine totally blocked iron uptake and at the same time depressed the amount of transferrin involved in the recycling process.
Amiloride (0.4mM) had no influence on transferrin recycling, but slightly inhibited iron uptake. The inhibitory

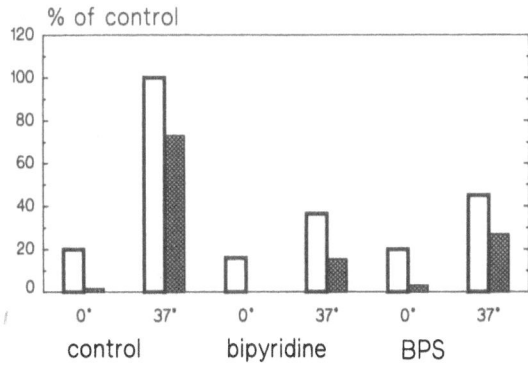

% of control

FIGURE 1. Effect of ferrous chelators on binding at $0°$ and uptake for 20min at $37°$ of ^{125}I-labelled transferrin into K562 cells. Clear area: total transferrin. Shaded area : intracellular label. 100% is used as control value for uninhibited total label.

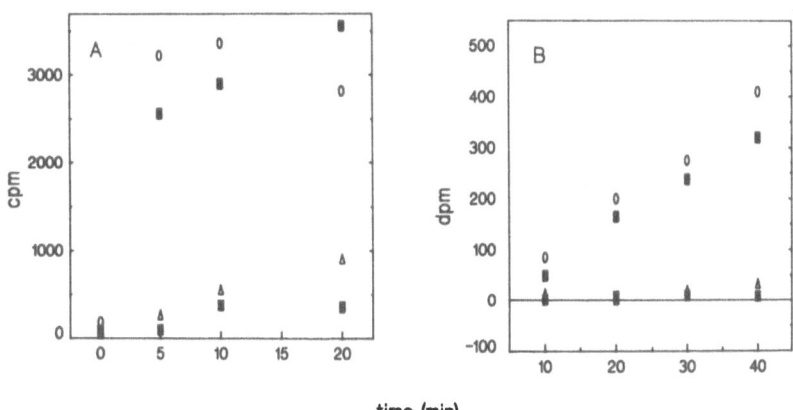

time (min)

FIGURE 2. Intracellular accumulation of transferrin (A) and iron (B) in K562 cells in absence and presence of 2mM bipyridine and 0.4mM amiloride, respectively. ● ‒ ●,control, △ ‒ △,bipyridine,■ ‒ ■,amiloride,■ ‒ ■,amiloride and bipyridine. Specific uptake is shown (total minus unspecific). Surface bound transferrin was removed by washing with DTPA (see METHODS).

effect of the chelator remained in presence of amiloride (Fig.2).
    75%(65%) respectively, of a single pulse of transferrin or iron were taken up into intracellular sites after five minutes at $37°$C. Washing of the cells with DTPA after binding equilibration at $0°$ removed 80% of the label. The total amount of transferrin and iron remained unchanged after 5min at $37°$, but after this time only 25%(35%) remained DTPA-accessible. In presence of bipyridine the total amount decreased to 70%(60%) of the uninhibited value, of which 50%(65%) were now DTPA-accessible (Fig.3).
    While intracellular transferrin protein was found in a low density intracellular compartment, presumably in endosomes, after Percoll gradient centrifugation of cell homogenates, iron had passed to the cytosol, as indicated by its uniform distribution in the gradient after washing with

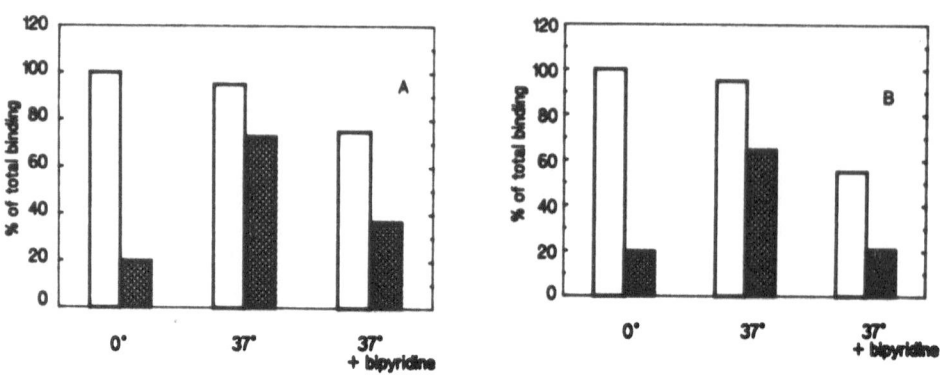

FIGURE 3. Total and intracellular transferrin (A) and iron
(B) after a single pulse of transferrin in K562 cells at 0°
and after 5min at 37°. Surface bound transferrin was removed
with DTPA. Clear area, total label. Shaded area, intracellular
label. The 100%-control values were 45000 molecules per cell
for transferrin and 83000 atoms per cell for iron.

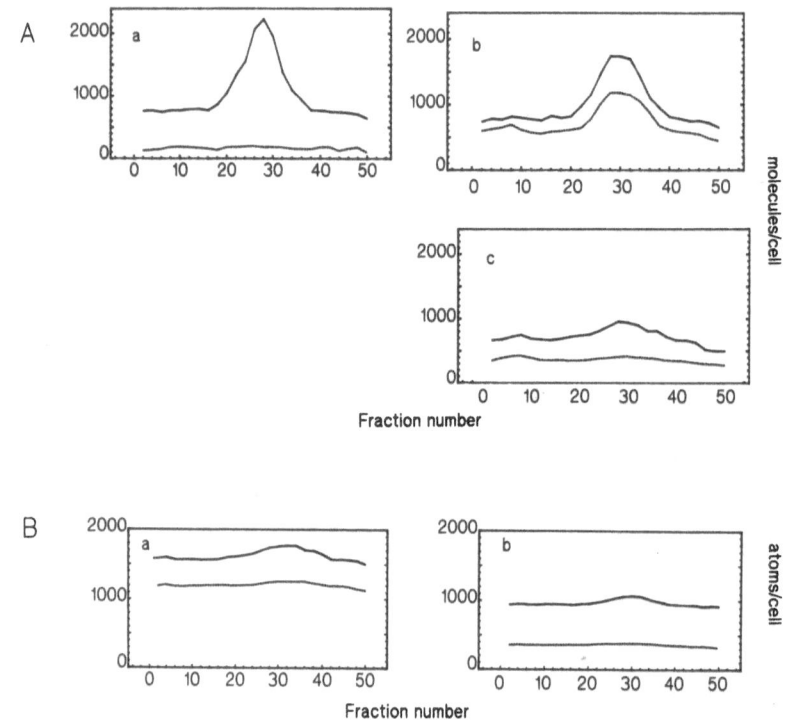

FIGURE 4A. Subcellular fractionation of a K562 cell homogenate
on a 10% Percoll gradient after incubation with [125]I-
transferrin. a, incubation at 0° only. b, single dose
incubation at 37° for 5min. c, incubation as in b, with 2mM
bipyridine present. Full line : total radioactivity. Broken
line: radioactivity after washing of the cells with DTPA
before homogenization.

FIGURE 4B. Experimental conditions as in Figure 4A, with [55]Fe
as marker. a, incubation at 37° without, b, with 2mM
bipyridine.

144

DTPA (Fig.4). Detachment of iron probably occurs very early after endocytosis (24). Comparison of the density gradient fractionations made after endocytosis in absence or presence of bipyridine clearly showed the disappearance of intracellular membrane bound transferrin in the latter case (Fig.4).

Intact K562 cells released iron from transferrin to bipyridine. This activity was dependent on specific receptor binding, because it was nearly completely inhibited by a surplus of unlabelled transferrin, and partly inhibited by the B3/25 monoclonal antitransferrin receptor antibody (27).

70% inhibition was observed with amiloride, and about 50% inhibition with ferricyanide and adriamycin (Fig.5). The latter two compounds had no effect on iron uptake at the concentrations used.

At higher transferrin concentrations (10uM) the formation of the ferrous - bipyridine complex could be followed directly by photometric assay. This activity was also inhibited by amiloride, B3/25 antibody, ferricyanide and adriamycin (Fig.5). Homogenates did not reduce transferrin iron under identical conditions. Reduction by homogenates was observed at pH 6 as an activity which was partly heat stable (Table 1).

An enzymatic activity dependent on reduced pyridine nucleotides could be shown in isolated total membranes by a similar release assay as used for intact cells (Table 2). Blank activity of membranes without pyridine nucleotide was heat stable, and may have been derived from residual cytosol trapped in membrane vesicles. At pH 6 only the blank reaction

Table 1

Formation of ferrous-bipyridine complex by intact K562 cells and by cell homogenates in nmoles/min.10^6 cells

Intact cells	0.106 ± 0.026
Homogenate at pH 7	0.019
Homogenate at pH 6	0.092
Homogenate at pH 6, boiled	0.046

Table 2

Reduction and release of transferrin iron by reduced pyridine nucleotides in absence and presence of cellular membranes from $8x10^6$ K562 cells (pmoles/h)

nucleotide	pH	membranes +	membranes −	boiled membranes
NADH	7	25.2	2.65	
	6.5	38.7	7.25	
	6	43.5	11.6	
NADPH	7	21.1	3.0	
	6.5	41.2	7.8	
	6	44.6	9.9	
none	7	10.9	1.5	7.4
	6.5	25.1	3.5	
	6	32.1	6	

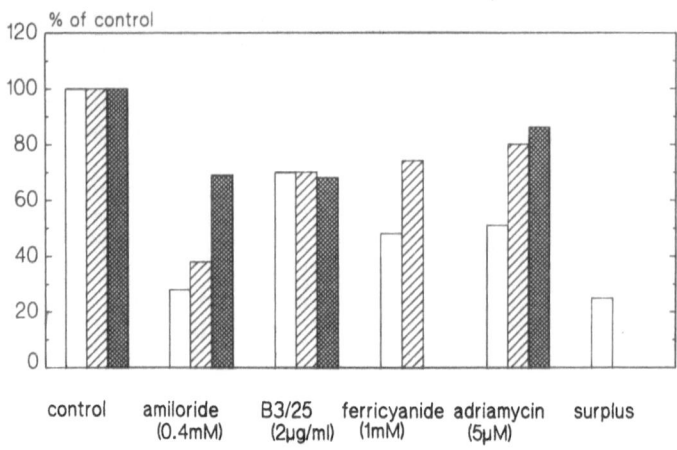

FIGURE 5. Effect of inhibitors on iron release by intact K562 cells (clear area), formation of ferrous-bipyridine complex by K562 cells (dashed area) and on NAD(P)H diferric transferrin reductase activity of membranes from K562 cells (shaded area). 100% was 10800 iron atoms per minute per cell ± 2300 for release, 0.106nmoles per minute per 10^6 cells ± 0.026 for chelate dye formation and 13600 iron atoms per cell per minute ± 2300 for enzymatic release of iron.

was observed. The NADH-dependent activity was seen after treatment of the membranes with CHAPS detergent, probably due to disruption of closed vesicles. Once again, inhibition by amiloride, B3/25 antibody and adriamycin could be demonstrated, though to a lesser degree (Fig.5).

DISCUSSION

It has been known for some time that ferrous chelators are effective inhibitors of iron uptake from transferrin. Morgan (8,9) as well as Nunez & Glass (10,16) have presented evidence that the lipophilic chelator 2,2´-bipyridine binds ferrous ions during their passage through the membrane of endocytic vesicles, where iron is detached from transferrin due to the low pH. Binding of ferrous ions by bipyridine needs a reduction step as a prerequisite, since iron is bound to transferrin in the ferric state (28).

Ferric ions bound to transferrin have a very low oxidation potential (14), so their enzymatic or chemical reduction by pyridine nucleotides, flavins or sulfhydryl compounds is highly unfavourable at pH 7, but readily occurs at lower pH due to loosening of the iron-protein binding. Presence of a ferrous chelator would , however, shift the equilibrium to reduction and detachment of iron due to formation of quaternary complexes (6,15).

According to the model presented by Nunez & Glass for reticulocytes (16) an oxidoreductase reduces iron after release from transferrin in endosomes, a pathway which would overcome the weak reducibility of transferrin iron.

Reduction of transferrin bound iron and formation of a ferrous chelate complex has also been ascribed to the activity

of a transplasma membrane NADH-dependent oxidoreductase
(13,29), which has been demonstrated in plasma membranes from
rat liver (13,30). The properties of extracellular
transferrin reduction in other cell types are very similar to
those of this enzymatic activity. A very prominent feature is
its connection to proton extrusion, probably via the Na^+/H^+-
channel (31), as also indicated by its sensitivity to
amiloride. Protonization could facilitate iron reduction and
release in presence of a chelator. Whether this reduction also
occurs in absence of the latter is not at all clear. It may
happen without detachment of iron and be reversed by the
intrinsic ferroxidase activity of transferrin (32), possibly
several times before iron is taken up by the cell. Iron uptake
itself almost certainly occurs via endocytosis together with
transferrin (3,33).

Data presented in this study may be taken as evidence
that in K562 cells the inhibitory effect of bipyridine on iron
uptake is in part due to reduction of surface bound
transferrin in the plasma membrane.

One point is that bipyridine also inhibited uptake of the
transferrin protein. Endocytosis and recycling themselves seem
unaffected by bipyridine, as shown by its lack of effect of
release of endocytosed transferrin (11). Thus the inhibitory
effect on transferrin endocytosis can be explained by a
partial loss of iron, which leads to cessation of transferrin
binding to its receptor (33). Figure 1 demonstrates a parallel
effect of bipyridine, but also of bathophenanthroline
sulfonate, on uptake of transferrin protein and iron, with the
marked difference that iron uptake is completely blocked by
the lipophilic bipyridine only.

Amiloride, which does not inhibit iron uptake in other
cell types (34), showed a slight (20%) inhibition of iron
uptake in K562 cells, while transferrin recycling was
unaffected (Fig. 2). This can be explained by its inhibitory
effect on transferrin iron reduction in the plasma membrane,
which may lower the amount of ferrous ions being able to cross
the endosomal membrane after release from transferrin. The
blocking effect of bipyridine nevertheless remained, because
this lipophilic chelator blocks the passage of ferrous ions in
every membrane.

Fractionation of cell homogenates after allowing for
endocytosis of a single 5min pulse of transferrin yielded a
similar picture. We chose this time point as reference from
data presented in the literature as a medium time point in the
recycling process in K562 cells (18,19,21-24). Bakkeren & al
(24) have shown that after 5min approximately 60% of the iron
has passed to the cytosol, while the rest is still membrane
bound.

Under the conditions used for gradient centrifugation,
plasma membrane and endosomes were not separated. Washing of
the cells with the impermeable chelator DTPA (20), however,
removed surface bound transferrin and iron and allowed
discrimination between extracelluar and intracellular label.

As seen from Figs 3 and 4, bipyridine inhibited
endocytosis of transferrin bound to the plasma membrane
without affecting surface binding itself. The difference
between total and intracellular transferrin or iron, as
expressed by DTPA-accessible label, remained constant. Taking
the $0°$ value as a reference, endocytosis of transferrin

protein was inhibited to 25% of the original, iron uptake into the cytosol was practically decreased to zero.

Further evidence derived from the effects of some inhibitors on the release of iron from transferrin to bipyridine by intact K562 cells (Fig 5). This reaction was dependent on specific receptor binding. It was blocked by a surplus of unlabelled transferrin and partly inhibited by the B3/25 antitransferrin receptor antibody.

0.4mM amiloride inhibited release of iron. This is evidence that proton transport from the cell is necessary for iron reduction.

Ferricyanide, a strictly extracellular oxidant, also inhibited iron release by intact cells. Ferricyanide is the unspecific general acceptor for the plasma membrane oxidoreductase and may compete with transferrin iron for electrons.

Adriamycin, which has been desribed as inhibitor of the transplasma membrane oxidoreductase (35), also decreased iron release to bipyridine.

These three inhibitory effects argue for a localization of the releasing effect in the plasma membrane.

Measuring formation of the ferrous-bipyridine complex by photometric assay, that is, kinetic measurement of iron release by intact K562 cells, yielded a similar picture (Fig 5). Cellular homogenates were much less effective at pH 7. This indicates the necessity of an intact membrane polarity, possibly for proton pumping (see above). At lower pH, homogenates did reduce transferrin iron. This activity was partly heat stable (Table 1) and may be due to activty of chemical reductants in the cell (36).

Release of iron from transferrin by isolated cellular membranes in presence of reduced pyridine nucleotide (Table 2) strongly argues for the action of a NAD(P)H diferric transferrin reductase in the releasing process (13,29). Of course, since plasma membranes were not purified, the subcellular localization of this enzyme activity remains unknown. It may be present in endosomes or in both plasma membrane and endosomes. The fact that this enzyme activity is inhibited by the same compounds as the release of iron by intact cells (Fig.5) is at least some evidence that both effects are due to the same activity. A dual localization is supported by the fact of less inhibition in isolated membranes by all effectors used.

Our findings are somewhat in contrast to what has been shown for reticulocytes (9,16). The enzymatic activity of NADH-diferric transferrin reductase in reticulocytes is not known. If this enzyme serves as a control element in cellular growth (37), as it may also be concluded from inhibition by adriamycin, it may not be present in significant amounts in non-growing cells.

None of the inhibitors of plasma membrane transferrin reductase except amiloride inhibited iron uptake. The purpose of this enzyme may therefore rest mainly in its ability to extrude electrons and protons from the cell. Reduced transferrin may be reoxidized by its own ferroxidase activity several times before it is freed from iron within the endocytic compartment. Endosomes may also contain a reductase

in all cells, which is mainly responsible for reduction of
transferrin iron for cellular uptake (16).

REFERENCES

1. J.A.Hanover and R.B.Dickson, Transferrin: Receptor
mediated endocytosis and iron delivery, in "Endocytosis",
I.Pastan and M.C.Willingham, eds., pp. 131 - 161. Plenum
Press, N.Y. (1985)
2. R.R.Crichton and M. Charloteaux - Wauters, Iron transport
and storage. Eur.J.Biochem. 164, 485 - 506 (1987)
3. J.van Renswoude and K.R.Bridges, J.B.Harford and
R.D.Klausner, Receptor -mediated endocytosis of transferrin
and the uptake of Fe in K562 cells: Identification of a
nonlysosomal acidic compartment. Proc.Natl Acad.Sci. 79, 6186
- 8190 (1982)
4. S.Paterson, N.J.Armstrong, B.J.Iacopetta, H.J.McArdle and
E.H.Morgan, Intravesicular pH and iron uptake by immature
erythroid cells. J. Cell. Physiol. 120, 225 - 232 (1984)
5. E.Ankel and D.H. Petering, Iron chelating agents and the
reductive removal of iron from transferrin. Biochem.
Pharmacol. 29, 1833 - 1837 (1980)
6. N.Kojima and G.W.Bates, The reduction and release of iron
from Fe^{3+} - transferrin. J.Biol.Chem. 254, 8847 - 8854 (1979)
7. B.P.Gaber and P.Aisen, Is divalent iron bound to
transferrin ? Biochim.Biophys.Acta 221, 228 - 233 (1970)
8. E.H.Morgan, A study of iron transfer from rabbit
transferrin to reticulocytes using synthetic chelating agents.
Biochim.Biophys.Acta 244, 103 - 116 (1971)
9. E.H.Morgan, Chelator mediated iron efflux from
reticulocytes. Biochim.Biophys.Acta 733, 39 - 50 (1983)
10. M.-T. Nunez, E.S. Cole and J.Glass, The reticulocyte
plasma membrane pathway of iron uptake as determined by the
mechanism of a,a -dipyridyl inhibition. J.Biol. Chem. 258,
1146 - 1151 (1983)
11. K.R.Bridges and A.Cudkowicz, Effect of iron chelators on
the transferrin receptor in K562 cells. J.Biol.Chem. 259,
12970 - 12977 (1984)
12. E.H.Morgan, Failure of a cell free system from rabbit
reticulocytes to remove iron from transferrin. Biochem.J. 158,
489 - 491 (1976)
13. H.Löw, I.L.Sun, P.Navas, C.Grebing, F.L.Crane and
D.J.Morré, Transplasmalemma electron transport from cells is
part of a diferric transferrin reductase system. Biochem.
Biophys. Res. Comm. 139, 1117 - 1123 (1986)
14. D.C. Harris, A.L.Rinehart, D.Hereld, R.W.Schwartz,
F.P.Burke and A.P.Salvador, Reduction potential of iron in
transferrin. Biochim. Biophys. Acta 838, 295 - 301 (1985)
15. G.W.Bates, Metal ion and anion exchange reactions of serum
transferrin: the role of quaternary complexes and
conformational transitions, in "The Biochemistry and
Physiology of Iron" pp. 1 - 18, P.Saltman and J.Hegenauer,
eds. Elsevier, Amsterdam (1982)
16. M.-T.Nunez and J.Glass, Iron uptake in reticulocytes:
Inhibition mediated by the ionophores monensin and nigericin.
J.Biol. Chem. 260, 14707 - 14711 (1985)

17. B.B.Lozzio and C.B.Lozzio, Properties and usefulness of the original K562 human myelogenous leukemia cell line. Leuk. Res. 3, 363 - 370 (1979).

18. B.S.Stein and H.H.Sussman, Demonstration of two distinct transferrin receptor recycling pathways and transferrin-independent receptor internalization in K562-cells. J.Biol. Chem. 261, 10319 - 10331 (1986)

19. A.Bomford, S.P.Young and R.Williams, Release of iron from the two iron binding sites of transferrin by cultured human cells: modulation by methylamine. Biochemistry 24, 3472 - 3478 (1985)

20. A.van der Ende, A. du Maize, C.F.Simmons, A.L.Schwartz and G.J.Strous, Iron metabolism in BeWo chorion carcinoma cells: Transferrin mediated uptake and release of iron. J.Biol.Chem. 262, 8910 - 8916 (1987)

21. C.A.Enns, J.W.Larrick, H.Suomalainen, J.Schroder and H.H.Sussman, Co-migration and internalization of transferrin and its receptor on K562 cells. J.Cell Biol.97, 579 - 585 (1983)

22. J.W.Larrick, C.Enns, A.Raubitschek and H.Weintraub, Receptor-mediated endocytosis of human transferrin and its cell surface receptor. J. Cell. Physiol. 124, 283 - 287 (1985)

23. R.D.Klausner, J.Van Renswoude, G.Ashwell, C.Kempf, A.N.Schechter, A.Dean and K.R.Bridges, Receptor mediated endocytosis of transferrin in K562 cells. J.Biol.Chem. 258, 4715 - 4724 (1983)

24. D.L.Bakkeren, C.M.H. de Jeu-Jaspars, M.J.Kroos and H.G.van Eijk, Release of iron from endosomes is an early step in the transferrin cycle. Int.J.Biochem. 19, 179 - 186 (1987)

25. G.Graham and G.W.Bates, Approaches to the standardization of serum unsaturated iron binding capacity. J.Lab.Clin.Med. 88, 477 - 486 (1976)

26. M. Hüttinger, W.J.Schneider, Y.K.Ho, J.L.Goldstein and M.S. Brown, Use of monoclonal anti-receptor antibodies to probe the expression of the Low Density Lipoprotein receptor in tissues of normal and Watanabe Heritable Hyperlipidemic rabbits. J. Clin. Invest. 74,1017 - 1026 (1984)

27. M.B.Omary, I.S.Trowbridge and J.Minowada, Human cell surface glycoprotein with unusual properties. Nature 286, 888 - 891 (1980)

28. P.Aisen and I.Listowski, Iron transport and storage proteins. Ann.Rev.Biochem. 49, 357 -393 (1980)

29. H.Löw, C.Grebing, A.Lindgren, M.Tally, I.L.Sun and F.L.Crane, Involvement of transferrin in the reduction of iron by the transplasma membrane electron transport system. J.Bioenerg.Biomembr. 19, 535 - 550 (1987)

30. I.L.Sun, P.Navas, F.L.Crane, D.J.Morré and H.Löw, NADH diferric transferrin reductase in liver plasma membrane. J.Biol.Chem. 262, 15915 -15921 (1987)

31. I.L. Sun, R.Garcia-Canero, W.Liu, W.Toole-Simms, F.L.Crane, D.J.Morré and H.Löw, Diferric transferrin reduction stimulates the Na^+/H^+-antiport of HeLa Cells. Biochem.Biophys.Res.Comm.145, 467 - 473 (1987)

32. N.Kojima and G.W.Bates, The formation of Fe^{3+}-transferrin-CO_3^- via the binding and oxidation of Fe^{2+}. J.Biol. Chem. 256, 12034 - 12039 (1981)

33. R.Klausner, G.Ashwell, J.Van Renswoude, J.Harford and K.R.Bridges, Binding of apotransferrin to K562 cells: Explanation of the transferrin cycle. Proc.Natl.Acad.Sci. 80, 2263 - 2266

34. J.Wang, J.Pouyssegur, M.C.Willingham and I.Pastan, The role of intracellular pH in ligand internalization. J.Cell.Physiol.128,18-22 (1986)

35. I.L.Sun, F.L.Crane, H.Löw and C.Grebing, Inhibition of plasma membrane NADH dehydrogenase by adriamycin and related anthracycline antibiotics. J.Bioenerg.Biomembr. 16, 209 - 221 (1984)

36. S.Pollack and J.Weaver, Iron release from transferrin: synergistic action between adenosine triphosphate and an ammonium sulfate fraction of hemolysate. J.Lab.Clin.Med. 108, 411-414 (1986)

37. F.L.Crane, I.L.Sun, M.G.Clark, C.Grebing and H.Löw, Transplasma membrane redox systems in growth and development. Biochim.Biophys.Acta 811, 233 - 264 (1985)

Acknowledgments: This work was supported by "Hochschuljubiläumsstiftung der Stadt Wien". We are grateful to Dr.I.Trowbridge (Salk Institute, LaJolla, Ca.), for donation of the B3/25 antibody

TRANSPLASMA MEMBRANE ELECTRON TRANSPORT FUNCTIONS

AS A FERRIC REDUCTASE

H. Löw, A. Lindgren, F.L. Crane, I.L. Sun,
W. Toole-Simms and D.J. Morré

Department of Endocrinology, Karolinska Hospital, Stockholm, Sweden
Department of Biology and Department of Medical Chemistry, Purdue University,
W. Lafayette IN 47907 USA

INTRODUCTION

We have previously presented evidence that ferric ions and iron in diferric transferric can be reduced at the cell surface by the transplasma membrane electron transport system acting in conjuction with the transferrin receptor [1,2]. In this paper we examine the reaction further especially in relation to the inhibition of both ferric citrate and diferric transferrin reduction by apotransferrin. We also consider the difference in activity found with different diferric transferrin preparations containing different amounts of loosely bound ferric iron. We will also consider the effects of argon, and cell transformation on the rate of ferric iron reduction in the presence and absence of diferric transferrin. The objective is to evaluate the extent to which loosely bound iron on transferrin can contribute to the formation of ferrous bathophenanthroline disulfonate ($Fe(BPS)_3$) through the transmembrane electron transport system. We also show that iron reduction at acidic pH by the transmembrane enzyme is not inhibited by apotransferrin which would favor its role for ferric reduction in acidic endosomes.

METHODS

Cell preparation and culture. Primary liver cells were prepared according to Clark et al.[3]. Cell culture was as previously described in media plus fetal calf serum and antibiotics. Ferric iron reduction was measured by following the formation of ferrous bathophenanthroline disulfonate with a dual beam spectrophotometer set at 535 minus 600 nm [1,4]. Ferricyanide reduction was measured in media A (130 mM NaCl, 5 mM KCl, 1 mM Mg Cl_2 and 2 mM Ca Cl_2 in 10 mM Hepes buffer pH 7.4) with 0.2 mM potassium ferricyanide [5] at 420 minus 500 nm. Diferric transferrin was prepared according to Karin and Mintz [6], or obtained from Miles Laboratories or Boehringer Mannheim. Apotransferrin was from Sigma. Diferric transferrin binding was performed with I^{125} labeled diferric transferrin with displacement with unlabeled diferric transferrin to determine specific binding.

Ascorbate-reducible iron in the diferric transferrin was measured at 535 nm by formation of ferrous bathophenanthroline sulfonate (BPS) with 3.4 µm diferric transferrin, 5 µm ascorbate pH 7.2 and 15 µm BPS in 2.5 ml media A. Absorbance increase at 535-600 nm was measured and an extinction coefficient for the Fe (BPS)$_3$ of 17.6 mM^{-1} cm^{-1} was used.

RESULTS

Ferric iron reduction by intact cells. The rate of formation of ferrous BPS by cells in the presence of ferric citrate varied from 5.8 to 0.2 nmoles min^{-1} 10^{-6} cells. The rate of ferrous BPS formation from diferric transferrin followed a similar pattern of activity except with L1210 cells. Reduction of ferricyanide was usually more rapid and was not always proportion at to the rate of ferric citrate reduction (Table 1).

Table 1. Typical rates of ferrous BPS formation with different cell types with diferric transferrin, ferric ammonium citrate or ferricynide as oxidants.

Celltype	Rate of iron reduction µmole min -1 cells -6			
	Diferric transferrin	Ferric ammonium citrate	Ferri- cyanide	% Fe2 Tf binding
Rat liver (primary)	8.2	5.8	9.0	
Swiss 3T3	2.6	2.9	3.2	7%
Sv40 Swiss 3T3	0.6	0.6	1.0	13%
Hep G2	0.85	0.9		
HeLa S3	0.65	1.2	4.6	22%
CC1-39 (day 1)	0.6	0.8		
CC1-39 (day 2)	0.3	0.4	0.2	
CC1-39 (day 9)	0.2	0.25	0.6	1%
HL60	0.2	0.2		
L1210	0.05	0.2	1.2	20%

Concentrations of oxidants: Diferric transferrin 17 µM, Ferric ammonium citrate 7.5 µM, Ferricyanide 0.2 mM. Incubation in media A. The diferric transferrin preparation contained 1 mole/mole extra, i.e. ascorbate reducible, iron.

The relative rate of Fe (BPS)$_3$ formation from diferric transferrin did not directly parallel the binding % for diferric transferrin especially in the case of CCl-39, swiss 3T3, and SV40 3T3 cells (Table 1).

Combined action of diferric transferrin and ferric ammonium citrate for formation of ferrous BPS. As shown previously[1] diferric transferrin can stimulate additional Fe (BPS)$_3$ formation in the presence of saturating concen-trations of ferric ammonium citrate with HeLa S3 cells (Table II). On the other hand diferric transferrin did not stimulate extra Fe (BPS)$_3$ formation when added to L1210 cells in addition to ferric ammonium citrate. With L1210 cells the extra diferric transferrin often inhibited the rate of ferric ammonium citrate reduction (Table 2).

Table 2.   Effect of diferric transferrin on Fe (BPS)3 formation in the presence of saturating concentrations of ferric ammonium citrate: comparison of effects with HeLa S3 cells and L1210 cells.

Cell type	Oxidants added	Fe(BPS)3 formation µmole min-1 cells-6
HeLa S3	FAC 7.5 µM	0.46
	FAC 7.5 µM + Fe2 Tf 7 µM	0.90
	FAC 7.5 µM + Fe2 Tf 14µM	1.2
	FAC 15 µM	0.61
	FAC 15 µM + Fe2 Tf 14 µM	1.4
L1210	FAC 7.5 µM	0.12
	Fe2 Tf 17 µM	0.03
	FAC 7.5 µM + Fe2 Tf 17 µM	0.04

FAC: ferric ammonium citrate. Incubation in media A. The Fe2 Tf preparation used contained 1 mole/mole ascorbate reducible iron.

Inhibition of Fe (BPS)$_3$ formation by apotransferrin. We have previously shown that apotransferrin inhibits Fe (BPS)$_3$ formation from both ferric citrate and diferric transferrin[1]. Evidence for apotransferrin binding to Hela cells[1] suggested a competitive inhibition at the transferrin receptor. (Table 3). The lack of apotransferrin binding by CCl-39 cells which still show apotransferrin inhibition of iron reduction indicates that the apotransferrin inhibits by binding free or loosely bound iron. This is further evidenced by the fact that apotransferrin inhibits Fe (BPS)$_3$ formation when ferrous iron is generated with ascorbate and ferric ammonium citrate, with no cells present (not shown).

Table 3. Inhibition of ferric ammonium citrate reduction and of ferric diferric transferrin reduction in different cells by cells by apotransferrin.

cell type	oxidant concentration µM	apotransferrin concentration µM	Fe(BPS)3 formation nmole min-1 cell-6
rat liver	Fe2 Tf* 3.4	0	5.7
rat liver	Fe2 Tf 3.4	1.7	0.7
HeLa S3	FAC 7.5	0	0.41
HeLa S3	FAC 7.5	7	0.04
CC1-39	Fe2 Tf 17	0	0.28
CC1-39	Fe2 Tf 17	14	0.06
CC1-39	FAC 7.5	0	0.41
CCI-39	FAC 7.5	14	0.09

Incubation in media A. FAC, ferric ammonium citrate.*
Note that CC1-39 cells show no apotransferrin binding at pH 7.4
*Fe2 Tf preparation used contained 1 mole/mole extra iron i.e. ascorbate reducible iron.

Relative transplasma membrane reductase activity with different prepations of diferric transferrin. The diferric transferrin preparations used in these studies have been prepared from apotransferrin or have been purchased as the holotransferrin (iron saturated) form. Most preparations had the expected ratio of absorbance at 465/280 nm of 0.043. There are significant differences in the activity observed with some preparations which is not correlated with the 465/280 ratio. The activity of the preparation with the lowest activity could be increased greatly by addition of ferric ammonium citrate whereas the most active preparation showed little increase with added ferric iron. This suggests the presence of excess loosely bound iron in the most active preparation. (Table 4).

Table 4. Relative transmembrane reductase activity with different preparations of diferric transferin with and without addition of ferric ammonium sulfate.

Diferric transferrin µM	FAC added µM	Fe(BPS)3 formation nmole min-1 10-6
Fe2 Tf II 17	0	0.11
Fe2 Tf I 17	0	0.01
Fe2 Tf II 17	7.5	0.11
Fe2 Tf I 17	7.5	0.20

Incubation in media A
FAC, ferric ammonium citrate
Fe2 Tf I, diferric transferrin preparation without extra iron
Fe2 Tf II, preparation containing 1 mole/mole extra iron.

Determination of loosely bound iron in the two preparations showed that the active preparation contains 1 µmole of ascorbate-reducible iron per µmol of diferric transferrin, (equivalent to triferric transferrin). The inactive preparation contained no ascorbate-reducible iron. The activity can therefore be primarily based on the presence of iron associated with a low affinity binding site on the diferric transferrin. The rapid reduction of the iron associated with diferric transferrin by liver cells allows the reduction to run to completion with a small amount of added diferric transferrin. Reduction of the iron in the diferric transferrin preparation which contains extra iron showed the formation of one µmole Fe(BPS)$_3$ per µmole added diferric transferrin which indicates that all of the loosely bound iron (1 µmole per µmole) is reduced by the cells.

Effect of argon on the formation of Fe(BPS)$_3$ from diferric transferrin. Since transferrin can act as a ferrous iron oxidase[7] and oxygen may compete with BPS for the ferrous iron formed by the redox system[8] the formation of Fe(BPS)$_3$ was measured under argon. With liver cells and HeLa S3 cells the rate of formation of Fe(BPS)$_3$ was increased under argon. With L1210 cells the rate was decreased (Table 5).

Table 5. Effect of argon on formation of Fe(BPS)3 in presence of diferric transferrin with different types of cells.

cell type	condition	Fe(BPS)2 formation nmole min-1 milj-1	per cent increase
HeLa S3	O2 (air)	0.22	
HeLa S3	argon	0.35	59
Liver cells	O2 (air)	7.0 ± 0.3	
Liver cells	argon	8.7 ± 0.7	25
L1210	O2 (air)	0.074	
L1210	argon	0.06	-20

Incubation in media A. Fe2 Tf preparation had 1 mole/mole extra iron.

DISCUSSION

Previous studies have shown that addition of diferric transferrin to cells leads to production of ferrous iron as measured with bathophenanthroline sulfonate (BPS),[4],[8]. Ferric ammonium citrate is also able to react with the redox site to produce ferrous BPS[1]. The evidence for the reduction of iron in diferric transferrin included direct measurement of the absorbance of the diferric transferrin complex at 465 nm. This change in absorbance of diferric transferrin is specific for transferrin iron reduction and is clear evidence for reduction of tightly bound iron by the redox enzyme. We have interpreted this to mean that the

157

transferrin reduction is not based on free iron complexes
in the solution. One aspect which we did not consider was
if extra iron associated with the transferrin at a less
tight binding site than the high affinity transferrin sites
could be reduced by the transmembrane enzyme. Loosely bound
iron on the diferric transferrin would remain inactive in
the presence of antitransferrin antibodies[2] since a close
association between membrane and iron carried on the transferrin would be blocked.

The dramatic difference in the rate of Fe $(BPS)_3$ formation with different preparations of diferric transferrin
which are equally iron saturated, 465/280=0.043, at the
high affinity sites [9] suggests that some stimulatory factor
is present in the active form or the inactive form may contain an inhibitor or be lacking in loosely bound iron. Measurement of ascorbate reducible iron by Fe$(BPS)_3$ formation
with the most active diferric transferrin preparation shows
one mole of loosely bound iron per mole of diferric transferrin. The diferric transferrin preparation with low activity had no ascorbate reducible iron. Since apotransferrin
inhibits ferric ammonium citrate reduction and is also
capable of tight binding of iron at alkaline pH, the partial inhibition by apotransferrin of Fe $(BPS)_3$ formation
from diferric transferrin may also be explained on the
basis of removal of loosely bound iron from the diferric
transferrin. If the apotransferrin inhibition was based on
competition for the transferrin receptor then kinetic analysis should show competitive effects. A Lineweaver-Burk
analysis of diferric transferrin reduction with HeLa cells
did not show competive inhibition by apotransferrin[10]. The
increase in the Fe$(BPS)_3$ formation rate observed when active diferric transferrin preparations were added to cells
together with ferric ammonium citrate[1] indicates that the
loose iron on diferric transferrin was reduced at a different site than the ferric ammonium ions or that the transferrin receptor was involved in the reduction reaction.
With Hela cells the ferric ammonium citrate reduction was
essentially saturated at 4 µM (Km 0.7 µM) (fig. 1,2). Addition of diferric transferrin then gave a 2 fold increase in
the reduction rate over the maximum rate with ferric ammonium citrate alone. On the other hand, with L1210 cells,
the addition of diferric transferrin with the ferric ammonium citrate inhibited the rate of Fe$(BPS)_2$ formation.

Ferrous iron in the presence of transferrin was oxidized by oxygen to ferric when the iron was rebound to the
transferrin[11].
The ferrous iron produced by the reductase may be partially
reoxidized by binding to transferrin when oxygen was present. This would decrease the rate of Fe$(BPS)_3$ formation.
In the presence of argon the full rate of Fe$(BPS)_3$ formation was observed. It was also possible that argon increased the level of cytosolic NADH to stimulate the ferric
reductase.

The uptake of diferric transferrin into endosome
resicles derived from the plasma membrane allows release of

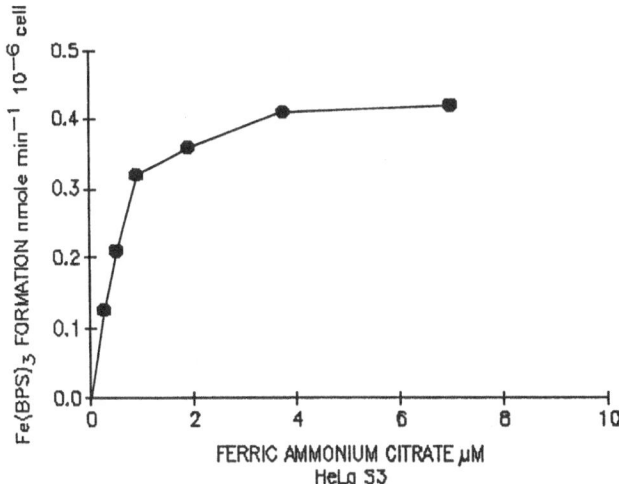

Fig. 1. Effect of ferric ammonium citrate
concentration on the rate of ferrous
BPS formation in the presence of
HeLa S3 cells.

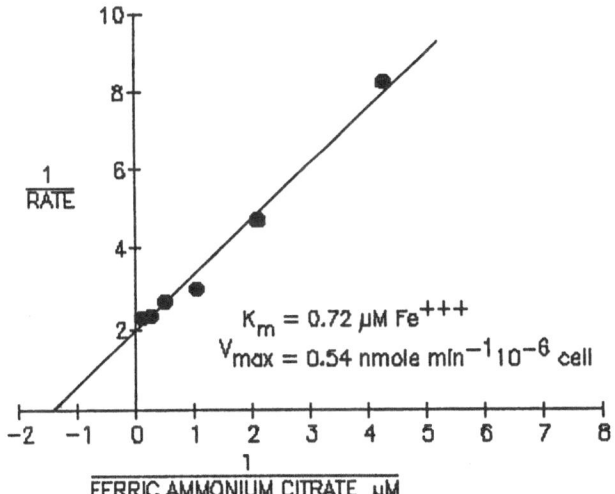

Fig. 2. Lineweaver-Burk analysis of the effect
of ferric ammonium citrate concentration
on the rate of ferrous BPS formation by
HeLa S3 cells.

the tightly bound iron in the acidic endosome[12]. The released iron can then be reduced to ferrous iron for passage
through the cell membrane by the transmembrane reductase.
This is shown by the fact that the reductase is still functional with a pH of 5.5 at the cell surface (Table 6).

**Table 6.   Effect of pH on ferric reductase activity in 3T3
              cells with and without apotransferrin present.**

pH	Electron acceptor, 7.5 μM	Apo Tf 7.5 μM	Rate of reduction nmol/min/10-6 cells
5.7	FAC,	-	0.38
5.7	Fe_2 Tf	-	0.47
5.7	FAC,	+	0.26
5.7	Fe_2 Tf	+	0.12
7.4	FAC	-	0.92
7.4	Fe_2 Tf	-	0.0
7.4	FAC	+	0.32

Incubation in media A at pH 7.4. At pH 5.7 HEPES in media A
was changed for MES, 2-(N-mospholino) ethanesulfonic acid,
FAC, ferric ammonium citrate. Fe_2 Tf contained no free
iron.

In conclusion, the transplasma membrane electron
transport system is very active as a ferric reductase[13] and
has a high affinity for ferric iron (0.7μm Km). It carries
out a very slow reduction of tightly bound iron at the two
high affinity sites in transferrin but shows excellent activity with iron bound to transferrin at low affinity sites
or iron released from transferrin at acid pH.

Supported in part by grants from the National Institutes of Health PO1 CA36761 and GMK6-21839.

References

1. H. Löw, C. Grebing, A. Lindgren, M. Tally, I.L. Sun and
   F.L. Crane. Involment of transferrin in the reduction of
   iron by the transplasma membrane electron transport system,
   J. Bioenerg. Biomemb. 19, 535-549 (1987).
2. I.L. Sun, P. Navas, F.L. Crane, D.J. Morré and H. Löw. NADH
   diferric transferrin reductase in liver plasma membrane,
   J. Biol. Chem. 262, 15915-15921 (1987).
3. M.G. Clark, E.J. Partick, G.S. Patten, F.L. Crane, H. Löw
   and C. Grebing. Evidence for extracellular reduction of
   ferricyanide by rat liver. Biochem. J. 200, 565-572 (1981).
4. H. Löw, I.L. Sun, P. Navas, C. Grebing, F.L. Crane and D.J.
   Morré. Transplasma membrane electron transport is part of a
   diferric transferrin reductase system. Biochem. Biophys.
   Res. Communic. 139, 1117-1123 (1986).

5. I.L. Sun, F.L. Crane, C. Grebing and H. Löw. Properties of a transplasma membrane electron transport system in HeLa cells. Bioenerg. Biomemb. 16, 583-595 (1984).
6. M. Karin and B. Mintz. Receptor-mediated endocytosis of transferrin in developmentally totipotent mouse teratocarcinoma stem cells. J. Biol. Chem. 256, 3245-3252 (1981).
7. D.C. Harris A.L. Rinehart, D. Hereld, R.W. Schwartz, F.P. Burke and A.P. Salvador. Reduction potential of iron in transferrin. Biochim. Biophys. Acta 838, 295-301 (1985).
8. K. Thorstensen and I. Romslo. Uptake of iron from transfer rin by isolated hepatocytes. Biochim. Biophys. Acta 804, 200-208 (1984).
9. P. Aisen and I. Listowski. Iron transport and storage pro teins. Ann. Rev. Biochem. 49, 357-393 (1980).
10. W. Toole-Simms. PhD Thesis Purdue University 1988 pp.
11. G.W. Bates, E.F. Workman and M.R. Schlabach. Does transferrin exhibit ferroxidase activity. Biochem. Biophys. Res. Communic. 50, 84-90 (1973).
12. J.N. Octave, Y.J. Schneider, A. Trouet and R.R Crichton. Iron uptake and utilization by mamalian cells: cellular uptake of transferrin and iron. Trends in Biochem. Sci. 8, 217-220 (1983).
13. M.D. Moody and H.A. Dailey. Ferric iron reductase. J. Bacteriol. 163, 1120-1125 (1985).

MODULATION OF TRANSFERRIN RECEPTOR GENE EXPRESSION BY PEPTIDE GROWTH FACTORS AND ROLE OF THE RECEPTOR IN HUMAN FIBROBLASTS

Paul Basset

Laboratoire de Génétique Moléculaire des Eucaryotes du CNRS, Unité 184 de Biologie Moléculaire et de Génie Génétique de l'INSERM, Faculté de Médecine, 11 rue Humann, 67085 Strasbourg Cédex, France*

ABSTRACT

The addition of serum or of the peptide growth factors PDGF, EGF and insulin in serum-free conditions to resting human HFL-1 fibroblasts stimulates cell proliferation and increases transferrin receptor mRNA levels and transferrin receptor numbers. The upregulation of transferrin receptor mRNA levels by growth factors is controlled at a level after transcription and is still observed when fibroblast proliferation is prevented by DNA synthesis inhibitors. Blocking of the transferrin receptor by 42/6 antibody does not influence the triggering of fibroblast proliferation by PDGF + EGF + insulin. Thus, in contrast to what has been observed with mitogen-stimulated human lymphocytes, transferrin receptor expression and upregulation in human diploid fibroblasts does not appear to be involved in the transduction of growth factor signals which trigger cell proliferation.

INTRODUCTION

Under normal circumstances cellular iron uptake is mediated by iron-transferrin through its interaction, at the cell surface, with a specific 180 kd glycoprotein receptor (for review see reference 1). Following internalization, transferrin receptor is not degraded but recycles to the cell surface together with apotransferrin, which is released [2-3]. Thus, given ample supplies of iron-transferrin in the extracellular milieu, the movement of iron within the cell is modulated by the total number and the function of transferrin receptors involved in the cycling process. In this regard, numerous studies have demonstrated that transferrin receptor expression is upregulated in proliferating cells[1], [4-7], an observation consistent with the knowledge that iron is an essential element for cell proliferation[8].

These observations have lead to the concept that the level of transferrin receptor expression may play a role in the control of cell proliferation. In this regard, it has been demonstrated that in mitogen-stimulated human T lymphocytes transferrin receptor expression is a prerequisite for DNA synthesis and is upregulated by interleukin-2 [9-11].

---

* Previous address : Pulmonary Branch, NHLBI, NIH, Bethesda, USA.

To evaluate whether comparable results are observed with cells other than lymphocytes, we have analyzed the relationships existing between transferrin receptor gene expression and cell proliferation in normal human diploid fibroblasts stimulated by growth signals.

METHODS

Fibroblast cultures

The normal human diploid fetal lung fibroblast strain HFL-1 (ATCC, CCL153) was cultured in DMEM supplemented with 10 % calf serum. Unless otherwise indicated, all studies were performed with density-arrested fibroblasts kept in serum-free medium (iron-free DMEM supplemented with 1 mg/ml bovine serum albumin and 10 mM HEPES). After 24 hr, fibroblasts were stimulated by growth signals (10 % calf serum or peptide growth factors in serum-free conditions, as indicated) and evaluated for transferrin receptor number, transferrin receptor mRNA level, transferrin receptor gene transcription rate, iron uptake and proliferation, as described below.

Transferrin Receptor Numbers

Total cell-associated transferrin receptor numbers were evaluated either in detergent cell extracts, as described by Rao et al[12], or in intact fibroblasts, as described by Ward et al[13]. In both conditions, receptor number was obtained by measuring the binding at saturation of diferric [I^{125}] transferrin.

Transferrin Receptor mRNA levels

Transferrin receptor mRNA levels per cell were quantified using cytoplasmic mRNA dot hybridization, according to White and Bancroft[14]. Filters were hybridized with a ^{32}P-labeled human transferrin receptor cDNA probe [pcD-TR-1, provided by A. McClelland and F. Ruddle[15]], followed by autoradiography.

In Vitro Nuclear Transcription Assays

To evaluate the relative rate of transcription of the transferrin receptor gene under conditions associated with growth signal-induced increase in transferrin receptor mRNA levels, nuclear transcription elongation assays were performed. In brief, nuclei from density arrested fibroblasts stimulated, or not, by serum were isolated as described by Mitchell et al. [16], and nuclear transcription elongation assays carried out according to Clayton and Darnell[17]. The ^{32}P-labeled RNAs were prepared as described by Rao et al. [18] and hybridized to specific cDNAs (transferrin receptor, γ-actin and $\alpha1$-antitrypsin) blotted onto nitrocellulose filters. After hybridization, filters were washed, treated with RNase A and proteinase K, and exposed for autoradiography.

Fibroblast Proliferation

DNA synthesis was evaluated using a 2 hr pulse of [^3H] thymidine (2 μCi/ml, 2.0 Ci/mmol) incorporation 16 to 18 hr after growth signal addition to density arrested fibroblasts. In this assay, 40 ng/ml PDGF + 30 μg/ml insulin or 10 ng/ml EGF + 30 μg/ml insulin gave a similar level of incorporation as 10 % serum. In some experiments, cell proliferation was evaluated by counting cell number. Fibroblasts were plated at low density (50000 cells/ml) and cell numbers determined after 5 days of culture in the presence of 10 % calf serum dialyzed against citrate-acetate (twice), followed by PBS (twice).

Iron uptake

Iron uptake by HFL-1 fibroblasts was evaluated by supplementing the culture medium with $[^{59}Fe]$ ferrous citrate (1 µCi/ml, 16 µCi/µg), in the presence, or not, of 2.5 µg/ml human apotransferrin.

Effects of Transferrin Receptor Blockade on HFL-1 Fibroblasts

The effects of transferrin receptor blockade were evaluated on fibroblasts using the 42/6 antibody, an IgA antibody to the human transferrin receptor raised by Trowbridge[19]. The effects of this antibody were tested on transferrin binding, iron uptake, DNA synthesis and cell division. In all experiments, the antibody (0-40 µg/ml) was added to fibroblast cultures 1hr before other test agents.

RESULTS

Expression of Transferrin Receptor and Transferrin Receptor mRNA in Quiescent and Proliferating Fibroblasts

Stimulation of fibroblasts by growth signals was associated with a significant increase of transferrin receptor mRNA level and of transferrin receptor number (Figure 1). The maximal upregulations were observed

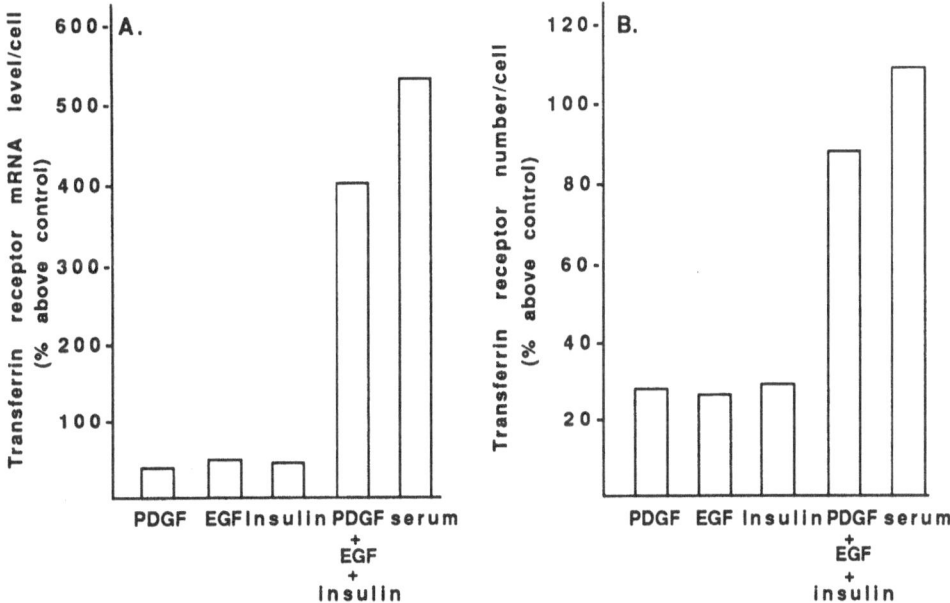

Fig. 1. Upregulation of transferrin receptor mRNA level and of transferrin receptor number by growth signals in HFL-1 fibroblasts. (A) Transferrin receptor mRNA levels. Density-arrested fibroblasts were cultured in the absence or presence of PDGF (40 ng/ml), EGF (10 ng/ml), insulin (30 µg/ml) or 10 % serum for 15 hr. Transferrin receptor mRNA levels were evaluated by cytoplasmic dot hybridization, using a ^{32}P-labeled transferrin receptor cDNA probe. (B) Transferrin receptor numbers. Density-arrested fibroblasts were cultured in the presence of growth signals, as indicated in panel A, for 24 hr. Transferrin receptor numbers were evaluated in detergent cell extracts by measuring the binding at saturation of human diferric ^{125}I-labeled transferrin.

respectively 15hr and 24hr following growth signal addition. Transferrin receptor mRNA levels rapidly declined after 15hr, while the number of transferrin receptors stayed at a higher level than in unstimulated fibroblasts, for at least an additional 24hr (not shown). In both cases, in serum-free conditions, PDGF, EGF and insulin individually each caused mild increases while together these growth factors demonstrated additive effects (Figure 1). Comparison of the upregulations induced by PDGF + EGF + insulin to the upregulations induced by serum demonstrated that the increases caused by the peptide growth factors represented approximately 80 % of the increases by serum. Taken together, these observations support the concept that a major part, if not all, of the increase of transferrin receptor numbers observed in the confluent fibroblasts after serum addition results from an increase of transferrin receptor mRNA levels by peptide growth factors present in serum. Furthermore, the upregulation of transferrin receptor mRNA levels by growth signals is still observed when cell proliferation is blocked by the addition of DNA synthesis inhibitor [20], indicating that this upregulation is not simply a consequence of fibroblast proliferation, but is induced by the growth factors per se.

Transcription Rate of Transferrin Receptor Gene in Fibroblasts

In vitro nuclei transcription assays were performed to evaluate the concept that the increase of fibroblast transferrin receptor mRNA levels induced by growth signals resulted from increased rates of transferrin receptor gene transcription. The transferrin receptor gene was transcribed both in quiescent and proliferating fibroblasts. However, its rate of transcription was not significantly different in the two conditions (Figure 2). As controls, the α1-antitrypsin gene, normally not expressed in fibroblasts, was not transcribed, while the γ-actin gene was transcribed at a higher rate than the transferrin receptor gene. These results suggest that the increase of transferrin receptor mRNA levels induced in HFL-1 fibroblasts by growth signals is likely controlled at a level after transcription, in contrast to the modulation of transferrin receptor mRNA level by intracellular iron in human leukemic cells[18].

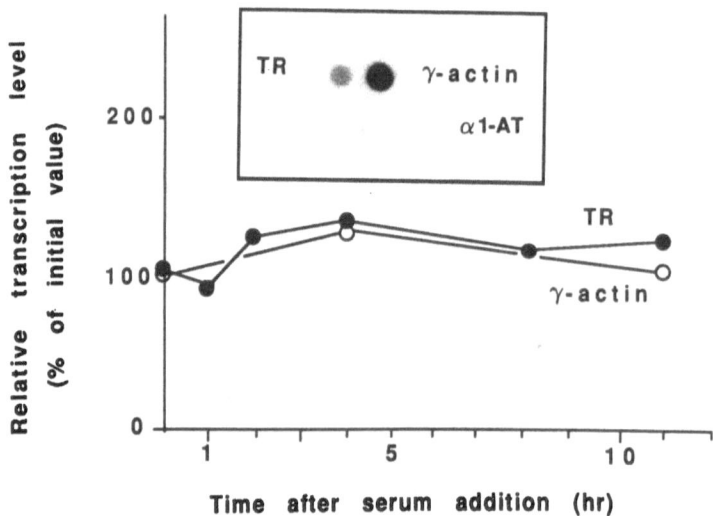

Fig. 2. Transcription rate of the transferrin receptor gene in HFL-1 fibroblasts. Density-arrested fibroblasts were cultured in the presence of 10 % serum for the times indicated. Nuclear transcription assays were performed as described in "Methods". Insert: an example of the dot blot autoradiographs for transcription rate analysis of the transferrin receptor gene (TR), the γ-actin gene and the α1-antitrypsin gene (α1-AT).

Effects of Transferrin Receptor Blockade on Fibroblast Proliferation

While transferrin receptor number increased together with the triggering of fibroblast proliferation by growth signals, several lines of evidence suggested that transferrin receptor upregulation and fibroblast proliferation might be independently modulated. First, the levels of fibroblast proliferation induced by PDGF + EGF + insulin were not significantly affected by the omission of exogenous transferrin in the culture medium (not shown). Second, 42/6 antibody to transferrin receptor,

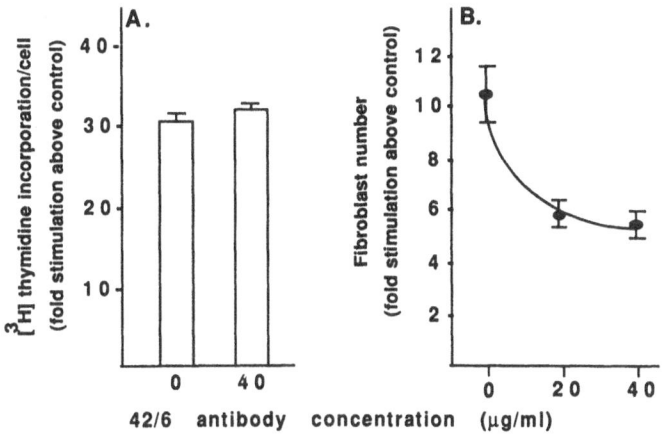

Fig. 3. Effects of transferrin receptor blockade on HFL-1 fibroblast proliferation. (A) Effect of 42/6 antibody on the triggering of DNA synthesis by peptide growth factors. Density-arrested fibroblasts, incubated alone or with 42/6 antibody (40 µg/ml) for 1 hr, were stimulated by PDGF (40 ng/ml)+EGF (10 ng/ml) + insulin (30 µg/ml). DNA synthesis was determined 16-18 hr after growth factor addition by a 2 hr pulse of [^{3}H] thymidine incorporation (B) Effect of 42/6 antibody on cell division. Fibroblasts, plated at low density (50 000 cells/ml) alone or in the presence of 42/6 antibody (20-40 µg/ml), were stimulated by 10 % dialyzed serum. Cell numbers were determined after 5 days of culture.

while blocking transferrin binding to HFL-1 fibroblasts, as has been demonstrated in other cell types [21], did not affect the triggering of DNA synthesis by peptide growth factors added to density-arrested fibroblasts (Figure 3A). Only a mild decrease (approximately 50 %) of proliferation rate was observed when 42/6 antibody was added to sparse fibroblast cultures (Figure 3B), in which several rounds of cell division were occuring. These results contrast with the strong inhibition of proliferation induced by 42/6 antibody in PHA-activated human T lymphocytes [10-11].

Iron Uptake by Fibroblasts

To evaluate whether transferrin-mediated iron uptake was the only possible source of iron for HFL-1 fibroblasts, iron uptake using [^{59}Fe] ferrous citrate were performed in the absence or in the presence of exogenous transferrin in the culture medium. Interestingly, a significant amount of ^{59}Fe was taken up by fibroblasts even in the absence of transferrin (Figure 4A). Furthermore, transferrin-mediated uptake was strongly decreased in the presence of 20 µg/ml 42/6 antibody, while transferrin-independent transport was not affected by the antibody (Figure 4B).

DISCUSSION

The present study demonstrates that the addition of PDGF, EGF and insulin to human diploid fibroblasts increases the number of transferrin receptors. This effect of PDGF, EGF and insulin is distinct from the control these growth factors exert on transferrin receptor internalization [22-24], which is a short term effect, characterized by a translocation of preformed intracellular receptors to the cell surface, and does not involve synthesis of new receptors. The observed increase of transferrin receptor number in HFL-1 fibroblasts is preceded by an increase of transferrin receptor mRNA levels, consistent with the concept that the

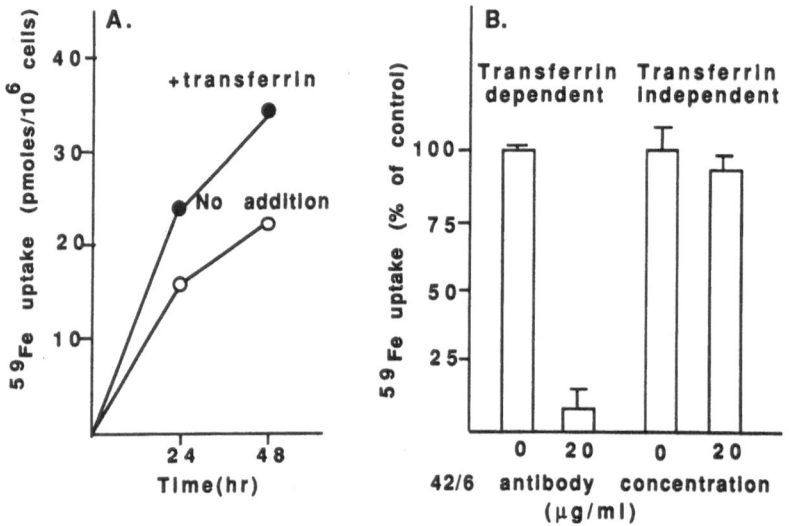

Fig. 4. Iron uptake by HFL-1 fibroblasts. (A) Density-arrested fibroblasts in serum-free conditions with 40 ng/ml PDGF + 10 ng/ml EGF + 30 µg/ml insulin, supplemented or not with apotransferrin (2.5 µg/ml), were cultured in the presence of [^{59}Fe] ferrous citrate (1 µCi/ml). At the times indicated, the fibroblasts were harvested and ^{59}Fe taken up by the cells determined. (B) Density arrested fibroblasts, supplemented or not with 42/6 antibody (20 µg/ml), were cultured for 24 hr and harvested, as indicated in panel A. Transferrin dependent iron uptake was evaluated as the difference between iron uptake observed in the presence and in the absence of apotransferrin.

upregulation of transferrin receptor expression at the protein level may be modulated by the level of transferrin receptor mRNA. Taken together, these results indicate that, as for the effects of interleukin-2 on transferrin receptor gene expression in lymphocytes [9-11,25,26], peptide growth factors known to signal fibroblasts to proliferate also upregulate transferrin receptor mRNA levels and transferrin receptor numbers in HFL-1-fibroblasts. However, the analogy between lymphocytes and fibroblasts cannot be further expanded. In lymphocytes, blocking of transferrin receptor by 42/6 antibody prevents interleukin-2 dependent DNA synthesis, while in fibroblasts the triggering of cell proliferation by peptide growth factors is not affected by 42/6 antibody. In HFL-1 fibroblasts,

42/6 antibody only slows proliferation when several rounds of cell division are involved. In this regard, diploid human fibroblasts appear to be similar to other non-hematopoietic human cells in which proliferation cannot be prevented by blocking of transferrin receptor [21],[27], in contrast to the potent inhibition which is observed in human hematopoietic cells [10],[11],[19],[28],[29].

In this context, the observation that HFL-1 fibroblasts are capable of taking up significant amounts of iron independent of transferrin might be important. Transferrin-independent iron transport systems have been characterized both in cultured cells [30],[31] and in animals [32],[33] and they can coexist with transferrin-mediated iron uptake [31]. However, the existence of transferrin-independent iron uptake is not sufficient to explain why some cells are capable of proliferation while transferrin receptor is blocked, since lymphocytes, in which proliferation is prevented by 42/6 antibody, can also acquire iron from low molecular weight chelates [34]. This suggests that the variation in the sensitivity to growth inhibition by blocking of transferrin receptor is more likely related to how some cells can efficiently utilize iron provided independently of transferrin [1], or to the fact that the role of diferric transferrin in some cells is not limited to iron transport [35].

Taken together, these results indicate that the role of transferrin receptor in non-hematopoietic cells should be envisaged independent of the modulation of cell proliferation. This role might be to supply, when needed, iron to cells in a non toxic form [32],[33], a concept consistent with the observation than in normal human fibroblasts transferrin receptor gene expression is a controlled process.

ACKNOWLEDGEMENTS

This work was carried out in the laboratory of Dr. R.G. Crystal, Chief of the Pulmonary Branch of NHLBI, at NIH, to whom we are indebted for support, advice and critical discussions. We are also grateful to Dr. I. Trowbridge, The Salk Institute, San Diego, for providing the 42/6 antibody used in this study.

REFERENCES

1. I. S. Trowbridge, and D. A., Shackelford, Structure and function of transferrin receptors and their relationship to cell growth, Biochem. Soc. Symp. 51:117 (1986).
2. A. Dautry-Varsat, A. Ciechanover, and H. F. Lodish, pH and the recycling of transferrin during receptor-mediated endocytosis, Proc. Natl. Acad. Sci. USA 80:2258 (1983).
3. R. D. Klausner, G. Ashwell, J. Van Renswoude, et al., Binding of apotransferrin to K562 cells: explanation of the transferrin cycle, Proc. Natl. Acad. Sci. USA 80:2263 (1983).
4. J. W. Larrick, and P. Cresswell, Modulation of cell surface iron transferrin receptors by cellular density and state of activation, J. Supramol. Struct. 11:579 (1979).
5. T. A. Hamilton, H. Garret Wada, and H. H. Sussman, Identification of transferrin receptors on the surface of human cultured cells, Proc. Natl. Acad. Sci. USA 76:6406 (1979).
6. G. M. P. Galbraith, R. M. Galbraith, and W. P. Faulk, Transferrin binding by human lymphoblastoid cell lines and other transformed cells, Cell. Immunol. 49:215 (1980).

7. I. S. Trowbridge, and M. Bishr Omary, Human cell surface glycoprotein related to cell proliferation is the receptor for transferrin, Proc. Natl. Acad. Sci. USA 78:3039 (1981).
8. R. J. Bergeron, Iron: a controlling nutrient in proliferative processes, Trends Biochem. Sci. 11:133 (1986).
9. T. A. Hamilton, Regulation of transferrin receptor expression in concanavalin A stimulated and Gross virus transformed rat lymphoblasts, J. Cell. Physiol. 113:40 (1982).
10. J. Mendelsohn, I. Trowbridge, and J. Castagnola, Inhibition of human lymphocyte proliferation by monoclonal antibody to transferrin receptor, Blood 62:821 (1983).
11. L. M. Neckers, and J. Cossman, Transferrin receptor induction in mitogen-stimulated human T lymphocytes is required for DNA synthesis and cell division and is regulated by interleukin 2, Proc. Natl. Acad. Sci. USA 80:3494 (1983).
12. K. K. Rao, D. Shapiro, E. Mattia et al., Effects of alterations in cellular iron on biosynthesis of the transferrin receptor in K562 cells, Mol. Cell. Biol. 5:595 (1985).
13. J. H. Ward, J. Jordan, J. P. Kushner, and J. Kaplan, Heme regulation of HeLa cell transferrin receptor number, J. Biol. Chem. 259:13235 (1984).
14. B. A. White, and F. C. Bancroft, Cytoplasmic dot hybridization. Simple analysis of relative mRNA levels in multiple small cell or tissue samples, J. Biol. Chem. 257:8569 (1982).
15. A. McClelland, L. C. Kuhn, and F. H. Ruddle, The human transferrin receptor gene: genomic organization, and the complete primary structure of the receptor deduced from a cDNA sequence, Cell 39:267 (1984).
16. R. L. Mitchell, C. Henning-Chubb, E. Huberman, and I. M. Verma, c-fos expression is neither sufficient nor obligatory for differentiation of monomyelocytes to macrophages, Cell 45:497 (1986).
17. D. F. Clayton, and J. E. Darnell, Changes in liver-specific compared to common gene transcription during primary culture of mouse hepatocytes, Mol. Cell. Biol. 3:1552 (1983).
18. K. Rao, J. B. Harford, T. Rouault, et al., Transcriptional regulation by iron of the gene for the transferrin receptor, Mol. Cell. Biol. 6:236 (1986).
19. I. S. Trowbridge, and F. Lopez, Monoclonal antibody to transferrin receptor blocks transferrin binding and inhibits human tumor cell growth in vitro, Proc. Natl. Acad. Sci USA 79:1175 (1982).
20. P. Basset, K. Yamauchi, and R. G. Crystal, Modulation of transferrin receptor mRNA levels in human fibroblasts by peptide growth factors, Fed. Proceed. 46:1995 (1987).
21. I. S. Trowbridge, and R. A. Newnan, Monoclonal antibodies to transferrin receptors, in: "Receptors and Recognition, Series B, vol. 17, M.F. Greaves, Ed. Chapman and Hall, London (1984).
22. H. S. Wiley, and J. Kaplan, Epidermal growth factor rapidly induces a redistribution of transferrin receptor pools in human fibroblasts, Proc. Natl. Acad. Sci. USA 81:7456 (1984).
23. R. J. Davis, and M. P. Czech, Regulation of transferrin receptor expression at the cell surface by insulin-like growth factors, epidermal growth factor and platelet-derived growth factor, EMBO J. 5:653 (1986).
24. R. J. Davis, and H. Meisner, Regulation of transferrin receptor cycling by protein kinase C is independent of receptor phosphorylation at serine 24 in Swiss 3T3 fibroblasts, J. Biol. Chem. 262:16041 (1987).
25. C. D. Pauza, J. D. Bleil, and E. S. Lennox, The control of transferrin receptor synthesis in mitogen-stimulated human lymphocytes, Exp. Cell. Res. 154:510 (1984).

26. J. M. Depper, W. J. Leonard, C. Drogula, et al., Interleukin 2 augments transcription of the interleukin 2 receptor gene, Proc. Natl. Acad. Sci. USA 82:4230 (1985).

27. R. Taetle, and J.M. Honeysett, Effects of monoclonal anti-transferrin receptor antibodies on in vitro growth of human solid tumor cells, Cancer Res. 47:2040 (1987).

28. R. Taetle, J. M. Honeysett, and I. Trowbridge, Effects of anti-transferrin receptor antibodies on growth of normal and malignant myeloid cells, Int. J. Cancer 32:343 (1983).

29. R. Taetle, K. Rhyner, J. Castagnola, et al., Role of transferrin, Fe, and transferrin receptors in myeloid leukemia cell growth, J. Clin. Invest. 75:1061 (1985).

30. J. A. Fernandez-Pol, Isolation and characterization of a siderophore-like growth factor from mutants of SV40-transformed cells adapted to picolinic acid, Cell 14:489 (1978).

31. P. Basset, Y. Quesneau, and J. Zwiller, Iron-induced L1210 cell growth: evidence of a transferrin-independent iron transport, Cancer Res. 46:1644 (1986).

32. T. L. Wright, P. Brissot, W. L. Ma, and R. A. Weisiger, Characterization of non-transferrin bound iron clearance by rat liver, J. Biol. Chem. 261:10909 (1986).

33. C. M. Craven, J. Alexander, M. Eldridge, et al., Tissue distribution and clearance kinetics of non-transferrin bound iron in the hypotransferrinemic mouse: a rodent model for hemochromatosis, Proc. Natl. Acad. Sci. USA 84:3457 (1987).

34. J. H. Brock, T. Mainou-Fowler, and L. M. Webster, Evidence that transferrin may function exclusively as an iron donor in promoting lymphocyte proliferation, Immunology, 57:105 (1986).

35. I. L. Sun, P. Navas, F. L. Crane, D. J. Morre and H. Löw, NADH Diferric Transferrin Reductase in Liver Plasma Membrane, J. Biol. Chem. 262:15915 (1987).

BASIC AND CLINICAL STUDIES OF TRANSFERRIN-ADRIAMYCIN CONJUGATES:

MECHANISMS FOR A NEW APPROACH TO DRUG TARGETING

W.P. Faulk, D.S. Torry, H. Harats, J.A. McIntyre
and C.G. Taylor[1]

Methodist Center for Reproduction and Transplantation
Immunology, Indianapolis, Indiana 46202
[1]Pembury Hospital, Tunbridge Wells, Kent, TN2 4Q5

INTRODUCTION

Research in drug targeting to cancer cells is almost exclusively limited to monoclonal antibodies. The clinical use of monoclonal antibodies is frought with problems. Three of the most serious problems are antigen shedding, sensitization of the patient to mouse immunoglobulins, and mutation of the antigenic target. Antigen shedding is the antibody-induced disappearance of antigens from cell membranes. Drugs cannot exert cytotoxic effects if their antibody carriers are shed with antigen. The second problem is sensitization of patients to mouse immunoglobulins. Mouse proteins are antigenically foreign to the human immune system. Some patients mount hypersensitivity reactions to mouse immunoglobulins and are in danger of developing anaphylactic responses. The magnitude of this problem is shown by the finding that 50% of leukemia and lymphoma patients treated with monoclonal antibodies develop hypersensitivity reactions.[1] The third problem with monoclonal antibodies relates to their restricted specificity. This is represented by cells that lose or mutate their surface antigens. Such cells are no longer recognized by cell-specific monoclonals and consequently acquire the ability to escape the anti-cancer effects of drugs conjugated to monoclonal antibodies.

What is needed in the field of drug targeting is a carrier that spares normal cells and reacts with neoplastic cells without encountering the above three problems. Transferrin seems to be a promising candidate. Transferrin conjugates would circumvent the problem of hypersensitivity, because it is a normal human plasma protein. In addition, receptors for transferrin are difficult to demonstrate on most resting normal cells, but are easily identified on extraembryonic[2], transformed[3], activated[4] and neoplastic[5,6] cells. The receptor is internalized following transferrin binding[7], thereby obviating the problem of antigen shedding. Finally, antigenic mutation would not be a problem because transferrin is metabolically required by all cells, so such a mutation of the receptor would be incompatible with survival.

The concept of using a receptor targeted drug delivery system prompted us to employ transferrin in the production of drug conjugates. Rationale for their use is the suggestion that the transferrin carrier will be bound by receptor and deliver drug to the targeted cell.[8] To test this concept, we chose to prepare transferrin-adriamycin conjugates.[9] The purpose of this paper is to introduce the idea of using transferrin receptors for drug targeting with conjugates of transferrin, and to present preliminary basic and clinical data obtained from our studies of transferrin-adriamycin conjugates.

STUDIES OF TRANSFERRIN-ADRIAMYCIN CONJUGATES

Conjugates of transferrin with adriamycin were prepared according to the method of Yeh and Faulk.[9] Glutaraldehyde (0.25%) was added dropwise to phosphate buffered saline (PBS) containing human apotransferrin and adriamycin-HCl. The coupling reaction was stopped by the addition of 1.0 M ethanolamine. The conjugate mixture was chromatographed through a 30 x 1.5 cm column of Sepharose CL-6B to separate transferrin aggregates from monomeric transferrin. Spectrophotometric readings of the collected column fractions were taken at 280 nm and 495 nm to identify protein and adriamycin peaks, respectively. Fractions corresponding to monomeric transferrin were pooled and dialyzed against PBS to remove unbound adriamycin.

In Vitro Studies of Conjugates

The conjugates were immunochemically shown to contain transferrin. They were immunocytologically shown to be bound by transferrin receptors on extraembryonic, transformed and malignant cells.[10] The biological activity of conjugates was evaluated by their ability to inhibit tritiated thymidine incorporation by human tumor cells.[9,10] This was done by using peripheral blood mononuclear (PBM) leukocytes from patients diagnosed as having acute myelocytic leukemia (AML). Conjugates were found to decrease tritiated thymidine uptake by 63.3 - 98.0% (Table 1). The effect of transferrin-adriamycin conjugates on normal cells was an order of magnitude less.

Table 1. Effects of Transferrin-Adriamycin Conjugates on [3]H Thymidine Uptake by AML PBM[a]

Patient #	% Inhibition[b]
PM	98
JK	85
JD	54.9
GL	63.3
PG	82.8
JF	69.5

[a] 5×10^5 cells were incubated with 100 μl of conjugate (40 μg transferrin labelled with 2 μg of adriamycin) or complete culture media for 3 hr. at 37°C in 5% CO_2/air, dispensed in triplicate as 100 μl aliquots into tissue culture wells and pulsed with ^{3}H thymidine for 16 hrs. at 37°C, harvested onto filter paper and counted for incorporated [3]H.

[b] % inhibition calculated as:

$$\frac{\text{CPM in conjugate treated}}{\text{CPM in culture media control}} \times 100\%$$

174

The reason for conjugate inhibition of tritiated thymidine utilization might be assumed to be that receptor-bound conjugate was endocytosed and deconjugated, and free drug diffused to nucleus where it intercalated with DNA. Indeed, we have identified adriamycin in nuclei of targeted but not in nuclei of control cells.[9] Recent studies, however, indicate that adriamycin inhibits a plasma membrane reductase which is associated with transferrin receptors.[11] In light of this, we asked Dr. Fredrick Crane at Purdue University to study our conjugates in his diferric transferrin reductase assay.[12] His results revealed that transferrin-adriamycin conjugates were potent inhibitors of HeLa cell diferric transferrin reductase as measured by the bathophenanthroline sulfonate assay. Details of this finding will be reported elsewhere.

In Vivo Study of Conjugates

A preliminary clinical trial was done with patients who were diagnosed by conventional peripheral blood and bone marrow criteria as having AML. Treatment was given for 5 days as a daily 10 minute intravenous push of 1 mg of adriamycin coupled to transferrin (transferrin:adriamycin ratio = 1:3.2). Peripheral blood counts and bone marrow aspirates were monitored before and after these procedures. The patients responded by showing some drop in the number of peripheral blast cells, but the degree of change was not uniform (Table 2). A drop in leukemic cells in peripheral blood counts could be detected within 36 hours of treatment. None of the patients suffered adriamycin toxicity, and no untoward effects were caused by the conjugates. Ten patients thus far have been treated, and a major phase 1 study of transferrin-adriamycin conjugates is being planned at Methodist Hospital in Indianapolis.

Table 2. Effects of in vivo Treatment with Transferrin-
Adriamycin Conjugates

Patients	% Blasts[a]		% PMN[a]	
	Before	After	Before	After
PM	76	30	8	46
BP	22	0	18	36
GL	98	93	1	2
PG	74	72	5	15

[a] These values represent counts done the day before treatment and on the 5th day of treatment.

STUDIES OF TRANSFERRIN RECEPTORS

Expression of transferrin receptors could represent proliferation. To test this, we studied dimethylsulfoxide controlled differentiation of HL-60 tumor cells and found no significant effect on transferrin receptor expression.[13] These results suggest that transferrin receptors on cancer cells are a manifestation of metabolic state and not of proliferation. This conclusion receives substantial support from the observation that transferrin receptors are found on syncytiotrophoblast, which is a non-proliferating tissue.[2] In light of this, we have studied trophoblast transferrin receptors in an attempt to determine if there is a relationship between receptor and diferric transferrin reductase activity.

Preparation of Trophoblast Transferrin Receptors

Human placentae are collected immediately after delivery, stripped of amniochorion and umbilical cord, and syncytiotrophoblast microvilli (STM) are prepared by differential centrifugation.[14] STM pellets are homogenized in PBS (pH 7.4) and stripped of endogenous transferrin by incubation in 0.1 M sodium citrate (pH 5.0) containing 50 μg/ml desferrioxamine followed by incubation in 1 mM Tris-HCl (pH 8.0) containing 1 M NaCl and 50 μg/ml desferrioxamine. Removing endogenous transferrin from the STM vesicles increases the availability of unbound transferrin receptors. Conventional chaotrophic agents (e.g., 3M KCl) remove transferrin from STM with only about 11% efficiency, but we find that employing pH shifts removes 69% of the transferrin (Table 3). This method of transferrin extraction is based on observations that holotransferrin binds strongly to the transferrin receptor at pH 7.4 but apotransferrin does not.[15-18] Thus, holotransferrin is converted to apotransferrin at pH 5 following which we adjust the STM vesicles to pH 8. This allows us to remove apotransferrin from the STM transferrin receptors in 1M NaCl. Our technique is a modification of other methods used to purify transferrin receptors.[19,20]

Table 3. Removal of Endogenous Transferrin from STM Vesicles[a]

Treatment Method[b]	STM (untreated)	STM (treated)	Efficiency
pH5/pH8	.628[c]	.197	69%
3M KCl	.428	.382	11%

[a] ELISA microtiter wells were coated with 100 μg/ml of STM membranes, air dried and fixed with 0.1% glutaraldehyde for 5 min, blocked with 2% bovine albumin, washed and reacted with rabbit anti-human transferrin (DAKO) 1:1000.
[b] pH5/pH8 = 0.1 M Na Citrate pH 5.0 + desferrioxamine for 60 min. followed by 1 mM Tris pH 8.0 + 1M NaCl + desferrioxamine for 30 min. 3M KCl = 16 hour treatment with 3M KCl.
[c] O.D. 450 nm

Transferrin extracted STM vesicles are solubilized in 1% NP-40, dialyzed against PBS and ultracentrifuged to remove nonsolubilized membrane. Solubilized STM preparations are passed through a human holotransferrin-coupled Sepharose column. The column is washed extensively with PBS, and bound transferrin receptors are eluted. Elution was originally done with 0.1M glycine-HCl, pH 3.0. The elution profile at OD 280 nm showed protein was eluted, but we were unable to immunologically detect transferrin receptor by means of using ELISA and monoclonal antibodies to transferrin receptors. We have subsequently found that treatment of intact STM preparations with pH 3.0 for as little as 30 minutes destroys antigenic determinants recognized by characterized monoclonal antibodies[21] to transferrin receptors (Table 4). The pH 3.0 eluted fraction, however, showed diferric transferrin reductase activity when assayed by Dr. Crane and his colleagues (F. Crane, personal communication, 1987). These results suggest an association between solid-phase affinity chromatography prepared transferrin receptors and diferric transferrin reductase.

Table 4.   Effects of pH 3.0 Treatment on STM Transferrin Receptors

MAb GB22[a]	Untreated STM	pH 3.0 treated[b] STM
1:10	0.513[c]	0.101
1:20	0.484	0.102
1:40	0.449	0.077
1:80	0.338	0.074

[a] GB22 — monoclonal antibody to transferrin receptor.
[b] STM coated wells were treated with 0.1 M glycine-HCl pH 3 for 30 min., washed with PBS and reacted with GB22.
[c] O.D. 450 nm

In light of the loss of receptor antigens at pH 3, we now elute transferrin receptors from columns of immobilized transferrin by using the method we developed to remove transferrin from STM vesicles. The column is loaded with solubilized membrane at pH 7.4, enabling soluble transferrin receptors to bind to the holotransferrin-Sepharose matrix. Unbound material is washed from the column at pH 7.4, following which the pH is changed to 5.0 and the washing process is continued until all iron is removed as evidenced by a color change of the column from salmon-pink to white. At this point the pH is changed from 5.0 to 8.0 and receptor is eluted from the column. The fractions from this column elution show the presence oftransferrin receptors by ELISA with monoclonal antibodies to transferrin receptors as well as the presence of diferric transferrin reductase activity as measured by Dr. Crane and colleagues at Purdue. These findings will form the basis of another communication. They support the possibility that transferrin receptors are closely associated with diferric transferrin reductase activity.

DISCUSSION

The ability to target adriamycin to cancer cells introduces 3 possible advances in the treatment of cancer. Firstly, most normal cells by virtue of not expressing transferrin receptors are not affected. Secondly, targeting of the drug enables one to treat patients with much less drug thereby reducing drug complications. In our preliminary clinical study, we injected the equivalent of about 80 times less adriamycin than is generally used in an unconjugated form for treating AML patients. This is clinically relevant because the side effects of adriamycin (e.g., cardiac toxicity) are dose related. Thirdly, the presence of transferrin receptors on other types of cancer cells might extend the use of receptor-targeted drug conjugates to other types of cancers.

One possible mechanism explaining the tumorcidal effect of the conjugate, aside from adriamycin intercalation with DNA, arises from recent biochemical research on cell growth has revealed an interdependent role for ferric ions, transferrin and transferrin receptors in energy metabolism which could contribute to adriamycin mediated cell death.[22] This is related to NADH oxidase and diferric transferrin reductase.[11] These enzymes are driven by the binding of diferric transferrin to transferrin receptors. This signals NADH oxidation which produces a proton and two electrons.[23]

Mechanism 1:

Inhibition of diferric transferrin reductase

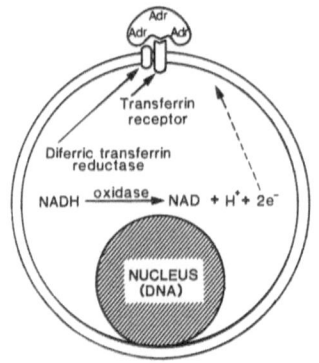

Mechanism 2:

Receptor occupancy by non-electron acceptor apotransferrin

Mechanism 3:

Intercalation with DNA base pairs

Figure 1. Mechanisms for Transferrin-Adriamycin Conjugates

The two electrons are transported across the plasma membrane by diferric transferrin reductase and reduce transferrin-bound ferric ions to diffusible ferrous ions.[24] The proton enters a hydrogen-sodium antiport that pumps out protons, opens calcium gates, elevates cytosol pH and initiates metabolic events that result in cell division.[25] The chain of events can be blocked by inhibiting the diferric transferrin reductase, and adriamycin is an excellent reductase inhibitor.[26]

Another possible mechanism by which the conjugate can inhibit cell growth arises with the finding that the reductase enzyme can also be blocked by substituting apotransferrin for holotransferrin in transferrin receptors.[27] The protein conjugates of adriamycin are prepared with apotransferrin and not holotransferrin[9], thus the conjugates can affect two pathways of membrane-mediated reductase inhibition: namely, adriamycin intoxication and apotransferrin blockade via the absence of electron acceptors. Both mechanisms stop energy metabolism and can lead to cell death. These plasma membrane systems are additive to the well known property of adriamycin to intercalate DNA base pairs and inhibit proliferation. These three mechanisms for targeted cell death by transferrin-adriamycin conjugates are depicted in the accompanying figure. They open a new approach to the possibility of receptor-mediated targeting of reagents to viable cells in vivo.

ACKNOWLEDGEMENTS

We thank Dr. C-J.G. Yeh and Dr. B-L. Hsi of INSERM U210 in Nice for antibody GB22. Dr. Yeh participated in early aspects of this research and produced data in Table 1 and assisted with Table 2. Our collaboration with Purdue has been through the generosity of Prof. D.J. Morre and Prof. F. Crane. Dr. Harats is supported in part by the Methodist Health Foundation.

REFERENCES

1. R. Levy and R.A. Miller, Tumor therapy with monoclonal antibodies, Fed. Proc. 42:2650, (1983).
2. W.P. Faulk and G.M.P. Galbraith, Trophoblast transferrin and transferrin receptors in the host-parasite relationship of human pregnancy, Proc. Roy. Soc. London B. 204:83, (1979).
3. G.M.P. Galbraith, R.M. Galbraith and W.P. Faulk, Transferrin binding by human lymphoblastoid cell lines and other transformed cells, Cell. Immunol. 49:215, (1980).
4. R.M. Galbraith and G.M.P. Galbraith, Expression of transferrin receptors on mitogen-stimulated human peripheral blood lymphocytes: relation to cellular activation and related metabolic events, Immunology 44:703, (1981).
5. C.J.G. Yeh, C.G. Taylor and W.P. Faulk, Transferrin binding by peripheral blood mononuclear cells in human lymphomas, myelomas and leukemias, Vox Sang. 46:217, (1984).
6. J.E. Shindelman, A.E. Ortmeyer and H.H. Sussman, Demonstration of the transferrin receptor in human breast cancer tissue. Potential marker for identifying dividing cells, Int. J. Cancer 27:329, (1981).
7. U. Testa, Transferrin receptors: structure and function, Curr. Top. Hematol. 5:127, (1985).
8. W.P. Faulk, B.L. Hsi and P.J. Stevens, Transferrin and transferrin receptors in carcinoma of the breast, Lancet ii:390, (1980).
9. C.J.G. Yeh and W.P. Faulk, Killing of human tumor cells in culture with adriamycin conjugates of human transferrin, Clin. Immunol. Immunopathol. 32:1, (1984).
10. C.J.G. Yeh, C.G? Taylor and W.P. Faulk, Targeting of cytotoxic drugs by transferrin receptors: Selective killing of acute myelogenous leukemia cells, Prot. Biol. Fluids 42:441, (1984).
11. I. Sun, P. Navas, F. Crane, D.J. Morre and H. Low, NADH diferric transferrin reductase in liver plasma membrane, J. Biol. Chem. 262:15915, (1987).
12. I. Sun, Crane F.L., C. Grebing and H. Low, Properties of a transplasma membrane electron transport system in HeLa cells, J. Bioenerget. Biomembr. 16:583, (1984).
13. C.J.G. Yeh, M. Papamichail and W.P. Faulk, Loss of transferrin receptors following induced differentiation of HL-60 promyelocytic leukemia cells, Exp. Cell Research 138:429, (1982).
14. W.P. Faulk, A. Temple, R.E. Lovins and N. Smith, Antigens of human trophoblasts: A working hypothesis for their role in normal and abnormal pregnancies, Proc. Natl. Acad. Sci. (USA) 75:1947, (1978).
15. B. Ecarot-Charrier, V.L. Grey, A. Wilcyznska and H.M. Schulman, Can. J. Biochem. 58:418, (1980).
16. E.H. Morgan, Biochem. Biophys. Acta 762:488, (1983).
17. A. Dautry-Varsat, A. Crechanover and H.F. Lodish, pH and the recycling of transferrin during receptor mediated endocytosis, Proc. Natl. Acad. Sci. (USA) 80:2258, (1983).
18. R.D. Klausner, G. Ashwell, J. Van Renswoude, J.B. Hartford and K.R. Bridges, Binding of apotransferrin to k562 cells: explanation of the transferrin cycle, Proc. Natl. Acad. Sci. (USA) 80:2263, (1983).
19. G.J. Anderson, A. Macherras, L.W. Powell and J.W. Halliday, Improved purification of the human placental transferrin receptor and a novel immunoradiometric assay for receptor protein, Biochimica et Biophysica Acta 884:225, (1986).
20. I.R. Van Driel, P.A. Stearne, B. Grego, R.J. Simpson and J.W. Goding, The receptor for transferrin on murine myeloma cells: One-step purification based on its physiology, and partial amino acid sequence, J. Immunol. 133:3220, (1984).

21. C.J.G. Yeh, B.L. Hsi, M. Samson, J.A. McIntyre, J.P. Breittmeyer and M. Fehlmann, Monoclonal antibodies (GB16, GB18, GB19, GB22) raised against human placental microvilli recognize the transferrin receptor, Placenta 8:627, (1987).

22. H. Low, C. Grebing, A. Lindgren, M. Tally, I. Sun and F. Crane, Involvement of transferrin in the reduction of iron by the transplasma membrane electron transport system, J. Bioenerget. Biomembr. 19:535, (1987).

23. H. Low, I. Sun, P. Navas, C. Grebing, F. Crane and D.J. Morre, Transplasmalemma electron transport from cells is part of a diferric transferrin reductase system, Biochem. Biophy. Res. Com. 139:1117, (1986).

24. F. Crane, I. Sun, M.G. Clark, C. Grebing and H. Low, Transplasma-membrane redox systems in growth and development, Biochem. Biophy. Acta 811:233, (1985).

25. I. Sun, R. Garcia-Canero, W. Liu, W. Toole-Simms, F. Crane, D.J. Morre and H. Low, Diferric transferrin reduction stimulates the Na+/H+ antiport of HeLa cells, Biochem. Biophy. Res. Com. 145:467, (1987).

26. I. Sun, P. Navas, F. Crane, D.J. Morre and H. Low, Differic transferrin reductase in the plasma membrane is inhibited by adriamycin, Biochem. Internatl. 14:119, (1987).

27. P. Navas, I. Sun, D.J. Morre and F. Crane, Decrease of NADH in HeLa cells in the presence of transferrin or ferricyanide, Biochem. Biophy. Res. Com. 135:110, (1986).

CONTROL OF TRANSPLASMA MEMBRANE DIFERRIC

TRANSFERRIN REDUCTASE BY ANTITUMOR DRUGS

I. L. Sun and F. L. Crane

Department of Biological Sciences, Purdue University
West Lafayette, IN 47907 U.S.A.

INTRODUCTION

It is well known that anthracycline antibiotics have been used successfully in the treatment of human cancer. It has long been thought that
DNA is the primary target for the cytotoxic action of these drugs on susceptible cells (Chandra, 1987; DiMarco, 1975; Kanter et al., 1979; Ross et
al., 1970). The DNA receptor hypothesis is attractive due to a reasonably
high affinity between some of these drugs and nucleic acid. However, this
hypothesis does not fully explain the antimitogenic effect of these drugs,
since significant mitotic inhibition by daunomycin and adriamycin are observed under conditions in which DNA synthesis is unaffected (Silvetrini,
1970,1973). The N-substituted derivatives, such as N-acetyl daunomycin
and N-trifluoro-acetyl adriamycin-14-valerate (AD32) have very low affinity toward DNA, yet they are capable of inhibiting cell mitosis (DiMarco
et al., 1965; DiMarco, 1975; Kanter et al., 1979). Similarly, bleomycin
and its chelate, Cu-bleomycin, inhibit mitosis equally well although
selective strong DNA binding affinity is shown only by bleomycin (Sun
and Crane, 1985a; Gutteridge and Chang, 1981). Furthermore, both cis and
trans diammine dichloroplatinum II bind equally well with DNA, yet the
trans analog is not effective as an antitumor agent (Macquet and Butour,
1985; Salles et al., 1983) and is much less cytotoxic than the cis analog
(Pascoe and Roberts, 1974). Therefore, there are reasons to suspect that
additional targets other than DNA are involved in killing cancer cells
with these drugs.

Effects of the drug on the surface properties of sarcoma cells (Murphree et al., 1976) and drug-cytoskeleton interaction (Na and Timasheff,
1977) have been described. In addition, anthracycline antibiotics affect
microsomal electron transport (Bachur et al., 1979). Recently, Tritton
and Yee have shown that adriamycin bound to agarose beads, which cannot
enter the cells, retains its cytotoxicity to sarcoma cells (Tritton and
Yee, 1982; Tritton et al., 1983). Likewise the cytostatic action of
adriamycin coupled to polyglutaraldehyde microspheres has been demonstrated on human leukemia cells and rat hepatocytes (Tokes et al., 1982).
Therefore, there is clear evidence that the killing effect of these drugs
may be achieved solely by action at the cell surface.

The transmembrane redox system has been shown directly to stimulate
the growth of melanoma cells (Ellem and Kay, 1983) and HeLa cells (Sun et
al., 1985b) by replacing the essential components of calf serum. In this
paper we describe the effect of anthracycline antibiotics and some other

anticancer drugs on plasma membrane redox enzymes at concentrations equivalent to those used in therapy and inhibition of cell division (Waring,
1981). It is concluded that plasma membrane is a promising target of the
cytotoxic action for these anticancer agents, due to its great importance
in regulating cellular metabolic functions.

METHODS

HeLa cells were grown in flasks with an α-modified minimal essential
medium containing 10% fetal calf serum, 100 μu of penicillin and 170 μg of
streptomycin per ml at pH 7.4 and maintained in a similar medium containing 2% fetal calf serum. Cells were prepared for study by pelleting the
trypsinized suspension cultures at 27,000 g. The pellet was diluted with
TD-Tris buffer (NaCl 8g/l, KCl 0.34g/l, Na_2HPO_4 0.1g/l and Trizma base
3g/l, pH 7.5) to a final concentration of 0.1 gm cells/ml.

A cell model system used in this study is a simian virus 40 (SV40)
temperature sensitive (ts) A209 virus-transformed fetal rat liver
RLA209-15 cell line and rat pineal gland RPNA209-1 cell line that are
temperature sensitive for growth and differentiation. They exhibit the
transformed phenotype at the permissive temperature (33°C) but mimic the
normal, nontransformed cells at the nonpermissive temperature (40°C).
Cell cultures were grown in flasks under conditions as described previously (Sun et al., 1986a,1986b; Chou and Schlegel-Haueter, 1981).

Reduction of iron in diferric transferrin by cells was measured as
described (Löw et al., 1986). The rate of ferricyanide reduction by cells
was determined by previously used methods (Sun et al., 1985b).

Ferricyanide induced proton generation was measured in a 2 ml cuvette
with an Orion 701 A pH meter and a Corning glass combination electrode.
Cells were suspended in a salt sucrose solution (10 mM KCl, 10 mM $CaCl_2$,
0.1 M sucrose and 5% TD-Tris buffer), to a final concentration of 0.005
grams wet weight per ml (g.w.w./ml). Sample was stirred continuously and
air was bubbled through the reaction mixture to remove CO_2. After the pH
came to an equilibrium, 0.15 mM ferricyanide was added. The proton generation was measured by the change in pH over the range from pH 7.0 to pH 7.4.

For drug treatment cells were harvested during the exponential growth
phase. The cells were suspended in TD-Tris buffer plus 2% fetal calf
serum and incubated with various concentrations of drugs at 37°C for 1
hour with shaking. After incubation the cells were chilled in an ice bath
and diluted 10-fold with ice-cold TD-Tris buffer to stop the drug reaction. Surviving cell fraction was measured immediately by eosin Y exclusion method. The colorless viable cells were counted (Mishell and Shrigi,
1980). NAD was extracted with perchloric acid and NADH with alkali, and
then quantitated with a cycling assay using alcohol dehydrogenase
(Matsumara and Miyachi, 1980).

RESULTS

Intact cells are known to reduce external impermeable electron acceptors. Diferric transferrin has been used as an electron acceptor and is
considered as a natural electron acceptor since it is required for maximum
cell growth. However, potassium ferricyanide shows the best activity
among all the artificial electron acceptors which have been tried (Crane
et al., 1982). We therefore use either ferricyanide or diferric transferrin in the measurement of the transplasmalemma redox enzymes activity.
Reduction of diferric transferrin by HeLa cell is observed with the formation of a pink chelate, $Fe-BPS_3$ externally since BPS is impermeable to
cells. There is an initial rapid rate of reduction followed by a slower

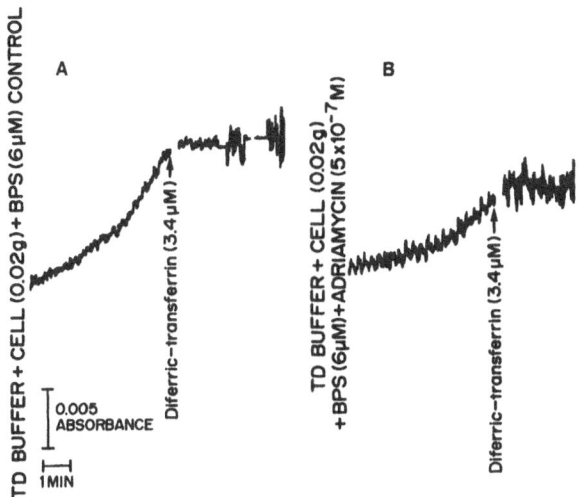

Fig. 1. Effect of adriamycin on diferric-transferrin reduction by HeLa cells. A. control; B. in the presence of adriamycin 10^{-7}M.

steady rate (Figure 1A). If adriamycin is incubated with the HeLa cells for three minutes before starting the reaction with diferric transferrin the rate of ferrous iron release is greatly decreased (Figure 1B). This reduction may depend on loosely bound iron on the transferrin. Similar two phase kinetics of ferricyanide reduction by cells has been demonstrated (Sun et al., 1985b). Effects of adriamycin and some other clinically used anthracyclines on the activity of ferricyanide reductase have also been reported (Sun et al., 1984c). Other anthracyclines, as indicated in Table 1, are also tested for effects on the plasma membrane oxidoreductase activity. The inhibition shows an order: doxorubicin (DX), 4-demothoxy DN > 4^1-deoxy-4^1-iodo-DX > 11-deoxydoxorubicin > 4-demethoxy-7R-9R-DN. The inhibition of diferric transferrin reduction with HeLa cells by some of these compounds is shown in Table 2.

Transmembrane redox activity is coupled to proton release in HeLa cells (Sun et al., 1985b). Under the influence of these cytostatic drugs, a significant decrease (30-77%) in ferricyanide induced proton generation was demonstrated at a concentration of 5 X 10^{-7}M (Table 3).

Table 1. Comparison of anthracycline effects on plasma membrane electron transport.

Compound	Inhibition of Electron Transport		
	HeLa Cell Fe(CN)$_6$ Reduct. SLOW ID 50 (μM)	HeLa Cell Fe$_2$Tf Reduct. ID 50 (μM)	Plasma Membrane NADH Fe(CN)$_6$ ID 50 (μM)
I. Adriamycin (Doxorubicin)	0.05	0.05	0.05
II. 11-deoxydoxorubicin	0.5	0.5	1.0
III. 4'-deoxy-4'-iodo-DX	--	0.1	0.05
IV. 4-demethoxy-DN	0.03	0.3	0.05
V. 4'-demethoxy-7R-9R-DN	>1.0	>1.0	1.0

Plasma membrane from pig erythrocyte; Fe$_2$Tf - diferric transferrin; DX - doxorubicin - adriamycin; DN - daunorubicin.

Table 2. The inhibition of anthracycline compounds on diferric transferrin reduction by HeLa cells.

Addition	Diferric transferrin reduction rate (n moles min^{-1} g.w.w.$^{-1}$)	% Inhibition
Control	10	--
Adriamycin (10^{-7}M)	3.2	68
11-deoxydoxorubicin (5 X 10^{-7}M)	4.6	52
4-demethoxy-DN (10^{-7}M)	7.0	30
4'-demethoxy-7R-9R-DN (10^{-7}M)	9.2	8

DN - daunorubicin.

Table 3. The effect of anthracycline antibiotics on ferricyanide induced proton extrusion in HeLa cells.

Additions	Activity [n moles H^{+}/min/g cell (w.w.)]	% Inhibition
Control 414 --	414	--
11-deoxydoxorubicin (5 X 10^{-7}M)	116	72
4'-deoxy-4'-iodo-doxorubicin (5 X 10^{-7}M)	104	75
4-demethoxy-daunorubicin (5 X 10^{-7}M)	77	82
Adriamycin (5 X 10^{-7}M)	416	43
Daunomycin (5 X 10^{-7}M)	168	60
Carminocyin (5 X 10^{-7}M)	96	77
Marcellomycin (5 X 10^{-7}M)	144	65
Aclacinomycin (5 X 10^{-7}M)	128	69
AD 32 (5.0 g/ml)	95	77

Ferricyanide induced a depolarization of membrane potential (Sun and Crane, 1981). This redox associated membrane depolarization is also inhibited by adriamycin and its analogs (Table 4). Ferricyanide induced membrane depolarization was measured in whole cells as described previously (Sims et al., 1974).

Table 4. Effect of anthracycline compounds of Fe(CN)$_6$ induced membrane depolarization in HeLa cells.

Additions	Fluorescence/0.01 g cells
Cells (0.1 ml) + Fe(CN)$_6$ (0.03 mM)	-0.25
Cells (0.1 ml) + Adriamycin (10^{-7}M) + Fe(CN)$_6$ (0.3 mM)	-0.08
Cells (0.1 ml) + Adriamycin (10^{-6}M) + Fe(CN)$_6$ (.03 mM)	-0.06
Cells (0.1 ml) + Daunomycin (10^{-7}M) + Fe(CN)$_6$ (0.3 mM)	-0.04
Cells (0.2 ml) + Carminomycin (10^{-7}M) + Fe(CN)$_6$ (0.3 mM)	-0.035
Cells (0.1 ml) + Marcellomycin (10^{-7}M) + Fe(CN)$_6$ (0.3 mM)	-0.02
Cells (0.1 ml) + Aclaicinomycin (10^{-7}M) + Fe(CN)$_6$ (0.3 mM)	-0.1

"-" indicates membrane depolarization.

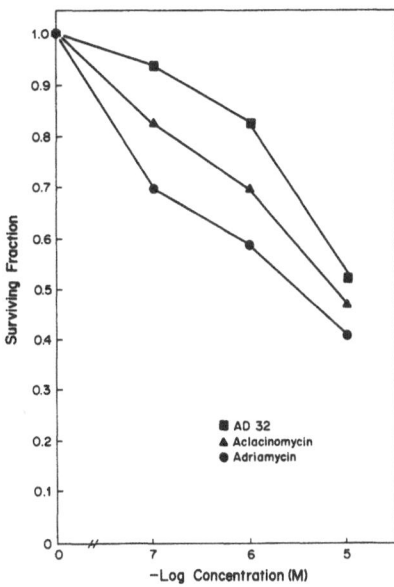

Fig. 2. The dose-response survival curve of HeLa cells after one hour drug treatment.

The cytotoxic activities of anthracycline antibiotics were compared by measuring the surviving fractions of cells treated with various concentrations of drugs for 1 hour. Figure 2 shows the dose-response survival curve. A significantly large portion of cells were sensitive to the three anthracycline compounds tested at a concentration of 1 X 10^{-7}M. However, the curve indicates that cells were a little more resistant to AD 32 than to aclacinomycin and adriamycin. Similar results were previously reported (Israel, 1975) for in vitro (human leukemia cells). AD 32 appeared to be 10-fold less potent than adriamycin as measured by effects on cell growth and cell survival. Other adriamycin analogs, such as 11-deoxydoxorubicin and 4-demethoxydaunorubicin also show 75% cytotoxicity (data not shown).

RLA209-15 fetal rat liver cells and RPNA209-1 rat pineal gland cells growth at 33°C (transformed phenotype) show a much slower rate of diferric transferrin reduction (Table 5) and ferricyanide reduction (Sun et al., 1983) in both the fast and slow phase than do the cells which have been

Table 5. Comparison of diferric transferrin reduction rates of cells with SV40 transformed and nontransformed phenotype.

Cell lines	Culture conditions	Diferric transferrin reduction rate nmoles min.$^{-1}$ g.w.w.$^{-1}$	
		fast rate	slow rate
RLA209-15*	33°C (transformed phenotype)	3.4	1.1
RLA209-15*	40°C (nontransformed phenotype)	8.7	2.7
RPNA209-1	33°C (transformed phenotype)	4.7	1.7
RPNA209-1	40°C (nontransformed phenotype)	11.1	3.2

Table 6. Effect of anticancer drugs on the transmembrane redox activities of HeLa Cells.

Addition	Diferric transferrin reduction	Ferricyanide reduction
	Specific Activity (nmoles/min/g.w.w.)*	
Control	13.5	480
Cis Diamminedichloro-Platinum (1 X 10^{-7}M)	9.3	216
Cis Diamminedichloro-Platinum (5 X 10^{-7}M)	7.4	182
Actinomycin D (7.5 X 10^{-6}M)	--	114
Bleomycin (50 µg/ml)	5.6	269
Bleomycin (75 µg/ml)	5.3	250
Bleomycin (100 µg/ml)	2.6	--
Adriamycin (1 X 10^{-6}M)	8.8	160
Adriamycin (5 X 10^{-6}M)	5.1	--
Adriamycin (1 X 10^{-5}M)	2.1	45
Anthramycin methyl ether (10^{-6}M)	4.5	--

*g.w.w. - grams wet weight.

grown at 40°C (nontransformed phenotype). In most cases, 33°C cultures show a rate about one-third of that of 40°C cultures. A similar difference has been observed with SV40 transformed 3T3 and nontransformed 3T3 cells (Löw, Grebing, Crane, unpublished). Hepatoma cells have a rate of ferricyanide reduction 30% less in the fast phase and 60% less in the slow phase than isolated fetal liver cells (Sun et al., 1983). High rates of ferricyanide reduction have also been observed with adult liver cells (Clark et al., 1981). This higher transmembrane redox activity of nontransformed cells also shows less sensitivity to anthracycline compounds than that of transformed cells (Sun et al., 1983).

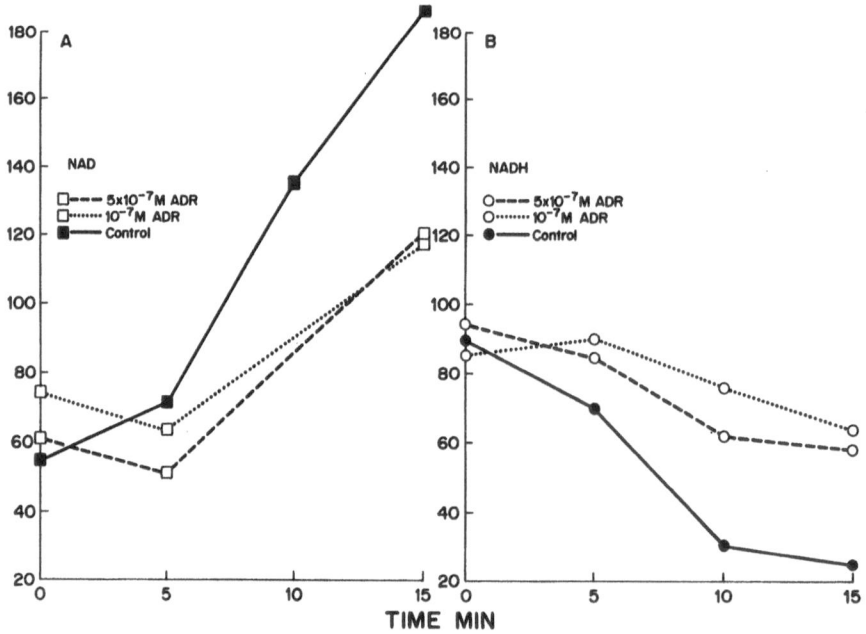

Fig. 3. Effect of adriamycin on the oxidation of internal NADH in HeLa cells by diferric transferrin. Cells incubated with adriamycin 5 min before addition of diferric transferrin (3.4 µM) at 0 time.

Besides anthracycline compounds, other types of antitumor drugs such as bleomycin, cis platin, actinomycin D, and anthramycin also show strong inhibition of transplasma membrane redox activities both in ferricyanide reduction and diferric transferrin reduction by HeLa cells (Table 6). Adriamycin still shows the strongest inhibition among these four types of drugs. Cytotoxicity of bleomycin, copper-bleomycin (Sun et al., 1985b) and cis platin but not trans analogs (Sun et al., 1984a) has been reported from our laboratory previously.

We have shown that reduction of external diferric transferrin by HeLa cells is accompanied by an oxidation of internal NADH with a subsequent increase of NAD (Navas et al., 1986). When adriamycin is incubated with the cells three minutes before starting the reaction there is less oxidation of internal NADH by addition of diferric transferrin and less increase of NAD (Figure 3). This is consistent with the inhibition of a transplasma membrane NADH diferric transferrin reductase by the adriamycin.

DISCUSSION

The reduction of iron in diferric transferrin on the outside of cells has been described (Sun et al., 1986b; Löw et al., 1986; Morley et al., 1982). Our studies present evidence that the reduction of diferric transferrin by HeLa cells is dependent on the transferrin receptor since monoclonal antitransferrin receptor antibody $B_3/25$ and apotransferrin inhibit diferric transferrin reduction as measured by the formation of ferrous BPS outside the cells (Löw et al., 1986). Thorstensen and Romslo (1984) have previously demonstrated BPS inhibition of iron uptake by liver cells and have proposed a role for transmembrane electron transport. It has also been shown that transmembrane electron transport to external electron acceptors including organic non-iron compounds can stimulate cell growth in the absence of diferric transferrin (Sun et al., 1984b). The diferric transferrin can be the natural electron acceptor for this growth stimulating electron transport in addition to providing iron for growth.

Effects of anthracycline compounds on plasma membrane redox reactions and their cytotoxicity related to the degree of inhibition of plasma membrane enzyme activities have been reported (Sun et al., 1984). Increasing evidence that adriamycin can be cytotoxic to cells by interacting with component of the plasma membrane (Tritton et al., 1982,1983; Tokes et al., 1982), together with the evidence of a vital function for the transmembrane redox system (Ellem and Kay, 1983) strongly suggests that the transmembrane enzyme could be the site for the cytotoxic effects of adriamycin. Inhibition of ferricyanide reductase by AD32, which does not intercalate with DNA (Waring, 1981), is another point in favor of anthracycline drug action on plasma membrane redox systems to produce cytotoxicity.

Study of the strong inhibitory effect of copper-bleomycin, which does not generate free radicals to react with DNA (Buettner and Oberley, 1979; Gutteridge and Chang, 1981; Burger et al., 1982) and the insignificant effect of trans platin, which has strong DNA binding affinity like cis platin indicate our proposal for a target on the plasma membrane could be right. The action of these drugs appears not to involve prior action at the level of DNA but is a direct effect on plasma membrane function.

We have previously shown that transmembrane electron transport is slower and more sensitive to antitumor drugs than the activity in non-transformed cells. RLA209-15 (33°C) and RPNA209-1 (33°C) show a much slower diferric transferrin reduction rate than cells grown at 40°C (non-transformed phenotype) (Table 5). The selective drug inhibition of transformed or tumor cells is further evidence for a change in their redox enzymes. Alteration of V_{max} and k_m of the enzyme has been described (Sun et al., 1986a,1986b). This may provide a basis for understanding selective effects of these antitumor drugs on cancer cells.

ACKNOWLEDGMENTS

Supported by grants from the NIH CA36761 (D.J.M.) and GM K6-21839 (F.L.C.). Adriamycin derivatives were provided by Dr. F. Arcamone, Farmitalia Carlo Erba, Milan, whose advice on these compounds was most helpful. Anthracycline methyl ether was kindly provided by Dr. A. Batcho, F. Hoffmann La Roache, Nutley, N.J.

REFERENCES

Bachur, N. R., Gordon, S. L., Gee, M. V., and Kon, H., 1979, NADPH cytochrome P-450 reductase activation of quinine anticancer agents to free radicals, Proc. Natl. Acad. Sci., 76:954.

Burger, R. M., Peisach, J., and Horwitz, S. B., 1982, Effect of O_2 on the reactions of activated bleomycin, J. Biol. Chem., 257:3372.

Buettner, G. R. and Oberley, L. W., 1979, The production of hydroxyl radical by tallysomycin and copper (II), FEBS Lett., 101:333.

Chandra, P., 1975, Role of chemical structure in biochemical activity of daunorubicin, adriamycin and some structure analogs: macromolecular interactions and their biological consequences, Cancer Res. Rep., 59:115.

Chou, J. W. and Schlegel-Haueter, S. E., 1981, Study of liver differentiation in vitro, J. Cell. Biol., 89:216.

Clark, M. G., Partick, E. J., Patten, G. S., Crane, F. L., Löw, H., and Grebing, C., 1981, Evidence for the extracellular reduction of ferricyanide by rat liver; a transplasma membrane redox system, Biochem. J., 200:565.

Cole, E. S. and Glass, J., 1983, Transferrin binding and iron uptake in mouse hepatocytes, Biochim. Biophys. Acta, 762:102.

Crane, F. L. and Löw, H., 1976, NADH oxidation in liver and fat cell plasma membranes, FEBS Lett., 68:153.

Crane, F. L., Crane, H. E., Sun, I. L., Mackellar, W. C., Grebing, C., and Löw, H., 1982, Insulin control of a transplasma membrane NADH dehydrogenase in erythrocyte membranes, J. Bioenerg. Biomemb., 14:425.

DiMarco, A., Silvetrini, R., DiMarco, S., and Dasdia, T., 1965, Inhibiting effect of the new cytotoxic antibiotic daunomycin on nucleic acids and mitotic activity of HeLa cells. J. Cell. Biol., 27:545.

DiMarco, A., 1975, Adriamycin (NSC - 123, 127): Mode and mechanism of action, Cancer Chem. Rep., 59:91.

Ellem, K. A. O. and Kay, G. F., 1983, Ferricyanide can replace pyruvate to stimulate growth and attachment of serum restricted human melanoma cells, Biochem. Biophys. Res. Commun., 112:183.

Gutteridge, J. M. C. and Chang, F. X., 1981, Protection of iron-catalyzed free radical damage to DNA and lipids by copper (II) - bleomycin. Biochem. Biophys. Res. Commun., 99:1354.

Israel, M., Modest, E. J., and Frei, E., 1975, N-trifluoracetyl-adriamycin-14-valerate, and analog with greater experimental antitumor activity and less toxicity than adriamycin. Cancer Res., 38:365.

Kanter, P. M. and Schwartz, H. S., 1979, Quantitative models for growth inhibition of human leukemia cells by antitumor anthracycline derivatives. Cancer Res., 39:3661.

Löw, H., Grebing, C., Navas, P., Sun, I. L., Crane, F. L., and Morré, D. J., 1986, Transplasmalemma electron transport from cells is part of a diferric transferrin reductase system, Biochem. Biophys. Res. Commun., 139:1117.

Maquet, J. P. and Butour, J. L., 1985, Avian erythroblastosis virus isolated from chick erythroblastosis induced by lymphatic leukemia virus
subgroup A, J. Nat. Cancer Inst., 70:899.

Matsumara, M. and Miyachi, S., 1980, Cycling assay for nicotinamide
adenine dinucleotides, Meth. Enzymol., 69:465.

Mishell, B. B. and Shrigi, S. M., 1980, Determination of viability by
eosin y exclusion, In: "Selected Methods in Cellular Immunology,"
W. H. Freeman Company, San Francisco.

Morley, C. D. G., Revers, K. and Bezkorvainy, A., 1982, "Biochemistry and
Physiology of Iron," (P. Saltman and Hagenauer, Eds.), Elsevier/North
Holland, Amsterdam.

Murphree, S. A., Cunningham, L. S., Hwang, K. M., and Sartorelli, A. C.,
1976, Effect of adriamycin on surface properties of sarcoma 180
ascites cells, Biochem. Pharmacol., 25:1227.

Na, C. and Timasheff, S. N., 1977, Physical-chemical study of daunomycin-
tubulin interactions, Arch. Biochem. Biophys., 182:147.

Navas, P., Sun, I. L., Morré, D. J., and Crane, F. L., 1986, Decrease of
NADH in HeLa cells in the presence of transferrin and ferricyanide,
Biochem. Biophys. Res. Commun., 135:110.

Pascoe, J. N. and Roberts, J. J., 1974, Interactions between mammalian
cell DNA and inorganic platinum compounds -I, Biochem. Pharmacol.,
23:1345.

Ross, W. E., Glaubiger, D., and Kohn, K. W., 1979, Qualitative and quantitative aspects of intercalator-induced DNA strand breaks, Biochim.
Biophys. Acta, 562:41.

Salles, B., Butour, J. L., Lesca, C., and Macquet, J. P., 1983, Cis-pt
$(NH_3)_2$ Cl_2 and trans-pt $(NH_3)_2$ Cl_2 inhibit DNA synthesis in cultured
L1210 leukemia cells, Biochem. Biophys. Res. Commun., 112:555.

Silvetrini, R., DiMarco, A., and Dasdia, T., 1970, Interference of daunomycin with metabolic events of the cell cycle in synchronized cultures of rat fibroblasts, Cancer Res., 60:603.

Silvetrini, R., Lenaz, L., and DiFronzo, C., 1973, Correlations between
cytotoxicity, biochemical effects, drug levels, and therapeutic
effectiveness of daunomycin and adriamycin on Sarcoma 180 ascites
in mice, Cancer Res., 33:2954.

Sims, P. J., Waggoner, A. S., Wang, C. H., and Hoffman, J. F., 1974,
Studies on the mechanism by which cyanine dyes measure membrane
potential in red blood cells and phosphatidylcholine vesicles.
Biochemistry, 13:3315.

Sun, I. L. and Crane, F. L., 1981, Evidence that a transplasma membrane
redox system is coupled to membrane potentials in HeLa cells, Abst.
Am. Soc. Microbiology, 81:165.

Sun, I. L., Crane, F. L., Chou, J. Y., Löw, H., and Grebing, G., 1983,
Transformed liver cells have modified transplasma membrane redox
activity which is sensitive to adriamycin, Biochem. Biophys. Res.
Commun., 112:183.

Sun, I. L and Crane, F. L., 1984a, The antitumor drug, cis-diamine-
dichloro-platinum, inhibits transplasmalemma electron transport
in HeLa cells, Biochem. Intern., 9:299.

Sun, I. L., Crane, F. L., Löw, H., and Grebing, C., 1984b, Transplasma
membrane redox stimulates HeLa cell growth, Biochem. Biophys. Res.
Commun., 125:649.

Sun, I. L., Crane, F. L., Löw, H., and Grebing, C., 1984c, Inhibition of
plasma membrane NADH dehydrogenase by adriamycin and related anthracycline antibiotics, J. Bioenerg. Biomemb., 14:425.

Sun, I. L. and Crane, F. L., 1985, Bleomycin control of transplasma membrane redox activity and proton movement in HeLa cells, Biochem.
Pharmacol., 34:617.

Sun, I. L., Crane, F. L., Grebing, C., and Löw, H., 1985, Transmembrane
redox in control of cell growth: Stimulation of HeLa cell growth by
ferricyanide and insulin, Exp. Cell. Res., 156:528.

Thorstensen, K. and Romslo, I., 1984, Uptake of iron from transferrin by isolated hepatocytes, _Biochim. Biophys. Acta_, 804:200.

Tokes, Z. A., Rogers, K. E., and Rembaum, A., 1982, Synthesis of adriamycin-coupled polyglutaraldehyde microspheres and evaluation of their cytostatic activity. _Proc. Natl. Acad. Sci. U.S.A._, 79:2026.

Tritton, R. R. and Yee, G., 1982, The anticancer agent adriamycin can be actively cytotoxic without entering cells, _Science_, 217:248.

Tritton, T. R., Yee, G., and Wingard, L. B. Jr., 1983, Immobilized adriamycin: a tool for separating cell surface from intracellular mechanisms, _Fed. Proc._, 42:284.

Waring, M. J., 1981, _Annu. Rev. Biochem._, 50:159.

A GENERAL INTRODUCTION TO THE LINK BETWEEN REDOX REACTIONS AND PROTON GRADIENTS

R.J.P. Williams

Inorganic Chemistry Laboratory, University of Oxford
South Parks Road, Oxford OX1 3QR, U.K.

INTRODUCTION

Since 1961 when Williams (1) and Mitchell (2) put forward the ideas concerning redox reactions in mitochondria and their relationship to proton gradients there has been an effort to resolve a series of related problems. (a) How do the reducing equivalents, i.e. initially hydride and then the electron, migrate in the membrane? (b) How and where do the redox reactions generate protons? (c) How do the protons migrate? (d) What are the coupling mechanisms? There has also been a major discussion concerning localised and delocalised events in this series of reactions to which I return later but this is only one mechanistic problem. I take the above questions in turn since they are all important at the plasma membrane (3), see Fig.1.

REDUCING EQUIVALENTS

There are a large number of enzymes which are concerned with the transfer of hydride from substrates to the coenzymes NAD and NADP. Such dehydrogenases are well studied. It is the next step in the chain which really begins to concern this meeting i.e. the transfer of hydride from NADH or NADPH to flavins. Whereas NAD and NADP are exchangeable coenzymes moving in the cytoplasm the flavin coenzymes are locked in proteins often on membranes. The reason is clear. Flavins are much more reactive especially toward dioxygen and could therefore generate free radicals (4). The direct use of flavin in this sense is well-known especially outside cells and in vesicles. A special use is in neutrophils to give a defensive supply of superoxide. The question now arises as to where are the flavoproteins relative to the cytoplasm? As far as this article is concerned I focus on the membranes of cells including eukaryotes, prokaryotes and even organelles related to prokaryotes e.g. mitochondria and chloroplasts. Typical well-known enzymes are the malate and succinate dehydrogenases but there are many others described in this symposium associated with cytochrome P-450 and so on. Most are associated with membranes.

DISLOCATED FLOW OF ELECTRONS AND PROTONS

The function of flavin is to convert the flow of hydride into the dislocated flow of electrons and protons

$$\text{Flavin H} \rightarrow \text{Flavin}^+ + \text{H}^+ + 2e$$

We see immediately that if the sink for the electron from this reaction is in a different direction in space from the pathway of release of protons then the redox energy of the flavin reaction will appear in the charge separation of the electron and the proton. This diffusion control was the basis of my 1961 article (1). It is not of necessary consequence that these activities should be in a sealed membrane.

The sink to which the electrons go is usually one or a series of Fe/S proteins and then in some cases to a quite different coenzyme, i.e. coenzyme Q, which to some extent resembles NAD (NADP) in that it is the hydride carrier to the next redox couple at a higher potential. Q diffuses in organic membrane phases, NAD diffuses in water. However while the water phase is perhaps poorly organised the membrane phase is highly organised as shown by its changes of curvature, see below. Flow is limited by organisation and becomes local flow. Q also resembles flavin in that at a particular membrane site it can do the reaction

$$QH_2 \rightarrow Q + 2H^+ + 2e$$

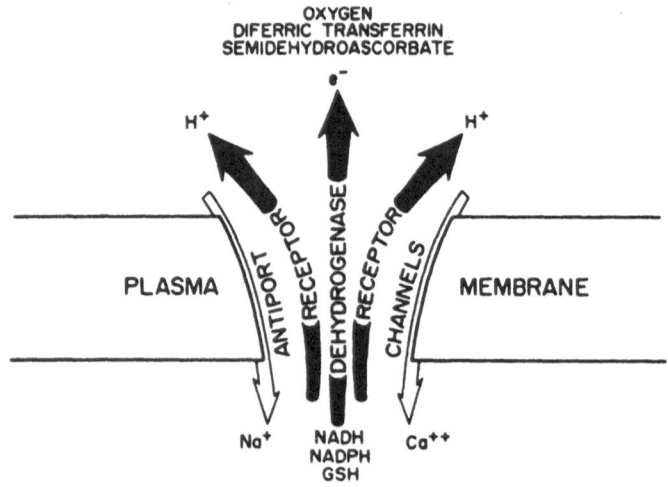

Fig. 1. The figure proposed to cover this meeting. N.B. It might be better if e^- and H^+ went to different sides of the membrane.

which gives the next diffusion controlled proton/electron charge gradient in the mitochondrial and thylakoid membranes. In this symposium we are probably not concerned with redox reactions other than those of flavins. Here the acceptor for the electrons after leaving the Fe/S proteins is not Q but must be an external oxidant such as Fe^{3+} or ascorbate (see articles in this symposium).

Now we have described the initial reaction but not the possible conformational changes which could help transfer energy from the redox reactions directly to proton movement if this happens at this stage.

Conformational Changes in Electron Transfer Proteins

In order to have safe knowledge of the conformational changes in redox enzymes in solution we in the Oxford Enzyme Group have carried out a very detailed analysis of the NMR proton spectra of three classes of proteins - cytochrome c, cytochrome b$_5$ and the copper plastocyanins. The results are clear-cut. All conformational changes accompanying electron transfer are small. It would appear that the proteins are protected against internal motion by their construction. Cytochrome c is cross-linked and contains a buried tryptophan. [Note that the apoprotein, which is not cross-linked is extremely flexible as is typical of a helical protein]. Plastocyanin is a β-barrel protein and all such proteins are known to be extremely immobile. Cytochrome b$_5$ is helical around the heme but is linked to a β-sheet structure. Its mobility is known to be limited but considerably greater than cytochrome c. Now all of these proteins must transfer electrons along membranes and must not have metal ion sites accessible to dioxygen or other small redox reagents, Fig.2. In fact they are "perfectly designed" to transfer electrons internally

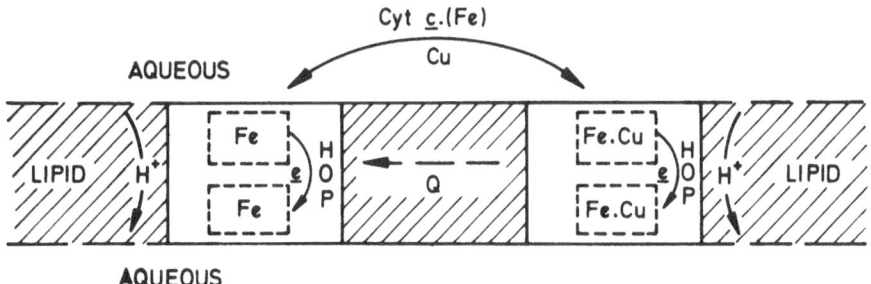

Fig. 2. A schematic mitochondrial membrane. The iron enzymes shown as hop electron conductors are proton pumps. Q carries $2H^+ + 2e$ in the membrane. Cyt.c and Cu carry electrons through water phases.

between themselves over a distance of say 15Å without movement of the protein as a whole and without giving access to dioxygen. The design principles are fully described elsewhere. However they can also transfer electrons over greater distances along membranes by molecular diffusion but in a specific manner since they only react with prescribed partners. [Note that while cytochrome c and plastocyanin travel from site to site by migration in aqueous phases, cytochrome b$_5$ travels along the surface of the membrane being held to it by a hydrophobic membrane segment.] The absence of conformational change on redox reaction means that they cannot be coupling devices however. In fact they transfer electrons from source to sink at constant potential so that there is no energy drop where there could be an energy coupling reaction. In one sense these proteins are just wires connecting sources and sinks of energy.

I turn now to the design of the Fe/S proteins listed in Table I. It is seen immediately that they are all related to β-sheets. Moreover it is expected that the redox reaction of Fe(II) to Fe(III) in a tetrahedral sulphide or thiolate matrix will generate but a relatively small change in bond lengths and no change in bond angle even though the metal ions are high-spin. This is due to the covalence of the bonding. Notice that FeS inorganic solids are good electronic conductors but FeO inorganic solids are relatively poor. Given that the β-sheet protein plastocyanin shows little or no conformation change on going from the large Cu(I) to the small Cu(II) we expect no conformation change in all the Fe$_n$/S$_n$ β-sheet proteins. This conclusion seems to be borne out by their observed fast electron transfer rates. Redox change occurs even in crystals without destroying the crystals. We must look at other members of the electron transfer systems in order to try to discover conformational changes suitable for coupling.

Table I   Structural Features of Fe/S Proteins

Protein	Structural Feature
Ferredoxin (Fe_2S_2)	Antiparallel β-sheet
Ferredoxin (Fe_4S_4)	Short antiparallel sheet
HIPIP (Fe_4S_4)	Three stranded β-sheet
Rubredoxin (Fe)	Small sheet segments

Flavodoxin

The structure of flavodoxin is typical of several water soluble electron transfer proteins. It is again a β-sheet protein and the flavin is deeply buried in a groove. We can expect but little back-bone conformational change on redox reaction despite the demand by this co-enzyme to undergo structural changes with redox state when free in solution. We shall now contrast the absence of conformational change in all the redox proteins known to be associated with the plasma membrane or particle I of mitochondria with the redox proteins of particles III and IV of mitochondria and similar particles in thylakoids.

Redox Membrane Proteins

Examination of the cytochromes shows that all these proteins are based on helices. As mentioned above the cytochromes c and the cytochromes c_1 are however cross-linked, relatively rigid, molecules which do not show high-spin to low-spin switches. They are not autoxidisable. Again as described above some cytochromes b are not cross-linked chemically but have associated with their helical regions extensive β-sheet segments which effectively cross-link the proteins, e.g. cytochromes b_5. These proteins show little tendency to go high-spin or to react with dioxygen. Finally the Fe_nS_n proteins of particle III would appear to be cross-linked by β-sheets e.g. HIPIP. By way of contrast cytochrome P-450 has a very low content of β-structure, readily goes high-spin, reacts with oxygen and undergoes conformational changes. Haemoglobin is similar to cytochrome P-450 and is virtually completely α-helical and undergoes conformational changes on redox reaction. Neither P-450 nor haemoglobin are cross-linked. These proteins show prefered order of reaction or allosteric switches. There is a parallel distinction between the rigid β-sheet copper-blue proteins and the mobile helical copper oxygen-carrier proteins. Somewhat tentatively we have concluded that purely helical proteins are conformationally mobile with respect to the movement of helical segments. If this is the case they alone can convert redox switches to mechanical switches of any size.

The discussion so far has concerned aqueous proteins. Turning to membrane proteins we have noted that most membrane proteins are helical (5). The argument has been extended to both cytochrome aa_3 and cytochrome b of the bc_1 and the bf complexes, Fig.3. Cytochrome oxidase contains both types of copper described above with the helical copper protein segment in the membrane, Fig.3. Now we know from spectral studies that these helical cytochromes show considerably conformational changes and are active as protein pumps for protons.

For some time we have supposed that it is the function of the motions of helices in proteins to transmit conformational change and to use this change to transfer ions (H^+, Na^+, Ca^{2+}) and molecules. Can we take over these ideas and use them to describe in a useful way the plasma membrane redox reactions? First we describe other known proton (and molecule) transfer proteins of membranes.

Proton (Molecule) Transfer

The basic unit we have supposed to form a proton channel is a set of helical protein segments. Modelling their dynamics on the observed dynamics of calmodulin we have described proton pumping through a channel as being due to (1) the reversible rotational/translational motion of helices on being protonated, (2) the reversible coupling of this motion to a (chemical) reaction connected to the helices with change of pK_a (3) the reversible release of protons from the new conformation. Such a mechanical device is easily coupled to different devices such that protons pump ions or molecules. Typical proteins are the channel peptides of the ATP synthases in organelles.

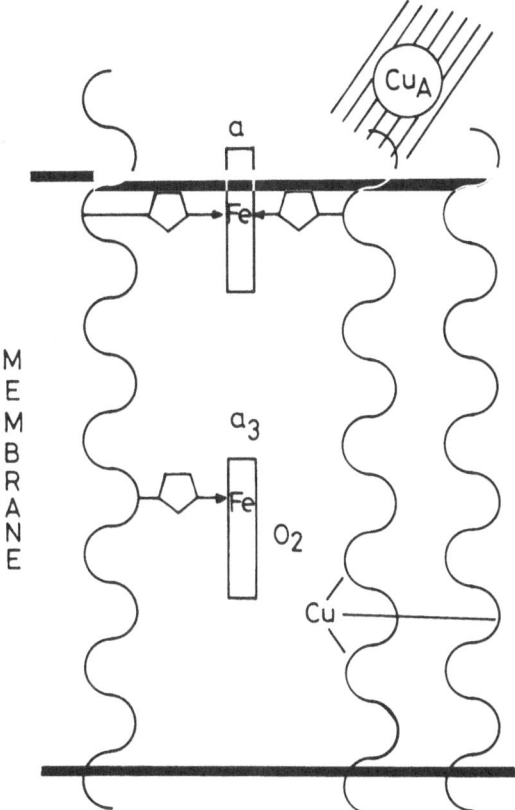

Fig. 3. Helical membrane proteins proposed in cytochrome oxidase as part of a proton pump and a β-sheet electron carrier Cu_A protein outside the membrane.

THE PLASMA MEMBRANE

It is quite generally true that the plasma membrane of a cell is not a spherical object. It has shape. The observed convolutions of the surface differ from cell type to cell type, e.g. from epithial kidney cells, Fig.4, to platelets and so on. Now shape requires that there are forces acting on the membrane which are of two kinds: (a) Constraints due to the fibrillar structures of the cell either running laterally and tangential to the plasma membrane surface or perpendicular to it: (b) Phase separation of components in the membrane due to size matching or binding. Now these two forces are not independent but are interactive through surface tension so that both external anisotropic physical forces and chemical interactions internal to the membrane cause observed curvature. For most cells then the plasma membrane is not laterally homogenous but a continuously changing series of localities following curvature changes, see Fig.4. Furthermore the localised proteins carry charge. Charge heterogeneity no matter how it arises laterally will force shape on to membrane. [It is not necessary for the curvature to be fixed for long periods of time and active zones can therefore move about].

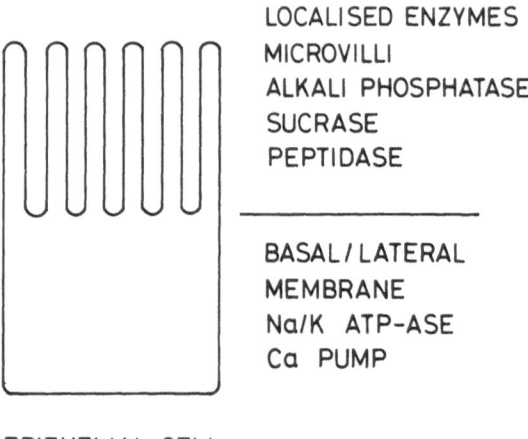

LOCALISED ENZYMES
MICROVILLI
ALKALI PHOSPHATASE
SUCRASE
PEPTIDASE

BASAL / LATERAL
MEMBRANE
Na/K ATP-ASE
Ca PUMP

EPITHELIAL CELL

Fig. 4. An epithial cell showing changes of membrane curvature and changes of protein locations.

Ion Currents

If we put together the impressions of the above paragraphs then the outward ion pumping enzymes of the plasma membrane are unlikely to be in exactly the same lateral zones of the membrane as the inward flowing channels. This is dramatically visualised in Fig.4. in the plasma membrane of the epithial cells of the kidney. Here there are no redox activities but we know that similar lateral sorting of redox enzymes occurs in thylakoid and mitochondria membranes. We also have good reason to believe that Chara cell plasma membranes set up local pH zones. All these local activities are associated with membrane cell shapes. Now let us return to the redox enzymes of plasma membranes.

Redox Activitity in Plasma Membranes

We know that some plasma membranes at least have redox enzymes associated with them. We know that the components are flavoproteins, cytochromes b_5, and

Fe/S proteins coupled to NADH or NADPH donors no matter what the acceptors. Now none of these proteins appear to be helical proteins in the membrane and we know of no direct evidence for proton pumps in the redox enzymes themselves. So far therefore we have no evidence for direct mechanical coupling to the electron/proton reaction. We do have evidence for electron transfer across the membrane presumably using the Fe/S proteins to some degree. Given that the flavoprotein will generate charge separation there will be an energised gradient. Questions of two kinds arise (a) Is there evidence of coupling to transport for example? (b) Is there evidence for local charge flow set up by these redox reactions? To deal with these questions, except in principle, we need to know the location of the enzymes along the membranes i.e. relative to the shape of the cell as seen in Fig.4.

My proposal is then that the electron/proton reactions of the plasma are not directly coupled to specific processes but are a general source of energy. In this way the proton gradients relate to those produced by irreversible ATP-hydrolysis at the plasma membrane by the ATP-ases which use phosphorylated intermediates. The coupling is then from these energy sources to a variety of transport mechanisms. It is given more selectivity by the positioning of proteins in local zones, Fig.4, so that chemicals are transfered into special parts of a cell.

Morphogenesis

There is the possibility that morphogenesis is related to the generation of patterns of fields around membranes. In such a case the fields which will be produced by redox reactions via the local sited enzymes will play a role in this process. I have given my views as to how such fields arise from hydrolytic reactions i.e. ATP as a source of ion-pumping. The direct use of redox pumps is also a possibility.

References

1.   Williams, R.J.P. J. Theoret. Biology, 1:1 (1961).
2.   Mitchell, P.D. Nature, 191:144 (1961).
3.   Crane, F.L.,   Sun, I.L., Clark, M.G., Grebning, C. and Laro, H.
Biochim. Biophys. Acta, 811:233 (1985).
4.   Williams, R.J.P. Chem. Scripta, 26:513 (1986).
5.   Williams, R.J.P. FEBS Letters, 226:1 (1987).
6.   Williams, R.J.P. J. Theoret. Biol. 121:1 (1986).

REGULATION OF PROTON DIFFUSION PATHWAYS IN

CHLOROPLAST THYLAKOIDS BY CA^{++} GATING

Richard A. Dilley and Gisela G. Chiang

Department of Biological Sciences, Purdue University
West Lafayette, IN 47907

INTRODUCTION

This chapter will present background information and some of the experiments which led to the hypothesis, and various tests of it, concerning
a Ca^{++} controlled gate for regulating H^{+} ion diffusion pathways between
localized and delocalized energy coupling gradients in thylakoid membranes. The arguments about whether photophosphorylation could be energized by localized proton gradients or only by delocalized gradients were
unresolved for some time and had proponents on each side (i.e., see Ferguson (1985) for thorough review of the question as of 1985). The issue
is clearer now, in light of recent results (see Dilley et al. (1987) for a
review), and our present view is that both types of energy coupling proton
gradients occur in thylakoids.

Our experimental approach for detecting localized or delocalized proton gradient energy coupling was to measure the lag time (number of sin-
gle-turnover flashes) required to reach the ATP formation threshold energization level, as influenced by the permeable buffer pyridine. Provided
the buffering range of the pyridine sufficiently overlaps the internal
acidic pH required to energize ATP formation [in the presence of valinomycin and K^{+} to collapse the $\Delta\psi$ component of the protonmotive force and with
the particular reactant and product concentrations we use, a $\Delta pH \approx 2.3$
units is required to reach the energization threshold], a delocalized
coupling mode should show a detectable and predictable increase in the
threshold energization flash number when 5 mM pyridine is present (pK$_a$ =
5.44). Localized proton gradient coupling should show no effect of pyridine on the length (number of flashes) of the ATP formation onset lag.
Similarly, for delocalized coupling, a post-illumination phosphorylation
ATP yield should show predictable permeable buffer effects, either increasing the ATP yield when the pK of the pyridine is near or slightly
more acidic than the threshold pH$_{in}$, or decreasing the ATP yield if the
energization threshold pH$_{in}$ is considerably below the pyridine pK (Beard
et al., 1988).

These predictions were thoroughly tested using spinach (Beard and
Dilley, 1986,1988a,b) or pea (Chiang and Dilley, 1987) thylakoids stored
in either a "high salt" (100 mM KCl) containing buffer, or a "low salt"
buffer (200 mM sorbitol and no KCl). The phosphorylation assay utilized
a real-time ATP detection system which measured ATP-dependent luminescence
using the LKB luciferin-luciferase kit. The assay utilized single turn-

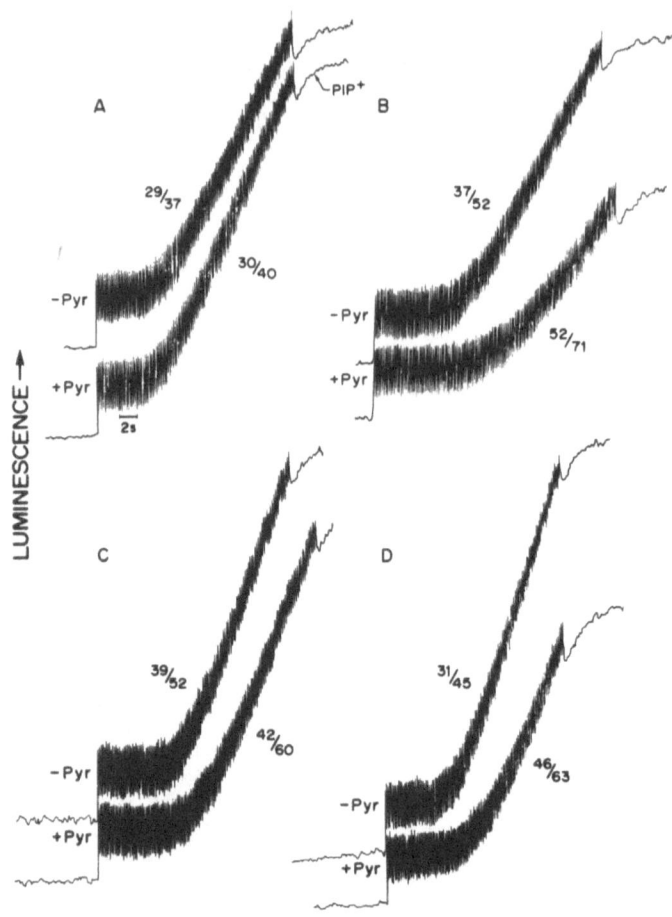

Figure 1. Effect of ionic composition and EGTA, with and without
pyridine, on the energization lag for ATP formation and on post-illumination phosphorylation. The thylakoid storage treatment, phosphorylation
medium and the luciferin-luciferase ATP assay were as described in Chiang
and Dilley (1987). The energization lag parameters are listed by each
trace; i.e., for A, top trace, the top number, 29, is the number of flashes to the first detectable rise in the luminescence signal, and 37 is the
flash number estimated by the intersection of the line given by the bottom
of rising signal and the extension of the horizontal line defined by the
bottom of the flashes occurring during the onset lag. In assays A-D, the
top trace was from a sample without pyridine and the bottom trace was from
a sample with 5 mM pyridine present, 3.5 min before beginning the flashes.
Thylakoid samples were stored in the following media prior to dilution
(approx. 5 μl added to 800 μl of reaction medium) into the phosphorylation
medium (identical for all samples): (A) Low salt (control) - 200 mM sucrose, 5 mM Hepes-KOH pH 7.5, 2 mM $MgCl_2$ and 0.5 mg ml^{-1} bovine serum albumin. (B) High salt - 100 mM KCl and 30 mM sucrose replaced the 200 mM
sucrose used in part A. (C) High salt + 1 mM $CaCl_2$ - the storage medium
was as in B plus 1 mM $CaCl_2$. (D) High salt + 30 mM $MgCl_2$ + 2 mM EGTA -
the medium was as in B plus 30 mM $MgCl_2$ and 2 mM EGTA. In (A), the bottom
trace identifies the post-illumination phosphorylation ATP yield (PIP$^+$).

TABLE 1.   EFFECT OF 100 mM KCl OR 200 mM SUCROSE IN THE THYLAKOID
           STORAGE MEDIA AND PYRIDINE ON THE ONSET OF ATP FORMATION AND
           PIP^+ POST-ILLUMINATION ATP YIELD.

Conditions	Number of flashes to the onset of ATP formation	ATP yield per flash nmol ATP· $(mg\ Chl\ flash)^{-1}$	Post-illumination ATP yield, PIP^+ [nmol ATP· $(mg\ Chl)^{-1}$]
pH 8			
SUCROSE STORAGE			
- pyridine	13±1/21±2	0.61±0.02	4.3±0.4
+ pyridine	14±1/25±1	0.64±0.08	4.3±0.4
KCl STORAGE			
- pyridine	30±3/42±3	0.56±0.05	5.4±0.5
+ pyridine	40±3/65±5	0.53±0.05	9.5±0.4

*Conditions were as in Fig. 1A and B. The lags for the onset of ATP for-
mation were determined as described in Fig. 1 and represent the actual/
extrapolated lags. The ATP yield per flash was determined from the
linear rise in luminescence (see Fig. 1). The results are the mean of 4
observations ± SE. PIP ATP yield was determined from the increase in
signal after the last flash in a flash sequence, while the remaining
rise in signal from the onset of ATP formation to the beginning of PIP
represents the total ATP yield from the flash train. The results are
the means of four observations ± SE (see Beard and Dilley, 1986,1987 for
details).*

over flashes, usually at 5 Hz, at 10°C, either pH 8 or 7 and with or
without 5 mM pyridine.

Experimentally Demonstrating Reversible Switching Between Localized and Delocalized Coupling Modes

Fig. 1 A and B shows typical pyridine effects on the phosphorylation
parameters at pH 8 for both the low and high KCl-stored thylakoids. The
low salt samples (Fig. 1A) showed phosphorylation parameters which were
insensitive to 5 mM pyridine, showing close to the same onset flash number
and similar post-illumination (PIP^+) ATP yields, consistent with energization by localized proton gradients. Table 1 gives a compilation of data
from several experiments. Chloroplasts stored in high salt, but assayed
in the same medium as used for the low salt-stored sample, showed a pyridine-induced increase in the ATP formation onset flash lag number of 15
flashes at pH 8 (Fig. 1B) and a much greater pyridine-induced lag at pH 7
(20 flashes, data not shown, but see Beard and Dilley, 1987). The pyridine-dependent increase in the energization flash number is the predicted
effect of a permeable buffer in the thylakoid lumen, interacting with a
delocalized proton gradient (val + K^+ present to keep Δψ suppressed). The
high KCl storage medium in some way induced a change, expressed in the
subsequent ATP formation assay, allowing the proton gradient to equilibrate with the lumen during the development of the energization threshold
ΔpH.

The much greater energization lag increase at pH 7 importantly affirms that the effects of the pK 5.44 pyridine buffer, are as predicted from theory for the delocalized coupling case for pH external of 7 compared to 8. At pH 7 outside, the lumenal pH of ≤ 4.7 is calculated to be required to overcome the thermodynamic driving force, ΔG_{ATP}, needed for ATP formation under these conditions, assuming 3 H^+ needed per ATP. For a buffer with a pK of 5.4, about 85% of the buffer capacity must be overcome to reach pH 4.7, whereas at pH 8 outside, only 35% of the pyridine buffering is encountered by pH 5.7, the calculated energetic threshold for that situation. Thus, the data for the high salt-stored thylakoids are exactly as expected for delocalized gradient coupling, but the low salt-stored thylakoids respond as expected for localized coupling.

The PIP^+ post-illumination phosphorylation ATP yield data were also completely confirmatory of the above conclusion (Table 1). In the high salt-stored samples pyridine increased the PIP^+ ATP yield by 4 nmol (mg chl)$^{-1}$, nearly a doubling. There was no effect of pyridine on the post-illumination ATP yield in the low salt-stored case. References (Beard and Dilley, 1986,1987,1988a,b) give details on various experiments designed to check for any membrane perturbations caused by either storage treatment, such as might result in changes in thylakoid volume, H^+ ion permeability, electron or H^+ transport. No trivial effects were noticed which could cause the results.

Earlier work from two other laboratories had reported permeable buffer effects consistent with delocalized coupling, and it is noteworthy that 50 mM or more KCl was used in those experiments for storing the thylakoids (Vinkler et al., 1980; Davenport and McCarty, 1980). In the work of Ort et al. (1976), which first reported permeable buffer-insensitive ATP formation onset lags, low salt media was used for thylakoid storage. Horner and Moudrianakis (1983) and Sigalat et al. (1985) have suggested that high KCl could cause a shift to a more delocalized coupling mode.

The question arises whether the high salt or the low salt treatments of the isolated thylakoids caused a type of state change of the membrane, shifting the phosphorylation responses irreversibly to different states. That this was not occurring was shown by washing and suspending high salt-stored thylakoids with the low salt medium. A subsequent phosphorylation assay showed no response to pyridine addition, by either phosphorylation criteria, thus a localized gradient coupling mode was reversibly restored due to the washing treatment (cf. Table IV of Beard and Dilley, 1988b). The reversibility of the localized-delocalized coupling responses argues in favor of there being a regulatable switch controlling the response, rather than the treatments causing irreversible membrane changes. This will be supported by reversible Ca^{++} ion effects described below. Before discussing the Ca^{++} effects, another aspect should be mentioned of the "switching" from apparent localized proton gradient coupling to delocalized gradients, which can be observed with the low salt-stored thylakoids.

While the low salt-stored thylakoids showed localized coupling responses in the assays described above, they can also show delocalized coupling under different energization conditions. That is, when the flash sequence was given in the absence of ADP and Pi and those components added immediately after the last flash, the "traditional" post-illumination ATP yield (termed PIP^-) was affected by pyridine exactly as predicted for a delocalized coupling pattern (Table 2; see Beard et al., 1988, for details). In agreement with theory, the relatively limited proton accumulation by 100 to 125 flashes results in pyridine reducing the PIP^- ATP yield for the pH 7 case, but increasing the PIP^- ATP yield for the pH 8 case. That is, in the pH 7 case the pK 5.44 pyridine buffered the proton accumulation sufficiently to keep the pH in the lumen from reaching as low a

TABLE 2. EFFECT OF PYRIDINE ON THE TRADITIONAL POST-ILLUMINATION PHOSPHORYLATION (PIP⁻) ATP YIELD[a]

pH	Number of Flashes	Pyridine	PIP⁻ ATP Yield (nmol ATP/mg chl)
7	125	-	21 ± 3
		+	14 ± 0.7
8	100	-	11.8 ± 0.4
		+	16.7 ± 0.5

[a]The reaction medium consisted of 50 mM Tricine-KOH (pH 8.0), 10 mM sorbitol, 3 mM $MgCl_2$, 400 nM valinomycin, 0.1 mM methylviologen, 5 mM DTT, with or without 5 mM pyridine and 10 μM diadenosine pentaphosphate. After the last flash, ADP and KH_2PO_4 were added to a final concentration of 0.1 mM and 2 mM, respectively. For the assays at pH 7, 50 mM MOPS-KOH replaced Tricine. The pH of the reaction mixture was adjusted at 10°C with and without 5 mM pyridine. The flash rate was 5 Hz. The values reported are the means ± S.E. of three determinations (cf. Beard and Dilley, 1988a,b for details).

value as occurs for the minus pyridine treatment. Hence, a significant amount of the protonated pyridine released protons in the dark, phosphory-lating stage, above the pH 4.7 threshold level, thus decreasing the PIP⁻ ATP yield.

Our interpretation is that the flash sequence builds up the proton accumulation in membrane-localized domains beyond the capacity reached when ADP and Pi are present, resulting in the "excess" proton concentra-tion opening a switch into the lumen with subsequent proton accumulation in the lumen. Pyridine present in the lumen would act as a buffer, giving increased proton accumulation during the flash train. As evidence for this, we have directly measured the pyridine-dependent increased proton uptake with low salt-stored thylakoids under basal (-ADP) conditions during a 5 Hz flash train at pH 8 outside to be 327 nmol H^+ (mg chl)$^{-1}$ when the minus pyridine control gave 220 nmol H^+ (mg chl)$^{-1}$. At pH 7 out-side with a 5 Hz, 125 flash sequence, pyridine also gave an increased H^+ accumulation from a control level of 368 up to 461 nmol H^+ (mg chl)$^{-1}$ with 5 mM pyridine present [Table III, Beard et al., 1988]. As shown in Table 2, in this "traditional" post-illumination phosphorylation experiment at pH 8.0, the addition of ADP, Pi after the last flash gave 11.8 nmol ATP (mg chl)$^{-1}$ for the control and 16.7 nmol ATP (mg chl)$^{-1}$ for the +5 mM pyr-idine case. At pH 7, pyridine inhibited the PIP⁻ATP yield, as expected if the 125 flashes did not drive the internal pH to as low a value when pyri-dine was present. The traditional post-illumination phosphorylation show-ing bulk phase protons contributing to the ATP yield is contrasted, using the same low salt-stored thylakoids, with the PIP⁺ type post-illumination ATP yield, mentioned above, which was not at all influenced by pyridine (Table 1). As discussed in more detail below we interpret the development of a delocalized gradient, in the traditional post-illumination phosphory-lation experiment, as due to excessive H^+ accumulation in the domains during the non-phosphorylating flash sequence causing protons to spill over into the lumen, where pyridine can act as a buffer giving predictable effects on both proton accumulation at pH 7 or 8 and the traditional (PIP⁻) post-illumination phosphorylation.

With the low salt-stored chloroplasts, when ADP and Pi are present during the flash train, the utilization of some of the H^+ gradient by the phosphorylation process would lead to less acidification in the domains than in the basal (-ADP, Pi) case, and the excessive spillover of H^+s into the lumen would not occur. That point was strongly supported by the finding that pyridine had little or no effect on H^+ accumulation during a phosphorylating flash train in low salt-stored thylakoids, but a pronounced effect with high salt-stored membranes (Table III, Beard et al., 1988). The above findings, taken together, strongly argue against the lumen being an obligatory part of the H^+ gradient in the low salt case, because for pH 8 outside with the energetic threshold pH of 5.7 required, pyridine could not have shared the H^+ source space and not have had an influence on the ATP onset lag length. The data can best be explained by a model wherein the energetic gradient can, in the low salt-stored case, build up to < pH 5.7 in a domain not in equilibrium with the lumen. In high salt-stored thylakoids such domains must equilibrate with the lumen, as suggested by the data.

The reversibility of the localized-delocalized coupling response by the washing treatment and the fact that the low salt-stored thylakoids showed delocalized gradient energy coupling properties in the traditional post-illumination phosphorylation experiment but not in the new, PIP^+, post-illumination mode, are consistent with the hypothesis that a regulatable switch must be present in thylakoids which controls the flux of protons. This set the stage for experiments designed to test for what factor(s) is responsible for the reversible switching between localized or delocalized proton gradient energy coupling modes.

Ca^{++} Ions Control a Proton Gating Function

Divalent cations, particularly Ca^{++}, were subsequently shown to be involved in an apparent switching function (Chiang and Dilley, 1987). This is shown in Fig. 1C, where it is clear that 1 mM $CaCl_2$ added to the 100 mM KCl storage media (compare Fig. 1C and 1B), had the effect of blocking what would otherwise have been the delocalizing action of the KCl and maintaining a localized coupling pattern (little or no effect of pyridine on the ATP formation onset lag). 30 mM $MgCl_2$ added to the 100 mM KCl storage medium gave the same effect as addition of 1 mM $CaCl_2$ (Table 3, compare treatments 3 and 4). That Ca^{++} is the agent affected by the 30 mM $MgCl_2$ is suggested by the fact that 2 mM EGTA (a Ca^{++} chelator) blocked the much higher Mg^{++} concentration effect (Fig. 1D). Furthermore, a membrane-permeable Ca^{++} chelator, TMB-8 (3,4,5-trimethoxybenzoic acid 8-(dimethyl-amino octyl ester) added to the low salt storage media caused the thylakoids to show pyridine effects on the ATP formation onset lags and post-illumination phosphorylation yield typical of the high salt-stored sample (Table 3, line 8). The proton gradient delocalizing effect of TMB-8 was blocked when 5 mM $CaCl_2$ was included in the storage buffer (Table 3, line 9). Table 3 shows statistically significant differences induced by the $MgCl_2$, $CaCl_2$ and chelator influences which switched on or off the pyridine-dependent effects on ATP formation onset lag and on the post-illumination ATP yield parameters.

A model expressing the concept of a Ca^{++}-controlled gate for regulating localized or delocalized proton gradient coupling (Chiang and Dilley, 1987) is shown in Fig. 2. The working hypothesis embodied in the model is not meant to imply that we know specific proteins that interact with the putative "gate" Ca^{++}; it is drawn as a speculative model, to stimulate the design of new experiments. Carboxyl groups associated with as-yet-unidentified membrane proteins are postulated as binding a Ca^{++} ion in the closed-gate configuration. Either K^+ ions in the 100 mM KCl storage medium, or H^+ ions in the case of high levels of H^+ accumulation from the

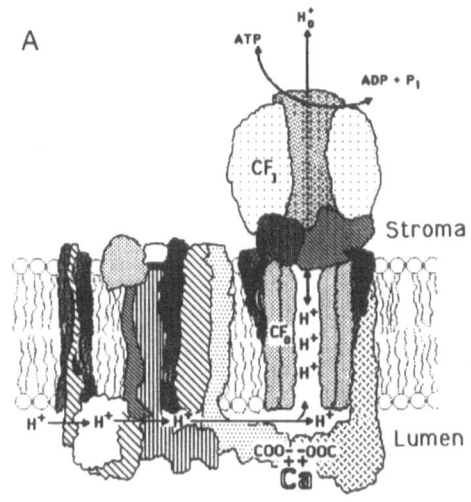

A

Localized coupling

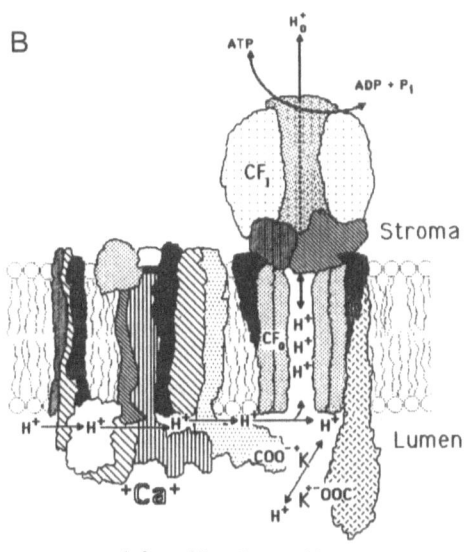

B

delocalized coupling

delocalized coupling

Figure 2. A model for a possible gating of proton fluxes between localized (A) or delocalized (B) energy coupling gradients. (A) A portion of a thylakoid membrane depicts, in a generalized and as yet speculative way, several intrinsic membrane proteins participating to form a localized proton diffusion domain from the proton releasing reactions in H_2O oxidation and plastoquinol oxidation into the CF_O channel (cf. ref. 2, Fig. 4 and accompanying discussion therein for details). Ca^{++} ions are hypothesized to form a cross-bridge between adjacent protein -COO⁻ groups to close a gated H^+ channel, although the cross-bridge could also form from tertiary structural parts of one polypeptide. (B) An open H^+ gate is shown owing to the putative Ca^{++} ligand being displaced by K^+ ion, producing an H^+ equilibration pathway between the localized domains and the lumen. The authors thank Dr. F. C. T. Allnutt for crafting the model with computer graphics.

TABLE 3. EFFECT OF SALTS AND Ca^{++} CHELATORS ON PHOSPHORYLATION PARAMETERS.[a]

Treatments	Energization Lag Number of Flashes		Lag Difference due to pyridine	PIP^+ ATP Yield nmol ATP $(mg\ Chl)^{-1}$	
	-Pyr	+Pyr	(+Pyr)-(-Pyr)	-Pyr	+Pyr
1. Control	29/37[b]	30/40	1/3	2.8	3.0
2. 100 mM KCl	37/52	52/71	15/19	3.3	6.7
3. 100 mM KCl 1.0 mM $CaCl_2$	39/52	42/60	3/8	3.4	4.7
4. 100 mM KCl 30 mM $MgCl_2$	32/47	35/53	3/6	2.4	2.6
5. 100 mM KCl 30 mM $MgCl_2$, 2 mM EGTA	31/45	46/63	15/18	3.7	7.8
6. 100 mM KCl 1 mM $CaCl_2$, 2 mM EGTA	37/54	49/71	12/17	4.1	7.0
7. 100 mM KCl 30 mM $MgCl_2$, 2 mM EGTA, 4 mM $CaCl_2$	41/54	43/62	2/8	3.6	5.8
8. low salt 25 μM TMB-8	33/43	45/54	12/11	3.9	6.2
9. low salt 25 μM TMB-8, 5 mM $CaCl_2$	32/39	34/43	2/4	3.9	3.8

[a]Thylakoids were resuspended after isolation in either the low-salt medium (control), the 100 mM KCl containing medium (high salt) as defined in the legend of Fig. 1, or medium with the addition of the various $CaCl_2$, $MgCl_2$, and EGTA combinations listed for treatments 3-7. The storage conditions and the subsequent dilution into the ATP formation assay medium are given under Materials and Methods. Where present, pyridine was added at 5 mM 3.5 min prior to starting the 5-Hz flash sequence. The two lag parameters are defined briefly in Fig. 1, and in more detail in Beard and Dilley (1986), and are the average of at least three independent assays. The PIP^+ ATP yield refers to post-illumination phosphorylation that occurred after the last flash of a sequence during which ATP was formed (ADP and P_i were present from the beginning of the flash sequence. [b]Average of three or more assays. Standard deviations for columns 1 and 2 were ±1 or ±2 in all cases except line 2 (-Pyr), where the data were 52 ± 3; for the PIP^+ data (columns 4 and 5) the standard deviations were similar to those shown in Table 1 (Chiang and Dilley, 1987).

redox reactions or from a sufficiently acidic storage medium (as in an acid-base phosphorylation protocol), could act to displace the Ca^{++} allowing protons in the localized domains to equilibrate freely with the lumen. The model can explain all the effects we have thus far observed, and it resolves the long-standing controversy about whether (and when) <u>localized</u> proton coupling occurs in a chloroplast system that can, in certain circumstances, so clearly exhibit <u>delocalized</u> coupling. Our experiments show, and the model conveys this notion, that either coupling mode can occur, depending on conditions. For the model to account for the data, the pK of the putative Ca^{++}-binding carboxyl groups is predicted to be below the pH reached in the localized domains under efficient phosphorylating conditions. Blocking ATP formation by withholding ADP, for instance, is predicted to lead to a sufficiently low pH in the domains to protonate the gate carboxyls, thus allowing domain protons to equilibrate with the lumen. The lumen could provide additional carboxyl buffering groups located on lumen-exposed portions of membrane proteins, effectively acting as an additional resevoir for protons that can contribute to energy coupling upon emptying the reservoir. Evidence for there being additional buffering groups in the lumen that contribute to the energetically competent buffering range near pH 5.5 is: a) the ATP formation onset lag, in the absence of pyridine, is more than doubled in the high salt-stored thylakoids (Table 1), understandable if additional carboxyl buffers are exposed to the developing ΔpH; b) with low salt-stored thylakoids the traditional post-illumination ATP yield (- pyridine) that occured after a flash train was also increased nearly twofold at pH 7.0 (outside) compared to pH 8 (21 compared to 11.8 nmol ATP (mg chl)$^{-1}$; Table 2). The latter effect, along with almost a two-fold increase in proton uptake (- pyridine) measured in a 5 Hz flash train at pH 7 compared to pH 8 (Table III, Beard et al., 1988), is consistent with greater lumen carboxyl buffering as the internal pH is driven into the pH $\approx$ 5 range.

Intact chloroplasts stored in either a 100 mM KCl-containing media or in a low salt, sucrose-containing media also showed Ca^{++} mediated switching between localized and delocalized responses (Chiang and Dilley, 1988). The intact chloroplasts were osmotically burst in the reaction cuvette immediately before running the ATP formation assay.

<u>Concluding Remarks</u>

The question arises as to what predicted effect the normally-occurring mono- and divalent cation concentrations found in intact chloroplasts would have on the coupling mode. Intact chloroplasts normally contain about 15 mM Ca^{++}, 30-35 mM Mg^{++} and 50 mM K^+ ion (Nakatani et al., 1979). With such high Mg^{++} and Ca^{++} levels, our results predict that the hypothesized Ca^{++} gate would normally be closed. Indeed, the data indicating localized coupling responses in the intact chloroplasts suspended in low salt media are consistent with this. Adding an additional 100 mM KCl to the intact chloroplast storage medium is apparently sufficient to induce opening of the gate in the phosphorylation assay, allowing pyridine buffering effects to be observed. Hence, <u>in situ</u>, conditions which would raise the stromal K^+ levels (especially if Mg^{++}, Ca^{++} were lowered) could lead to delocalized H^+ gradient formation. We would predict that electron-proton transport in the intact system occurring in the absence of full coupling to ATP formation, would also shift the proton flux to the lumen through the effect of protonating the carboxyl groups that bind the Ca^{++} ion. Thus, there seems to be a good basis to suggest that chloroplasts in the leaf have the capacity to express either localized proton gradient energy coupling or delocalized gradient coupling.

We have previously discussed (Dilley et al., 1987) possible reasons for a physiological benefit that might accrue to the system by having a localized proton gradient coupling mode. It may be that energizing ATP formation via localized gradients would reduce the tendency - that a more acidic lumen may promote - for $H^+_{in} \rightleftharpoons$ cation out (K^+, Na^+) exchanges to occur with subsequent Cl^- and water uptake. It is known that such exchange-driven salt uptake can occur in energized thylakoids (Murakami and Nobel, 1967; Nishida et al., 1966; Packer et al., 1965), and the resulting high amplitude swelling can be deleterious. One way to modulate this sort of osmoregulatory stress would be to maintain proper thylakoid salt content and volume, partly through the ability of the system to keep the energy coupling proton gradients localized as much as possible.

Acknowledgements

       This work was supported in part by grants from N.S.F. and U.S.D.A. The authors thank Janet Hollister for excellent assistance in preparation of the manuscript.

REFERENCES

Beard, W. A. and Dilley, R. A., 1986, FEBS Lett., 201:57-62.
Beard, W. A. and Dilley, R. A., 1987, Prog. in Photosyn. Res., III:165.
Beard, W. A. and Dilley, R. A., 1988a, J. Bioenerg. Biomemb., 20:85.
Beard, W. A. and Dilley, R. A., 1988b, J. Bioenerg. Biomemb., 20:129.
Beard, W. A., Chiang, G., and Dilley, R. A., 1988, J. Bioenerg. Biomemb., 20:107.
Chiang, G. and Dilley, R. A., 1987, Biochemistry, 26:4911.
Chiang, G. and Dilley, R. A., 1988, Plant Physiol., submitted.
Davenport, J. W. and McCarty, R. E., 1980, Biochim. Biophys. Acta, 589:353.
de Kouchkovsky, Y., 1985, Biochim. Biophys. Acta, 809:403.
Dilley, R. A., Theg, S. M., and Beard, W. A., 1987, Annual Rev. Plant Physiol., 38:348.
Ferguson, S. J., 1985, Biochim. Biophys. Acta, 811:47.
Horner, R. and Moudrianakis, E. N., 1983, J. Biol. Chem., 258:11643.
Murakami, S. and Nobel, P. S., 1967, Plant Cell. Physiol., 8:657.
Nakatani, H. Y., Barber, J., and Minski, M. J., 1979, Biochim. Biophys. Acta, 545:24.
Nishida, K., Tamai, N., and Ryoyama, K., 1966, Plant Cell Physiol., 7:415.
Ort, D. R., Dilley, R. A., and Good, N. E., 1976, Biochim. Biophys. Acta, 449:108.
Packer, L., Siegenthaler, P. A., and Nobel, P. S., 1965, Biochem. Biophys. Res. Comm., 18:474.
Sigalat, C., Haraux, F., de Kouchkovsky, F., Hung, S. P. N., and de Kouchkovsky, Y., 1985, Biochim. Biophys. Acta, 809:403.
Vinkler, C., Avron, M. and Boyer, P. D., 1980, J. Biol. Chem., 255:2263.

ENERGY-DEPENDENT TRANSPORT OF URATE AND XANTHINE IN THE UNICELLULAR GREEN ALGA *CHLAMYDOMONAS REINHARDTII*

Manuel Pineda, Rafael Pérez and Jacobo Cárdenas

Departamento Bioquímica y Biología Molecular y Fisiología, Fac. Ciencias, Univ. Córdoba, Avda San Alberto Magno s/n, 14071-Córdoba (Spain)

INTRODUCTION

Purines and purine derivatives are utilized by many organisms as sources of nitrogen and energy or as precursors in nucleic acids synthesis. Prior to their utilization these compounds must enter the cells through different transport systems depending on the type of cells. Up to now it has been very hard to distinguish between the transport process proper and the subsequent enzymatic oxidation of the transported substrate, since purines and other nucleic acids breakdown products have been observed to be utilized immediately after their transport into the cells. Thus, it is small wonder that there are so few reports in which these transport systems are distinguished from the subsequent metabolism of these compounds (Syrett, 1981; Munch-Petersen and Mygind, 1983).

Apart from our previous characterization of urate uptake and urate and xanthine metabolism in Chlamydomonas reinhardtii (Pineda et al., 1984, 1987; Pineda and Cárdenas, 1985; Pérez-Vicente et al., 1987, 1988), studies on purine uptake and metabolism in algae are rather scarce and centered mainly upon the ability of these compounds to be used as nitrogen sources for growth (Syrett, 1981; Devi Prasad, 1983), whereas the mechanisms by which these compounds are translocated into the cells have been but poorly studied.

In this paper we review our present knowledge about the uptake of purines by Chlamydomonas reinhardtii cells. Data on transport of urate and xanthine in this unicellular green alga strongly support the conclusion that both purines are translocated through inducible mediated transport systems, exhibiting Michaelian kinetics, lacking a diffusion component, energy-dependent, and distinguishable from the enzyme-catalyzed intracellular substrate oxidation.

MATERIALS AND METHODS

Growth Conditions and Preparation of Enzyme Extracts

Cells of Chlamydomonas reinhardtii 6145c (from the collection of Dr R. Sager, Sydney Farber Center, New York) were cultured phototrophically as previously described (Pineda et al., 1984). Cells were collected by centrifugation at 20 000 g, 10 min, and broken by freezing at -40°C and thawing with gentle stirring in 0.1 M Tris-HCl buffer, pH 8.5 (xanthine dehydrogenase) or 9.0 (urate oxidase). The homogenates were centrifuged at 27 000 g, 10-20 min, and the resulting supernatants used as source of enzymes.

Enzyme Assays

Xanthine dehydrogenase (EC 1.2.1.37) activity was measured spectrophotometrically by following NAD^+ reduction at 340 nm in a reaction mixture containing in a final volume of 1 ml 100 μmol Tris-HCl, pH 8.5, 0.6 μmol hypoxanthine, 1 μmol NAD^+ and the adequate amount of enzyme extract (Pérez-Vicente et al., 1988). Urate oxidase (EC 1.7.3.3) activity was assayed spectrophotometrically by following the decrease in absorbance at 292 nm due to urate oxidation in a reaction mixture containing in a final volume of 1 ml 100 μmol Tris-HCl, pH 9.0, 50 nmol urate and an adequate amount of enzyme extract (Pineda et al., 1984). One unit of enzyme is defined as the amount of enzyme which catalyzes the reduction of 1 μmol NAD^+ (xanthine dehydrogenase) or the oxidation of 1 μmol urate (urate oxidase) per min under optimal conditions of assay.

Xanthine and Urate Uptake

Xanthine or urate uptake was determined by following their disappearance from the media after removal of the cells by centrifugation. For spectrophotometric measurements, 1 ml aliquots of the culture media were centrifuged at 13 500 rpm, 1-2 min, in a Beckman Microfuge 11. In radioactivity experiments, samples of 0.2 ml were centrifuged (13 500 rpm, 30 s) in polypropylene microcentrifuge tubes (0.4 ml) containing 0.1 ml of a 3:2 mixture of silicone DC-550 and bis(3,5,5-trimethyl hexyl)phthalate. Absorbance or radioactivity was measured in the supernatants.

Kinetic experiments were performed by following the disappearance of purines in culture media at 30-60 s periods and plotting the corresponding progress curves ([S] versus t). Parameters were calculated by means of the integrated Michaelis-Menten equation as described by Cornish-Bowden (1981).

Analytical Procedures

Cell growth was measured turbidimetrically at 660 nm. Xanthine was determined enzymatically with milk xanthine oxidase at 292 nm (Krenitsky et al., 1986) or directly at 268 nm (millimolar extinction coefficient 9.15). Urate was measured enzymatically by the colorimetric assay of Fossati et al., (1980) or directly at 292 nm (millimolar extinction coefficient 12.2). When necessary (6-^{14}C)xanthine or (2-^{14}C)urate was used and radioactivity was measured in a Beckman LS 3801 liquid scintillation counter. Protein was determined according to Bradford (1976), using bovine serum albumin as standard. All spectrophotometric assays and determinations were performed in a Bausch & Lomb Spectronic 2000.

RESULTS AND DISCUSSION

Chlamydomonas cells grown on ammonium and transferred to either xanthine or urate media took up these purines after a lag phase needed to induce the corresponding transport systems (Fig. 1). Neither xanthine nor urate were consumed significantly in the presence of 10 mM chloral hydrate, a reported inhibitor of protein synthesis in Chlamydomonas (McMahon and Blaschko, 1971), which indicates that both systems are repressed in algal cells cultured in ammonium. Very recently we have also

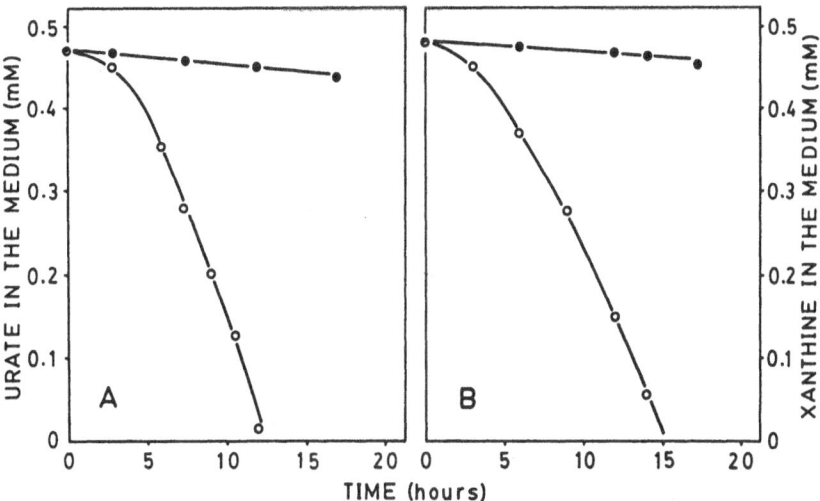

Fig. 1. Time course of urate and xanthine uptake by C. reinhardtii cells growing phototrophically. Cells grown on
ammonium until mid-logarithmic phase of growth were
harvested, washed and transferred to media with urate
(A) or xanthine (B) as sole nitrogen sources in the
light. Purine concentration in the media in absence (O)
or in presence (●) of 10 mM chloral hydrate was measured at the indicated times.

found that ammonium, besides acting as co-represor, inhibits the urate
uptake system of C. reinhardtii (Pineda et al., 1987).

The growth rate of algal cells on urate was very similar to that observed with ammonium and slightly faster than that seen with xanthine
(Table 1). High amounts of xanthine dehydrogenase and urate oxidase, enzymes responsible for the biological oxidation of xanthine and urate,
respectively, were found in cells cultured with either xanthine or urate,
whereas in ammonium grown cells the levels of both enzymes were negligible. Synthesis of protein de novo is involved in the induction of both
enzyme activities, since they did not appear in the presence of inhibitors
cycloheximide or chloral hydrate (Pineda et al., 1987; Pérez-Vicente et

Table 1. Growth rate and xanthine dehydrogenase and urate
oxidase activities of C. reinhardtii cells grown
on different sources of nitrogen

Nitrogen source	Doubling[a] time (h)	Xanthine dehydrogenase (mU/mg protein)	Urate[a] oxidase (mU/mg protein)
Ammonium	8.5	0	5
Xanthine	11.0	35	31
Urate	8.0	30	39

Enzyme activities were measured in cells harvested at the
mid-exponential phase of growth. Nitrogen compounds were
used at a concentration 4 mM in N. [a]Taken from Pineda et
al., 1984.

Table 2. Kinetic parameters for transport and enzymatic
oxidation of urate and xanthine in C. reinhardtii
cells

Substrate	Transport		Oxidation	
	K_t (µM)	Optimum pH	K_m (µM)	Optimum pH
Xanthine	1.1	6.8	70.0[b]	8.5[b]
Urate	0.9[a]	6.2[a]	12.0[c]	9.0[c]

[a]Taken from Pineda and Cárdenas, 1985; [b]Pérez-Vicente et
al., 1988; [c]Pineda et al., 1984.

al., 1988). Unlike fungi, where glutamine seems to be the metabolite responsible for nitrogen catabolite repression (Marzluf, 1981), in C. reinhardtii cells this role appears to be played by ammonium as demonstrated
in several routes of nitrogen metabolism in this alga (Syrett, 1981;
Franco et al., 1984; Pineda et al., 1987; Pérez-Vicente et al., 1988). On
the other hand, neither xanthine dehydrogenase nor urate oxidase specifically required xanthine or urate to be induced, since high levels of both
activities were also detected in algal cells cultured with other purines
or purine derivatives (Pineda et al., 1984; Pérez-Vicente et al., 1988).
Similar results have been reported in Neurospora crassa (Marzluf, 1981).

Both urate (Pineda and Cárdenas, 1985) and xanthine accumulated
against a concentration gradient inside C. reinhardtii cells by means of
mediated transport systems lacking diffusion components, which permits
appropriate distinction between the transport process in itself and the
intracellular assimilation step of these purines. Intracellular accumulation in this non-vacuolated alga is due to the fact that transport
occurs at a faster velocity than assimilation, probably because of a
higher affinity of molecular carriers for their substrates as deduced from
the different values of K_t and K_m of transport and oxidation (Table 2).
This difference in the affinities for their substrates in the processes
of transport and enzymatic assimilation is common in green algae and has
been also observed with other nitrogen sources such as ammonium, nitrate,
nitrite, urea, amino acids, pyrimidines, etc. (Syrett, 1981; Córdoba et
al., 1986). It appears probable that the mechanism of urate and xanthine
transport in Chlamydomonas is energy-dependent and not attributable to
metabolic drag. Similar purine uptake has been reported in several yeasts
(Roush and Domnas, 1956; Roush et al., 1959), bacteria (Rouf and Lomprey,
1968; Salas and Ellar, 1985), N. crassa (Tsao and Marzluf, 1976) and
Chlorella pyrenoidosa (Ammann and Lynch, 1964). In C. reinhardtii a ratio
of 3 000 between urea concentration inside and outside the cells has also
been found (Dagestad et al., 1981).

Another feature of an active and mediated transport is its dependence
on pH. Optimum pH values for transport of urate and xanthine were 6.2 and
6.8, respectively (Table 2), whereas maximum activity for urate oxidase
was found at pH 9.0 (Pineda et al., 1984) and that of xanthine dehydrogenase was at pH 8.5 (Pérez-Vicente et al., 1988). These differences
between optimum pH values for transport and oxidation corroborate the
distinction between both processes.

Xanthine and urate transport in C. reinhardtii was light-dependent.
When cells actively consuming either xanthine or urate in the light were
transferred to dark conditions the transport velocity was halved (Table 3).

Table 3. Effect of light-dark transitions on transport
and enzymatic oxidation of urate and xanthine
in C. reinhardtii cells

Treatment	Transport (%)		Enzyme activities (mU/mg protein)	
	Urate[a]	Xanthine	XDH	UO[a]
Light	100	100	23	19
Dark	55	49	24	17

Cells actively consuming urate or xanthine in the
light were darkened and, after 1 h, urate and xanthine
transport and oxidation were measured as indicated in
Materials and Methods. [a]Adapted from Pineda and Cárdenas, 1985. XDH = xanthine dehydrogenase; UO = urate
oxidase.

A similar reduction independent from terminal electron aceptor (NO_3^-, NO_2^-,
N_2O or O_2) and pH has been observed for the active transport of proline
in Rhodopseudomonas sphaeroides f. sp. denitrificans (Kundu and Nicholas,
1986). This effect of light on proline transport has been related to a
proton motive force generation and an acceleration of acceptor reduction
by light. Our data for xanthine and urate transport in C. reinhardtii are
better explained by assuming a light-dependent proton motive force generation than an acceleration of the intracellular oxidation of either urate
or xanthine, since neither urate oxidase nor xanthine dehydrogenase were
affected to a significant extent by light-dark transition of cells (Table
3). Chemiosmotic mechanisms have been proposed for amino acid transport

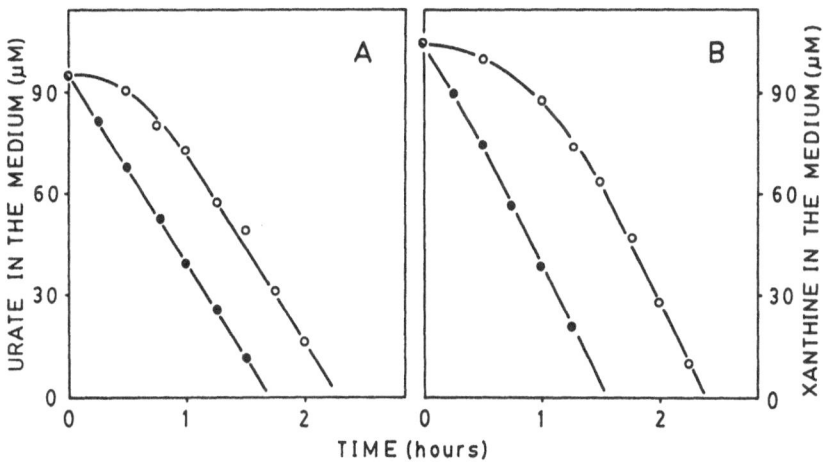

Fig. 2. Time course of urate and xanthine uptake by N-starved
C. reinhardtii cells in the light. Cells were grown
phototrophically on ammonium until the nitrogen source
was exhausted. Then, cells were transferred either
immediately of after 10 h in nitrogen-free media to
media containing urate (A) or xanthine (B). Uptake of
purines by untreated (O) or N-starved (●) cells was
measured at the indicated times.

Table 4. Effect of different metabolic inhibitors on both
xanthine and urate uptake in C. reinhardtii cells

Inhibitor	Concentration (mM)	Xanthine uptake (%)	Urate uptake (%)
None	-	100	100
2,4-DNP	0.1	54	16[a]
DCMU	0.1	50	45
Cyanide	0.1	40	0
Azide	0.5	15	0
CCCP	0.01	77	57
	0.1	0	0

Cells actively consuming urate or xanthine were divided into
glass flasks and the listed reagents were added at the indicated final concentrations. After 30 min, the uptake was
determined during 1 h at 15 min intervals. DNP: dinitrophe-
nol; DCMU: 3-(3,4-dichlorophenyl)-1,1-dimethyl urea; CCCP:
carbonylcyanide m-chlorophenylhydrazone. [a]Taken from Pineda
and Cárdenas, 1985.

in algae (Cho and Komor, 1983), other eukaryotic organisms (Horák, 1986)
and phototrophic bacteria (Knaff, 1986).

Transport of both urate and xanthine was sensitive to nitrogen starvation. When Chlamydomonas cells grown on ammonium were subjected to a
period of nitrogen starvation, the uptake systems for both purines were
derepressed (Fig. 2). Nitrogen deprivation can be associated with energy
processes, since under nitrogen starvation conditions algal cells show
an accumulation of carbon compounds such as polysaccharides and fats
and an increase in enzyme levels such as glutamine synthetase, glutamate
synthase, glutamate dehydrogenase, and nitrate and nitrite reductases
(Syrett, 1981). Nitrogen deprivation also enhanced activity of uptake
systems for nitrate, nitrite, ammonium, amino acids (Syrett, 1981), urate
and xanthine as well as that of oxidizing activities urate oxidase and
xanthine dehydrogenase (Pineda et al., 1987; Pérez-Vicente et al., 1988).

The effect of different metabolic inhibitors on the transport of
urate and xanthine corroborated that the process is energy-dependent. Both
transport systems were affected by energy poisons although that of urate
was more severely inhibited (Table 4), which indicates either that these
purines are transported by means of different systems or energy requirements
are different for each purine. Both systems were completely inhibited by
0.1 mM carbonylcyanide-m-chlorophenylhydrazone (CCCP), a very efficient
uncoupler that acts on the two components (ΔpH and Δψ) responsible for
the proton motive force (Clayton, 1980; Moreland, 1980).

Although urate and xanthine transport systems exhibited similar kinetic characteristics (Table 2) and both were also similarly affected by
light-dark transitions (Table 3) and inhibited, albeit to a different
extent, by several metabolic poisons (Table 4), it remains to be elucidated whether or not both purines share a common transport system. Xanthine inhibited urate uptake (Fig. 3) and vice versa (data not shown).

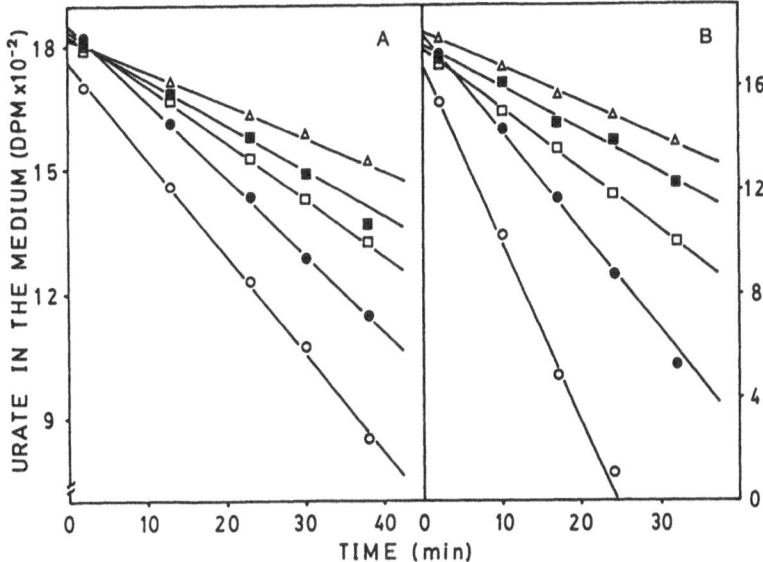

Fig. 3. Effect of xanthine on urate uptake by C. reinhardtii
cells in the light. Cells linearly consuming urate
were distributed in flasks containing 120 μM (A) and
40 μM (B) urate. At zero time 0.2 μCi (2-^{14}C)urate
(SA = 51 μCi/μmol) and different concentrations of
xanthine (0 (O); 50 (●); 100 (□); 150 (■) and 200 (△)
μM) were added to each flask. (2-^{14}C)urate concentration in the media was measured at the indicated
times.

When inhibition data were adjusted to the Dixon's equation

$$\frac{1}{v} = \frac{K_m \ [I]}{V_{max} \ [S] \ K_i} + \frac{1}{V_{max}} \left(\frac{1 + K_m}{[S]} \right)$$

it was found that each purine acted as alternative competitive inhibitor
of the other's transport, which is considered by some as evidence of the
presence of a common transport system (Tsao and Marzluf, 1976; Beaman et
al., 1983). Nevertheless, to reach this conclusion further investigations
are required, since the possibility exists that we are dealing with two
different systems induced by both urate and xanthine with different
energy requirements (Table 3) and each one capable of transporting both
purines. As for the mammalian systems, the number of transporters and
their specificity can only be resolved unequivocally by isolating and
characterizing appropriate mutants or by isolating the transporters and
assessing their substrates specificities in reconstituted systems, or by
both (Puziss et al., 1983).

ACKNOWLEDGEMENTS

This work was supported by grants of the CAICYT nº PB86-0167-CO3-01
of Ministerio de Educación y Ciencia and nº 111 of Junta de Andalucía
(Spain). The authors acknowledge the participation of Pedro Barrera Bueno
in the preliminary stages.

REFERENCES

Ammann, E.C.B., and Lynch, V.H., 1964, Purine metabolism by unicellular algae. II. Adenine, hypoxanthine and xanthine degradation by Chlorella pyrenoidosa, Biochim. Biophys. Acta, 87:370.

Beaman, T.C., Hitchins, A.D., Ochi, K., Vasantha, N., Endo, T., and Freese, E., 1983, Specificity and control of uptake of purines and other compounds in Bacillus subtilis, J. Bacteriol., 156:1107·

Bradford, M.M., 1976, A rapid and sensitive method for the quantitation of microgram quantities of protein utilizing the principle of protein-dye binding, Anal. Biochem., 72:248·

Cho, B-H., and Komor, E., 1983, Mechanism of proline uptake by Chlorella vulgaris, Biochim. Biophys. Acta, 735:361.

Clayton, R.K., 1980, "Photosynthesis: physical mechanisms and chemical patterns", Cambridge University Press, London.

Córdoba, F., Cárdenas, J., and Fernández, E., 1986, Kinetic characterization of nitrite uptake and reduction by Chlamydomonas reinhardtii, Plant Physiol., 82:904.

Cornish-Bowden, A., 1981, "Fundamentals of Enzyme Kinetics", Butterworths, London.

Dagestad, D., Lien, T., and Knutsen, G., 1981, Degradation and compartmentalization of urea in Chlamydomonas reinhardtii, Arch. Microbiol., 129:261.

Devi Prasad, P.V., 1983, Hypoxanthine and allantoin as nitrogen sources for the growth of some freshwater green algae, New Phytol., 93:575.

Fossati, P., Prencipe, L. and Berti, G., 1980, Use of 3,5-dichloro-2-hydroxybenzenesulfonic acid/4-aminophenazone chromogenic system in direct enzymatic assay of uric acid in serum and urine, Clin Chem., 26:227.

Franco, A.R., Cárdenas, J., and Fernández, E., 1984, Ammonium(methylammonium) is the co-repressor of nitrate reductase in Chlamydomonas reinhardtii, FEBS Lett., 176:453.

Horák, J., 1986, Amino acid transport in eucaryotic microorganisms, Biochim. Biophys. Acta, 864:223.

Knaff, D.B., 1986, Active transport in phototrophic bacteria, Photosynth. Res., 10:507.

Krenitsky, T.A., Spector, T., and Hall, W.W., 1986, Xanthine oxidase from human liver: purification and characterization, Arch.Biochem. Biophys., 247:108.

Kundu, B., and Nicholas, D.J.D., 1986, Active transport of proline into washed cells of Rhodopseudomonas sphaeroides f. sp. denitrificans grown with nitrate, Arch. Microbiol., 144:237.

Marzluf, G.A., 1981, Regulation of nitrogen metabolism and gene expression in fungi, Microbiol. Rev., 45:437.

McMahon, D., and Blaschko, W., 1971, Chloral hydrate inhibits protein synthesis in vivo, Biochim. Biophys. Acta, 238:338.

Moreland, D.E., 1980, Mechanisms of action of herbicides, Ann. Rev. Plant Physiol., 31:597.

Munch-Petersen, A., and Mygind, B., 1983, Transport of nucleic acid precursors, in: "Metabolism of nucleotides, nucleosides and nucleobases in microorganisms", A. Munch-Petersen, ed., Academic Press, London.

Pérez-Vicente, R., Pineda, M., and Cárdenas, J., 1987, Occurrence of an NADH diaphorase activity associated with xanthine dehydrogenase in Chlamydomonas reinhardtii, FEMS Microbiol. Lett., 43:321.

Pérez-Vicente, R., Pineda, M., and Cárdenas, J., 1988, Isolation and characterization of xanthine dehydrogenase from Chlamydomonas reinhardtii, Physiol. Plant., 72:101.

Pineda, M., and Cárdenas, J., 1985, The urate uptake system in Chlamydomonas reinhardtii, Biochim. Biophys. Acta, 820:95.

Pineda, M., Cabello, P., and Cárdenas, J., 1987, Ammonium regulation of urate uptake in Chlamydomonas reinhardtii, Planta, 171:496.

Pineda, M., Fernández, E., and Cárdenas, J., 1984, Urate oxidase of Chlamydomonas reinhardtii, Physiol. Plant., 62:453.

Puziss, M.B., Wohlhueter, R.M., and Plagemann, P.G.W., 1983, Adenine
transport and binding in cultured mammalian cells deficient in
adenine phosphoribosyltransferase, Mol. Cell. Biol., 3:82.

Rouf, M.A., and Lomprey Jr, R.F., 1968, Degradation of uric acid by
certain aerobic bacteria, J. Bacteriol., 96:617.

Roush, A.H., and Domnas, A.J., 1956, Induced biosynthesis of uricase in
yeast, Science, 124:125.

Roush, A.H., Questiaux, L.M., and Domnas, A.J., 1959, The active transport and metabolism of purines in the yeast, Candida utilis, J.
Cell. Comp. Physiol., 54:275.

Salas, J.A., and Ellar, D.J., 1985, Uric acid and allantoin uptake by
Bacillus fastidiosus spores, FEBS Lett., 183:256.

Syrett, P.J., 1981, Nitrogen metabolism of microalgae, in: "Physiological
basis of phytoplankton ecology", Can. Bull. Fisheries and Aquatic
Sci., 210:182.

Tsao, T-F., and Marzluf, G.A., 1976, Genetic and metabolic regulation of
purine base transport in Neurospora crassa, Mol. Gen. Genet., 149:
347.

MECHANISMS OF ION TRANSPORT IN PLANTS:

K+ AS AN EXAMPLE

William J. Lucas* and Leon V. Kochian

Department of Botany*, University of California,
Davis, CA 95616 and United States Plant, Soil
and Nutrition Laboratory , USDA-ARS, Cornell
University, Ithaca, NY 14853

INTRODUCTION

Following Epstein's pioneering research on the kinetics of K^+ uptake into barley roots, in which he proposed that specific carriers are responsible for the movement of K^+ across the plasma membrane, an extensive literature has developed in the general area of K^+ transport in plant systems (for a review, see Kochian and Lucas, 1988b). The development of hypotheses on the actual mechanism(s) of K^+ transport was strongly influenced by the advent of Mitchell's chemiosmotic hypothesis. These models will be reviewed in an attempt to provide an appropriate background for an analysis of the involvement of transplasma membrane redox systems in influencing K^+ transport.

CHEMIOSMOTIC MODELS OF K^+ TRANSPORT

Although it is generally accepted that K^+ uptake is dependent on H^+ efflux, no consensus exists as to the exact nature of this dependence. The two most widely accepted forms of coupling are illustrated in Figure 1. Pathway 1 involves an indirect, electrophoretic coupling of the K^+ flux to the electrical component produced by the electrogenic nature of the plasma membrane H^+-ATPase (see Marré, 1979). Pathway 2 involves a direct chemical coupling via a K^+/H^+-ATPase (Poole, 1974; Leonard and Hanson, 1972; Leonard and Hodges, 1973; Leonard, 1984; Briskin, 1986). However, the existence of a direct or an indirect coupling between K^+ influx and H^+ efflux has been challenged by Glass and coworkers. They suggest that the primary role for H^+ efflux is the maintenance of charge balance, when the uptake of cations exceeds anionic uptake (Glass and Siddiqi, 1982; Siddiqi and Glass, 1984).

Analysis of the K^+/H^+-ATPase Hypothesis

The concept of a K^+-transporting ATPase developed from the work conducted on corn, wheat, oat, and barley roots in which K^+ influx was correlated with K^+-stimulated ATPase activity in a microsomal preparation isolated from these roots (Fisher and Hodges, 1969; Fisher et al., 1970). Characterization of this putative plasma membrane ATPase indicated that it had an acidic pH optimum, required Mg^{2+}, and its activity was stimulated by monovalent cations, particularly K^+ and Rb^+ (see Hodges, 1976).

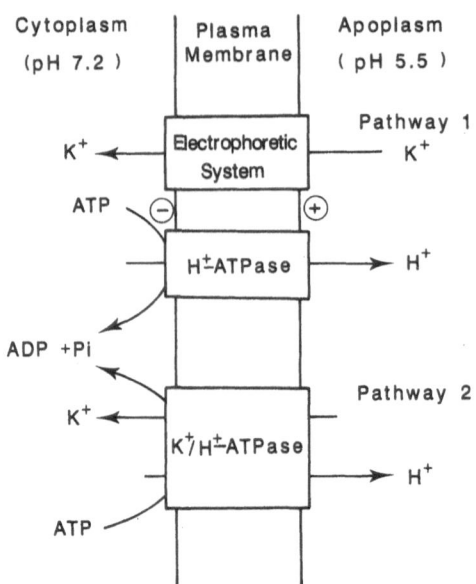

Cytoplasm
(pH 7.2)

Plasma
Membrane

Apoplasm
(pH 5.5)

Fig. 1. Possible mechanisms for K^+ transport into
plant cells. The plasma membrane H^+-ATPase
produces a pH gradient and a membrane potential during ATP hydrolysis. Pathway 1
represents an electrophoretic system for K^+
movement, and Pathway 2 represents K^+ transport by direct coupling to ATP hydrolysis via
a K^+/H^+ ATPase. (Both systems may function
across the same membrane.)

More recent work with isolated plasma membranes from corn roots
(Briskin and Leonard, 1982a,b), oat roots (Vara and Serrano, 1983) and red
beet storage tissue (Briskin and Poole, 1983a,b)‘, has established that an
ATPase, present in this membrane fraction, forms a covalently-bound,
phosphorylated intermediate during ATPase hydrolysis. This result is
similar to the Na^+/K^+ ATPase of animal cells which appears to transport K^+
by direct coupling (Cantley, 1981).

Briskin and Leonard (1982a,b) reported that in their corn root
microsomal fraction, K^+ caused a slight stimulation in the breakdown of
this phosphorylated intermediate. Vara and Serrano (1983), on the other
hand, could find no effect of K^+ on the phosphorylation/dephosphorylation
of the oat root plasma membrane ATPase. This result, and other inconsis-
tencies between the response of the plant and animal plasma membrane
ATPases, led Serrano (1984) to conclude that the plant system is not a
K^+/H^+-translocating ATPase.

Another way to test the K^+/H^+ ATPase hypothesis is to examine the
stoichiometry of K^+ influx and H^+ efflux to see if the predicted 1:1 ratio
is always observed. Most studies directed along these lines involved
measurements on bulk root tissue, where bathing medium pH and K^+ depletion
(or radioisotope uptake) were used to obtain the appropriate parameters.
Newman et al. (1987) overcame the limitations associated with this approach
by employing ion-specific microelectrodes for K^+ and H^+. By simultaneously
measuring the radial activity gradients of K^+ and H^+ near the surface of
intact corn roots, these workers were able to apply a diffusion-analysis
approach to obtain values for net K^+ influx and net H^+ efflux. The point

Table 1. Net K[+] influx and net H[+] efflux calculated by diffusion analysis of the activity gradients generated near the surface of low salt-grown corn roots.

Time (min)	Net K[+] Influx[a]	Net H[+] Efflux
0	2.92[b]	1.87[b]
3	3.18	0.76
8	2.52	0.00
12	2.89	0.00
20	2.27	1.12

[a] Data from Kochian and Lucas (1988a).
[b] Values given as μmol g fr wt^{-1}h^{-1}.

of importance with this approach is that the values reflect the activity of only a few cells, rather than an average of all the cells in a root. Typical results, obtained on low salt-grown corn roots using this micro-electrode approach, are presented in Table 1. Data obtained in this manner demonstrated that no fixed stoichiometry exists between net K[+] influx and net H[+] efflux. Although the magnitude and direction of change in these fluxes were often correlated, there were many situations in which no correlation existed. Thus, results obtained from these studies also indicate that K[+] influx may not be mediated by a K[+]/H[+] ATPase.

K[+]-H[+] Cotransport Hypothesis

In a biophysical study on K[+] transport into corn roots, Newman et al. (1987) observed that the membrane potential of the cortical cells was extremely sensitive to small changes in external K[+] within the micromolar range. An increase from 0.1 μM K[+] to 20 μM K[+] caused a rapid 70 mV depolarization of the potential, from -190 mV to -120 mV; however, beyond this level of K[+], the membrane potential exhibited a marked reduction in K[+] sensitivity (Figure 2). By making simultaneous measurements of the membrane potential and net K[+] influx (using K[+] ion specific microelec-trodes), as a function of external K[+] concentration, Newman et al. (1987) were able to compare the apparent $K_{\frac{1}{2}}$ of the two parameters. Both the kinetics for net K[+] influx and the K[+]-induced depolarization of the membrane potential yielded similar apparent $K_{\frac{1}{2}}$ values in the 6 to 9 μM range. Thus, low salt-grown corn roots possess a very high-affinity K[+] uptake system that is extremely electrogenic (Newman et al., 1987).

Slayman and coworkers have recently described a high affinity (apparent $K_{\frac{1}{2}}$ = 1-10 μM) K[+] transport system, in K[+]-starved *Neurospora* cells, that is also highly electrogenic in nature. An analysis of the electrochemical potential for K[+] across the plasma membrane of this fungal system indicated that this high affinity system may mediate active K[+] uptake. This system appears to be coupled to the *Neurospora* plasma membrane H[+]-ATPase, in that a 1:1 stoichiometry of K[+] influx to H[+] efflux was found. Flux and current voltage studies indicated that the K[+]-associ-ated inward current was twice the magnitude of the measured net K[+] influx (Rodriguez-Navarro et al., 1986). These results are consistent with the second charge coming in with the K[+] being a H[+], which suggests that this high-affinity K[+] uptake system may be a K[+]-H[+] cotransport system (Rodriguez-Navarro et al., 1986; Blatt et al., 1987).

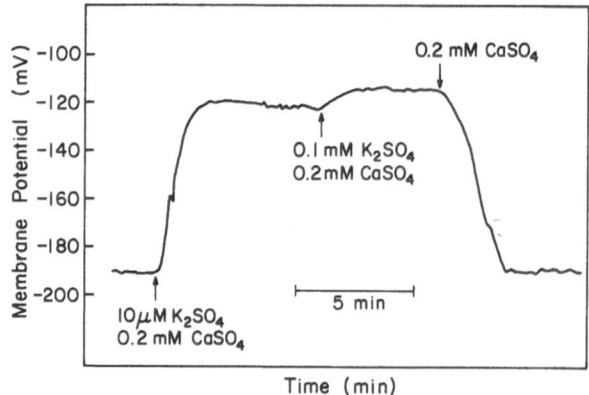

Fig. 2. Response of the cortical membrane potential
of intact, low salt-grown corn roots to
changes in the K^+ concentration in the
bathing medium. From Newman et al. (1987).

The similarities between the K^+ uptake systems seen in K^+-starved
Neurospora cells and low salt-grown corn roots, suggests that such a K^+-H^+
cotransport system may also be operating in higher plants. On a kinetic
basis, a K^+-H^+ cotransport system should reflect a sensitivity to both
extracellular and cytoplasmic pH values (see Blatt et al., 1987). However,
Kochian et al. (1987) found that varying the external pH from 4 to 8 had no
effect on either net K^+ influx (measured with K^+ ion selective microelectrodes), unidirectional K^+ ($^{86}Rb^+$) influx or the K^+-induced depolarization
of the membrane potential. Either the corn root K^+-H^+ cotransport system
has an extremely high affinity for H^+ ($K_{\frac{1}{2}} < 10^{-8}$ M) or K^+ influx is
mediated by a different molecular mechanism; e.g. a K^+-ATPase. Preliminary
evidence in support of such a system in red beet storage tissue was
recently reported by Giannini et al. (1987).

Electrophoretic K^+ Uptake

Based on electrophysiological, kinetic and inhibitor studies, performed on root tissues, we can conclude that under low external K^+ concentrations, uptake of this ion is a thermodynamically active process (see
Pitman, 1976; Cheeseman and Hanson, 1980). Uptake at concentrations above
0.5 mM appears to be passive (Marré, 1979). It is interesting to note that
Cheeseman and Hanson (1979a,b) suggested that both the active and the
passive fluxes of K^+ could be mediated by the same transport entity.
However, they later modified this concept to suggest that a separate
passive system operates at all external K^+ concentrations, but becomes
dominant at higher K^+ concentrations.

This interpretation is consistent with most of the kinetic data
obtained on corn roots. Kochian and Lucas (1982) have shown that the
kinetics for K^+ influx follow a smooth non-saturating profile, which
appears to be the combination of a saturable- and a first-order-kinetic
system (see Figure 3A). The nature of this passive, electrophoretic system
has been established, to some degree, by kinetic and biophysical studies.
Kochian et al. (1985) found that the K^+-channel blocking agent, tetraethyl-
ammonium chloride, caused a dramatic (75%) and specific inhibition of the
first-order kinetic component of K^+ influx into high salt-grown corn roots
(see Figure 3B). Also, in anion substitution studies on corn roots,
Kochian et al. (1985) obtained results consistent with earlier work
conducted on barley roots (Epstein et al., 1963) in that they found the

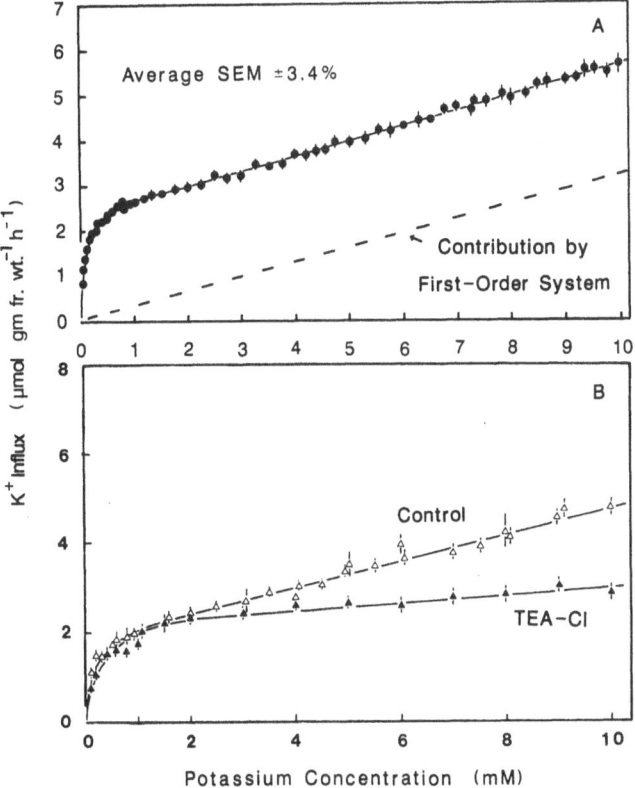

Fig. 3. Kinetic profiles for K^+ ($^{86}Rb^+$) influx into corn root segments that were grown in 5 mM KCl + 0.2 mM $CaSO_4$ (high-salt status). A: Control experiment; linear regression performed on data points from 1 to 10 mM K^+ gave a value of 0.334 μmoles g fresh $wt^{-1}h^{-1}mM^{-1}$ for the first-order rate coefficient (regression coefficient was 0.999). B: Influence of 10 mM TEA-Cl on K^+ influx (data from Kochian et al., 1985).

first-order kinetic system of K^+ to be extremely sensitive to the nature of the anions present in the bathing medium. These results led Kochian et al. (1985) to conclude that "the linear component of K^+ influx may occur by a passive process involving transmembrane K^+ channels," with the fluxes through these channels being partly coupled (in some way) to the mechanism of Cl^- transport.

Support for this hypothesis has been provided by patch-clamp studies conducted on various higher plant cells, including guard cells (Schroeder et al., 1984, 1988) and corn root cells (Ketchum et al., 1987). Future studies using the patch-clamp technique should provide a full characterization of the type(s) of K^+ channel(s) present in a specific cell-type, including the mechanism by which it is regulated. Information of this type may then provide an adequate explanation as to why, on occasion, we obtain unusual kinetic profiles in which the first-order kinetic component does not contribute to K^+ influx over the low K^+ concentration range (see Figure 4).

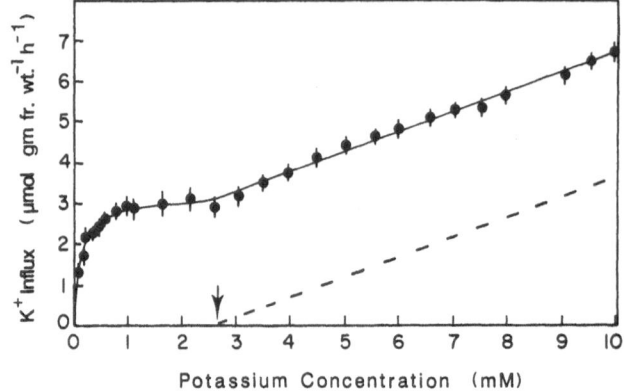

Fig. 4. Unusual kinetic profiles obtained for K^+
influx into high salt-grown corn root
segments. The saturable system was normal,
but the first-order kinetic system did not
contribute to the flux until the external K^+
concentration was raised above 2.5 mM.

Summary of K^+ Transport Mechanisms

Figure 5 presents a summary of the possible mechanisms by which K^+ may
enter the symplasm of plant roots. Uptake at low external concentrations
most likely occurs via a K^+-translocating ATPase or, as in *Neurospora*, via
a K^+-H^+ cotransport system coupled to a H^+-translocating ATPase. One
should not exclude the possibility that both types of active transport may
occur in parallel. Either system could be subject to kinetic and/or
thermodynamic control as well as allosteric regulation from internal K^+
levels. Uptake at high external K^+ levels appears to be a passive process
that may involve movement via channels that are coupled to anion (Cl^-)
transport systems. In this situation, net uptake will only occur when the
membrane potential is more negative than the Nernst potential for K^+.

REDOX-COUPLED PLASMA MEMBRANE TRANSPORT OF K^+

Several workers have investigated the effects of electron donors and
acceptors in terms of an involvement of plasma membrane redox systems in
the energetics of solute transport into plants (see Møller and Lin, 1986).
Crane and coworkers (Craig and Crane, 1980, 1981; Misra et al., 1984)
investigated the effects of exogenously added NADH and ferricyanide on
membrane transport and cell growth in suspension cultures of carrot. From
these studies, it would appear that carrot suspension cells have the
ability to oxidize exogenous NADH, and that this oxidation results not only
in a 30% increase in O_2 consumption, but also a 60% stimulation of K^+
influx. Experiments conducted on protoplasts prepared from the cortex of
corn roots also showed that K^+ influx and H^+ efflux could be markedly
stimulated by exogenous NADH (Lin, 1982a,b). These results suggested that
a plasma membrane redox system may operate in close association with the
H^+-ATPase to exert an influence on K^+ transport (Misra et al., 1984).

Influence of Exogenous NADH on K^+ Influx

Although a stimulatory effect of exogenous NADH has been reported on
K^+ uptake, there is a discrepancy between these findings and the results
obtained by Kochian and Lucas (1985) and Thom and Maretzki (1985).

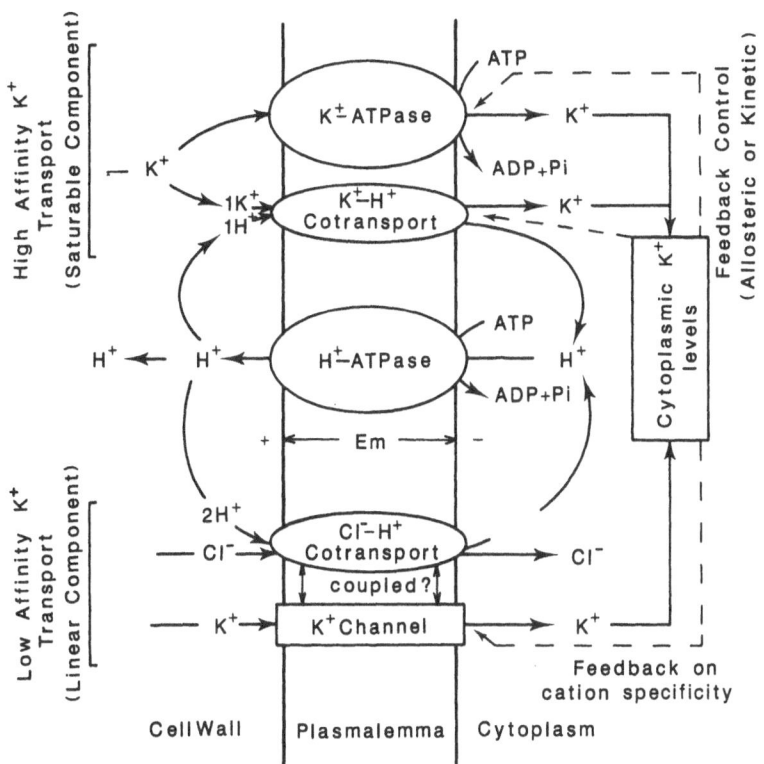

Fig. 5. Schematic representation of possible K^+ transport systems operating at the plasmalemma to facilitate K^+ uptake at both low and high external K^+ concentrations. At low K^+ levels, a high-affinity system is hypothesized to be either a K^+-ATPase, or a H^+-K^+ cotransport system, coupled to the H^+-ATPase. Both systems would be subject to feedback control by internal K^+. Under high K^+ levels, non-saturating K^+ uptake involves a K^+ channel which is, in some way, coupled to anion (Cl^-) transport. The degree of K^+-specificity exhibited by this channel may be influenced by cytoplasmic K^+ levels. (From Kochian and Lucas, 1988b.)

Treatment of high salt-grown corn roots with 1.5 mM NADH resulted in a significant inhibition of K^+ influx (see Figure 6). A similar effect was obtained on K^+ influx into protoplasts derived from sugar cane suspension cells; leucine, arginine and 3-0-methyl glucose influx into these same protoplasts were also inhibited by NADH application (Thom and Maretzki, 1985). It is noteworthy that K^+ influx into corn roots was also inhibited by addition of ferricyanide to the experimental medium (Kochian and Lucas, 1985; Rubinstein and Stern, 1986).

Analysis of the type of inhibition produced by exogenous NADH led Kochian and Lucas (1985) to propose that this reductant elicits a "wound" response within the exposed tissue. The general characteristics of such a response are a reduction in K^+ influx and a stimulation of K^+ efflux, a reduction in net H^+ efflux and a depolarization of the membrane potential. A recovery period of about 4 hours is normally required for "wounded" tissue to regain their normal functions.

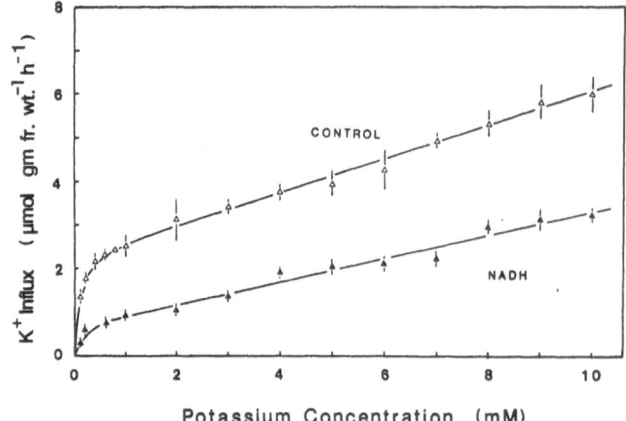

Fig. 6. Influence of 1.5 mM NADH on $^{86}Rb^+$ influx
into high salt-grown corn root segments.
(A similar response was observed in NADH
experiments conducted on low-salt corn
roots.) From Kochian and Lucas (1985).

Kochian and Lucas (1985) showed that NADH did not affect K^+ influx if
the corn root segments had not recovered from excision-associated wounding.
Also, if cycloheximide was included in the recovery medium, K^+ influx did
not return to the control level, and again NADH had no effect on K^+ influx.
These results indicate that activation of a wound response blocks the NADH
effect. Kochian and Lucas (1985) demonstrated that in this state, no NADH-
stimulated O_2 consumption could be detected. Only in the recovered state
could they measure a 30% stimulation of O_2 uptake upon addition of 1.5 mM
NADH. Interestingly, although addition of NADH perturbed K^+ influx (and
efflux), it continued to be oxidized by these perturbed root segments.

As mentioned earlier, when the redox system in corn roots (or protoplasts) was stimulated by NADH, net apparent H^+ efflux was reported to
increase (Lin, 1984). However, Kochian and Lucas (1985) found that NADH
elicited a significant decrease in H^+ efflux in both washed corn root
segments and intact roots (see also Lucas and Kochian, 1988). In some
cases an alkalinization of the external medium was observed after addition
of NADH, which implies net apparent H^+ influx. A similar situation has
also been reported for sugarcane protoplasts (Thom and Maretzki, 1985).

Transferring NADH-treated roots to fresh control solution resulted in
an almost immediate recovery of net apparent H^+ efflux to pretreatment
values (Kochian and Lucas, 1985). This imbalance between recovery of H^+
efflux and K^+ influx has also been observed with roots that are recovering
from excision-wounding (Gronewald and Hanson, 1982). Wounding the root
elicits a depolarization of the membrane potential, a response which is in
direct contrast to the NADH-induced hyperpolarization of the potential
reported by Lin (1982a). In recent studies conducted by Lucas and Kochian
(1988), NADH was found to cause a significant depolarization of the corn
root membrane potential. In some cases the potential remained in this
depolarized state, while in others it repolarized very slowly towards the
pre-NADH resting potential. However, consistent with the response of H^+
efflux discussed above, removal of NADH always resulted in a repolarization
of the potential. Certainly, then, the NADH influence on K^+ influx, H^+
efflux, and the membrane potential is consistent with the hypothesis that
exogenous application of this redox reagent elicits some form of wound
response in corn root tissue.

Exogenous NADH-Elicited Wound Response

Even though conflicting data are available on the effects of exogenous NADH on plasma membrane transport systems, little doubt exists as to the bona fide nature of the redox systems located in the plant plasma membrane (see Chapters by Buckhout and Pupillo, this volume). The challenge is going to be in the development of a complete understanding of the manner in which these redox systems interact with the various transport systems that function across the plasma membrane. In this final section we will restrict our attention to the special case of the exogenous NADH syndrome.

Application of various physical shock treatments to corn roots results in an increase in the phosphorylation level of various putative plasma membrane proteins, including the H^+-ATPase (Zocchi et al., 1983). Protein phosphorylation is thought to result in modified activity, with a blockage of H^+ efflux being the outcome for the H^+-ATPase. The important question that we need to address is how this tissue injury is "sensed" by the cells and is then transduced into protein phosphorylation.

Several possible control systems appear to be tenable. Calcium influx has been shown to increase upon wounding (Zocchi and Hanson, 1982; De Quintero and Hanson, 1984) and the resultant change in the cytosolic Ca^{2+} level may activate a Ca^{2+}-calmodulin system. One consequence of this activation could be the stimulation of a specific protein kinase. Alternatively, the turnover of phosphatidylinositol may function as a biochemical signal transduction system in roots in a manner analogous to its function in animal tissues (Berridge, 1984; Nishizuka, 1984). In animal cells, various external stimuli activate a phospholipase C on the inner plasma membrane surface, which then hydrolyses phosphatidylinositol-4,5-bisphos-phate to release myo-inositol-1,4,5-triphosphate (IP_3) into the cytosol and diacylglyceride (DG) which remains within the membrane. IP_3 can stimulate numerous biochemical and biophysical processes, including the release of internally sequestered Ca^{2+} and activation of certain kinase systems and membrane channels.

Recent studies on plant tissues have demonstrated the presence and metabolic turnover of IP_3 (Boss and Massel, 1985; Morse et al., 1987; Sandelius and Sommarin, 1986; Rincon and Boss, 1987). These reports, along with the central role played by IP_3 in regulatory phenomena in animal tissues, provided the impetus for Lucas and Kochian (1988) to incorporate the IP_3 cycle as a central component in their NADH "wound" model (see Figure 7). When the alternate agonist membrane receptor (MR) in this model has been stimulated by a wound "signal," the membrane transmitter (MT) activates the signal transduction system (STS; phospholipase C) and this produces an increase in the level of IP_3. As outlined by Lucas and Kochian (1988), the predicted consequences of this increase in IP_3 are:

(i) Activation of a protein kinase which results in phosphorylation of a protein subunit of the H^+-ATPase near the P_i release site. This causes a reduction in H^+ efflux; the effect will depend on the extent of phosphorylation and the dephosphorylation capability of the cell.

(ii) Ca^{2+} release, which may or may not exert its effects on cellular metabolism via Ca^{2+}-calmodulin (CaCM).

(iii) Direct or indirect effects of IP_3 on the gating (G) of ionic channels in both the plasma membrane and tonoplast.

An important consequence of this wound activation of MT is that in this state the transmission system may be incapable of accepting alternative stimuli. As indicated in Figure 7, an external NADH oxidizing site

227

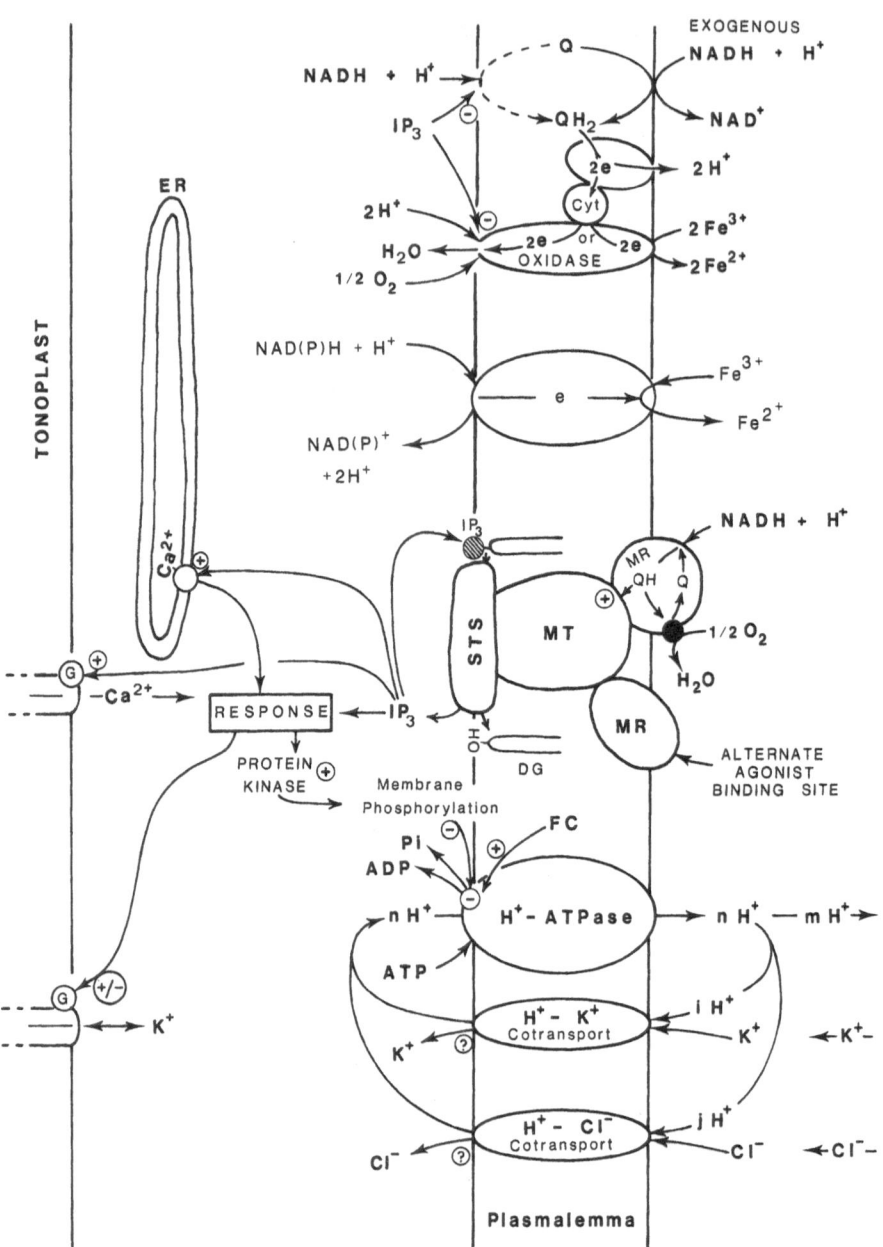

Fig. 7. Model summarizing the various effects of NADH on K^+ influx across the plant plasma membrane. Two putative transplasma-lemma NADH redox systems are shown. The Lin (1984) model (top) utilizes endogenous or exogenous NADH and transport both electrons and H^+. The Rubinstein and Stern (1986) model transfers only electrons across the plasmalemma. In this model, generation of H^+ in the cytoplasm, by the oxidation of NAD(P)H, is compensated for by transport out of the cell by the H^+-ATPase. In both models, stimulation of K^+ uptake would occur by H^+-K^+ cotransport. Lucas and Kochian (1988a) proposed that the NADH-induced "wound response," exhibited by corn, is mediated by the coupling of a membrane bound signal transduction system (STS) to the IP_3 cycle. From Kochian and Lucas (1988b).

may function to stimulate the MT system. Evidence for the presence of this NADH oxidation system comes from a reinterpretation of Lin's trypsin data (Lin, 1982b; see also Buckhout and Hrubec, 1986). If endomembrane contamination (mitochondria and endoplasmic reticulum) was absent (but see Komor et al., 1987), external NADH oxidation may occur by the 42 kD polypeptide released by mild trypsin action (Lin, 1982b, 1984). This protein appears to be able to oxidize NADH and reduce O_2 (provided O_2 is present); these properties are inconsistent with the NADH-redox model proposed by Lin (1984). According to the model developed by Lucas and Kochian (1988), exogenous NADH will result in the following:

(i) A stimulation of the MT and STS, with the consequences outlined above.

(ii) A continued stimulation of O_2 consumption, even though the cell has received a wound signal.

(iii) Proton consumption at the outer surface of the plasmalemma; depending on the degree to which the total H^+-ATPase system is inhibited, this may give rise to alkalinization of the root surface.

(iv) Continual exogenous NADH oxidation is required to maintain the enhanced IP_3 level; removal of the exogenous NADH results in a decline in IP_3 with the rate being dependent on the levels and/or activation states of the enzymes involved in its breakdown and resynthesis to phosphatidylinositol-4,5-bisphosphate.

These properties would account for all of our experimental observations on the perturbative influence of exogenous NADH on K^+ fluxes in corn roots and corn root segments (Kochian and Lucas, 1985; Lucas and Kochian, 1988). How, then, is it possible to explain the NADH-mediated stimulation of K^+ influx observed in corn root segments, corn root protoplasts (Lin, 1982a, 1984) and carrot suspension cells (Misra et al., 1984)? Based on recent reports of plasma membrane-bound electron transport systems, we have now included two forms of NADH redox systems in our model. We propose that these systems would function only when the protoplast, cell or tissue is not in a "wound" status. The protoplast is the easiest system to rationalize, because lysis during preparation could release proteases that actually remove the NADH MR protein. Buckhout and Hrubec (1986) also reported that washing their isolated plasma membrane preparation (with or without salt) resulted in a partial loss of NADH reductase as well as oxidase activity. In this state, the protoplasts may oxidize NADH (either exogenously supplied or via the cytosol) to cause an increase in the electrochemical potential gradient for H^+ across the plasma membrane either by direct transport of protons (Lin, 1984; Böttger and Lüthen, 1986), or by the production of H^+ in the cytosol which then stimulates the H^+-ATPase (Rubinstein and Stern, 1986).

CONCLUSIONS

A detailed understanding of the role played by plasma membrane redox systems, in terms of regulating/energizing transport processes, like K^+ uptake, must await a complete elucidation of the specific mechanism(s) involved in the transfer of each substrate. K^+ as a case history clearly illustrates this point, in that if active uptake occurs via a K^+-ATPase, then it would be difficult to envision how any of the putative redox systems could contribute <u>directly</u> to the movement of this ion. Alternatively, a K^+-H^+ cotransport system could be influenced either through an effect on cytoplasmic pH or on the magnitude of the electrochemical potential gradient for protons. Clearly, if the redox systems influence the magnitude of the membrane potential, one could predict the effect on the first-order kinetic system for K[+], provided this mechanism does involve K[+] channels.

Purification of these putative redox systems of the plasma membrane should greatly assist the elucidation of their function. Questions of sidedness, direction of electron transport, nature of the substrates transported, and natural substrates (donors and acceptors) are all important issues that need to be resolved. Reconstitution of these proteins into liposomes may then provide us with an unambiguous answer to the question of whether or not they are important in energizing K[+] transport into roots.

ACKNOWLEDGEMENTS

Work on K[+] transport in our laboratories has been supported by grants from the United States National Science Foundation (W.J.L.) and the United States Department of Agricultural Research Service (L.V.K.).

REFERENCES

Berridge, M.J., 1984, Inositol trisphosphate and diacylglycerol as second messengers. *Biochem. J.*, 220:345-360.

Blatt, M.R., Rodriguez-Navarro, A., and Slayman, C.L., 1987, Potassium-proton symport in *Neurospora*: Kinetic control by pH and membrane potential. *J. Membr. Biol.* 98:169-189.

Boss, W.F., and Massel, M.O., 1985, Polyphosphoinositides are present in plant tissue culture cells. *Biochem. Biophys. Res. Commun.*, 132:1018-1023.

Böttger, M., and Lüthen, H., 1986, Possible linkage between NADH-oxidation and proton secretion in *Zea mays* L. roots. *J. Exp. Bot.*, 37:666-675.

Briskin, D.P., 1986, Plasma membrane H[+]-transporting ATPase: role in potassium ion transport? *Physiol. Plantarum*, 68:159-163.

Briskin, D.P., and Leonard, R.T., 1982a, Partial characterization of a phosphorylated intermediate associated with the plasma membrane ATPase of corn root. *Proc. Nat. Acad. Sci. USA*, 79:6922-6926.

Briskin, D.P., and Leonard, R.T., 1982b, Phosphorylation of the adenosine triphosphatase in a deoxycholate-treated plasma membrane fraction from corn roots. *Plant Physiol.*, 70:1459-1464.

Briskin, D.P., and Poole, R.J., 1983a, Plasma membrane ATPase of red beet forms a phosphorylated intermediate. *Plant Physiol.*, 71:507-512.

Briskin, D.P., and Poole, R.J., 1983b, Role of magnesium in the plasma membrane ATPase of red beet. *Plant Physiol.*, 71:969-971.

Buckhout, T.J., and Hrubec, T.C., 1986, Pyridine nucleotide-dependent ferricyanide reduction associated with isolated plasma membranes of maize (*Zea mays* L.) roots. *Protoplasma*, 135:144-154.

Cantley, L.C., 1981, Structure and mechanism of the (Na,K)-ATPase. *Curr. Topics Bioenerg.*, 11:201-237.

Cheeseman, J.M., and Hanson, J.B., 1979a, Mathematical analysis of the dependence of cell potential on external potassium in corn roots. *Plant Physiol.*, 63:1-4.

Cheeseman, J.M., and Hanson, J.B., 1979b, Energy-linked potassium influx as related to cell potential in corn roots. *Plant Physiol.*, 64:842-845.

Cheeseman, J.M., and Hanson, J.B., 1980, Does active K[+] influx to roots occur? *Plant Sci. Lett.*, 18:81-84.

Craig, T.A., and Crane, F.L., 1980, Evidence for a trans-plasma membrane electron transport system in plant cells. *Proc. Indiana Acad. Sci.*, 90:150-155.

Craig, T.A., and Crane, F.L., 1981, Hormonal control of a transplasma membrane electron transport system in plant cells. Proc. Indiana Acad. Sci., 91:150-154.

De Quintero, M.R., and Hanson, J.B., 1984, Reactions of corn root tissue to calcium. Plant Physiol., 76:403-408.

Epstein, E., Rains, D.W., and Elzam, O.E., 1963, Resolution of dual mechanisms of potassium absorption by barley roots. Proc. Nat. Acad. Sci. USA, 49:684-692.

Fisher, J., and Hodges, T.K., 1969, Monovalent ion stimulated adenosine triphosphatase from oat roots. Plant Physiol., 44:385-395.

Fisher, J.D., Hansen, D., and Hodges, T.K., 1970, Correlation between ion fluxes and ion-stimulated adenosine triphosphatase activity of plant roots. Plant Physiol., 46:812-814.

Giannini, J.L., Gildensoph, L.H., Ruiz-Cristin, J., and Briskin, D.P., 1987, Isolation and characterization of sealed plasma membrane vesicles from red beet (Beta vulgaris) storage tissue. Plant Physiol., 83:S-55.

Glass, A.D.M., and Siddiqi, M.Y., 1982, Cation-stimulated H^+ efflux by intact roots of barley. Plant Cell Environ., 5:385-393.

Gronewald, J.W., and Hanson, J.B., 1982, Adenine nucleotide content of corn roots as affected by injury and subsequent washing. Plant Physiol., 69:1252-1256.

Hodges, T.K., 1976, ATPases associated with membranes of plant cells, in: "Encyclopedia of Plant Physiology," New Series, Vol. 2A, U. Lüttge and M.G. Pitman, eds., pp. 260-283, Springer-Verlag, Berlin.

Ketchum, K.A., Cooper, E., Shrier, A., and Poole, R.J., 1987, Voltage-dependent whole cell currents observed in corn protoplasts. Plant Physiol., 83:S-112.

Kochian, L.V., and Lucas, W.J., 1982, Potassium transport in corn roots. I. Resolution of kinetics into a saturable and linear component. Plant Physiol., 70:1723-1731.

Kochian, L.V., and Lucas, W.J., 1985, Potassium transport in corn roots. III. Perturbation by exogenous NADH and ferricyanide. Plant Physiol., 77:429-436.

Kochian, L. V., and Lucas, W.J., 1988a, Investigating root ion transport processes: an integrated experimental approach, in: "Redox Functions of the Eukaryotic Plasmalemma," J. Ramirez, ed., C.S.I.C., Madrid, in press.

Kochian, L. V., and Lucas, W.J., 1988b, Potassium transport in roots. Adv. Bot. Res., 15:in press.

Kochian, L. V., Shaff, J.E., and Lucas, W.J., 1987, Electrogenic K^+ uptake in corn roots: a lack of correlation with H^+ efflux. Plant Physiol., 83:S-111

Kochian, L.V., Xin-Zhi, J., and Lucas, W.J., 1985, Potassium transport in corn roots. IV. Characterization of the linear component. Plant Physiol., 79:771-776.

Komor, E., Thom, M., and Maretzki, A., 1987, The oxidation of extracellular NADH by sugarcane cells: coupling to ferricyanide reduction, oxygen uptake and pH change. Planta, 170:34-43.

Leonard, R.T., 1984, Membrane-associated ATPases and nutrient absorption by roots, in: "Advances in Plant Nutrition," Vol. 1, P.B. Tinker and A. Läuchli, eds., pp. 209-240, Praeger Scientific, New York.

Leonard, R.T., and Hanson, J.B., 1972, Induction and development of increased ion absorption in corn root tissue. Plant Physiol., 49:430-435.

Leonard, R.T., and Hodges, T.K., 1973, Characterization of plasma membrane-associated adenosine triphosphatase activity of oat roots. Plant Physiol., 52:6-12.

Lin, W., 1982a, Responses of corn root protoplasts to exogenous reduced nicotinamide adenine dinucleotide: oxygen consumption, ion uptake, and membrane potential. Proc. Nat. Acad. Sci. USA, 79:3773-3776.

Lin, W., 1982b, Isolation of NADH oxidation system from the plasmalemma of corn root protoplasts. Plant Physiol., 70:326-328.

Lin, W., 1984, Further characterization on the transport property of plasmalemma NADH oxidation system in isolated corn root protoplasts. Plant Physiol., 74:219-222.

Lucas, W.J., and Kochian, L.V., 1988, Influence of exogenous NADH on K^+ and H^+ transport in corn roots, in: "Redox Functions of the Eukaryotic Plasmalemma," J. Ramirez, ed., C.S.I.C., Madrid, in press.

Marré, E., 1979, Fusicoccin: a tool in plant physiology. Annu. Rev. Plant Physiol., 30:273-288.

Misra, P.C., Craig, T.A., and Crane, F.L., 1984, A link between transport and plasma membrane redox system(s) in carrot cells. J. Bioenerg. Biomembr. 16:143-152.

Møller, I.M., and Lin, W., 1986, Membrane-bound NAD(P)H dehydrogenases in higher plant cells. Annu. Rev. Plant Physiol., 37:309-334.

Morse, M.J., Crain, R.C., and Satter, R.L., 1987, Phosphatidylinositol cycle metabolites in Samanea saman pulvini. Plant Physiol., 83:640-644.

Newman, I.A., Kochian, L.V., Grusak, M.A., and Lucas, W.J., 1987, Fluxes of H^+ and K^+ in corn roots: characterization and stoichiometries using ion-selective microelectrodes. Plant Physiol., 84:1177-1184.

Nishizuka, Y., 1984, The role of protein kinase C in cell surface signal transduction and tumor promotion. Nature, 308:693-698.

Pitman, M.G., 1976, Ion uptake by plant roots, in: "Encyclopedia of Plant Physiology," New Series, Vol. 2B, U. Lüttge and M.G. Pitman, eds., pp. 95-128, Springer-Verlag, New York.

Poole, R.J., 1974, Ion transport and electrogenic pumps in storage tissue cells. Can. J. Bot., 52:1023-1028.

Rincon, M., and Boss, W.F., 1987, myo-Inositol trisphosphate mobilizes calcium from fusogenic carrot (Daucus carota L.) protoplasts. Plant Physiol., 83:395-398.

Rodriguez-Navarro, A., Blatt, M.R., and Slayman, C.L., 1986, A potassium-proton symport in Neurospora crassa. J. Gen. Physiol., 87:649-674.

Rubinstein, B., and Stern, A.I., 1986, Relationship of transplasmalemma redox activity to proton and solute transport by roots of Zea mays. Plant Physiol., 80:805-811.

Sandelius, A., and Sommarin, M., 1986, Phosphorylation of phosphatidyl-inositols in isolated plant membranes. FEBS Lett., 201:282-286.

Schroeder, J.I., Hedrich, R., and Fernandez, J.M., 1984, Potassium-selective single channels in guard cell protoplasts of Vicia faba. Nature, 312:361-362.

Schroeder, J.I., Raschke, K., and Neher, E., 1988, Voltage-sensitive K^+ channels in guard cell protoplasts. Proc. Nat. Acad. Sci. USA, in press.

Serrano, R., 1984, Plasma membrane ATPase of fungi and plants as a novel type of proton pump. Curr. Topics Cell Reg., 23:87-126.

Siddiqi, M.Y., and Glass, A.D.M., 1984, The influence of monovalent cations upon influx and efflux of calcium in barley (Hordeum vulgare). Plant Sci. Lett., 33:103-114.

Thom, M., and Maretzki, A., 1985, Evidence for a plasmalemma redox system in sugarcane. Plant Physiol., 77:873-876.

Vara, F., and Serrano, R., 1983, Phosphorylated intermediate of the ATPase of plant plasma membranes. J. Biol. Chem., 258:5334-5336.

Zocchi, G., and Hanson, J.B., 1982, Calcium influx into corn roots. Plant Physiol., 70:318-319.

Zocchi, G., Rogers, S.A., and Hanson, J.B., 1983, Inhibition of proton pumping in corn roots is associated with increased phosphorylation of membrane proteins. Plant Sci. Lett., 31:215-221.

ELECTRON TRANSPORT AT THE PLASMA MEMBRANE AND ATP-DRIVEN H$^+$ EXTRUSION IN

Elodea densa LEAVES

E. Marrè, M.T. Marrè, F.G. Albergoni, V. Trockner
and A. Moroni

Department of Biology
University of Milano
Via Celoria 26, 20133 Milano, Italy

INTRODUCTION

In a number of plant materials the reduction of extracellular ferricyanide is associated with a marked acidification of the extracellular
medium first demonstrated by Craig and Crane, 1982. A preliminary description of the phenomenon as well as an approach to the problem of its
physiological significance implicate the recognition of: 1) the biochemical nature of the plasmalemma-located system mediating the flow of
electrons from the cytoplasm to the extracellular acceptor; 2) the nature of the initial intracellular reducing agent; 3) the nature of the
eventual natural electron or hydrogen acceptor (O_2?) in the absence of
artificial extracellular oxidants; 4) the origin of the protons which
appear in the external medium when ferricyanide (or other electron acceptor) is reduced.

The present paper discusses the last of these four points, formu-
lated as follows: is the plasma membrane redox system transporting both
electrons and protons outward, or it only transports electrons, while
the export of H$^+$ depends on the activity of a different mechanism, activated by the effects of the electron flux to the exterior? Which of these two alternatives is valid is still debated among the workers in the
field. These two main alternatives lead to two models. According to a
first model (Böttger and Lüthen, 1986) the redox system would
transport both electrons and protons to the outside. With ferricyanide
(or similar electron acceptors) the protons would accumulate in the
medium. With oxygen as a potential physiological oxidant, <u>and if it is
reduced in the cytoplasm</u>, the protons extruded by the redox system would
accumulate in the medium, while the protons necessary for the reduction
of O_2 to H_2O would come from the cytosol. In this case the system would
function as a proton extruding redox energy-driven electrogenic pump. An
easy prediction is that the activation of such a pump would induce an at
least transitory hyperpolarization of E_m and an alkalinization of the
cytoplasm. In the alternative case , if O_2 is the acceptor of both electrons and protons <u>in the apoplast</u>, neither E_m nor cytoplasmic and apoplastic pH would be influenced.

According to a second model, (Rubinstein and Stern, 1986) the redox
system would export electrons only, and the protons appearing in the
medium when ferricyanide is reduced would be extruded by the well known
and ubiquitous ATP-driven, vanadate-inhibited plasmalemma H$^+$-ATPase. The
coupling between the redox system and the ATP-driven H$^+$ pump might de-

pend on the depolarization and/or the changes of pH and metabolism induced by the efflux of electrons by the redox system. According to this
model, and with O_2 as the physiological acceptor in the cytoplasm neither cytosolic and apoplastic pH nor E_m should be influenced. In contrast, if O_2 is reduced within the apoplast one would expect a depolarization of Em, an acidification of the cytoplasm and an alkalinization
of the apoplast. It is clear that in this case the redox system would
pump H^+ from the apoplast to the cytoplasm, rather then viceversa! (Marrè M.T. et al., 1988).

The discrimination between these two models seems fundamental not
only for the understanding of the mechanism and of the regulation of the
system but also for the obvious metabolic and physiological implications. In fact, the two models implicate different effects of the operation of the redox chain on intracellular pH, apoplastic pH and E_m, with
presumably long reaching consequences on solute transport, metabolism,
and various cell functions (see, for the situation in animals, Busa and
Nucitelli, 1984, and, in plants, Marrè M.T. et al., 1986).

The purpose of the present article is a coordinate presentation of
the results obtained by our group while working on Elodea densa leaves,
a material apparently well fit for this type of study, inasmuch as all
cells are exposed to the medium and the activities of both the ferricyanide-reducing system and of ATP-driven H^+ extrusion are particularly
high.

MATERIALS AND METHODS

Young, well developed Elodea densa leaves were treated as described
in previous papers (Marrè E. et al., 1987; Marrè M.T. et al., 1988 and
Trockner and Marrè E., 1988) also describing the experimental details
concerning the measurements of E_m, H^+ extrusion ($-\Delta H^+$), cell sap pH
cytoplasmic and vacuolar pH, K^+ uptake ferricyanide reducton, oxygen uptake and ATP, Glu-6-P and malate levels. All experiments were run in
triplicate or quadruplicate, and repeated at least 3 times.

RESULTS AND DISCUSSION

The ATP-driven H^+ pump of Elodea densa leaves

The demonstration of an ATP-driven H^+ pump in Elodea densa leaves
is based on the finding that H^+ extrusion in exchange with K^+: i) is
increased by fusicoccin (FC), a toxin shown to stimulate ATP-dependent,
electrogenic H^+ transport both "in vivo" and in isolated plasmamembrane
vesicles (Marrè E. et al., 1987; Rasi-Caldogno et al., 1986); ii) is
almost completely suppressed by the H^+ transporting plasmamembrane
ATPase inhibitors orthovanadate and erythrosine B (Albergoni et al.,
1987 and 1988; Marrè E. et al., 1987); iii) a vanadate-,
erythrosine-inhibited ATPase is present in plasmalemma-enriched vesicle
preparations obtained from Elodea leaves, and able to carry out
ATP-dependent, vanadate-inhibited H^+ transport (Marrè M.T. et al.,
unpublished).

As is known for other materials, vanadate-sensitive H^+ extrusion in
Elodea leaves strictly requires the presence of K^+ (or other permeant Em
depolarizing cation) in the medium, and is stimulated by treatments
acidifying the cytoplasm (Marrè E. et al., 1986 and 1987; Romani and
Beffagna, unpublished). This is interpreted as indicating that both Em
and cytosolic pH regulate the activity of the ATP-driven H^+ pump in this
material. The activation of H^+ extrusion by K+ or, more efficiently, by
K^++FC is associated with a significant alkalinization of the cell sap
including both the cytoplasm and the vacuole (Marrè E. et al., 1986 and

1987). Also photosynthesis-inducing light strongly stimulates in <u>Elodea</u> leaves K^+-dependent vanadate- and erythrosine-inhibited electrogenic H^+ extrusion (Marrè M.T. et al., 1987).

The significance of these results to the problem here investigated is that, taken together, they provide some means for the discrimination between ATP-driven H^+ extrusion and other H^+ transporting processes.

Ferricyanide reduction-associated changes of H^+ and K^+ fluxes in the presence of either FC or H^+-ATPase inhibitors

As previously reported by other authors (Ivankina et al.,1984; Elzenga and Prins, 1987) ferricyanide reduction (e^-) by <u>Elodea</u> leaves is associated with a marked acidification of the medium. In our experiments the $e^-/-\Delta H^+$ ratio was strongly influenced by treatments known to either stimulate or to inhibit the ATP-driven H^+ pump. In fact, the ratio changed from 2.1, in the controls, to 1.2, in the presence of FC and to about 5 and 6 in the presence vanadate and erythrosine, respectively.

Thus, the increase of negative charges due to ferricyanide reduction was electrically almost balanced by the efflux of the positive protons when the ATP-driven pump was activated by FC, while a deficit in positive charges was already evident in the controls and became very large in the presence of the H^+-ATPase inhibitors. This suggested that other ions should be excreted in order to ensure electroneutrality of the medium. In fact, determinations of net K^+ fluxes showed that ferricyanide reduction was associated with a net efflux of K^+, much larger in the presence of the inhibitors; so that the $e^-/-\Delta (H^++K^+)$ ratio remained close to 1 under all conditions (Marrè M.T. et al., 1988; see also, for similar results in <u>Lemna</u>, Lass et al., 1986)

Interestingly, the FC-induced enhancement of H^+ extrusion, strictly dependent on the presence of K^+_o in the absence of ferricyanide, was still fully evident in the presence of the electrons acceptor, but in this case it became K^+_{out} independent. As discussed below, this is interpreted as indicating that K^+_{out} activates ATP-driven H^+ pumping by depolarizing E^m, a factor which becomes no longer limiting in the presence of the very marked depolarization induced by ferricyanide.

Ferricyanide-induced changes of E_m

As already reported by other authors (Ivankina et al., 1987; Elzenga and Prins, 1987,see also Sijmoms et al., 1984),we observed that ferricyanide reduction by <u>Elodea</u> leaves is associated with an important depolarization of E_m. The addition of 0.5 mM K^+, which strongly depolarizes E_m in the controls had little effect on the E_m of the already strongly depolarized ferricyanide-treated leaves. This supports the interpretation proposed above, i.e. that the requirement of FC-induced H^+ extrusion for K^+ extrusion depends on the depolarizing action of this permeant cation, which removes the inhibition of the H^+ pump by E_m; an action no longer necessary in the presence of the strongly depolarizing ferricyanide.

Intracellular pH changes associated with ferricyanide reduction

No intracellular pH change should be observed, if the plasmalemma redox chain simultaneously transports electrons and protons to ferricyanide and to the medium. Instead, an at least transitory acidification should be expected if the redox chain were transporting electrons only, and the activation of H^+ extrusion were mediated by a distinct, secondarily activated system. Moreover, if the H^+ are extruded by the ATP-driven pump, H^+-ATPase inhibitors should increase the eventual cytoplasmic acidification associated with ferricyanide reduction. Determinations of intracellular pH as measured in the cell sap or,

Table 1. Effects of FC, erythrosine B (EB), isobutyric acid, diethyldithiostilbestrol (DES) on ferricyanide reduction and on cell sap pH in Elodea densa leaves.

| | Ferricyanide Reduction μmol / g FW 90 min | | Cell Sap pH (units) | |
	Δ from Control	Δ from + FC	Δ from Control	Δ from + FC
FC	+ 39%		+ 0.24	
EB	− 42%		− 0.37	
EB + FC	+ 4%	− 25%	− 0.02	− 0.3
IBA	− 21%		− 0.20	
IBA + FC	− 24%	− 45%		
DES	− 22%		− 0.05	
DES + FC	− 50%	− 60%	+ 0.05	− 0.2

separately, in the cytoplasm and in the vacuole (weak acid and base distribution method and NMR method) show that indeed ferricyanide reduction is accompanied in Elodea leaves by a consistent acidification of all cell compartments, which is larger in the presence of the H^+-ATPase inhibitors vanadate and erythrosine B (Table 1 and Fig. 1., see also Marre M.T. et al., 1988). In another material, acer cells in suspension culture (see poster by Guern et al., this workshop), cytoplasmic pH measurements with the ^{31}P-NMR method showed that ferricyanide (although influencing pH very little when alone) induced a marked, rapid cytoplasmic acidification when it was added together with erythrosine B.

The simplest interpretation of these data on the ferricyanide-induced changes of E_m, H^+ and K^+ fluxes and on the effects of inhibitors on $- \Delta H^+$ is that: a) the (depolarizing) electron transport to ferricyanide is only loosely linked to the (hyperpolarizing) H^+ extrusion; b) the H^+ transporting system is heavily inhibited by H^+-ATPase inhibitors, while the electron transporting system is only slightly affected; c) E_m depolarization by electron efflux might play a role in linking ferricyanide reduction to K^+ net efflux due to an E_m-regulated mechanism (Marrè M.T. et al., 1988).

Metabolic changes associated with ferricyanide reduction

If the redox system tranfers electrons to ferricyanide and releases H^+ in the cytosol, while H^+ extrusion is carried out by the secondarily activated ATP-driven pump, one would expect ferricyanide reduction to be associated with metabolic changes counteracting those induced by the direct FC-induced activation of the cytoplasm alkalinizing ATP-pump. In fact, we found that malate and glucose-6-P levels, while they are characteristically increased (together with cytosolic pH) by the treatment with FC and K^+, in contrast are significantly depressed by the presence of the cytoplasm acidifying ferricyanide. Measurements of the incorporation of C-1 and C-6 of ^{14}C-labeled glucose into RNA ribose also indicated that the FC-induced inhibition of the pentose phosphate oxidative pathway was completely reversed, and transformed into a stimulation when ferricyanide was present (Trockner and Marrè, 1988). The finding that the depressing action of ferricyanide on glucose-6-P level was larger than that on malate was in agreement with the proposal by Sijmons et al. (1984) that NADPH is the physiological electron donor for ferricyanide reduction.

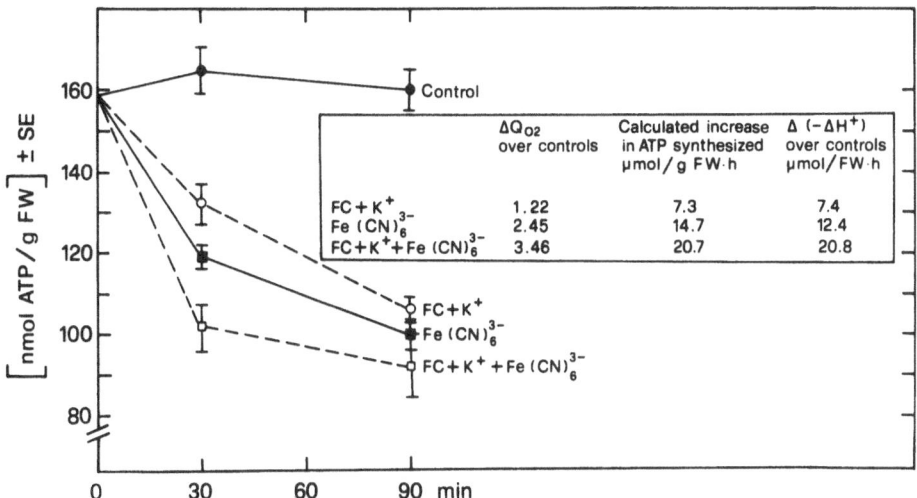

Fig. 1. Effects of ferricyanide and of FC on ATP level and
on Q_{O2} in _Elodea_ _densa_ leaves.

Measurements of respiration showed that ferricyanide reduction is associated with a marked increase in Q_{O2}, approximately corresponding to one mol O_2 evolved per 6 mol of protons extruded. This stoichiometry is that expected with a P/O ratio of 3, and an ATP molecule hydrolized per proton extruded by the H^+ ATPase (Marrè E. and Ballarin-Denti, 1985). On the other hand, in _Elodea_ leaves ferricyanide induced a marked decrease of ATP level, presumably dependent on the ATP utilization by the H^+ extruding mechanism.

Both the $\Delta Q_{O2}/-\Delta H^+$ ratio value and the decrease in ATP level presented very similar features in the presence of either FC alone, or ferricyanide alone, or the two agents together, thus providing further support to the view that a single mechanism, i.e. the ATP-driven pump, mediates H^+ extrusion in the absence as well as in the presence of ferricyanide (Trockner and Marrè E., 1988).

Relationships between the ferricyanide reducing redox chain and the ATP-driven H^+ pump

Convincing evidence exists that the H^+ pump is activated both by a decrease of cytosolic pH and by a depolarization of E_m (Bellando et al., 1979; Cerana et al., 1981; Marrè M.T. et al. 1982; Romani et al., 1985). This provides a satisfactory interpretation of the stimulation of ATP-driven H^+ extrusion by the cytosolic acidification and the depolarization associated with ferricyanide reduction. On the other hand, an increase in the rate of ferricyanide reduction following treatment with FC has been repeatedly reported (Crane and Craig, 1982; Rubinstein and Stern, 1986; Federico and Giartosio, 1983), and is also evident (although with a large variability) in _Elodea_ leaves. As a possible interpretation of this effect we propose that the activity of the redox system might be to some extent limited by the associated cytoplasm acidification and E_m depolarization and that FC, by increasing cytoplasmic pH and hyperpolarizing E_m, antagonizes the opposite effects of ferricyanide reduction on these two parameters. Some support to this interpretation comes from the finding that cytoplasm acidification induced by treatment with isobutyric acid inhibits ferricyanide reduction by about 30% in _Elodea_ leaves. Whatever be the mechanism of FC action on ferricyanide reduction, the possibility that the activity of the redox system may be regulated by that of the ATP-driven pump (just as the latter system may be by that of the former) seems of importance as far as the physiological role of the two systems is concerned.

Interactions between photosynthesis, ATP-driven H^+ extrusion and ferricyanide reducing system

Previous work showed that light strongly stimulates in _Elodea_ leaves electrogenic H^+ extrusion associated with K^+ uptake. This effect is maximal for the light frequencies inducing photosynthesis, depends on the presence of. CO_2 and is suppressed by DCMU. It also shows a strict requirement for the presence of K^+ in the medium and is strongly inhibited by the inhibitors of plasmalemma H^+-ATPase, vanadate and erythrosine B (Marrè M.T. et al., 1986; Albergoni et al., 1987). It appears therefore that photosynthesis markedly activates the ATP-driven proton pump by some yet unknown mechanism, presumably related to a product of photosyntetic CO_2 fixation.

If this conclusion is valid, and on the basis of the results presented in the above sections, the interaction of the ferricyanide reducing system response to light should present the same features as its interaction with FC-stimulated H^+ extrusion. The results of Fig. 2 show that this indeed is the case. In fact, as shown in Fig. 2, light was found to stimulate ferricyanide reduction to about the same extent as FC, the effects of the two treatment being practically non additive. Moreover, the effect of light on intracellular pH were very similar to that of FC: both treatments induced an alkalinization, partially counteracting the acidification associated with ferricyanide reduction in the light as well as in the dark.

These results are interpreted as indicating that photosynthetic activity may regulate the operation of both the redox system and the ATP-driven H^+ pump.

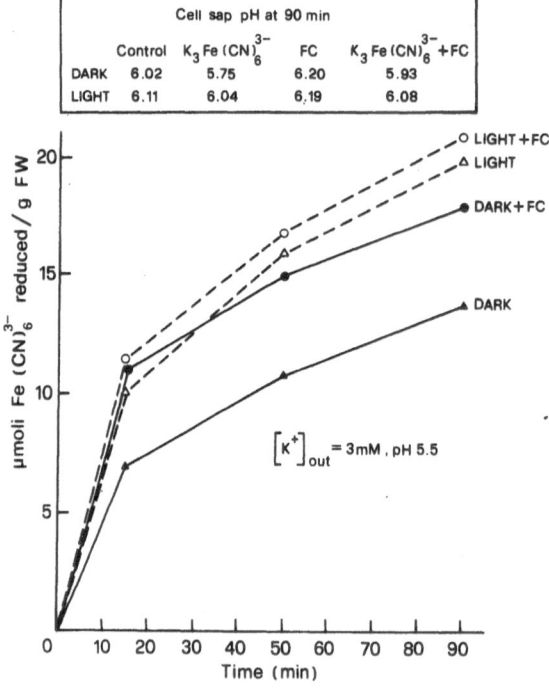

Fig. 2. Effects of light and FC on $Fe(CN)_6^{3-}$ reduction and on cell sap pH (inset) in _Elodea densa_ leaves.

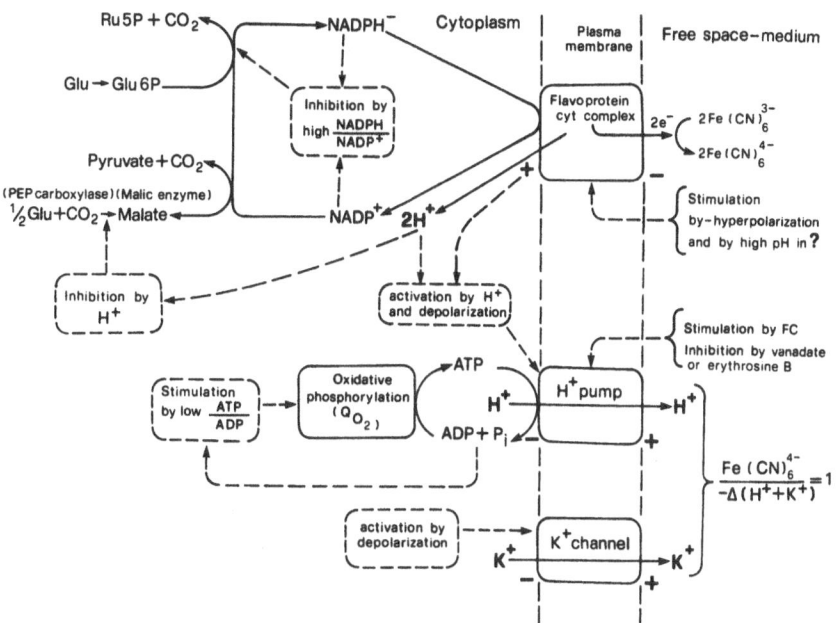

Fig. 3. Model for the relationships between the ferricyanide-reducing redox system and the ATP-driven H^+ pump and for the effects of H^+ transport activators (light, fusicoccin) and inhibitors on electron, proton and H^+ fluxes and metabolism.

Proposed model for the relationship between redox system, ATP-driven H^+ pump and metabolism

On the basis of the results obtained in Elodea, integrated with those reported by other authors for other materials, we propose as a working hypothesis the model presented in Fig. 3. According to this model, the redox system transports electron only, 2 protons being released in the cytoplasm per two electrons transported to ferricyanide. The main intracellular reductant "in vivo" would be NADPH, even if a participation of NADH is probable as suggested by the stimulating effect of ethanol on ferricyanide reduction (Craig and Crane, 1982). The H^+ extrusion associated to ferricyanide reduction would depend on the stimulation of the ATP-driven pump by both the cytoplasmic acidification and the depolarization due to the electron efflux through the redox system. The decrease in ATP and the increase in Q_{O_2} would be due to ATP hydrolysis by the ATP-driven pump. The changes in malate and glucose-6-P levels would depend on the cytosolic pH changes induced, in opposite directions, by the operations of the redox system and of the ATP-driven pump, respectively. The reciprocal interactions between the two systems would be mediated by their opposite effects on intracellular pH and on E_m. Thus, factors influencing the ATP-driven pump (such as FC, probably hormones such as auxin and absisic acid, and photosynthesis products) would influence the redox chain, and, conversely, factors influencing the latter would influence the former system. Obviously, the model is incomplete, and its only value is that of covering most of the evidence available up to now, and to open a way to further investigations.

Physiological significance

What is known concerning the plasmamembrane redox chain and its relationship with proton extrusion is sufficient to postulate an important physiological role for this system. It seems clear that its rate of operation can influence a number of important cell parameters. Among these are: a) the pH values in the cytoplasm and in the apoplast, with serious implications for metabolism, transport, growth and differentiation; b) the transmembrane electrical potential difference, thus a number of transport processes; c) the oxidation/reduction state of catalytic proteins in the plasmamembrane; d) the oxidation/reduction state and, in general, the biochemistry of a variety of compounds in the apoplast. Furthermore, and if the results suggesting that the redox chain may influence the NADPH/NADP ratio (SiJmons et al., 1984; Trockner and Marrè E., 1988) are confirmed, then changes in activity of the redox chain might influence the rate of pentose phosphate synthesis by the oxidative pentose phosphate pathway, the reduction state of important sulfhydryl-containing regulators such as thioredoxins, etcetera.

On the other hand, any more definite speculation on the physiological significance of the plasmamembrane redox chain in the absence of artificial extracellular electron acceptors (i.e. with O_2 as the presumable terminal acceptor) depends on the answer to a number of questions. Among these, a) what is the rate of the electron flux, and how is it regulated?; b) which is (are) the initial metabolite reductant(s) and which are the intermediates on the way towards O_2?; c) where is O_2 reduced, in the cytoplasm or in the apoplast (i.e., is there an apoplast respiration?). Finding the answers to these questions represents an highly stimulating challenge to plant physiologists.

AKNOWLEDGEMENTS

This work was supported by the Italian National Research Council (C.N.R.)

REFERENCES

Albergoni, F. G., Marrè, M. T., Trockner, V., Beffagna, N., Romani, G , and Marrè, E., 1987, Changes of vacuolar and cytoplasmic pH associated with the activation of the H^+ pump in Elodea leaves, in: "Plant Vacuoles. Their Importance in Plant Cell Compartmentation", B.P. Marin, ed., NATO ASI Series, Plenum Publ. Corporation, in press.

Albergoni, F. G., Marrè, M.T., and Marrè, E., 1988, Photosynthesis-induced activation of electrogenic H^+ extrusion in Elodea densa leaves, VII International Workshop on Plant Membrane Transport, Sydney, August 1986, in press.

Bellando, M., Trotta, A., Bonetti, A., Colombo, R., Lado, P., and Marrè, E., 1979, Dissociation of H^+ extrusion from K^+ uptake by means of lipophilic cations, Plant Cell Environm., 2:39.

Böttger, M., and Lüthen, H., 1986, Possible linkage between NADH-oxidation and proton secretion in Zea mays L. roots, J. Exp. Bot., 37:666.

Busa, W. B., and Nuticelli R., 1984, Metabolic regulation via intracellular pH. Am. J. Physiol., 246:409.

Cerana, R., Bonetti A., Colombo, R., and Lado P., 1981 Tributylbenzylammonium ($TBBA^+$)-dependent fusicoccin (FC)-induced H^+ extrusion in maize roots: relationship between the stimulating effects of $TBBA^+$ on H^+ extrusion and on Cl^- efflux, Planta, 152:202.

Craig, T. A., and Crane, F. L., 1982, Hormonal control of a transplasmamembrane electron transport system in plant cell, Proc. Ind. Acad. Sci., 91:150.

Elzenga, J.T.M., and Prins, H.B.A., 1987, Light induced polarity of redox reactions in leaves of Elodea canadensis michx, Plant

Physiol., 85:239.

Federico, R., and Giartosio, C.E., 1983, A transplasmamembrane electron transport system in maize roots, Plant Physiol., 73:182.

Ivankina, N. G, Novak, V. A., and Miclashevich A.I., 1984, Redox reactions and active H^+-transport in the plasmalemma of Elodea leaf cells, in "Membrane Transport in Plants", W.J. Cram, K. Janacek, R. Rybova, S. Sigler, eds., J. Wiley & Sons, England.

Lass, B., Thiel, G., and Ullrich-Eberius, C. I., 1986, Electron transport across the plasmalemma of Lemna gibba G1, Planta, 169:251.

Marrè. E., and Ballarin-Denti, A., 1985, The proton pumps of the plasmalemma and the tonoplast of higher plants, J. Bioenerg. Biomembr., 17:1.

Marrè, E., Romani, G., Beffagna, N., and Trockner, V., 1986, Intracellular pH alkalinization associated with the fusicoccin- and K^+-induced activation of the H^+ pump in Elodea densa leaves, VII International Workshop on Plant Membrane Transport, Sydney, August 1988, in press.

Marrè, E., Romani, G., and Beffagna, N., 1987, Potassium transport and regulation of intracellular pH in Elodea densa leaves, Botan. Acta, 101:17.

Marrè, M. T., Albergoni, F. G., and Marrè, E., 1987, Preliminary observations on the integration of photosynthesis with solute transport. II. Photosynthesis-induced activation of the vanadate-sensitive electrogenic proton pump in Elodea densa planch leaves, Rend. Accad. Naz. Lincei, 80:233.

Marrè, M.T., Moroni, A., Albergoni, F.G., and Marrè, E., 1988, Plasmalemma redox activity and H^+ extrusion. I. Activation of the H^+-pump by ferricyanide-induced potential depolarization and cytoplasm acidification, Plant Physiol., in press.

Marrè, M. T., Romani, G., Cocucci, M., Moloney, M. M., and Marrè, E., 1982, Divalent cation influx, depolarization of the transmembrane electric potential and proton extrusion in maize root segments. in "Plasmalemma and Tonoplast: Their Functions in the Plant Cell", D. Marmè, E. Marrè, R. Hertel, eds., Elsevier Biomedical Press, Amsterdam.

Marrè, M. T., Romani, G., Bellando, M., and Marrè, E., 1986, Stimulation of weak acid uptake and increase in cell sap pH as evidence for fusicoccin-induced cytosol alkalinization, Plant Physiol., 82:316.

Rasi-Caldogno, F., De Michelis, M. I., Pugliarello, M. C., and Marrè, E., 1986, H^+ pumping driven by the plasma membrane ATPase in membrane vesicles from radish: stimulation by fusicoccin, Plant Physiol., 82:121.

Romani, G., Marrè, M. T., Bellando, M., Alloatti, G., and Marrè, E., 1985, H^+ etrusion and potassium uptake associated with potential hyperpolarization in maize and wheat root segments treated with permeant weak acids, Plant Physiol., 79:734.

Rubinstein, B., and Stern, A. I., 1986, Relationship of transplasmalemma redox activity to proton and solute transport by roots of Zea mays., Plant Physiol., 80:805.

Sijmons, P. C., Lanfermeijer, F. C., de Boer, A. H., Prins, H. B. A., and Bienfait, H. B. F.,1984, Depolarization of cell membrane potential during trans-plasmamembrane electron transfer to extracellular electron acceptors in iron-deficient roots of Phaseolus vulgaris L., Plant Physiol., 76:943.

Trockner, V., and Marrè, E., 1988, Plasmalemma redox chain and H^+ extrusion. II. Respiratory and metabolic changes associated with fusicoccin-induced and with ferricyanide-induced H^+ extrusion, Plant Physiol., in press.

THE RELATIONSHIPS BETWEEN FERRICYANIDE REDUCTION AND

ASSOCIATED PROTON EFFLUX IN MECHANICALLY ISOLATED

PHOTOSYNTHETICALLY COMPETENT MESOPHYLL CELLS

Alan Bown, Lesley Crawford and Damian Rodriguez

Department of Biological Sciences
Brock University
St. Catharines, Ontario L2S 3A1, Canada

INTRODUCTION

The plasma membrane of plant cells contains an outwardly directed, proton translocating, electrogenic ATPase. The activity of this enzyme establishes both an electrical gradient and pH gradient across the plasma membrane (Sze 1985). The resulting proton electrochemical gradient favours proton entry into the cell and this downhill flux of protons is coupled to and drives the accumulation of sugars and amino acids (Reinhold and Kaplan 1984). Substantial evidence also suggests that a plasma membrane redox enzyme can reduce exogenous electron acceptors using an endogenous electron donor. Associated with this redox activity is proton efflux (Crane et al 1985). The mechanism coupling the reduction of an exogenous oxidant to proton efflux is contentious. In particular it is not clear whether the redox enzyme transfers both electrons and protons to the external medium or whether the redox activity transfers electrons only and thereby activates the proton-translocating ATPase. Some workers suggest that the redox enzyme catalyses an electrogenic transfer of electrons, which depolarizes the plasma membrane reducing the proton electrochemical gradient and activating the plasma membrane ATPase (Rubinstein and Stern 1986, Lass et al 1986). Others have noted an apparent 1:1 stoichiometry between electron and proton movements and the apparent insensitivity of redox activity dependent proton efflux to ATPase inhibitors. They suggest that a single redox system is responsible for movement of both electron and protons (Crane et al 1985, Craig and Crane 1985, Neufeld and Bown 1987, Bown and Crawford 1988).

Light stimulates the reduction of an exogenous electron acceptor by the photosynthetic tissue of Lemna gibba (Lass et al 1986) and Elodea canadensis (Elzenga and Prins 1987) and by mesophyll cells from oat (Dharmawardhane et al 1987). In isolated mesophyll cells of Asparagus sprengeri light stimulates both reduction of exogenous ferricyanide and proton efflux (Neufeld and Bown 1987). These light stimulated processes appear to be mediated by photosynthesis since DCMU (3-(3,4-dichlorophenyl)-1, 1-dimethylurea) an inhibitor of non-cyclic electron flow inhibits ferricyanide reduction and associated proton efflux. It has been suggested that stimulation by photosynthesis of events at the plasma membrane involves the export of triose phosphate from the chloroplast and its subsequent oxidation to generate reduced NAD or NADP (Dharmawardhane et al 1987, Neufeld and Bown 1987). However, DCMU did not completely eliminate white light stimulated redox reactions in Asparagus mesophyll cells and the possibility of other light stimulated processes being involved in redox activity has not been excluded. Blue light activates numerous physiological activities, stimulates the reduction of a plasma membrane located cytochrome b protein (Widell et al 1982) and induces proton efflux from guard cells (Shimazaki et al 1986).

In the present study mechanically isolated photosynthetically competent mesophyll cells from Asparagus sprengeri were employed. The possible involvement of the proton translocating plasma membrane ATPase in redox activity dependent proton efflux was investigated. The ATPase was inhibited and subsequent addition of exogenous ferricyanide was employed to determine whether proton efflux was still observed. In addition, the role of light and photosynthesis in redox dependent proton efflux was investigated using a pH electrode and O_2 electrode to simultaneously measure changes in rates of proton efflux and oxygen evolution.

METHODS

Asparagus sprengeri Regel was grown under greenhouse conditions. Mesophyll cells were isolated using a technique of gentle mechanical disruption (Colman et al 1979). Cells were prepared daily and counted using a light microscope and a hemocytometer. Chlorophyll was determined by the method of Arnon (1949).

Cells were suspended in 10 or 20 ml of 1 mM $CaSO_4$, aerated and stirred at 30°C. Acidification rates were determined with a Radiometer recording pH apparatus. To reduce light stimulated rates of redox activity- dependent acidification to measurable levels 1 to 3 mM MES-Ca(OH)$_2$ pH 6.2 was added to the medium. When ATP dependent proton efflux was invest-igated the rate was stimulated by addition of 10 mM K_2SO_4. The buffering capacity of the medium was determined at the beginning and end of experiments to determine whether any of the experimental procedures changed the buffering capacity. Net rates of proton efflux in nmol $H^+/10^6$ cells.min were calculated using the rate of acidification, the appropriate buffering capacity and the number of cells (Bown 1982).

Ferricyanide concentrations were determined spectrophotometrically. Approximately 15×10^6 cells were suspended in 20 ml of medium. At 5 min intervals 2.5 ml aliquots of the cell suspension were removed, centrifuged to remove the cells and the difference in the absorbance of the supernatant fluid at 420 and 500 nm determined. Using an extinction coefficient of 1.0 mM^{-1} cm^{-1} at 420 nm the concentration of ferricyanide was obtained and rates of reduction calculated as nmol ferricyanide/10^6 cells.min. Where indicated the supernatant fluid was also analysed for ferrocyanide concentration using a colorimetric method and an extinction coefficient of 10.8 mM^{-1} cm^{-1} (Avron and Shavit 1969).

Rates of proton efflux and O_2 evolution were recorded simultaneously using a Radiometer combination pH electrode and an O_2 electrode (Clark YSI

4004) connected to two strip chart recorders set at a rate of 2 min per cm. The two electrodes were inserted through a tapered polyethylene cone which contained a third aperture for a Hamilton syringe needle. The polyethylene cone slid into an open water jacketed vessel with a tapered inside wall (taper stopper 34/45) producing an enclosed space for the stirred 10 ml cell suspension. Prior to measurements the cell suspension was bubbled with nitrogen to remove oxygen and carbon dioxide.

Cell suspensions were illuminated with a 300W reflector lamp (Sylvania) which gave an irradiance at the surface of the glass incubation vessel of 1.7×10^4 μWcm^{-2}. For blue light the value was 5.3×10^1 $\mu W\,cm^{-2}$. Stock solutions of potassium ferricyanide and DABS (diazobenzenesulfonate) were prepared daily. Stock solutions of DES (diethylstilbestrol) and DCCD (dicyclohexyl-carbodiime) were prepared using 80% (v/v) ethanol. All concentrations quoted are the final value after dilution in the cell suspension.

RESULTS AND DISCUSSION

General properties of the redox activity

The rate of disappearance of ferricyanide from the illuminated cell suspension medium was linear over a 30 min period, and was accompanied by a stoichiometric accumulation of ferrocyanide (Fig. 1). These results establish that the decline in ferricyanide concentration in the medium was not due to ferricyanide breakdown or uptake into the cell.

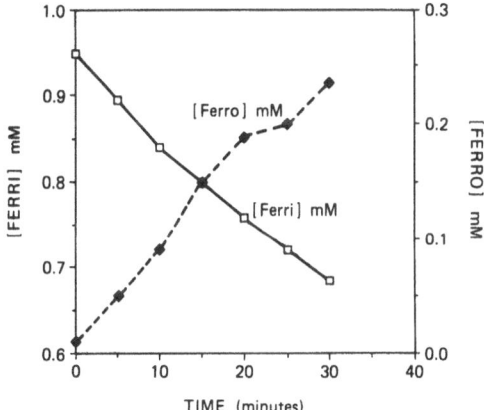

Fig. 1. Rates of ferricyanide disappearance and ferrocyanide appearance in the medium of illuminated cell suspensions. Each point is the mean of 4 determinations obtained with different cell preparations.

The reduction of ferricyanide was also accompanied by a stoichiometric efflux of protons. These two processes were stimulated considerably by white light. However, the molar ratios of ferricyanide disappearance, ferrocyanide appearance and proton efflux approached 1:1:1. in both light and dark. The non-permeant DABS when present at a concentration of 1 mM eliminated ferricyanide reduction but inhibited white light stimulated H^+ efflux by 64% and the rate in the dark by 56% (Table 1). The results confirm recent reports of

TABLE 1. Rates of ferricyanide disappearance, ferrocyanide appearance, and H+ efflux in the presence or absence of illumination.

CONDITIONS	FERRICYANIDE DISAPPEARANCE	FERROCYANIDE APPEARANCE	H+ EFFLUX
	nmol/min.10^6 cells ± SD		
+ ILLUMINATION - DABS	27.2 (5.0)	27.6 (7.4)	22.5 (5.8)
+ ILLUMINATION + DABS	0	0	8.1
- ILLUMINATION - DABS	0.8 (0.1)	0.7 (0.1)	0.9 (0.2)
- ILLUMINATION + DABS	0	0	0.4 (0.3)

Rates were determined over a 30 min period using a 20 ml cell suspension. DABS and ferricyanide were added to give a final concentration of 1 mM. In the light 10 to 15 x 10^6 cells were suspended in 3 mM Ca (OH)$_2$ pH 6.2. In the absence of light 15 to 20 x 10^6 cells were suspended in 1 mM MES Ca(OH)$_2$ pH 6.2. Each figure is the mean of 3 determinations.

light stimulated redox activity at the plasma membrane of photosynthetic cells (Dharmawardhane et al 1987, Elzenga and Prins 1987, Neufeld and Bown 1987) and reports of a 1:1 stoichiometry between ferricyanide reduction and H+ efflux (Crane et al 1985, Ivankina et al 1984). The inhibition of redox activities by DABS has been reported previously. However, in contrast to the present data, the percentage inhibition of reduction and proton efflux in carrot cells appeared to be equal (Craig and Crane 1985).

Ferricyanide reduction and proton production was not measurable when ferricyanide was added to medium which had been used to suspend illuminated cells for 20 min. In addition cell suspensions which had been illuminated for 20 min prior to ferricyanide addition did not stimulate rates above those rates obtained when the ferricyanide was added immediately (Table 2). These two experiments indicate that redox activity involves neither secretion of a reductant, nor secretion of a reductant which reduces ferricyanide in the presence of an enzyme located in the cell wall or on the outer face of the plasma membrane. Thus in contrast to results obtained with soybean roots no evidence for secretion of malate and subsequent redox activity in the cell wall space was obtained (Tipton and Thowsen 1985).

The absence of data supporting the secretion of a reductant but data demonstrating the light stimulated reduction of an exogenous electron acceptor indicates the presence of a light stimulated redox process across the plasma membrane of Asparagus mesophyll cells.

Coupling of ferricyanide reduction to proton efflux

An ATP-dependent ferricyanide-independent proton efflux mechanism in Asparagus mesophyll cells has been documented (Bown 1982). This process

246

TABLE 2. Light stimulated rates of ferricyanide disappearance, ferrocyanide appearance and H+ efflux with respect to the possible release of a reductant to the medium.

CONDITIONS	FERRICYANIDE DISAPPEARANCE	FERROCYANIDE APPEARANCE	H+ EFFLUX
	$nmol/min.10^6$ cells $\pm$ S D		
A CELLS +	10.4	10.7	11.8
MEDIUM	12.3	12.9	13.1
	31.8	31.4	33.5
B CELLS +	11.6	11.1	10.2
CONDITIONED	9.8	9.5	10.2
MEDIUM	29.3	30.3	30.1
C MEDIUM	0	0	0
ALONE	0	0	0

In A, ferricyanide was added to illuminated cell suspensions and measurements made over a 20 min period.

In B, the cell suspension was illuminated for 20 min then ferricyanide added and measurements made over the subsequent 20 min.

In C, the cell suspension was illuminated for 20 min, it was then centrifuged and ferricyanide added to the supernatant fluid. Measurements were made over the subsequent 20 min.

is stimulated optimally by 20 mM potassium (data not shown) and is stimulated after the addition of 0.01 mM fusicoccin (FC) by a further 340% (Fig. 2). Similar rates were obtained in the light or dark. These results indicate that, in common with other tissues, Asparagus cells have a plasma membrane located proton translocating ATPase which is significantly stimulated when FC is added to tissue exposed to a high potassium concentration (Marrè 1979).

To investigate the possible involvement of ATP-dependent proton efflux in redox activity-dependent proton efflux inhibitors of the plasma membrane ATPase were employed. The criterion for complete inhibition of the enzyme was a zero rate of proton efflux which could not be reactivated by application of FC. With non-illuminated cell suspensions these conditions were obtained with 0.1 mM DES or 0.3 mM DCCD. Subsequent addition of ferricyanide, however, stimulated significant rates of proton efflux suggesting that redox activity-dependent proton efflux did not involve the ATPase (Fig. 2). In addition, when the inhibitors were added after ferricyanide, rates of proton efflux were reduced from 1.62 ± 0.38 SD (n=4) nmol H+/million cells.min to 1.43 ± 0.51 SD (n=4) nmol H+/million cells.min in the case of DES and to 0.30 ± 0.18 (n=5) nmol H+/million cells.min in the case of DCCD. Thus DCCD, unlike DES, appeared to inhibit a significant component of the ferricyanide-dependent proton efflux process. DES and DCCD were also used to eliminate FC stimulated ATP-dependent proton efflux in illuminated cells. Subsequent application of ferricyanide resulted in light stimulated rates of proton efflux. Thus light stimulated redox activity also appears to be independent of the ATPase (Bown and Crawford 1988). In the absence of cells DES, DCCD or ferricyanide had no influence on the pH of the medium. The phosphate analogue vanadate inhibits in vitro plasma membrane ATPase. A

0.1 mM concentration inhibited the acidification of the medium without significantly inhibiting the in vivo rate of proton efflux. This resulted from the increase in the buffering capacity of the medium on vanadate addition.

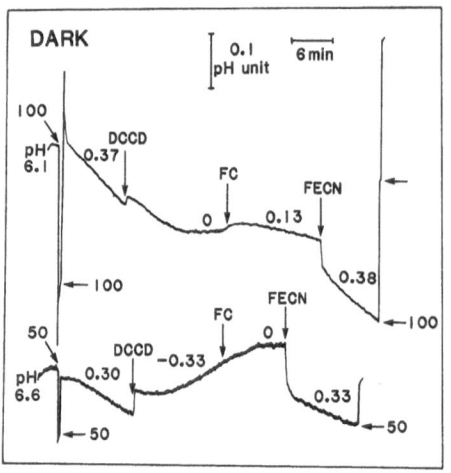

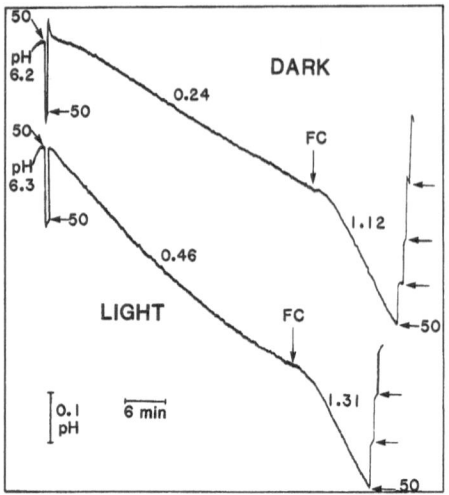

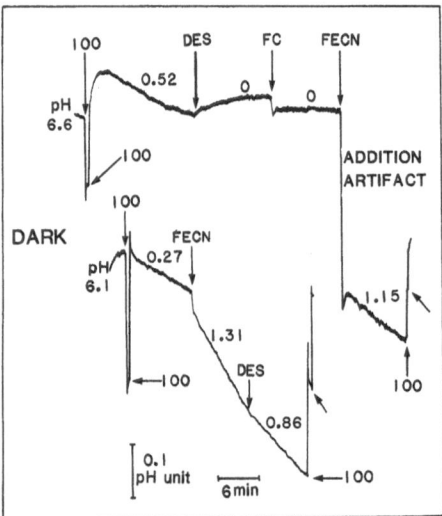

Fig. 2. Ferricyanide-dependent proton efflux in cell suspensions treated with plasma membrane ATPase inhibitors. Cells were suspended in a medium containing 1 mM CaSO₄ and 10 mM K₂SO₄. Additions of 50 or 100 nmol of H⁺ (downward deflection) or OH⁻ (upward deflection) are indicated by arrows. FC (0.01 mM), DES (0.1 mM), DCCD (0.1 mM upper curve, 0.3 mM lower curve) and ferricyanide 1 mM) were added where indicated. Net rates of proton efflux are indicated in nmol H⁺/million cells.min. All experiments were performed with non-illuminated cells, unless indicated otherwise. Published with the permission of Physiol. Plant.

The data suggest that redox activity-dependent proton efflux can be observed in the absence of active proton-translocating ATPase. Thus they are not consistent with proposals that redox activity involves electrogenic transfer of electrons to the exogenous acceptors, depolarization of the plasma membrane and consequent activation of the ATPase (Lass et al 1986, Rubinstein and Stern 1986). They are consistent with suggestions that the plasma membrane redox enzyme transfers both electrons and protons to the external medium (Craig and Crane 1985, Neufeld and Bown 1987). Further evidence is consistent with the tightly coupled transfer of electrons and protons by a single enzyme. The stoichiometry of electron and proton transfer approaches 1 (Table 1 Craig and Crane 1985). This ratio is not predicted by indirect coupling between electron and proton movements. Light appears to stimulate both redox-dependent electron and proton movements (Table 2) whilst having little or no effect on the ATP-dependent proton efflux rate (Bown 1982). Addition of ferricyanide stimulates proton efflux without any significant lag period for depolarization and activation of the ATPase (Neufeld and Bown 1987). In contrast, several authors have demonstrated membrane depolarization on the addition of exogenous electron acceptors (Lass et al 1986, Elzenga and Prins 1987 and Rubinstein and Stern 1986). Depolarization is not predicted by a stoichiometric coupled transfer of electrons and protons.

Light stimulated redox activity

Addition of ferricyanide to illuminated cells resulted in an immediate stimulation of oxygen evolution and proton efflux. Subsequent addition of DCMU completely eliminated oxygen evolution and reduced proton efflux by 90%. The remaining rate of proton efflux, although not associated with oxygen evolution, appeared to be light stimulated since removal of the light resulted in an immediate proton influx process (data not shown). A similar discrepancy between the influence of darkness and DCMU has been observed previously (Neufeld and Bown 1987). The data suggest the existence of more than one mechanism of light stimulated ferricyanide dependent proton efflux. The possible involvement of a blue light stimulated process was investigated using simultaneous measurements of proton and oxygen fluxes.

In the absence of DCMU and illumination ferricyanide stimulated proton efflux without a detectable effect on the rate of oxygen uptake. Blue light (400 to 460 nm) application stimulated proton efflux after a characteristic 2 to 3 min lag period without stimulating oxygen evolution. A subsequent switch from blue light to white light resulted in the large stimulation of both proton efflux and oxygen evolution without measurable lag periods (Fig. 3). To separate the blue light effect from photosynthesis the experiment was repeated in the presence of DCMU. Again blue light stimulated proton efflux was detected in the absence of oxygen evolution. White light however did not stimulate oxygen evolution, and stimulated proton efflux to a much smaller extent. The absence of any change in the buffering capacity of the cell suspension during the experiment indicate that changes in the rate of medium acidification result from changes in the rate of proton efflux (Fig. 4). The blue light effect was dependent on the presence of ferricyanide in the cell suspension. The lack of both oxygen evolution and inhibition by DCMU suggest that blue light stimulated proton efflux is not mediated by photo-synthesis.

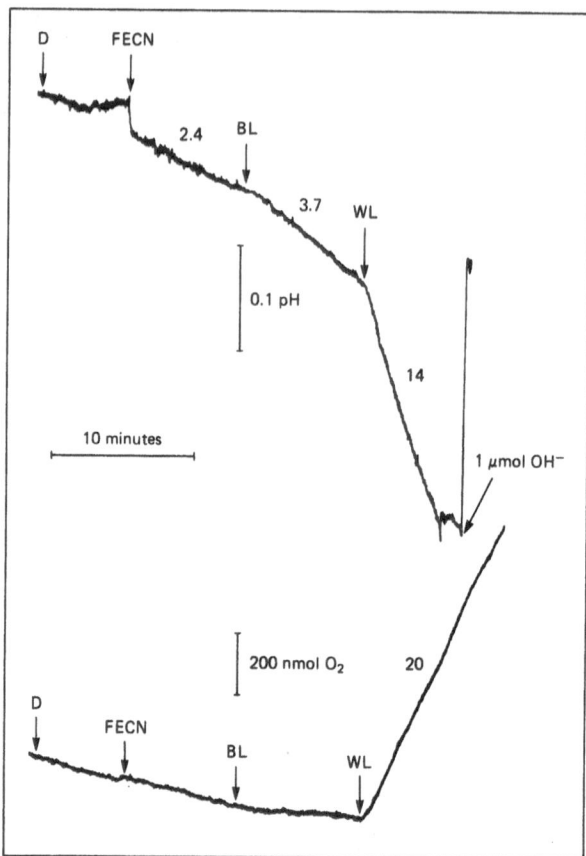

Fig. 3. Blue light stimulated proton efflux. Ferricyanide
(1 mM) was added where indicated. D, BL and WL
indicate darkness, blue light and white light respectively.
Ten ml of 1 mM Ca(OH)$_2$ pH 6.2 contained 11 million
cells. Rates of proton efflux are expressed as nmol H$^+$/
million cells.min. For oxygen rates are expressed as
micromol O$_2$/mg Chl.hr. The upper recording indicates
pH, the lower oxygen concentration.

Blue light stimulated proton efflux by guard cell protoplasts of Vicia
faba also exhibits a 2 to 3 min lag period (Shimazaki et al 1986). In contrast
with the present study, however, the stimulation was obtained in the absence
of an added exogenous reductant. In addition DES completely inhibited blue
light or FC stimulated proton efflux. Vanadate was not inhibitory. The
authors suggest that the plasma membrane ATPase is involved in blue light
stimulated proton efflux. The mechanism and possible physiological
functions of a blue light stimulated plasma membrane redox process remain
to be characterized.

The rate of white light stimulated ferricyanide dependent proton efflux
is much greater than that obtained with blue light. In addition it is mediated
by photosynthesis as indicated by a simultaneous rate of oxygen evolution
(Fig. 3) and its inhibition by DCMU (Fig. 4). The data are consistent with the
stimulation by photosynthesis of electron transport across the plasma

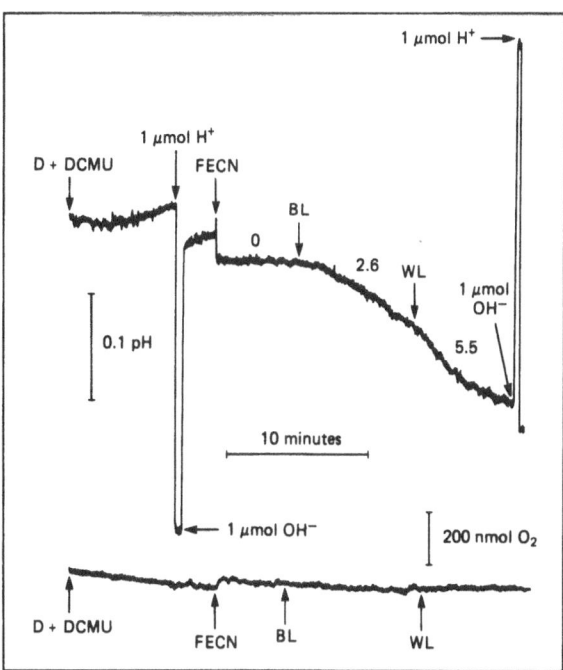

Fig. 4. Blue light stimulated proton efflux in the presence of DCMU. Ferricyanide (1 mM) and DCMU (0.01 mM) were added where indicated. D, BL, and WL indicate darkness, blue light, and white light respectively. Ten ml of 1 mM MES CA(OH)$_2$ pH 6.2 contained 10 million cells. Rates of proton efflux are expressed as micromol O$_2$/mg Chl.hr. The upper recording indicates pH, the lower oxygen concentration.

membrane. This view is supported by DCMU inhibition of ferricyanide reduction in oat mesophyll cells (Dharmawardhane et al 1987) and the light stimulated precipitation of a reduced tetrazolium salt at the plasma membrane of Elodea leaf cells (Elzenga and Prins 1987). Similarities in the kinetic characteristics of the redox activity occurring in illuminated and non-illuminated cells suggests that photosynthesis stimulates a process which operates in the dark (Dharmawardhane et al 1987). The evolution of oxygen in response to reduction at the plasma membrane indicates the presence of a redox chain leading to the water splitting process. However, the possibility that a component of photosynthetically stimulated oxygen evolution, ferricyanide reduction and proton efflux derives from the Hill reaction should not be overlooked. This could arise if tissue or cell preparations were damaged to the extent that non-permeant electron acceptors could rapidly penetrate the plasma and chloroplast envelope membrane barriers to the thylakoid membrane.

ACKNOWLEDGEMENTS

Supported by a grant from the Natural Sciences and Engineering Research Council of Canada.

REFERENCES

Arnon, D.I., 1949. Copper enzymes in isolated chloroplasts. Polyphenol-oxidase in *Beta vulgaris*. Plant Physiol. 24:1-15.

Avron, H. and Shavit, N., 1969. A sensitive and simple method for the determination of ferrocyanide. Anal Biochem 6:549-554.

Bown, A.W., 1982. An investigation into the roles of photosynthesis and respiration in H^+ efflux from aerated suspensions of *Asparagus* mesophyll cells. Plant Physiol 70:803-810.

Bown, A.W. and Crawford, L.A., 1988. Evidence that H^+ efflux stimulated by redox activity is independent of plasma membrane ATPase activity. Physiol. Plant. In press.

Colman, B., Mawson, B.T. and Espie, G.S., 1979. The rapid isolation of photosynthetically active mesophyll cells from *Asparagus* cladophylls. Can. J Bot. 57:1505-1510.

Craig, T.A. and Crane F.L., 1985. The transplasma membrane redox system in carrot cells: inhibited by membrane-impermeable DABS and involved in H^+ release from cells. In Current Topics in Plant Biochemistry and Physiology (D.D. Randall, D.G. Blevins and R.L. Larson eds.), Vol. 4 p. 247, Univ, Missouri, Columbia.

Crane, F.L., Sun, I.L., Clark, M.G., Grebing, C. and Low, H., 1985. Transplasma-membrane redox systems in growth and development. Biochim Biophys. Acta 811:233-264.

Dharmawardhane, S., Stern, A.I. and Rubinstein, B., 1987. Light-stimulated transplasmalemma electron transport in oat mesophyll cells. Plant Science 51: 193-201.

Elzenga, J.T.M. and Prins, H.B.A., 1987. Light induced polarity of redox reactions in leaves of *Elodea canadensis*. Plant Physiol. 85:239-242.

Ivankina, N.G., Novak, V.A., Miclashevich, A.I., 1984. Redox reactions and active H^+ transport in the plasmalemma of *Elodea* leaf cells. In W. J. Cram, K. Janacek, R. Rybova, S. Sigler, eds, Membrane Transport in Plants. J. Wiley & Sons, England, pp. 404-405.

Lass, B., Thiel, G. and Ullrich-Eberius, C.I., 1986. Electron transport across the plasmalemma of *Lemna gibba* G1. Planta, 169:251-259.

Marrè, E., 1979. Fusicoccin: a tool in plant physiology. Ann. Rev. Plant Physiol. 30:273-288.

Neufeld, E. and Bown, A.W., 1987. A plasmamembrane redox system and proton transport in isolated mesophyll cells. Plant Physiol. 83:895-899.

Reinhold, L. and Kaplan, A., 1984. Membrane transport of sugars and amino acids. Annu Rev Plant Physiol 35:45-83.

Rubinstein, B., Stern, A.I., 1986. Relationship of transplasmalemma redox activity to proton and solute transport by roots of *Zea mays*. Plant Physiol. 80:805-811.

Shimazaki, K., Iino, M. and Zeiger, E., 1986. Blue light-dependent proton extrusion by guard-cell protoplasts of *Vicia faba* . Nature 319:324-326.

Sze, H., 1985. H^+ translocating ATPases: advances using membrane vesicles. Annu Rev. Plant Physiol. 36:175-208.

Tipton, C.I. and Thowsen, J., 1985. Fe(III) reduction in cell walls of soybean roots. Plant Physiol. 79:432-435.

Widell, S., Lundborg, T. and Larsson, C., 1982. Plasma membranes from oats prepared by partition in an aqueous polymer two-phase system. On the use of light-induced cytochrome *b* reduction as a marker for the plasma membrane. Plant Physiol. 70:1429-1435.

THE FERRICYANIDE-DRIVEN REDOX SYSTEM AT THE PLASMALEMMA OF PLANT CELLS :

ORIGIN OF THE PROTON PRODUCTION AND REAPPRAISAL OF THE STOICHIOMETRY e^-/H^+

Jean Guern[1] and Cornelia I. Ullrich-Eberius[2]

Laboratory of Plant Cell Physiology, CNRS-INRA, 91198
Gif sur Yvette Cedex, France[1] and Botanisches
Institut, Technische Hochschule, D-6100 Darmstadt,RFA[2]

INTRODUCTION

An external acidification has been found associated with the activity of the ferricyanide-driven redox system at the plasmalemma of animal and plant cells [see 1, 2, 3 for reviews]. This has been interpreted as evidence that the transfer of electrons across the plasmalemma is accompanied by a release of protons from the cells [4, 5]. This conclusion has been formalized by assuming that the redox system at the plasmalemma of plant cells functions as a redox H^+ pump [6, 7].

More recently, the question of the H^+ pumping activity of the redox system has been discussed and evidence has been provided for *Lemna gibba* fronds [8] *Zea mays* roots [9] and *Elodea densa* leaves [10] showing that the redox system probably transfers only electrons across the plasmalemma, protons being excreted by the proton pump ATPase. However, this hypothesis, mainly based on differential effects of inhibitors of the proton pump ATPase on the rates of proton excretion and ferricyanide reduction, is still a matter of controversy [5].

The stoichiometry between the electrons transferred to the external ferricyanide and the apparent proton excretion associated with the ferricyanide reduction appears to vary widely. In *Asparagus* cell suspensions the stoichiometry e^-/H^+ of one measured in different conditions was considered as an argument for a tight coupling between electron and proton transfer [11]. However, the e^-/H^+ ratio appears to vary from 0.8 to 2.5 according to the material studied [1] and no clear cut hypothesis has been provided to account for these variations. Furthermore, the stoichiometry appears more variable with time (from 20 to about 2) during the course of ferricyanide-induced effects in *Zea mays* roots [9].

Aside from the external acidification, a depolarization of the membrane potential with an associated leakage of K^+ ions has been described for different materials [6, 8, 10, 12]. The release of K^+ is most likely to contribute to the equilibrium of charges transferred through the plasmalemma [8, 10].

Here we contribute to the debate about the origin of the external acidification linked to ferricyanide reduction by considering the possibility that the reduction of $[Fe^{3+}(CN)_6]^{3-}$ to $[Fe^{2+}(CN)_6]^{4-}$ anions could result in a scalar proton production. We report a strong modulation of the stoichiometry between the rate of electron transfer and the rate of the apparent proton excretion by the extent of K^+ release induced by the membrane depolarization. We also provide an interpretation for the occurence of the lag period observed between the onset of ferricyanide reduction and external acidification. We finally discuss several models describing tentative relationships between the redox system and the plasmalemma ATPase.

MATERIALS AND METHODS

Lemna gibba fronds were cultivated in aseptic conditions and preincubated for several days on 1 X medium consisting of 1 mM KCl, 1 mM $Ca(NO_3)_2$, 0.25 mM $MgSO_4$, 1 mM NaH_2PO_4, pH 5.7. The fronds were harvested from their culture medium (about 2 g FW for each experiment), quickly rinsed with 1 mM $CaSO_4$, and infiltrated under vacuum during 45 min in a 1 mM $CaSO_4$ solution. They were then packed in a perfusion chamber linked to a reservoir beaker by plastic tubing and perfused in the dark with 1 mM calcium sulfate at flow rates of 15 to 30 ml. min^{-1}, for a total perfusion volume of 45 ml.

The pH of the perfusion medium was maintained at 6.0 by titrating the medium in the reservoir with 5 mM NaOH (Radiometer titrator TTT8Q, Copenhagen, Denmark) to measure the rate of net H^+ efflux. The activity of the redox chain at the plasmalemma was studied by injecting the impermeant external electron acceptor Na_3 or $K_3[Fe^{3+}(CN)_6]$ into the perfusion medium.

The rate of ferricyanide reduction was determined by measuring at intervals the optical density of the perfusion medium at 405 or 420 nm using the corresponding absorbance coefficients.

The release of K^+ associated with the reduction of ferricyanide was measured by injecting $Na_3[Fe^{3+}(CN)_6]$ into the perfusion medium. K^+ concentration was monitored continuously by using a K^+-specific electrode (Radiometer F2312K potassium ions electrode). The response of the electrode was calibrated by flame photometry of aliquots taken at intervals. In parallel experiments, the influence of ferricyanide on the transmembrane electrical potential difference was measured essentially as described previously [8].

RESULTS AND DISCUSSION

<u>A scalar proton production can be associated with ferricyanide reduction</u>

Ferricyanide, as revealed by the neutral character of $K_3[Fe^{3+}(CN)_6]$ solutions and by titrating its aqueous solutions, can be considered as a strong acid bearing three anionic charges in the pH range used in biological experiments. Its reduction to ferrocyanide generates an extra anionic charge with a strong acidic character which is neutralized by a fourth K^+ in the neutral salt $K_4[Fe^{2+}(CN)_6]$. In a solution of K-ferricyanide with calcium sulfate, the concentration of protons is determined by the balance between the concentrations of cations and anions of strong bases and acids i.e. by the strong ion difference SID [13].

$$H^+ - OH^- + K^+ + 2 \times Ca^{2+} - 2 \times SO_4^{2-} - 3x [Fe^{3+}(CN)_6]^{3-} - 4x[Fe^{2+}(CN)_6]^{4-} = 0$$

Equation (1) shows that the reduction of ferricyanide to ferrocyanide decreases SID and consequently acidifies the solution.

Such a behaviour was observed when ferricyanide was reduced by diphenylcarbazide: the ferricyanide reduction monitored by the decrease of the absorbance at 420 nm was associated with a stoichiometric acidification (Fig. 1A). Reciprocally, the oxidation of ferrocyanide to ferricyanide consumes one anionic charge, increases the SID and consequently alkalinizes the solution (Fig. 1B). Fig. 1A and 1B demonstrate that, in agreement with the strong acid character of the anionic charges involved, one proton equivalent was either produced or consumed per electron transferred.

This raises the possibility that scalar productions or consumptions of protons can be associated with redox shifts of the couple ferri and ferrocyanide. More specifically, if electrons alone are transferred across a membrane to the electron acceptor ferricyanide a scalar acidification would result with a ratio electrons fixed to protons produced of one. The scalar production of protons has not been considered as a mechanism accounting for the ferricyanide-induced acidifications reported in the literature for a number of materials. These acidifications have been systematically interpreted as the result of a vectorial proton transfer from the intracellular spaces to the external medium, catalyzed either by the redox system itself or by the proton pump ATPase. However, such a scalar acidification can occur if the redox system transfers only electrons across the membrane or if the electron donor NAD(P)H and the acceptor ferricyanide are used on the same side of the membrane. As a matter of fact, some evidence has been described recently showing that the NAD(P)H-driven ferricyanide reduction could occur without creating a transmembrane proton gradient; the associated acidification likely corresponding to a scalar proton production [14, 15].

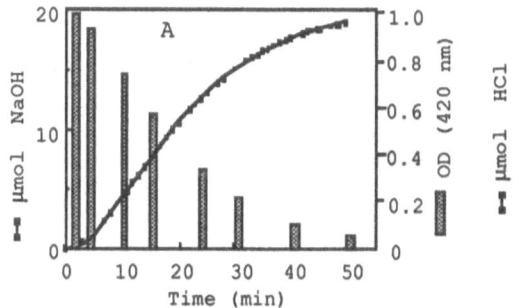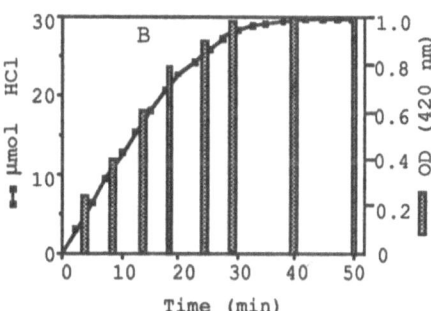

Figure 1. Acidification linked to the reduction of ferricyanide and alkalinization linked to the oxidation of ferrocyanide.

A- Diphenylcarbazide (5 mg) was added to a 20 ml 1 mM $K_3[Fe^{3+}(CN)_6]$ solution in 1 mM $CaSO_4$. The reduction of ferricyanide was monitored by measuring the absorbance decrease at 420 nm (O.D. 1 mM solution: 1.0). The acidification linked to the appearance of one anionic charge per ferrocyanide formed was measured by maintaining the pH of the solution to its initial value (5.5) by titration with 5 mM NaOH.

B- A 30% H_2O_2 solution was added (about 3 mM final concentration) to an aqueous solution of K^+-ferrocyanide (30 ml-1mM). The appearance of ferricyanide was monitored by measuring the absorbance increase at 420 nm (O.D. 1 mM solution: 1.0). The alkalinization linked to the disappearance of one anionic charge per ferrocyanide oxidized was measured by maintaining the pH of the solution to its initial value (6.63) by titration with 5 mM HCl.

Ferricyanide reduction induces membrane depolarization and potassium release

As previously described [8], *L. gibba* cells were depolarized (by about 100mV) when treated with 2.0 mM ferricyanide (fig. 3). The maximum depolarization was reached within about 1 min and was, in some conditions, followed by a slow recovery as already observed for the same material [8] and also in bean roots [12]. This strong depolarization of the membrane potential reflects the transfer of charges linked to the electron flow and indicates a strong electrogenicity of the redox system, at least during the first minutes of ferricyanide reduction.

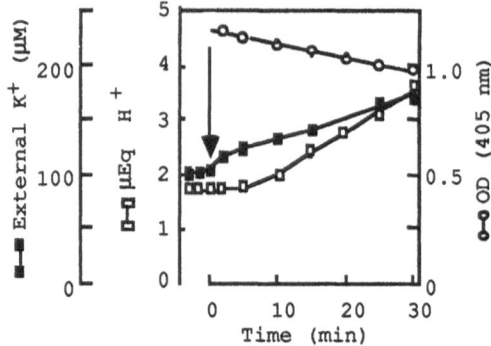

Figure 2. Time dependence of ferricyanide reduction, apparent proton excretion and potassium efflux by *L. gibba* fronds.

L. gibba fronds were first incubated for 96 h on 1X medium in continuous light; then rinsed and vacuum-infiltrated for 45 min with 1mM $CaSO_4$. Infiltrated fronds (1.5 g original fresh weight in 45 ml 1 mM $CaSO_4$) were then perfused in the dark at a flow rate of 30 ml.min^{-1}. After equilibration, 1.2 mM Na^+ ferricyanide was added (arrow). The reduction of ferricyanide was followed by measuring the absorbance at 405 nm (O.D. 1 mM solution: 0.975). The apparent rate of proton excretion was measured by continuously titrating the perfusion medium with 5mN NaOH for an end point pH of 6.0. The K^+ concentration in the perfusion medium was continuously monitored with a K^+ specific electrode and the values obtained were corrected according to flame photometry of the K^+ concentration of aliquots taken at intervals.

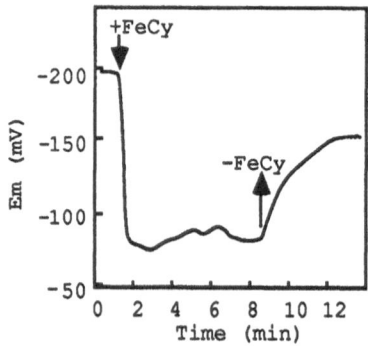

Figure 3. Membrane depolarization of *L. gibba* cells linked to ferricyanide reduction.

Plants pretreated for 96 h on 1 X medium in continuous light, were perfused with 1 X medium pH 5.7 at a flow rate of 15 ml. min^{-1}. Arrows indicate addition and removal of 2.0 mM ferricyanide.

Table I - Comparison of the rates of ferricyanide reduction (e^-), apparent proton excretion (H^+) and potassium efflux (K^+) for L. gibba fronds preincubated on $CaSO_4$ or 1 X medium.

Mean rates were calculated for steady-state situations and expressed as μmol or μEq $.h^{-1}.g^{-1}$ FW (± SE).

	e^-	H^+	K^+	e^-/H^+	$e^-/(H^++K^+$
$CaSO_4$	10.6 ± 0.8	8.6 ± 0.5	1.2 ± 0.4	1.16 ± 0.04	0.98 ± 0.04
1 X medium	10.5 ± 0.9	5.7 ± 0.7	3.4 ± 0.5	1.97 ± 0.14	1.18 ± 0.08

L. gibba frond perfused with a calcium sulfate solution as described in materials and methods reduced ferricyanide at a rate of 6 to 15 μmol. $h^{-1}.g^{-1}$ FW. The disappearance of ferricyanide was associated with an acidification of the external medium and a release of K^+ from the fronds. The intensity of the acidification was low when the fronds were preincubated for several days on 1X medium prior to the perfusion and higher when the fronds were preincubated on 1 mM $CaSO_4$. Conversely, K^+ release was higher for 1X preincubated fronds.

The net efflux of K^+ is most likely driven by the membrane depolarization. Accordingly, the time course of K^+ efflux was characterized by a fast release during the first min of ferricyanide reduction (Fig. 4) followed by a slower efflux at a constant rate, lasting for 30 to 60 min (Fig. 2). When the rate of K^+ efflux calculated for the first 2 min of ferricyanide reduction was compared to the rate of electron transfer, a ratio close to 1 was obtained (Table 2).

Table II - Comparison of the rates of ferricyanide reduction (e^-) to the initial (K^+_i) and steady-state (K^+_s) net rates of potassium efflux.

K^+_i was estimated from the extent of K^+ release during the first 2 min of ferricyanide reduction. K^+_s corresponds to the steady-state rate of efflux reached 10 to 30 min after the onset of ferricyanide reduction. Mean rates are expressed as μmol or μEq $.h^{-1}.g^{-1}$ FW (± SE).

e^-	K^+_i	e^-/K^+_i	K^+_s	e^-/K^+_s
8.5 ± 0.5	9.0 ± 0.9	0.99 ± 0.12	2.8 ± 0.5	3.94 ± 0.98

The stoichiometry between the transfer of electrons and the apparent proton excretion depends on the physiological status of the material and varies with time

As shown in Fig. 2, the ferricyanide-induced acidification of the external medium and release of K^+ proceeded at constant rates only after 5 to 20 min depending on the experiment. Therefore, the rates of these ionic changes were determined when they reached their steady value. Table 1 shows a comparison of the rates of ferricyanide reduction, apparent proton excretion and potassium efflux for a number of experiments. _L. gibba_ fronds preincubated on $CaSO_4$ displayed a mean ratio e^-/H^+ close to 1 while this ratio was close to 2 for fronds preincubated on 1X medium. Correlatively, the rate of K^+ efflux was about 2 fold higher for 1X preincubated fronds. In both conditions the ratio $e^-/(H^+ + K^+)$ was close to 1.

To interpret these different stoichiometries, we consider here the hypothesis that the ferricyanide driven redox system at the surface of plant cells only transfers electrons across the plasmalemma with the following initial consequences (i) the reduction of external ferricyanide (ii) a scalar acidification of the external medium corresponding to the extra anionic charge of the ferrocyanide ion (iii) a strong depolarization of the membrane potential due to the transfer of charges linked to the electron flow (iv) a net efflux of K^+ driven by the membrane depolarization.

Starting from a basal stoichiometry of one corresponding to the initial scalar proton production, the external acidification linked to the production of ferrocyanide ions is counteracted by the efflux of K^+ with a resultant increase of the e^-/H^+ ratio. From equation (1) one can predict that the intensity of the apparent proton excretion should be negatively correlated with the extent of K^+ efflux. This is quite well verified by the results reported in Table 1 and fig. 4 showing that the rate of acidification is low and the ratio e^-/H^+ is high when an important efflux of K^+ is observed. The results obtained here are also in good agreement with those already published on the same material [8] showing that the ratio $e^-/(H^+ + K^+)$ was 1.0 for an e^-/H^+ ratio of 1.72. The same result has been obtained recently with _Elodea densa_ leaves [10] for which the stoichiometry e^-/H^+ was 1.63 while the ratio $e^-/(H^+ + K^+)$ was 1.19.

It is interesting to see that according to the physiological status of _L. gibba_ fronds (i.e. likely their K^+ status prior to the ferricyanide treatment) the stoichiometry can vary widely. When tentatively extended to the different situations reported in the literature one can think that the intensity of K^+ release modulates the stoichiometry e^-/H^+ : values close to 1 [11, 16] corresponding to negligible K^+ effluxes while values of 2.5 reported for corn roots [4] could be linked to strong K^+ release as observed here for _L. gibba_ fronds preincubated on 1X. Stoichiometries lower than 1 such as the value of 0.8 reported for carrot cells [1] are more difficult to interpret with the model proposed here.

A lag period can be observed in some conditions between the beginning of ferricyanide reduction and the onset of external acidification

A clear-cut lag of 10 to 20 min was observed between the injection of ferricyanide and the onset of the external acidification when _L. gibba_ fronds preincubated on 1X medium were used (Fig.4A and B).

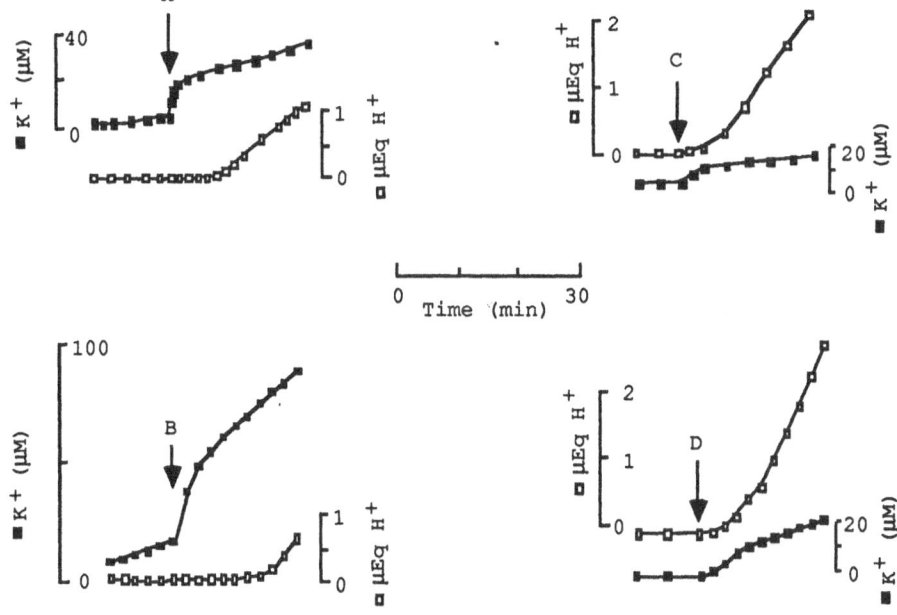

Figure 4. Influence of the extent of K^+ efflux on the duration of the lag phase observed between the onset of ferricyanide reduction and the apparent net H^+ excretion.
Open symbols represent the time course of the apparent proton excretion titrated with 5 mM NaOH (μEq). Black symbols represent the K^+ concentration with time in the perfusion medium (μM). *L. gibba* fronds were preincubated in continuous light, infiltrated with 1 mM $CaSO_4$ and perfused with 40 to 45 ml of the same solution. A : 1.5 g FW preincubated on 1X for 96 h. B : 1.8 g FW preincubated on 1X for 30 h. C : 1.1 g FW preincubated on $CaSO_4$ for 72 h. D : 1.4 g FW preincubated on $CaSO_4$ for 48 h.

The lag period was much shorter for fronds preincubated on $CaSO_4$ (Fig. 4 C and D). Furthermore, long lag periods appeared associated with large K^+ releases.

The existence of a lag period between the beginning of ferricyanide reduction and the onset of extracellular acidification has already been described [9] for Corn roots. Its occurence has not received up to now any interpretation and its existence has even been discussed [11]. The results we report here demonstrate that the lag period corresponds to the large initial burst of K^+ release which counteract the scalar acidification. According to the intensity and duration of this burst, the net acidification of the external medium starts 1 to 20 min after the onset of ferricyanide reduction. Furthermore, from the model proposed here one can predict that the absence of lag should be correlated with a low K^+ release and consequently with an e^-/H^+ ratio close to 1.

Possible interactions between the proton pump ATPase and redox system activities.

This problem is a matter of debate and opposite views have been proposed [5, 9, 10] as to the interactions between the ATPase and the redox system (Fig. 5).

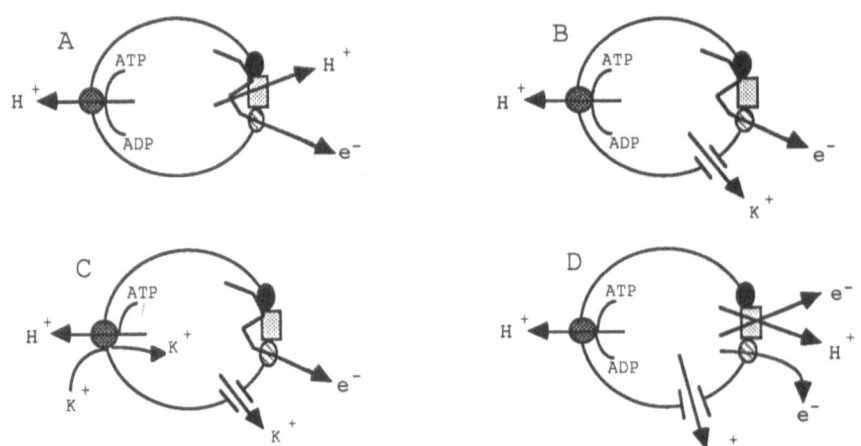

Figure 5. Comparison of different models describing possible relationships between the proton pump ATPase and the ferricyanide-driven redox system.

A- The redox system transfers both electrons and protons across the plasmalemma and acts as a proton pump with a ratio e⁻/H⁺ of 1.

B- The redox system transfers only electrons across the plasmalemma. The associated depolarization induces a K^+ efflux through voltage activated K^+ channels and activates the ATPase.

C- The redox system transfers only electrons across the plasmalemma. The associated depolarization induces a K^+ efflux and stimulates the ATPase. An influx of K^+ is chemically coupled to the efflux of protons.

D- The redox system initially transfers only electrons across the plasmalemma. The associated depolarization induces a K^+ efflux, stimulates the ATPase and modifies the spatial arrangement of the redox system. Ferricyanide can now accept electrons from a different component of the redox chain resulting in the transfer both of electrons and protons.

The first model (Fig. 5A) admits that the redox system transfers both electrons and protons with a stoichiometry of one and considers the redox system as a redox proton pump [5, 6, 7]. It accounts neither for the membrane depolarization, nor for the K^+ release and the variability of the ratio e⁻/H⁺ observed for different materials.

The second model (Fig. 5B) assumes that only electrons are transferred across the redox system. Such a flux induces membrane depolarization, efflux of K^+, cytoplasmic acidification and ATPase activation [9, 10]. The external acidification is assumed to correspond, from the beginning, to the protons excreted by the pump. This model is mainly based on the use of inhibitors of the ATPase which, to a variable extent, inhibit the ferricyanide-induced acidification more than the ferricyanide reduction [9, 10]. This strategy which relies on the assumption that the inhibitors used are really specific gave rather ambiguous results for *L. gibba* fronds preincubated on calcium sulfate and incubated on 1 mM CaSO4 + 50 mM glucose. DES (100 µM) inhibited almost totally the ferricyanide-induced acidification but also blocked ferricyanide reduction.

DCCD (100 μM) suppressed the basal and FC-induced acidifications and, in the same conditions, almost totally blocked the ferricyanide-induced acidification after 20 min while ferricyanide reduction was inhibited by about 75 %. Erythrosin B (100 μM) inhibited both the ferricyanide-induced acidification and ferricyanide reduction by about 60 % after 20 min.

More support is given to this second model by the finding that the cytoplasm is acidified when the redox system is driven by external ferricyanide [10 and our own results not published]. Furthermore, the slow hyperpolarization observed in different circumstances after a few min of ferricyanide treatment [8, 12] is most likely the result of the pump activation triggered by the cytoplasmic acidification.

However, this model does not consider the possibility that at least one part of the acidification observed could correspond to a scalar proton production and it does not give any immediate interpretation of the kinetics of K^+ release and external acidification observed here.

The large ferricyanide-induced depolarization of the membrane measured here and already described on different materials [6, 8, 10, 12] fits well, at first sight, with the proposal that only electrons are transferred through the plasmalemma by the redox system. However simple calculations, assuming a membrane capacitance of $1 \mu F.cm^{-2}$, shows that only a small amount of electrons need to be transferred in order to induce a 100 mV depolarization. Thus, the flux of electrons should be balanced in terms of charges by an efflux of cations or an influx of anions. According to model B, the balance of charges is insured by K^+ efflux and H^+ excretion through the pump.

The results illustrated by fig. 4 show that, during the first minutes of ferricyanide reduction where a large depolarization of the membrane is observed, the efflux of K^+ is particularly important. The near equilibrium of charges transferred across the membrane during the initial activity of the redox system is apparently due to the K^+ efflux driven by the membrane depolarization (table 2). After a few min, the efflux of K^+ was reduced by about 3 (table 2 and fig. 4) and no longer equilibrated the flux of electrons. According to model B, H^+ excretion through the pump can compensate for the charge deficit corresponding to the lower K^+ release.

Models C and D provide alternative proposals to account for the stoichiometry e^-/H^+ and charge compensation during the first and second periods of ferricyanide reduction and more specifically for the decrease of K^+ efflux with time. According to model C (i) the large initial depolarization induces the opening of voltage-activated K^+ channels [17] through which K^+ is released. (ii) The membrane depolarization and (or) the drop of cytoplasmic pH linked to the decrease of cytoplasmic K^+ and (or) the increase of external K^+ activate the proton pump ATPase at the plasmalemma. (iii) The pump catalyzes a K^+ influx, chemically coupled to the proton excretion in a slightly electrogenic (i.e. hyperpolarizing) manner which is responsible for the decreased measured net K^+ efflux (K^+_{net}). (iv) The external acidification is determined by the evolution of $[Fe^{2+}(CN)_6]^{4-}$ and K^+ concentrations in the medium, i.e. by the difference ($e^- - K^+_{net}$). The critical point of this model concerns the postulated influx of K^+ linked to the activation of the pump. This point, which has not been tested up to now in this study, is not supported by the observation that ^{86}Rb influx in Corn roots is inhibited by ferricyanide [9, 18].

According to model D (i) the redox system initially only transfers electrons to ferricyanide with a resultant membrane depolarization, K^+ efflux and pump activation (ii) the membrane depolarization and change of the electric field in the membrane modify the topography of the redox chain in such a way that ferricyanide can accept electrons from a different component of the redox system, now exposed at the outer face of the plasmalemma, which transfers both electrons and protons across the membrane. The reduction of ferricyanide is then less electrogenic, with a lower net negative charge transfer and consequently a lower K^+ efflux. This model is purely speculative as no experimental evidence supports the ad hoc hypothesis that membrane properties such as electrical field, local protonation, changes in the redox state of critical groups, etc, which can differ from one material to the other, determine which component of the chain transfers electrons to ferricyanide.

The simplest model which fits most of the experimental evidence is model B. According to the kinetics of ferricyanide-induced acidification and K^+ release it appears that time is needed to activate the ATPase which indicates that the cytoplasmic acidification more than the membrane depolarization could be the triggering factor. Furthermore, different levels of pump activity prior to the ferricyanide addition could explain the different intensity of the ferricyanide-induced membrane depolarization (measurements made in light, dark or dark + glucose; ref. 8) and the different stoichiometries e^-/H^+ observed.

ACKNOWLEDGEMENTS

Prof. J. GUERN gratefully acknowledges the award donated by the Alexander Von Humboldt foundation. The authors are indebted to Prof. U. Lüttge for stimulating discussions and to H. Schipp von Branitz for his assistance in the computerized acquisition and treatment of data.

REFERENCES

[1]   Crane, F.M., Sun, I.L., Clark, M.G., Grebing, C. and Low H. (1985) Biochim. Biophys. Act. 811, 233-264.
[2]   Moller, I.M. and Lin W. (1986) Annu. Rev. Plant Physiol. 37, 309-334.
[3]   Lüttge, U. and Clarkson D.T. (1985) Progress in Botany 47, 73-86.
[4]   Federico, R. and Giartoso C.E. (1983) Plant Physiol. 73, 182-184.
[5]   Bown, A.W. and Crawford L. (1988) Physiol. Plant., in the press.
[6]   Ivankina, N.G., Novak, V.A. and Miclashevich A.I. 1984 Redox reactions and active H^+-transport in the plasmalemma of Elodea leaf cells. In WJ Cram, K Janacek, R Rybova, S Sigler, eds, Membrane Transport in Plants,J. Wiley & Sons, England, pp 404-405.
[7]   Novak, V.A. and Miklashevich A.I. (1984) Fiziologiya Rastenii, 31, 489-495.
[8]   Lass, B., Thiel, G. and Ullrich-Eberius C.I. (1986) Planta 169, 251- 259.
[9]   Rubinstein, B. and Stern A.I. (1986) Plant Physiol. 80, 805-811.
[10]   Marre, M.T., Moroni, A., Albergoni, F.G. and Marre E. (1988) Plant Physiol., in the press.
[11]   Neufeld , E. and Bown A.W. (1987) Plant Physiol. 83, 895-899.
[12]   Sijmons, P.C., Lanfermeijer, F.C., de Boer, A.H., Prins H.B.A. and Bienfait H.F. (1984) Plant Physiol. 76, 943-946.
[13]   Stewart, P.A.(1981) In How to understand acid-base. A quantitative acid-base primer for biology and medicine.E. Arnold, London.
[14]   Macri, F. and Vianello A. (1986) Plant Science 43, 25-30.
[15]   Komor, E., Thom M. and Maretzki A. (1987) Planta 170, 34-43.
[16]   Blein, J.P., Canivenc, M.C., De Cherade, X., Bergon, M., Calmon, J.P and Scalla R. (1986) Plant Science 46, 77-85.
[17   Schroeder, J.I., Raschke, K. and Neher E. (1987) Proc. Natl. Acad. Sci. USA, 84, 4108-4112.
[18]   Kochian, L.V. and Lucas W.J. (1985) Plant Physiol. 77, 429-436.

FERRICYANIDE CHANGES THE TRANSPORT PROPERTIES OF LAMPROTHAMNIUM

PAPULOSUM PLASMALEMMA

G. Thiel and G.O. Kirst

Fachbereich Biologie, Universität Bremen NW II
2800 Bremen 33, W. Germany

ABSTRACT

External perfusion of the charophyte <u>Lamprothamnium papulosum</u> with medium containing 0.5 mM ferricyanide induced depolarization of the transmembrane potential. The depolarization was associated with a decrease in membrane resistance, presumably due to an increase in K^+ conductance.

During the depolarization cytoplasmic streaming slowed down transiently. The decrease of streaming velocity could be overcome by lowering the external Ca^{2+} concentration or preconditioning the cells with the Ca^{2+} channel blocker La^{3+}. Since cytoplasmic streaming is sensitive to the concentration of cytoplasmic free Ca^{2+} it seems as if $Fe^{3+}Cy$ stimulates Ca^{2+} influx. As a sequence of events we propose: The transmembrane reduction is a depolarizing current. The depolarization triggers the opening of voltage gated Ca^{2+} and/or K^+ channels.

However, the $Fe^{3+}Cy$ stimulated net proton efflux was not a result of the membrane depolarization.

INTRODUCTION

A transmembrane electron current can be elicited across the plasmalemma of the brackish water Characeae <u>Lamprothamnium papulosum</u> by addition of extracellular ferricyanide ($Fe^{3+}Cy$). Similar to redox systems of higher plants extracellular $Fe^{3+}Cy$ reduction by <u>L.papulosum</u> is accompanied by a net proton efflux (Thiel and Kirst 1988). In some plants a stimulated K^+ efflux (<u>Lamprothamnium</u>: Thiel and Kirst 1988, <u>Lemna gibba</u>: Lass et al. 1986, <u>Phasaeolus vulgaris</u>: Sijmons et al. 1984) and an inhibition of K^+ influx (maize roots: Rubinstein et al. 1986) upon $Fe^{3+}Cy$ addition was reported. It is still controversial if the operation of the redox system generates the driving force for the stimulated H^+ and K^+ efflux and if special proteins associated with the redox system serve as a conduit for these fluxes. H^+ and K^+ may, in contrast also move via transporters independent of the redox system. The driving force for H^+ and K^+ fluxes could then originate from secondary effects of the redox reaction. The depolarization of the transmembrane potential (Em) generally observed after addition of $Fe^{3+}Cy$ (Lass et al. 1986, Sijmons et al. 1984) may result in an indirect impact on ion movement across the membrane. Membrane depolarization is reported to alter the transport properties of ion conducting channels in the outer membrane (K^+: Lühring 1986, Ca^{2+}: Shiina and Tazawa 1987, Cl^-: Tyerman et al. 1986).

In this study we provide evidence that $Fe^{3+}Cy$ changes the transport properties of the membrane for K^+ and Ca^{2+}.

Abbreviations: DCCD = dicyclohexylcarbodiimide, Em = electrical membrane potential, Rm = electrical membrane resistance, $(K^+)o(Ca^{2+})o$ = extracellular K^+,Ca^{2+} concentration.

MATERIALS AND METHODS

Lamprothamnium papulosum (Wallr) J.Gr. was grown in tanks with artificial sea water (280–290 mosmol kg^{-1}) as described by Kirst and Wichmann (1987). One day before the experiment young vegetative internodes were isolated from plants and preincubated in medium of the same composition as was used throughout as experimental medium (Table 1). In experiments with an altered $(K^+)o$ the Na^+ concentration was changed to keep the cation concentration constant.

$Fe^{3+}Cy$ reduction was measured photometrically as a decrease of absorption at 420 nm (A420) in a constant flow circuit.

To measure H^+ efflux 10 internodes were incubated in 3 ml unbuffered experimental medium and the decrease of external pH was recorded with a pH meter (Radiometer, Copenhagen). The amount of NaOH used to balance the pH drop was used as direct measure of the net H^+efflux.

The electrical properties of the membrane were measured as recently described (Thiel and Kirst, 1988). Em was estimated by the K^+-anesthesia methode (Shimmen et al. 1976) and Rm by the external electrode method (Okazaki et al. 1984).

Cytoplasmic streaming was observed under the microscope at a magnification of 160. The velocity of the fast moving particles was timed with a stop watch.

Table 1. Experimental media

	control	$Fe^{3+}Cy$	Anesthesia
NaCl (mM)	150	150	45
KCl (mM)	3	3	110
$MgCl_2$(mM)	9	9	9
$MgSO_4$(mM)	7.5	7.5	7.5
$CaCl_2$(mM)	3	3	3
$NaHCO_3$(mM)	2	2	2
$K-Fe^{3+}Cy$(mM)	–	0.5	–
Hepes(mM)	5	5	5
pH	8	8	8

RESULTS

Effect of the $Fe^{3+}Cy$ on the membrane properties

$Fe^{3+}Cy$ (0.5 mM) was introduced into the medium with internodal cells of L.papulosum to test its effect on the membrane potential and membrane resistance. In all cells tested 0.5 mM $Fe^{3+}Cy$ depolarized the membrane and considerably decreased the membrane resistance (Fig. 1). Em depolarized from a resting value of -148 ± 15 mV to a new stable depolarized level of -60.6 ± 9 mV (Table 2).

At the resting state the membrane was only slightly sensitive to changes in $(K^+)_o$ up to 30 mM. In the depolarized state Em changed as a function of $(K^+)_o$. After $Fe^{3+}Cy$ treatment Em decreased with increasing $(K^+)_o$ from 3 mM to 30 mM approximately as predicted by the Nernst equation (Fig. 2).

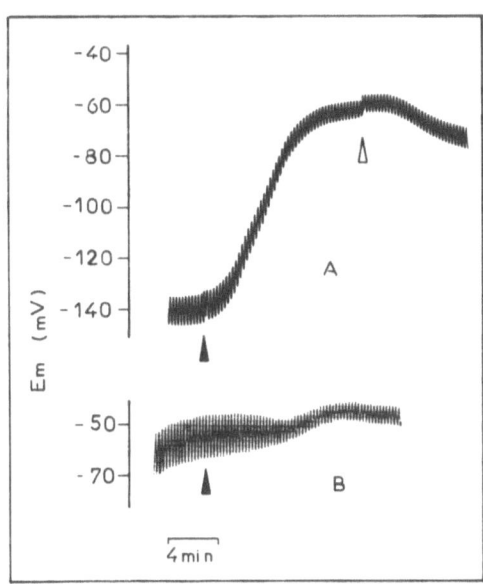

Fig. 1. Characteristic recording of the membrane potential (Em) after addition of $Fe^{3+}Cy$ (solid arrowhead) to the outside perfusion medium of L. papulosum internodes (A). Cell B was preincubated in 50 μm DCCD for 2 h. Sine form current pulses with an amplitude of 23 nA were constantly applied during the measurement to monitor concomitant Rm changes.

The maximum depolarization also in good approximation displayed a $(K^+)o$ dependence. Em levelled off at values more positive for increased $(K^+)o$. The increase in Em with increased $(K^+)o$ followed the predictions of the Nernst equation (Fig. 3).
Rm decreased from 1.0 ± 0.2 m² at the resting state to 0.22 ± 0.2 m² in the depolarized state (Fig.4).

Table 2. Effect of different preconditioning treatments on the transmembrane potential of L. papulosum internodes and on the magnitude of $Fe^{3+}Cy$ induced depolarizations. Mean $\pm$ SD. Number in brackets denotes number of tested cells.

Preconditioning treatment:	$Fe^{3+}Cy$ (0.5 mM)		Em (mV)
	(−) EM (mV)	(+) Em (mV)	
no	−148±15.1 (33)	−60.6±8.9 (26)	87.4
2h 50 μM DCCD	−69.4±5.4 (5)	−58.6±5 (5)	10.8
3 min Anesth. med. 3 min Control	−69.0±15 (4)	−62.1±7 (4)	6.9

Effect of DCCD on the membrane properties

When internodes were incubated for 2 h in medium containing 50 μM of the ATPase inhibitor DCCD Em depolarized to −69.4±5.4 mV and Rm slightly increased (Table 2, Fig. 1). After DCCD removal from the medium and cell perfusion with medium containing 0.5 mM $Fe^{3+}Cy$ Em depolarized only slightly from 69.4±5.4 mV to 58.6±6 mV. Rm similarly decreased as in non-poisoned cells from 1.5±0.7 to 0.4±0.2 m² (Fig. 1,4).

Effect of membrane depolarization on H^+ efflux

To investigate if the stimulated H^+efflux is due either to the trans-membrane redox reaction or generated by the membrane depolarization and the concomitant decrease in membrane resistance we estimated the H^+ efflux after cells have been depolarized by a high external K^+ concentration. With in-creasing K^+ a threshold concentration was reached which changed the mem-brane properties of L.papulosum similar to that produced by Fe^{3+}Cy. Em de-polarized to a K^+ sensitive state and Rm decreased to lower values (unpub-lished data). To assure depolarization into the diffusion state internodes were incubated in anesthesia medium (Table 1) for three min. (Fig. 5 insert, arrow head 1-2). Then cells were transferred to experimental medium causing within 3 min a resting diffusion

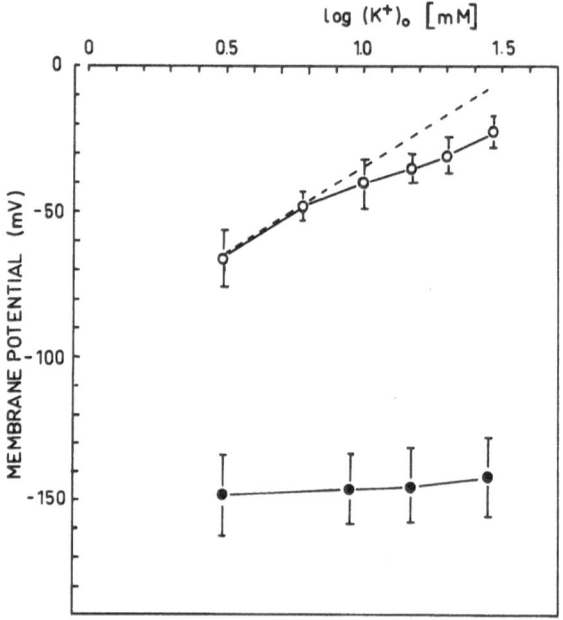

Fig. 2. Effect of $(K^+)o$ on Em before (●) and after (○) depolari-zation by 0.5 mM Fe^{3+}Cy. The dotted line shows the change of Em as predicted by the Nernst equation for increasing K^+ beyond the control value (3 mM). Mean ± SD, (5). Data from Thiel and Kirst 1988, modified.

potential of -70 mV (arrow head 2-3). After this preconditioning H^+ efflux was estimated in fresh experimental medium (control) or in a medium contain-ing Fe^{3+}Cy. H^+ efflux from cells in the depolarized state did not exceed from the cells in the resting state. However, external perfusion of the depolariz-ed cells with solution containing Fe^{3+}Cy, stimulated net proton efflux more than that of non-depolarized cells (Fig. 5).

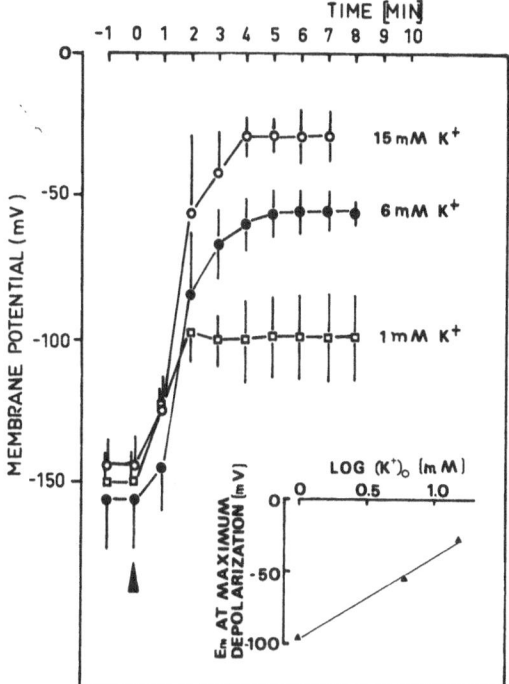

Fig. 3. Time course of $Fe^{3+}Cy$-induced membrane depolarization.
0.5 mM $Na-Fe^{3+}Cy$ was added (arrow head) to the external perfusion medium at different $(K^+)o$.
Insert: Em value of maximum $Fe^{3+}Cy$ induced depolarization as a function
of $(K^+)o$. Mean $\pm$ SD, (5).

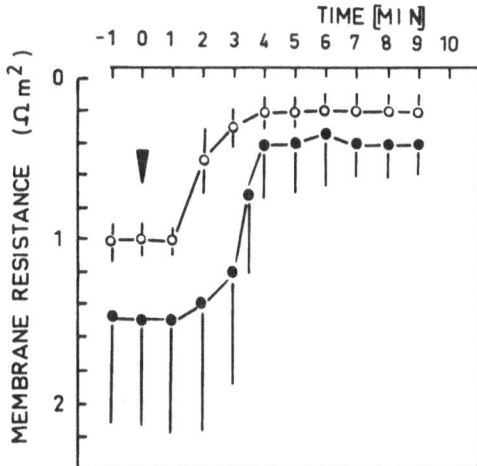

Fig. 4. Change of membrane resistance (Rm) after addition of 0.5 mM
$Fe^{3+}Cy$ (arrow head) to untreated internodes (o) and internodes which
were preincubated in 50 μm DCCD for 2h (●). Mean $\pm$ SD, 6-7 cells
measured per treatment.

Effect of $Fe^{3+}Cy$ on cytoplasmic streaming

$Fe^{3+}Cy$ was added to the external perfusion medium. Within one min the
cytoplasmic streaming slowed down to approximately 10% of the control
value. With time the streaming rate recovered and after 15 min the rate
was back to the original value (Fig. 6A). Since the cytoplasmic streaming
is sensitive to cytoplasmic free Ca^{2+} and ceases in Characeae upon an influx of Ca^{2+} we investigated the effect of $(Ca^{2+})o$ on $Fe^{3+}Cy$ induced
streaming inhibition. Fig. 6 B shows that by reducing the external $(Ca^{3+})o$
cytoplasmic streaming remained uneffected by $Fe^{3+}Cy$ treatment.

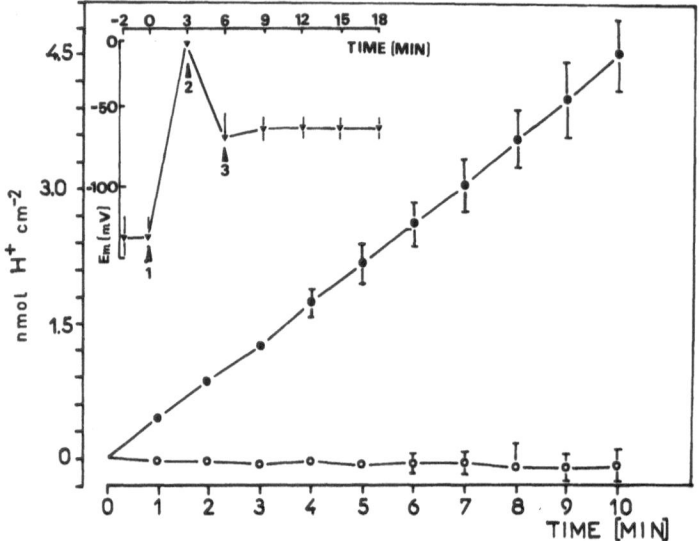

*Fig. 5. Net H^+-efflux into unbuffered medium from L. papulosum
cells in a depolarized state. H^+-efflux was first determined in
a control medium. Then cells were treated as shown in the insert:
3 min in medium of high $(K^+)o$ and 3 min in control medium. After
this treatment H^+-efflux into a unbuffered control medium was
measured with (●) or without (○) addition of 0.5 mM $Fe^{3+}Cy$. Data
are presented as increase of H^+-efflux in respect to the control
efflux. Insert: Increasing the (K^+) in the external medium to
110 mM arrow 1 (Na^+ was proportionally lowered) depolarized cells
to an Em of 0 mV. In the control medium arrow 2 cells repolarized
to -69 mV within 3 min. At arrow 3 the medium was exchanged with
fresh control medium (course of Em not shown) or medium containing
0.5 mM $Fe^{3+}Cy$. Mean ± SD, (5) (insert (5)).*

Internodes were preincubated for 1 h in experimental solution containing 0.5 mM of the Ca-channel blocker LaCl . Following then the streaming
rate in experimental medium, ($-LaCl_3$) $Fe ^+Cy$ did not affect the velocity
of cytoplasmic particles (Fig. 6C).

Low Ca^{2+} did not effect the reduction rate of extracellular $Fe^{3+}Cy$.
After $LaCl_3$ treatment the reduction rate was slightly lower. With respect
to the control the reduction rate was 102% ± 13 (4) in the low Ca^{2+} medium and 92% ± 7 (4) after La-treatment.

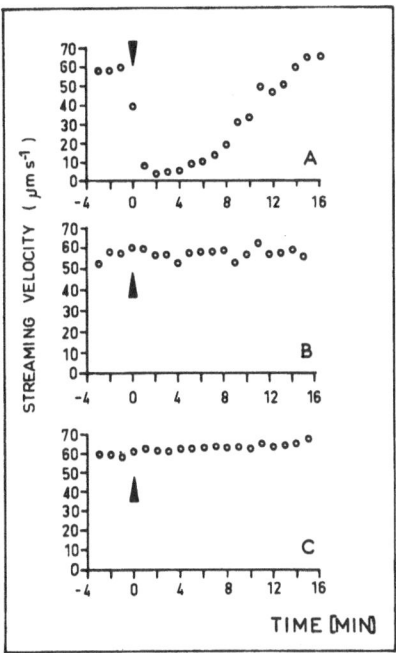

Fig. 6. Effect of 0.5 mM $Fe^{3+}Cy$ (arrow head) on the
cytoplasmic streaming rate of L.papulosum internodes in
different outside media.
(A) Cell in control medium. (B) Cell in medium of low
$(Ca^{2+})o$ (0.01 mM). The cell was pretreated for 5 min in Ca
free medium containing 1 mM (EDTA). (C) Cells in control medium. The
cell was pretreated for 1 h in control medium containing 0.5 mM $LaCl_3$.
Representative examples from 5-7 cells shown in each case.

DISCUSSION

The membrane properties of Characeae are under the control of several
transport systems. The operation of different channels and pumps depends
on the conditions in the outside medium. In Chara corallina at pH less
than Em is dominated by an electrogenic proton pump (Bisson and Walker,
1982). Increasing $(K^+)_o$ above 2.0 mM (Beilby 1985) or depolarizing the
transmembrane potential (Em) to a voltage range less negative than -60 mV
(Ohkawa et al., 1986) shifts the membrane into a state which is dominated
by passive K^+-diffusion, called the K-state.

$Fe^{3+}Cy$ decreased the Em of L.papulosum to a stable potential which is,
with respect to Em and Rm comparable to the K^+-state of Chara corallina.
Rm in the $Fe^{3+}Cy$-induced depolarized state was decreased to 0.2 m^2.
(Fig. 4). In comparison Rm of the Chara membrane drops to 0.2-0.05 m^2
when it is shifted from the pump state into the K^+ diffusion state
(Beilby, 1986).

Depolarizing the membrane of Chara into the K^+-diffusion state by
means of a constant technical voltage resulted in different depolarized
voltage plateaus for different $(K^+)o$. The higher $(K^+)o$ was, the larger
was the depolarization (Ohkawa et al. 1986). The same was found in L.papu-
losum when Em was depolarized by $Fe^{3+}Cy$ in medium with different $(K^+)o$.
With increasing $(K^+)o$ the depolarization resulted in more positive Em
plateaus (Fig.3). The Em values of maximum depolarization were a function
of $(K^+)o$ and obeyed the Nernst equation (Fig.3 Insert). From this we conclude that $Fe^{3+}Cy$ elicits a depolarizing current across the plasmalemma

which is strong enough to shift the membrane from the pump dominated state into the state dominated by passive K^+-diffusion.

A membrane state in L.papulosum characterized by K^+-diffusion was also generated by poisoning the ATPase with DCCD (Thiel and Kirst 1988) or exposing cells to high concentrations of $(K^+)o$ (Fig. 6 and unpublished data). Addition of $Fe^{3+}Cy$ to cells in the inhibitor produced diffusion state caused depolarizations in the range of only 10 mV. A comparable small depolarization was found by addition of $Fe^{3+}Cy$ to cells which were forced into the diffusion state by high $(K^+)o$ (Table 2). We suppose that in the diffusion state the current carried by the excreted electrons is masked by the high membrane conductance. The depolarizing electron current contributes considerably to the total membrane conductance only in the hyperpolarized pump state.

While Rm in the diffusion state produced by high K^+ was low (unpublished data) the membrane resistance increased after DCCD treatment (Fig.2,4). A similar increase in Rm was found in Chara coralline after DCCD treatment (Moriyasu et al. 1984). The rise in Rm is thought to originate in the DCCD blocking of H^+-channels (Kishimoto et al. (1984). In addition an increase in K^+ permeability has been discussed (Keifer and Spanswick 1978). From our data it remains unclear whether $Fe^{3+}Cy$ decreases Rm of the inhibitor treated cells by further increasing the number of open K^+-channels or by activating H^+-channels associated with the redox reaction (Neufeld and Bown 1987).

For the activation of K^+ channels in Chara two mechanisms are discussed: voltage activation and activation by increasing $(Ca^{2+})_i$ (Shimmen and Tazawa 1983). It is likely that $Fe^{3+}Cy$ stimulates Ca^{2+} influx. A rise in $(Ca^{2+})_i$ may than trigger the opening of K^+ channels (Shiina and Tazawa 1987). $Fe^{3+}Cy$ stimulation of Ca^{2+} influx can be deduced from the influence on cytoplasmic streaming. The streaming rate of cytoplasmic particles in charophytes is inhibited by an increase of cytoplasmic free Ca^{2+} (Okazaki and Tazawa 1986) or a decrease of ATP (Reid and Walker, 1983). The observaion that cytoplasmic streaming remained uneffected by $Fe^{3+}Cy$ after preconditioning cells with the Ca^{2+} channel blocker or lowering $(Ca^{2+})o$ can be explained by assuming that Ca^{2+} influx was inhibited. This makes Ca^{2+} a likely candidate to cause the $Fe^{3+}Cy$ induced streaming inhibition.

The presence of a high Mg^{2+} concentration in the experimental medium clearly shows that Mg^{2+} cannot substitute for the Ca^{2+} effect. However, the high Mg^{2+} concentration may mask the inhibitory effect of low Ca^{2+} on the reduction of extracellular $Fe^{3+}Cy$. (Marigo and Belkouera, 1986).

Even a small Ca^{2+} influx into Chara (Shiina and Tazawa 1987) can activate Cl^- channels. Therefore $Fe^{3+}Cy$ could also elicit Cl^- fluxes.

In summary our results indicate that a calculation of stoichiometries upon $Fe^{3+}Cy$ addition becomes complex. Rather a cascade of fluxes seems to be elicited by $Fe^{3+}Cy$.

From our data we cannot decide whether the plasmalemma H^+-ATPase continues to pump or is stalled in the depolarized state. For the depolarization induced by high K_O in Chara it is proposed that the pump continues to extrude protons (Beilby 1986) or even increases its pump activity (Kishimoto et al. 1985). In this context the $Fe^{3+}Cy$ induced depolarization also may be considered as origin of the observed acidification. This would be an alternative to the common view that the redox reaction alone is responsible for the measured net H^+-efflux in media with $Fe^{3+}Cy$ (Neufeld and Bown 1987). H^+ efflux during the resting state reflects the balance of H^+ extrusion by the electrogenic pump(s) and H^+ influx through leaks and H^+ coupled co-transport processes. Depolarization, however, decreases the electrical driving force for co-transport processes. In fact $Fe^{3+}Cy$ was found to inhibit co-transport linked uptake to sulphate (Lass et al. 1986). Therefore the observed acidification upon $Fe^{3+}Cy$ addition (i.e. an increase in net H^+ efflux) could very well originate from a decreased return rate in the cell.

When L.papulosum cells were depolarized by high $(K^+)o$ no additional acidification of the external medium was detected (Fig. 6). However, addition of $Fe^{3+}Cy$ to cells already in the depolarized state caused an increase in net H^+efflux. Hence it is not the $Fe^{3+}Cy$ induced depolarization of Em itself which stimulates net proton efflux. Further investigations have to show whether the acidificaion (net H^+ efflux) is caused by a $Fe^{3+}Cy$ action on H^+efflux or H^+return pathways.

Acknowledgement

This work was supported by the Deutsche Forschungsgemeinschaft. We thank Dr.C.I. Ullrich-Eberius (TH Darmstadt) for critically reading the manuscript.

LITERATURE CITED

Beilby,M.J., 1985. J.Exp.Bot.36,228-239.
Beilby,M.J.,1986. J.Membr.Biol.89,241-249.
Bisson,M.A.,N.A. Walker, 1982.J.Exp.Bot.33,520-532.
Keifer,D.W.,R.M.Spanswick, 1978. Plant Physiol. 62, 653,659.
Kirst,G.O., F.Wichmann, 1987. J. Plant Physiol. 131, 413-422.
Kishimoto,U., N.Kami-ike,Y.Takeuchi,T.Ohkawa, 1984. Membr. Biol. 80, 175-183.
Kishimoto,U., Y.Takeuchi,T.Ohkawa,N.Kami-ike, 1985. J.Membr.Biol.86, 27-36.
Lühring,H. 1986. Protoplasma 133, 19-28.
Lass, B., G.Thiel,C.I.Ullrich-Eberius, 1986. Planta 169, 251-259.
Moriyasu,Y., T.Shimmen,M.Tazawa, 1984. Cell Structure and Function 9, 235-246.
Marigo,G., Belkoura,M., 1985. Plant Cell Rep. 4, 311-314.
Neufeld, E., A.W.Bown, 1987. Plant Phyisol. 83, 895-899.
Okazaki,Y., T.Shimmen,M.Tazawa,1984. Plant and Cell Physiol. 25, 573-581.
Okazaki,Y., M.Tazuawa, 1986. Plant Cell Environ. 9, 491-494.
Ohkawa,T., T.Tsutsui, U.Kishimoto, 1986. Plant Cell Physiol. 27, 1429-1438.
Reid,R.J., N.A. Walker, 1983. Aust. J.Plant Physiol. 10, 373-383.
Rubinstein, B., A.I. Stern, 1986. Plant Physiol. 80, 805-811.
Shiina,T., M.Tazawa, 1987. J.Membr.Biol. 99, 137-146.
Shimmen,T., Kikujama, M.Tazawa, 1976. J.Membr.Biol. 30, 249-270.
Shimmen,T., M.Tazawa, 1983. Plant Cell Physiol. 24, 1511-1524.
Sijmons,P.C., F.C. Lanfermeijer,A.H.de Voer, H.B.A. Prins, H.F.Bienfait, 1984. Plant Physiol. 76, 943-946.
Thiel, G., G.O.Kirst, 1988. J.Exp.Bot. 39, in press.
Tyerman,S.D.,G.P.Findley, G.J. Patterson, 1986. J.Membr. Biol. 89, 139-152.

b-TYPE CYTOCHROMES, LIGHT- AND NADH-DEPENDENT OXIDO-REDUCTASE ACTIVITIES IN PLANT PLASMA MEMBRANES

Roland Caubergs, Han Asard and Jan A. De Greef

University of Antwerpen, R.U.C.A.
Department of Biology
Groenenborgerlaan, 171, B-2020 Antwerpen, Belgium

INTRODUCTION

Blue light induces many important biological phenomena in plants and microorganisms (Senger, 1980, 1984a; Senger and Briggs, 1981; Senger and Schmidt, 1986). The characterization of the sequence of events from the photoperception to the final respons is the ultimate aim in the area of photobiological research. These responses are often extensively described at the macrophysiological level but the molecular perception mechanism(s) remain obscure. In higher plant phototropism it seems justified to propose an alteration in membrane properties (Firn, 1986). The basic discussion centers obviously around the primary reactions as a consequence of the photochemical reaction of the receptor. Changes in transmembrane ion gradients are considered as an early step in the reaction chain and occur by alterations in the existing electrochemical potential differences (Evans, 1985). According to chemiosmotic principles, changes in plasma membrane ATPase activity are directly or indirectly responsible for ion exchange (Serrano, 1985). Another system also working as a proton pump is provided by an electron transfer chain. The latter system is well described for mitochondria and chloroplasts but now increasing evidence supports the presence of a proton translocating redox mechanism in the plasma membrane (Crane et al., 1985; Møller and Lin, 1986). It is precisely the aim of this symposium to discuss the possible involvement of these systems in membrane transport and growth phenomena. Several contributions will describe the various oxido-reductases located on the plasma membrane.

This paper deals with a blue light induced cytochrome reduction in highly purified plant plasma membranes, mainly isolated from cauliflower inflorescences. The eventual relationship with NAD(P)H oxido-reductases is discussed.

OXIDO-REDUCTASE REACTIONS

LIAC

A blue light inducible absorbance change (LIAC) clearly shows the photoreduction of a b-type cytochrome. Reoxidation occurs in the subsequent dark period (Fig. 1).

The discovery of this light inducible redox system was made by W.L. Butler and associates in several fungi (Munoz and Butler, 1975; Poff and Butler, 1974). The action spectrum of this in vivo process resembles those of various other blue light inducible phenomena which are called "cryptochrome" responses. The term "cryptochrome" is however an operational name (Senger, 1984b). The similarity of action spectra indeed is not a guaranty for the involvement of only one photoreceptor in the "cryptochrome" responses. Flavins and carotenoids are likely candidates (Horwitz and Gressel, 1986). For phototropism circumstantial evidence points in the direction of flavoproteins as photoreceptor (Schmidt et al., 1977; Vierstra and Poff, 1981). Flavins were also proposed as photoreceptors for the LIAC activity by Butler and associates, mainly on the basis of their close association with b-type cytochromes in other systems where they contribute in redox activities (Munoz and Butler, 1975). Carotenoids do not participate in redox reactions as a consequence of their specific molecular structure (Song and Moore, 1974). Butler concluded that the LIAC is the reflection of the primary step in the photocontrol of "cryptochrome" phenomena. To explain the diversity of blue light phenomena, a multitude of flavins and flavoproteins are possibly involved as photoreceptors.

The position of the LIAC as a representative signal for "cryptochrome" phenomena comparable for instance to the photoreversibility test for phytochrome, is disputed and the physiological relevance is questioned (Schmidt, 1987). The notion that flavins and cytochromes in vitro show a variety of mutual redox reactions, including blue light induced reduction of cytochromes, implies the possibility for an artificial nature of the LIAC, especially because of the unnatural conditions (anaerobiosis) needed to generate the light induced reduction.

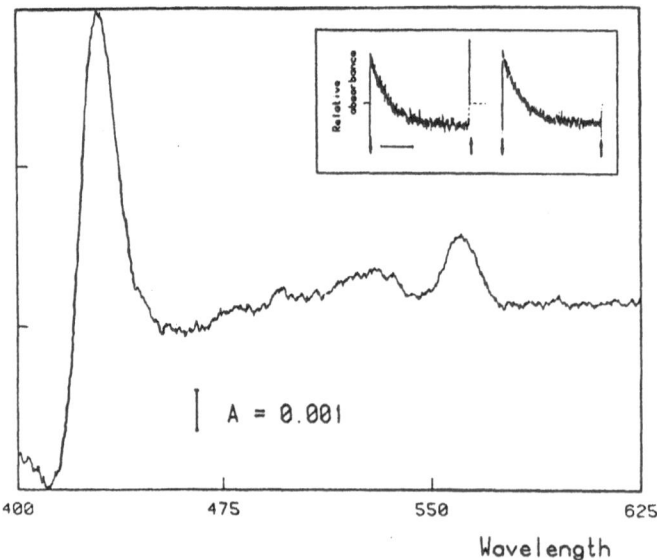

Fig. 1.   Light induced absorbance changes in cauliflower PM's purified through aqueous two phase partitioning. The light (1 min blue light 370 W.m^{-2}) minus dark difference spectrum shows the reduction of a b-type cytochrome. Inset : decay kinetics (at 430 nm with reference at 410 nm) after actinic radiation.

Some points however are in favour of the physiological relevance of the LIAC in the sequence of events upon blue light perception. Firstly, one particular b-type cytochrome is involved in the LIAC reaction and secondly, the elements producing LIAC activity appear to be intrinsic constituents of the plasma membrane. Evidence supporting these statements will be presented in the next section of this presentation.

In addition to biochemical evidence various tests of a physiological nature support the relevance of LIAC's. For instance, inhibitors, with high specificity for LIAC or "cryptochrome" response have been used (Leong and Briggs, 1982; Vanden Driessche and Caubergs, 1985). Also mutants of fungi with reduced blue light sensitivity were examined for correlated changes in LIAC activity (Trad and Lipson, 1987; Klemm and Ninnemann, 1978, 1979; Horwitz et al., 1986).

The characteristics of the LIAC and a H^{+}-ATPase in plant membranes

Because of their well known sensitivity to blue light, coleoptiles were an obvious choise for LIAC studies in higher plants. Corn coleoptiles were used to optimize the conditions for obtaining reproducible LIAC activities. It is important to mention that an O_2-scavenging system (glucose-glucose oxidase) was added to the sample to create partial anaerobiosis.

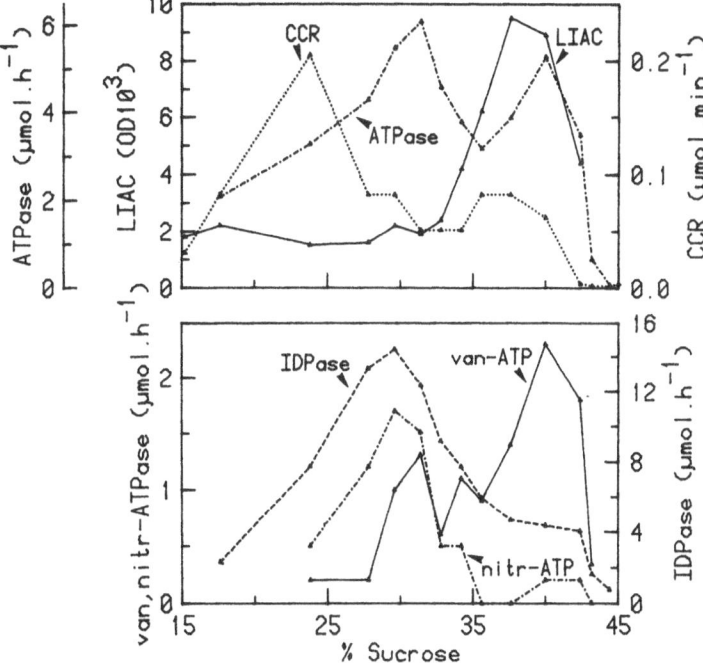

Fig. 2. Distribution of LIAC and marker enzyme activities after fractionation of cauliflower microsomes on a sucrose density gradient (adapted from Caubergs et al., 1986 with permission). Enzyme activities (on a 1 ml basis) respectively indicate ER (CCR), tonoplast (nitrate sensitive ATPase), Golgi (latent IDPase) and PM (vanadate sensitive ATPase).

EDTA in general is used as a one electron donor (Goldsmith et al., 1980). Based on the results obtained by Leong and Briggs (1981) gradient centrifugation techniques were used to identify the LIAC containing membrane fraction derived from cauliflower inflorescences. Figure 2 shows LIAC and marker enzyme profiles of a microsomal fraction on a linear sucrose gradient. It is clear that the distribution of the light sensitive cytochrome b reduction coincides with one of the two peaks of total ATPase activity. The fact that sodium orthovanadate (Na_3VO_4) preferentially inhibits this ATPase activity, indicates the presence of plasma membrane vesicles.

The ATPase activity coinciding with LIAC was more extensively studied in membranes obtained by an aqueous polymer two phase separation method. This method has been described in full detail several times (Yoshida et al., 1983; Widell and Larsson, 1987). By this method highly purified plasma membrane vesicles are obtained, although significant latent IDPase (Golgi-marker) and CCR activity (ER-marker) are detectable. The characteristics of the ATPase of this plasma membrane fraction matches those of the typical proton extruding pump, with Km and Vmax values as described for other systems (Serrano, 1985; Sze, 1985). The notion that the proton ATPase is associated with the same membrane as the presumptive blue light photoreceptor, leads to speculations about their mutual interaction. Model systems describing a direct or indirect relationship have been designed and are mentioned later.

The b-type cytochrome

Work of Rich and Bendall (1975) on the presence of electron transport components in microsomal fractions of higher plants, as from cauliflower buds, shows the presence of various b-type cytochromes. In spite of considerable variation in presence and amounts, the b_5 and P-450/420 components were always detectable. Frequently an ascorbate reducible b-type cytochrome was measured. Because of the multitude of membrane types the subcellular localization of the cytochromes was unknown.

It is certain now that cytochrome b_5 and P-450/420 are constituents of microsomal material (ER) (Hendry et al., 1981). It is reasonable to assume on the basis of the existing developmental relationship between ER and plasma membrane that common cytochromes are present in both membrane types. This assumption however is debatable.

One of the earliest reports in which a separation was achieved between ER and plasma membrane by sucrose gradient centrifugation, mentions the presence of a b-type cytochrome, not reducible by NADH in the plasma membrane of corn coleoptiles (Jesaitis et al., 1977). The similarity in characteristics with the cytochrome related to the LIAC system of fungi was readily recognized. LIAC research concerning the cytochrome of corn plasma membranes revealed a b-type cytochrome with a midpoint potential of -65mV (Leong et al., 1981). The authors suggest that this cytochrome participates in LIAC, although the presence of ascorbate reducible cytochrome(s) is mentioned. For plasma membranes of cauliflower P-450 was assigned as the cytochrome participating in LIAC activity on the basis of low temperature spectroscopy (Widell et al., 1983). Recently carbonmonoxide difference spectra and pyridine binding spectra support the presence of P-450 in cauliflower plasma membranes (Kjellbom et al., 1985). No clear correlation between P-450 and LIAC however was presented.

The true nature of the b-type cytochrome associated with the plasma membrane LIAC activity remains uncertain. It was already mentioned that high potential cytochromes are present in plasma membranes of corn coleoptiles. Best fit analysis of data obtained by potentiometrical titration combined with simultaniously recorded spectra of the alpha-band region of the cytochromes (Van Wielink et al., 1982), revealed the existence of at least 2 b-type cytochromes in the plasma membrane of cauliflower inforescences (Caubergs et al., 1986). The major component has a midpoint potential of +160mV (Fig. 3), and is consequently reducible by ascorbate. It fits most likely the describtion of a cytochrome b reported by Rich and Bendall (1975) in a cauliflower microsomal fraction showing absorption maxima at 559.5, 527 and 429nm at room temperature. The second minor component has a negative midpoint potential and is not ascorbate reducible. It has to be mentioned that the titration range has a span from -200 to 200mV and very low potential cytochromes are consequently not detected. This could mean that the presence of P-450/420 with a E'o lower than -200mV is overlooked. Carbon-monoxide difference spectra show in our hands the presence of P-450 in a microsomal fraction, but only small amounts were detectable in the plasma membrane fraction after two phase purification. It should be mentioned that purified plasma membranes clearly show TMBZ (3,3',5,5'-tetramethylbenzidine) staining, indicative for the presence of haem proteins with peroxidase activity. In addition there is evidence that the LIAC in plasma membranes reflects the reduction of the high potential cytochrome.

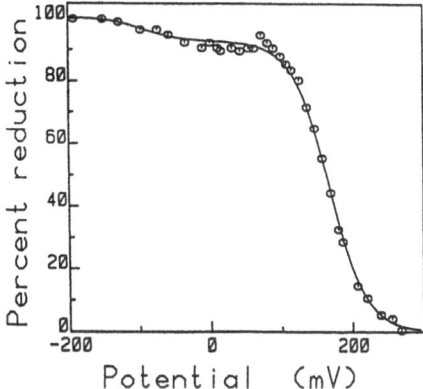

		%	E'o (mV)
2 comp.	74		159
	24		- 7
3 comp.	68		164
	18		56
	14		-85

Fig. 3. Potentiometric redox titration of cytochromes in purified cauli-flower PM's. Statistical analysis reveals two (solid line) or three components (table). Results are the mean of two experiments.

Firstly, the amount of cytochrome b reduced by light is roughly of the same order of magnitude as the total amount of the ascorbate insensitive component. So that all cytochrome of this type needs to be used in the light reduction (Table 1). Theoretically such a highly efficient system is not excluded since the quantum efficiency for the photoreduction of the cyto-chrome is close to unity (Lipson and Presti, 1980). However with the addi-tion of external photosensitizers such as riboflavin, the amount of cyto-chrome reduced by light increases three fold. This effect is only explain-able if both types of cytochromes or the ascorbate reducible cytochrome is involved. Double reciprocal plots however, showing the effect of varying concentrations of several flavins on cytochrome reduction by blue light, reveal that the maximum absorbance change is similar for added riboflavin, FMN and FAD which suggests that all flavins interact competitively at the same site (Goldsmith et al., 1980). Only one type of cytochrome is respon-sible for the increased LIAC activity upon addition of external photo-sensitizers and only the ascorbate reducible cytochrome is present in sufficient amount to fulfil this role.

Table 1. Cytochrome (b-type) reduction in cauliflower PM fractions.
(units = $\Delta A_{430} \cdot 10^{-3} \cdot mg^{-1}$ protein ; n = number of experiments)

	Absorbance + S.E.	n	%
Dithionite	33.2 + 1.0	(2)	100
Ascorbate	27.0 + 1.4	(2)	81
Blue light	4.6 + 0.4	(5)	14
BL + riboflavin (3 μM)	13.5 + 1.7	(4)	41
BL + duroquinone (3 μM)	9.4 + 2.2	(2)	28
BL + catalase (8300 U/ml)	20.6	(1)	62

Secondly, the increased LIAC activity upon catalase addition, probably resulting in a decrease in the dark reoxidation rate, is hardly explainable on a quantitative base if the participation of the ascorbate reducible cytochrome is excluded.

Thirdly, it is clear that in the presence of ascorbate all LIAC activity is inhibited. The obvious interpretation is that the chemically reduced cytochrome is functionally disabled to participate in any light reaction. In view of the complexity of the flavin cytochrome interaction it is cautious not to exclude other explanations.

Quinone stimulated NADH oxidation and b-type cytochromes

The presumptive physiological significance of flavoproteins and b-type cytochromes are of common interest for the photobiologists and researchers dealing with NADH dependent oxidation-reduction reactions at the plasma membrane. From the various model systems proposed from both research areas, there are several examples in which extrusion of redox equivalents are considered as fundamental (Lüttge and Clarkson, 1985; Schmidt, 1984). The participation of b-type cytochromes is envisaged and whether these electron carriers are eventually common components for both LIAC and NADH oxidoreductases needs sorting out.

A first approach to gain some basic information relates to the distribution of the LIAC and the NADH dehydrogenase activity upon gradient centrifugation (Asard et al., 1987). Gradient profiles reveal, as expected, a predominant LIAC activity at the plasma membrane although some activity can be designated to the ER region (Fig. 4). NADH oxidase activity is detected in both ER and plasma membrane. The bimodal distribution allows no conclusion about the resemblance or difference of both systems. An observation that might suggest the presence of common components is the stimulation upon quinone addition of both LIAC (Table 1) and oxido-reductase activity. This interpretation is contradicted by measurements of NADH-oxido-reductase activity in the presence of ascorbate. Neither chemical nor light induced reduction of the high potential cytochrome, resulted in any alteration of the NADH oxidation. This seems to deny any direct interaction between the two redox systems.

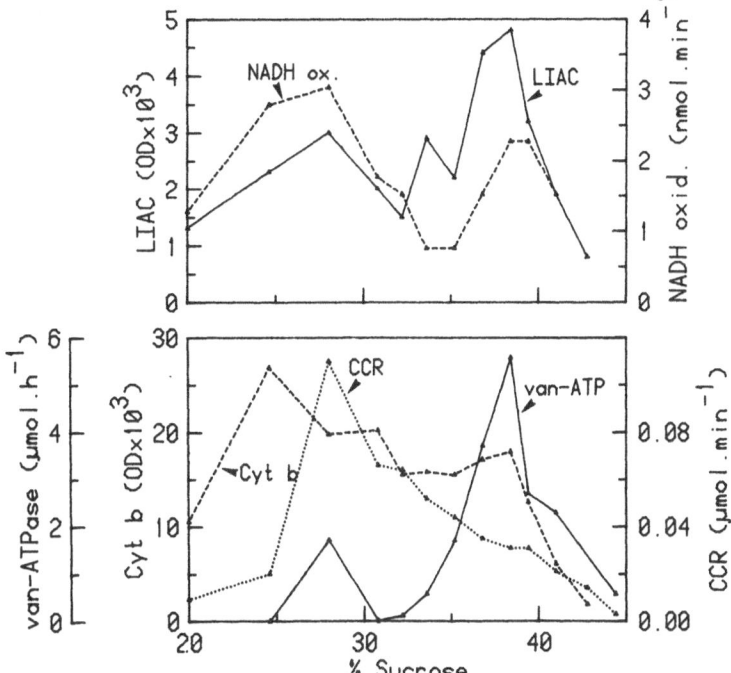

Fig. 4.   Distribution of DQ-stimulated NADH oxidase, LIAC, b-type cytochromes and marker enzyme activities in cauliflower microsomes
after gradient fractionation. Total amounts of b-type cytochromes
are estimated after dithionite reduction (adapted from Asard et
al., 1987 with permission).

CONCLUSIONS

     The availability of highly purified plasma membranes offers in prin-
ciple the opportunity to characterize the various proteinacious and lipo-
phylic components.

     In view of the increasing importance of redox systems in physiological
phenomena such as transport, - both for ions and hormones, - and photorecep-
tion, it seems promising to concentrate further immediate research on a
thorough characterization and quantification of the haemo- and flavoproteins
of the plant plasma membrane. Also further study concerning the ATPase
system(s) is needed. There are indications that oxido-reductase(s) and H^+
extruding ATPase(s) operate simultaniously (Crane et al., 1982; Rubinstein
and Stern, 1986) and that an electrogenic H^+ ATPase is a component of the
photosensory transduction system in blue light perception (Cooke et al.,
1983; Shimazaki et al., 1986).

     A strict analytical approach alone will probably be unable to resolve
the diverse mechanisms linked to the physiological events performed by this
membrane type. Clear determination of the constituents will however contribute to a better understanding of the dynamic entity.

ACKNOWLEDGEMENTS

The excellent technical assistance of H. Van Hemelrijk and W. Reynders
is greatfully acknowledged. H.A. is a research assistant at the National
Science Foundation (N.F.W.O.).

REFERENCES

Asard, H., Caubergs, R., Renders, R. and De Greef, J. A., 1987, Duroquinone-
    stimulated NADH oxidase activity and b-type cytochromes in the plasma
    membrane of cauliflower and mung beans. Plant Science, 53:109-119.
Caubergs, R. J., Asard, H. H., De Greef, J. A., Leeuwerik, F. J. and Oltmann,
    F. L., 1986, Light-inducible absorbance changes and vanadate-sensitive
    ATPase activity associated with the presumptive plasma membrane frac-
    tion from cauliflower inflorescences. Photochem. Photobiol.,
    44:641-649.
Cooke, T. J., Racusen, R. H. and Briggs, W. R., 1983, Initial events in the
    tip-swelling response of the filamentous gametophyte of Onoclea
    sensibilis L. to blue light., Planta, 159:300-307.
Crane, F. L., Roberts, H., Linnane, H. and Löw, A.W., 1982, Transmembrane
    ferricyanide reduction by cells of the yeast Saccharomyces cerevisiae.,
    J. Bioenerg. Biomembr., 14:191-205.
Crane, F. L., Sun; I. L., Clark, M. G., Grebing, C. and Löw, H., 1985,
    Transplasma membrane redox systems in growth and development. Biochem.
    Biophys. acta, 811:233-264.
Evans, R. D., 1985, The action of auxin in plant cell elongation. in: "CRC
    Critical Reviews in Plant Sciences", CRC Press, Inc., Boca Raton,
    Florida, vol. 2, issue 4, p. 317.
Firn, R. D., 1986, Phototropism. in: "Photomorphogenesis in Plants", R. E.
    Kendrick and C. H. M. Kronenburg, eds., Martinus Nijhoff/Dr. W. Junk
    Publishers, Dordrecht, The Netherlands, p. 367.
Goldsmith, M. H. M., Caubergs, R. J. and Briggs, W. R., 1980, Light-
    inducible cytochrome reduction in membrane preparation from corn
    coleoptiles. Plant Physiol., 66:1067-1073.
Hendry, G. A. F., Houghton, J. D. and Jones, O. T. G., 1981, The cytochromes
    in microsomal fractions of germinating mung beans. Biochem. J.,
    194:743-754.
Horwitz, B. A. and Gressel, J., 1986, Properties and working mechanisms of
    the photoreceptors. in: "Photomorphogenesis in Plants", R. E. Kendrick
    and C. H. M. Kronenburg, eds., Martinus Nijhoff/Dr. W. Junk Publishers,
    Dordrecht, The Netherlands, p. 159.
Horwitz, B. A., Trad, C. H. and Lipson, E. D., 1986, Modified light-induced
    absorbance changes in dim photoresponse mutantes of Trichoderma.,
    Plant Physiol., 81:726-730.
Jesaitis, A. J., Heners, P. R., Hertel, R. and Briggs, W. R., 1977,
    Characterization of a membrane fraction containing a b-type cytochrome.
    Plant Physiol., 59:941-947.
Kjellbom, P., Larsson, C., Askerlund, P., Schelin, C. and Widell, S., 1985,
    Cytochrome P-450/420 in plant plasma membranes : a possible component
    of the blue light reducible flavoprotein cytochrome complex. Photochem.
    Photobiol., 42:779-783.
Klemm, E. and Ninnemann, H., 1978, Correlation between absorbance changes
    and a physiological response induced by blue light in Neurospora.,
    Photochem. Photobiol., 28:227-230.
Klemm, E. and Ninnemann, H., 1979, Nitrate-reductase - a key enzyme in blue
    light promoted conidiation and absorbance change of Neurospora,
    Photochem. Photobiol., 29:629-632.

Leong, T. Y. and Briggs, W. R., 1981, Partial purification and characteri-
    zation of a blue light sensitive cytochrome flavin complex from corn
    membranes. Plant Physiol., 67:1042-1046.
Leong, T. Y. and Briggs, W. R., 1982, Partial purification and characteri-
    zation of a blue light sensitive cytochrome flavin complex from corn
    membranes. Plant Physiol., 67:875-881.
Leong, T. Y., Vierstra, R. D. and Briggs, W. R., 1981, A blue light-sensi-
    tive cytochrome flavin complex from corn coleoptiles. Further
    characterization. Photochem. Photobiol., 34:697-703.
Lipson, E. D. and Presti, D., 1980, Graphical estimation of cross sections
    from fluence-response data. Photochem. Photobiol., 32:383-391.
Lüttge, U. and Clarkson, D.T., 1985, Mineral nutrition : plasmalemma and
    tonoplast redox activities. Prog. Bot., 47:73-86.
Møller, I. M. and Lin, W., 1986, Membrane-bound NAD(P)H dehydrogenases in
    higher plant cells. Ann. Rev. Plant Physiol., 37:309-334.
Munoz, V. and Butler, W. L., 1975, Photoreceptor pigments for blue light in
    Neurospora crassa., Plant Physiol., 55:421.
Poff, K. L. and Butler, W. L., 1974, Absorbance changes induced by blue
    light in Phycomyces blakesleanus and Dictyostelium discoideum., Nature
    (London), 248:799-801.
Rich, P. R. and Bendall, D. S., 1975, Cytochrome components of plant micro-
    somes. Eur. J. Biochem., 55:353-341.
Rubinstein, B. and Stern, A.I., 1986, Relationships of trans plasmalemma
    redox activity to proton and solute transport by roots of Zea mays.,
    Plant Physiol., 80:805-811.
Schmidt, W., 1984, Blue light-induced, flavin-mediated transport of redox
    equivalents across artificial bilayer membranes. J. Membr. Biol.,
    82:113-122.
Schmidt, W., 1987, Primary reactions and optical spectroscopy of blue light
    photoreceptors., in: "Blue Light Responses Phenomena and Occurence in
    Plants and Microorganisms", H. Senger, ed., CRC Press, Inc., Boca
    Raton, Florida, p. 20.
Schmidt, W., Hart, J., Filner, P. and Poff, K. L., 1977, Specific inhibition
    of phototropism in corn seedlings. Plant Physiol., 60:736-738.
Shimazaki, K., Iino, M. and Zeiger, E., 1986, Blue light-dependent proton
    extrusion by guard-cell protoplasts of Vicia faba., Nature,
    319:324-326.
Senger, H., 1980, "The Blue Light Syndrome", Springer-Verlag, Berlin.
Senger, H., 1984a, "Blue Light Effects in Biological Systems", Spinger-
    Verlag, Berlin.
Senger, H., 1984b, Cryptochrome, some terminological thoughts. in: "Blue
    Light Effects in Biological Systems", Spinger-Verlag, Berlin, p. 72.
Senger, H. and Briggs, W. R., 1981, The blue light receptor(s) : primary
    reactions and subsequent metabolic changes. in: "Photochemical and
    Photobiological Review", K. C. Smith, ed., Pleunm Press, New York,
    vol. 6, chapt. 1.
Senger, H. and Schmidt, W., 1986, Diversity of photoreceptors. in: "Photo-
    morphogenesis in Plants", R. E. Kendrick and C. H. M. Kronenburg, eds.,
    Martinus Nijhoff/Dr. W. Junk Publishers, Dordrecht, The Netherlands,
    p.137.
Serrano, R., 1985, Plasma membrane ATPase of Plants and Fungi. CRC Press,
    Inc., Boca Raton, Florida.
Song, P.-S. and Moore, T. A., 1974, On the photoreceptor pigment for photo-
    tropism and phototaxis : is a carotenoid the moste likely candidate ?,
    Photochem. Photobiol., 19:435-441.
Sze, H., 1985, H$^+$ translocating ATPase : advances using membrane vesicles.
    Ann. Rev. Plant Physiol., 36:175-208.
Trad, C. H. and Lipson, E. D., 1987, Biphasic fluence-response curves and
    derived action spectra for light-induced absorbance changes in
    Phycomyces mycelium. J. Photochem. Photobiol., B. Biology, 1:169-180.

Vanden Driessche, T. and Caubergs, R., 1985, Inhibiting the transduction of blue light signals in Acetabularia., in: "Acetabularia 1984", S. Bonotto, F. Cenelli, R. Billiau, eds., Belgian Nuclear Center, C.E.N.-S.C.K., Mol, Belgium, p. 91.

Van Wielink, J. E., Oltmann, L. F., Leeuwerik, F. J., De Hollander, J. A. and Stouthamer, A. H., 1982, A method for in situ characterization of b- and c-type cytochromes in E. coli and in complex III from beef heart mitochondria by combined spectrum deconvolution and potentiometric analysis. Biochim. Biophys Acta., 681:177-190.

Vierstra, R. D. and Poff, K. L., 1981, Mechanism of specific inhibition of phototropism by phenylacetic acid in corn seedlings. Plant Physiol., 67:1011-1015.

Widell, S. and Larsson, C., 1987, Plasma membrane purification., in: "Blue Light Responses Phenomena and Occurence in Plants and Microorganisms", H. Senger, ed., CRC Press, Inc., Boca Raton, Florida, p. 99-107.

Widell, S., Caubergs, R. J. and Larsson, C., 1983, Spectral characterization of light-reducible cytochrome in a plasma membrane-enriched fraction and in other membranes from cauliflower inflorescences. Photochem. Photobiol., 38:95-98

Yoshida, S., Uemura, M., Niki, T., Sakai, A. and Gusta, L. V., 1983, Partition of membrane particles in aqueous two-polymer phase system and its practical use for purification of plasma membranes from plants. Plant Physiol., 72:105-114.

KINETICS OF REDOX STATE AS RELATED

TO PLANT GROWTH AND DEVELOPMENT

E. Wagner[1], M. Bonzon[2], S. Frosch[1], S. Ruiz
Fernández[1], S. Kiefer[2] and H. Greppin[2]

[1] Institut für Biologie II, Universität
Freiburg, D-7800, Freiburg
[2] Physiologie Végétale, Université de Genève,
CH-1211 Genève

Growth and differentiation in most living systems are tightly coupled to seasonal changes in thermo- and photoperiod. Photo- and thermo-periodic phenomena show a rhythmic change in sensitivity which is endogeneous in character and reflects an endogeneous circadian rhythm in metabolic activity. The network of eucaryotic energy metabolism might be conceived as an evolutionary adaptation to a cyclic energy supply from the environment. It is probable that, in the course of evolution, circadian metabolic acitvty became the innate timer controlling growth, differentiation and behaviour in eukaryotes.

In plants, photoperiodic behaviour has been used successfully to define operationally a photoreceptor system – phytochrome –. This system distinguishes between light and dark through the action of the active form of the pigment on metabolism, membranes, and gene activity. The changes in the pigment and the timing of the availability of the "substrate" on which it acts possibly represent the essentials of photoperiodic control. Circadian rhythmicity in energy transduction seems to be a necessity for control of the subsystems of energy metabolism in the eukariotic cell for the benefit of the whole organism (1-4).

Circadian changes in energy metabolism would be expected to be reflected in a corresponding rhythm in the state of energy-transducing biomembranes which might control the binding sites for photoreceptors and other effectors. The interaction of effectors and binding sites could therefore be controlled by energy metatolism. Stimulation of the receptors could lead to transitory modulation (phase setting) of oscillatory energy metabolism and the switching of metabolic controls. The energy dependent state of membranes in turn may determine the sensitivity of membrane-bound receptors and the activity of membrane-linked enzyme systems (5,6).

The experimental results support the idea that the different subsystems of energy metabolism (glycolitic sequences, oxidative pentose-phosphate cycle, mitochondria, chloroplasts) in eukariots are synchroniced and coordianted in time. This is demonstrated by systemic changes in the redox- and phosphorylation state and by modulation of the enzymatic capacity at the cell subsystems (7-9). Under constant conditions, a circadian rhythm in reduction- and energy charge is observed. Under light-dark cycles, the NAD and NADP pools are characterized by reciprocal changes in size (10-13). The transition from vegetative growth to flower initiation is marked by a characteristic change in the catabolic and anabolic reduction charges (13). Photomodulation of the phytochrome system results in transient changes in pyridine nucleotide pool size levels which are most likely mediated by NAD-kinase and the Ca^{2+}-calmodulin system (14).

It is assumed that nucleotide ratios are involved in the stoichiometric coupling of metabolic compartments at the cellular level. The response to environmental and internal signals depends on the state of the membranes bearing the receptor systems. This view is supported by the observation of photomodulation of a daily rhythm in peroxidase binding to microsomes (15,16).

REFERENCES

1. WAGNER, E. (1976): Endogeneous rhythmicity in energy metabolism: Basis for timer-photoreceptor-interactions in photoperiodic control. In: Dahlem Workshop on the Molecular Basis of Circadian Rhythms, J.W. Hastings and H.G. Schweiger (eds.), Dahlem Konferenzen, Berlin Aabkon-Verlagsgesellschaft, pp. 215-238.
2. WAGNER, E. (1977): Molecular basis of physiological rhythms. In: Integration of Activity in the Higher Plant, D.H. Jennings, ed., Sociuenty for Experimental Biology, Symposium 31, Cambridge, Univerity Press, pp. 33-72.
3. WAGNER, E., BONZON, M. and GREPPIN, H. (1985): Membrane-oscillator hypothesis of metabolic control in photoperiodic time measurement and the temporal organization of development and behaviour in plants. In: New Developements and Methods in Membrane Research and Biological Energy Transduction, L. Packer, ed., Plenum Pub. Co., pp. 525-546.
4. WAGNER, E. and FUKSHANSKY, L. (1985): Die zeitliche Organisation des eukaryoten Stoffwechsels als Grundlage für die Signalverarbeitung in Photo- und Thermodperiodismus. Ber. Dtsch. Bot. Ges. 98, 35-52.
5. WAGNER, E. (1983): Circadian rhythms: The basis for information processing in eukaryotes during adaptation to seasonal changes in photo- and thermoperiods, In: Molecular Models of Photoresponsiveness. G. Montagnoli and B.F. Erlanger (eds.), Plenum Pub. Co., pp. 197-202.
6. WAGNER, E., HAERTLE, U., KOSSMANN, I. and FROSCH, S. (1983): Metabolic and development adaptation of eukaryotic cells as related to endogeneous and exogeneous control of translocators between subcellular compartments. In: Edocytobiology II, W. Schwemmler and H. Schenk (eds.), Walter de Gruyter & Co., Berlin, New York, pp. 341-352.

7. WAGNER, E., TETZNER, J., HAERTLE, U. and DEITZER, G.F. (1974): Endogeneous rhythmicity and energy transduction. VIII. Kinetics in enzyme activity, redox state and energy charge as related to photomorphogenesis in seedlings of Chenopodium rubrum L., Ber. Dtsch. Bot. Ges. 87, 291-302.

8. WAGNER, E., DEITZER, G.F., FISCHER, S., FROSCH, S., KEMPF, O. and STOEBELE, L. (1975): Endogeneous oscillations in pathways of energy transduction as related to circadian rhythmicity and photoperiodic control. BioSystems 7, 68-76.

9. BONZON, M., SIMON, P., DEGLI AGOSTI, R., GREPPIN, H. and WAGNER, E. (1987) Activity of glyceraldehyde-3-phosphate dehydrogenase isozymes during photoperiodic floral induction in spinach leaves. Physiol. Plant. 70, 577-582.

10. WAGNER, E., STOEBELE, L. and FROSCH, S. (1974): Endogeneous rhythmicity and energy transduction. V. Rhythmicity in adenine nucleotides and energy charge in seedlings of Chenopodium rubrum L. J. Interdiscipl. Cycle Res. 5, 77-88.

11. WAGNER, E. and FROSCH, S. (1974): Endogeneous rhythmicity and energy transduction. VI. Rhythmicity in reduced and oxidized phyridine nucleotide levels in seedlings of Chenopodium rubrum L. J. Interdiscipl. Cycle Res. 5, 231-239.

12. BONZON, M., HUG, M., WAGNER, E. and GREPPIN, H. (1981): Adenine nucleotides and energy charge evolution during the induction of flowering in spinach leaves. Planta 152, 189-194.

13. BONZON, M., SIMON, P., GREPPIN, H. and WAGNER, E. (1983): Pyridin nucleotides and redox charge evolution during the induction of flowering in spinach leaves. Planta 159, 254-260.

14. WAGNER, E., FROSCH, S. and KEMPF, O. (1974): Endogeneous rhythmicity and energy transduction. VII. Phytochrome-modulated rhythms in pyridine nucleotide levels in seedlings of Chenopodium rubrum. Plant. Sci. Lett. 3, 43-48.

15. KIEFER, S., KLINGLER, H., PENEL, C., GREPPIN, H. and WAGNER, E. (1986): Photoperiodic and phytochrome control of peroxidase binding to micorsomes from Pharbitis nil cotyledons. 5th Congress of FESPP, Hamburg, FRG, Book of Abstracts 5-25.

16. KIEFER, S., PENEL, C., GREPPIN, H. and WAGNER, E. (1987): Association de peroxydases aux membranes de courgettes, de raifort et de pharbitis nil. Arch. Sc. Genève 40, 369-378.

REGULATION OF TRANSPLASMALEMMA ELECTRON TRANSPORT BY CALCIUM AND LIGHT

IN OAT MESOPHYLL CELLS

Suranganee Dharmawardhane, Bernard Rubinstein
and Arthur I. Stern

Botany Department
University of Massachusetts
Amherst, Massachusetts USA

INTRODUCTION

The transport of electrons across the plasmalemma from a cytosolic donor to an extracellular acceptor can provide an energy source capable of regulating cellular events. Calcium ions and light, either alone or interacting in some way, control important activities in the cell and evidence exists that both effectors also influence rates of transplasmalemma electron flow.[1,2,3,4] We have been attempting to characterize further calcium and light effects on cell surface redox activity. If redox activity and a particular cellular process are both regulated by calcium and/or light, then the possibility exists that redox activity mediates the action of these effectors. An indication that such a relationship may exist with respect to blue light has been obtained in vitro. In this case, the same wavelengths of blue light that lead to absorbance changes characteristic of a reduced b-type cytochrome, are also active in phototropism.[5,6]

Evidence will be presented using the Ca^{2+} ionophore, A23187 and substances which influence Ca^{2+} channels, that a relationship exists between Ca^{2+} levels and rates of transplasmalemma electron transport. Furthermore, we will show that a phorbol ester analog of diacylglycerol, which stimulates the Ca^{2+}-requiring enzyme protein kinase C in animal cells, has similar effects in mesophyll cells in the dark. One modulator of protein kinase C, the sphingoid base phytosphingosine, has the unusual property of stimulating transmembrane electron transport in the light.

MATERIALS AND METHODS

The plant material and the assay procedure used to detect ferricyanide reduction were similar to those already reported.[2] Briefly, 7-day-old light-grown oat leaves were peeled, cut into two, 1 cm segments, 5 mm from the tip, and washed in two changes of distilled water. Fifteen to twenty segments were then incubated on 2 ml of the desired media which eventually contained 80 mM K_3FeCN_6. After the incubation period, an aliquot of the solution was removed, and the ferrocyanide present was measured colorimetrically.[7]

Manipulations for dark treatments were carried out under a dim green safe light. When a sphingoid base was used, the medium was filtered through two layers of coarse Fisher brand filter paper before assay. No corrections were made for any chemical interactions between ferricyanide or ferrocyanide and the sphingoid base.

Stock solutions of the agents used in this study were dissolved in 95% ethanol except for DCMU which was in methanol. After A23187 or sphingoid bases were diluted into the aqueous medium, they were sonicated in a Branson B-220 bath-type sonicator for 2 min. Equivalent amounts of solvent were added to controls and preliminary experiments determined that the solvents had no effect on redox activity at the concentrations used.

Oxygen evolution or consumption was measured at a constant temperature of 25C with a YSI Clark-type oxygen electrode as previously described.[2]

RESULTS

A role for cytosolic Ca^{2+} as a regulator of transmembrane electron transport is suggested by the data in Fig. 1. The Ca^{2+} ionophore, A23187, stimulates ferricyanide reduction at concentrations of 10 uM and above; as expected, the effect is seen only when the ionophore is added in the absence of $CaCl_2$ and the $CaCl_2$ is introduced later (data not shown).

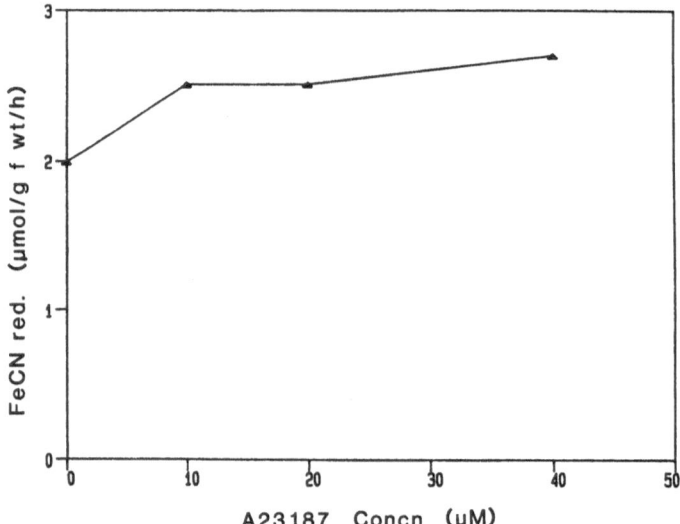

Fig. 1. Effect of A23187 concentration on ferricyanide reduction in the dark. Leaf segments were incubated on buffered medium
containing A23187 at the concentrations
indicated for 10 min before adding 1 mM
$CaCl_2$. Ferricyanide was added after a
further 10 min to a concentration of
0.8 mM. The amount of ferricyanide reduced was assayed after 15 min.

Table 1. Effect of Modulators of Calcium Channels on Transplasmalemma Electron Transport in Oat Mesophyll Cells

| | 3 mM KCl | | 50 mM KCl | |
	$Fe(CN_6)^{-3}$ reduced[a]	Δ	$Fe(CN_6)^{-3}$ reduced[a]	Δ
	μmoles/g f wt/h	%	μmoles/g f wt/h	%
Control	3.4	100	3.3	100
(+)202-791	3.4	100	4.3	130
(-)202-791	3.1	92	2.7	85

[a]Ferricyanide reduction was measured for leaf segments incubated for
30 minutes in 0.1 mM (+) or (-) 202-791 in buffer solution containing
either 3 mM KCl or 50 mM KCl, in the dark.

Table 2. The Effect of DCMU and Phytosphingosine on Redox Activity
in the Light and Dark.

| | Ferricyanide reduced | | | |
| | -DCMU | | +DCMU | |
	Dark	Light	Dark	Light
	μmoles/g f wt/h			
Control[a]	3.5	5.9	3.6	3.7
Sphingosine[a]	1.8	18.9	1.7	3.4

[a]Leaf segments were preincubated in the dark with 0.5 mM phytosphingosine with or without 50 μM DCMU for 15 minutes. Ferricyanide
(0.8 mM) was added at the end of this incubation period to the same
medium and the amount of ferricyanide reduced was measured after
15 minutes in dark or 20 W/m^2 white light.

The involvement of Ca^{2+} is further indicated by the use of the Ca^{2+}
channel opener, (+)202-791, and channel blocker, (-)202-791.[8] As shown in
Table 1, neither form of the drug has an effect in the presence of 3 mM KCl,
but in the presence of 50 mM KCl, ferricyanide reduction is stimulated 30%
by the channel opener and inhibited slightly by the blocker. The higher
concentration of KCl depolarizes the membrane potential of mesophyll cells[9]
and the drugs are able to act more effectively, since the Ca^{2+} channel is
thought to be voltage regulated.[8]

One of the possible Ca^{2+}-requiring reactions which may regulate transmembrane redox activity is the phosphorylation of certain proteins by a
protein kinase C-like enzyme. To test for the involvement of such an enzyme, the phorbol ester, phorbol myristate acetate (PMA), was added together
with ferricyanide to leaf segments for 30 min. This drug mimics the action
of diacylglycerol, which is required, along with Ca^{2+}, for protein kinase C
activity.[10] As shown in Fig. 2, PMA stimulates ferricyanide reduction at
concentrations above 25 ng/ml.

Next, compounds reputed to specifically inhibit protein kinase C action were investigated. H-7 inhibits phytoalexin production in carrot cells at concentrations above 1 μm,[11] but 0.1 to 0.5 mM H-7 had no effect on ferricyanide reduction by mesophyll cells (data not shown). On the other hand, sphingoid bases, which appear to inhibit transmembrane electron transport in neutrophils via protein kinase C,[12] are very effective regulators of transmembrane electron transport in mesophyll cells. The data in Table 2 (first two columns) show that phytosphingosine, which is structurally similar to sphingosine of animal cells, appears to inhibit ferricyanide reduction in the dark. Surprisingly, however, phytosphingosine markedly stimulates reduction in the light.

The enhancement of transmembrane redox activity in the light by phytosphingosine is also seen with other sphingoid bases whose activities parallel those reported for inhibiting protein kinase C in neutrophils.[12] These observations may account for the effects of the fatty acid chain elongation inhibitor, cerulenin, which stimulates ferricyanide reduction in the light by 1.75 to 2.0 times the rate of cerulenin-treated tissue in the dark. This may be due to the inhibitor being a long chain amide, a molecular structure related to sphingoid bases, rather than to its inhibitory action on fatty acid chain elongation.

The light-induced stimulation of ferricyanide reduction in the presence of sphingoid bases is not related to leakage of reductants. As shown in Table 3, structural analogs of sphingoid bases cause some leakage of substances from the cells which reduce ferricyanide, but, except for octylamine, the rates are less than 10% of those seen in the light without DCMU when the tissue is present.[13] These effects appear to be unrelated to

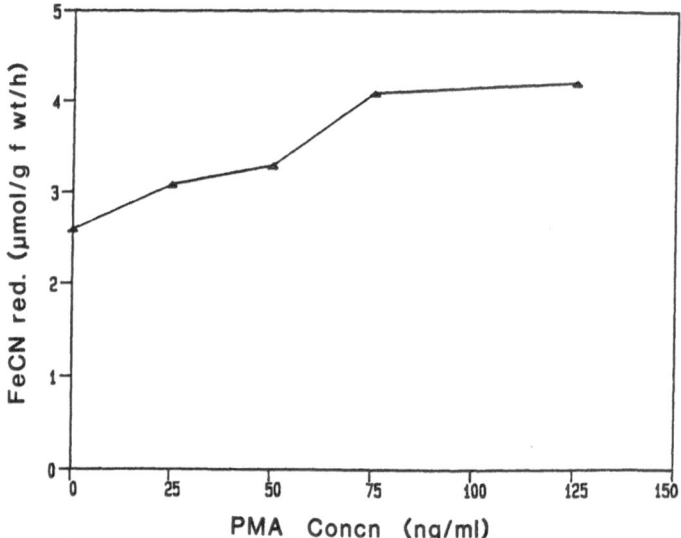

Fig. 2.   Effect of phorbol myristate acetate (PMA)
on ferricyanide reduction in the dark.
Leaf tissue was preincubated on PMA at
the concentrations indicated for 30 min.
Ferricyanide was then added to a concentration of 0.8 mM and the amount reduced
was assayed after 15 min.

Table 3. Leakage of Reductant from Oat Leaf Segments in the Presence of Sphingoid Bases.

	Ferricyanide reduced[a]
	µmoles/g f wt/h
Control	0.05
Threo-dihydrosphingosine	0.42
Erythro-dihydrosphingosine	0.53
Phytosphingosine	0.68
Stearylamine	0.5
Octylamine	3.5

[a]Peeled leaf segments were incubated in the indicated compounds at 0.5 mM for 15 minutes in the light. The segments were removed and the amount of ferricyanide reduced was measured after a subsequent 15 minute period in 0.8 mM ferricyanide.

changes in rates of respiration or photosynthesis, since sphingoid bases were found to have little or no effect on O_2 exchange in dark or light (data not shown).

Based in part on previously reported results using DCMU, photosynthesis appears to be totally responsible for light-induced ferricyanide reduction observed in the absence of phytosphingosine.[2] However, when DCMU and phyto-sphingosine are present together, a two-fold stimulation of ferricyanide reduction by light is still observed when compared to the corresponding dark control (Table 2, columns 3 and 4); this remaining increment of transplasma-lemma electron transport in the presence of sphingoid bases and DCMU, is due solely to blue light.[13] The ten-fold stimulation of ferricyanide reduction in the light by phytosphingosine in the absence of DCMU (Table 2, columns 1 and 2), suggests that in this case photosynthetic processes are also involved.

DISCUSSION

Indirect evidence that cytosolic Ca^{2+} is an important regulator of transmembrane electron transport is presented in Fig. 1 and Table 1. Both A23187 and (+)202-791 in the presence of exogenous Ca^{2+} may lead to elevated levels of Ca^{2+} within the cell. This could be due to the ionophoric property of A23187 which allows Ca^{2+} to diffuse down its electropotential gradient; the (+)202-187 is reported to open voltage regulated Ca^{2+}-specific channels.[12] Both of these agents stimulate ferricyanide reduction by mesophyll cells. (-)202-187 is reported to block the Ca^{2+}-specific channels and this agent slightly inhibits redox activity. The rather small effects of these drugs may indicate that factors other than cytosolic Ca^{2+} levels are limiting for transmembrane electron transport.

One possible factor controlling electron transport at the cell surface is the activity of protein kinase C, an important regulator of many hormonal and light-mediated processes;[14] it is significant for our studies that this enzyme controls the activity of the transmembrane NADH oxidase of human neutrophils.[15] While there is still no unequivocal evidence for protein kinase C in plants, we can show that substances which modify the activity of

291

this enzyme in certain animal systems can affect rates of ferricyanide reduction by mesophyll cells. For example, the phorbol ester, PMA, which stimulates protein kinase C activity in animal cells,[15] stimulates redox activity (Fig. 2). The inhibitor, H-7, had no effect on ferricyanide reduction under our conditions, but H-7 is also inactive on the transplasmalemma redox-mediated superoxide release in human neutrophils.[16]

What is difficult to explain, however, is how phytosphingosine, a close structural analog of the sphingoid bases found to inhibit protein kinase C,[12] markedly stimulates ferricyanide reduction in the light (Table 2). The responses are most pronounced using sphingoid bases which are the most effective inhibitors of protein kinase C.[13] Cerulenin, whose mode of action has been thought to be primarily an inhibitor of fatty acid chain elongation,[17] could also be considered a structural analog of sphingoid bases; its action on ferricyanide reduction is consistent with this suggestion.

The stimulations of ferricyanide reduction in the light by sphingoid bases are not due to leakage of reducing substances (Table 3); the effects are also not directly related to a corresponding increase in the rate of photosynthesis, since changes in oxygen evolution are not detectable under similar conditions. Photosynthesis is important, however, since DCMU at concentrations which eliminate oxygen evolution in mesophyll cells[2] markedly reduce the stimulatory effects of phytosphingosine (Table 2).

The light-induced transmembrane redox activity still observed in the presence of DCMU (Table 2) may provide a clue to the nature of the phytosphingosine stimulation. This result indicates that a photoreaction other than photosynthesis is activated when a sphingoid base is added. The pigment responsible for this effect absorbs light primarily in the blue region,[13] and may act in conjunction with a photosynthetically-regulated process to give the observed stimulations.

A tentative hypothesis is that a redox system, perhaps partially inhibited by sphingoid bases, operates slowly in the dark and is the system which is stimulated by photosynthesis.[2] However, a second redox system exists which is activated solely by blue light. This second system is normally maintained in an inactive state by a mechanism which is inhibited by sphingoid bases such as phytosphingosine. Thus, in the absence of photosynthesis, net stimulations of ferricyanide reduction would be observed in blue light only when a sphingoid base is added.

It is tempting to speculate that the blue light reaction discovered here may represent at least part of a primary photochemical system responsible for certain morphogenic events. And, because of its presence in plant cells, phytosphingosine might be a hitherto unrecognized regulator of these events. It appears rewarding, therefore, to enlarge this study to characterize the blue light response further as well as to investigate the effects of sphingoid bases on other tissues known to exhibit blue light effects. We also intend to study the effects of sphingoid bases on _in vitro_ systems in which transplasmalemma electron transport is known to occur.[18]

REFERENCES

1. R. Barr, B. Stone, T. A. Craig, and F. L. Crane, Evidence for Ca^{2+}-calmodulin control of transplasmalemma electron transport in carrot cells, Biochem. Biophys. Res. Commun. 126:262 (1985).
2. S. Dharmawardhane, A. I. Stern, and B. Rubinstein, Light-stimulated transplasmalemma electron transport in oat mesophyll cells, Plant Sci. 51:193 (1987).

3. B. Lass, G. Theil, and C. I. Ulrich-Eberius, Electron transport across the plasmalemma of _Lemna Gibba Gi._, Planta 169:251 (1986).

4. E. Neufeld and A. W. Bown, A plasmamembrane redox system and proton transport in isolated mesophyll cells, Plant Physiol. 83:895 (1987).

5. T. Y. Leong and W. R. Briggs, Partial purification of a blue light sensitive cytochrome-flavin complex from corn membranes, Plant Physiol. 67:1042 (1981).

6. S. Widdell and C. Larsson, Distribution of cytochrome b photoreduction mediated by endogenous photosensitizer or methylene blue in fractions from corn and cauliflower, Physiol. Plant. 57:196 (1983).

7. M. Avron and N. Shavit, A sensitive and simple method for the determination of ferrocyanide, Anal. Biochem. 6:549 (1963).

8. S. Kongsamut, T. J. Kamp, R. J. Miller, and M. C. Sanguinetti, Calcium channel agonist and antagonist effects of the sterioisomers of the dihydropyridine 202-791, Biochem. Biophys. Res. Commun. 130:141 (1985).

9. B. Rubinstein and T. A. Tattar, Regulation of amino acid uptake into oat mesophyll cells: a comparison between protoplasts and leaf segments, J. Exp. Bot. 31:269 (1980).

10. M. Castagna, Y. Takai, K. Kaibuchi, K. Sano, U. Kikkawa, and Y. Nishizuka, Direct activation of calcium-activated phospholipid-dependent protein kinase by tumour-promoting phorbol esters, J. Biol. Chem. 257:7847 (1982).

11. F. Kurosaki, Y. Tsurusawa, and A. Nishi, Breakdown of phosphatidylinositol during elicitation of phytoalexin production in cultured carrot cells, Plant Physiol. 85:601 (1987).

12. E. Wilson, M. C. Olcott, R. M. Bell, A. H. Merrill, and J. D. Lambeth, Inhibition of the oxidative burst in human neutrophils by sphingoid long-chain bases, J. Biol. Chem. 261:12616 (1986).

13. S. Dharmawardhane, Light-stimulated transplasmalemma electron transport in oat mesophyll cells, Ph.D. Thesis, University of Massachusetts, Amherst (1988).

14. M. J. Berridge, Inositol triphosphate and diacylglycerol: two interacting second messengers, Annu. Rev. Biochem. 56:159 (1987).

15. A. I. Tauber, D. B. Bretter, C. A. Kennington, and P. A. Blumberg, Relation of human neutrophil phorbol ester occupancy and NADPH-oxidase activity, Blood 60:333 (1982).

16. F. Rossi, The O_2-forming NADPH oxidase of the phagocytes: nature, mechanisms of activation and function, Biochim. Biophys. Acta 853:65 (1986).

17. S. Omura, The antibiotic cerulenin, a novel tool for biochemistry as an inhibitor of fatty acid synthesis, Bacteriol. Rev. 40:681 (1976).

18. M. Hassidim, B. Rubinstein, H. R. Lerner, and L. Reinhold, Generation of a membrane potential by electron transport in plasmalemma-enriched vesicles of corn and radish, Plant Physiol. 85:872 (1987).

THIOREDOXIN AND GLUTAREDOXIN: FUNCTIONS IN GROWTH-DEPENDENT DNA

SYNTHESIS AND CELLULAR REGULATION VIA THIOL REDOX CONTROL

Arne Holmgren

Department of Physiological Chemistry
Karolinska Institutet
S-104 01 Stockholm, Sweden

INTRODUCTION

Oxidation-reduction systems involving thiols and disulfides are of importance for many cellular processes. Among reactions required for cell growth, protein thiol-disulfide redox reactions are important for: (i) the mechanism of some biosynthetic enzymes, (ii) as a posttranslational event in the formation of native disulfides in proteins or their reduction, and (iii) for the regulation of the activity of certain enzymes and receptor by thiol redox control.

Studies of the enzymatic synthesis of deoxyribonucleotides, which are the precursors of DNA,[1,2] resulted 1964 in the isolation of thioredoxin[3] and in 1976 in the discovery of glutaredoxin.[4] these small proteins, both contain two vicinal redox-active half-cystine residues in the active site (-Cys-Xxx-Yyy-Cys-), forming a fourteen-membered disulfide ring in their oxidized form.[3,4] Thioredoxin and glutaredoxin have been isolated from Escherichia coli as well as mammalian tissues like rat liver or calf thymus and are now known to be involved in a number of reactions (for a recent summary see ref. 5). Studies of the structure of thioredoxin and glutaredoxin[3,4] have revealed two proteins with different primary structures but similarities in folding and three-dimensional structures.[6] This may reflect the evolution of thioredoxin to catalyze dithiol-dependent reactions and of glutaredoxin to catalyze monothiol or glutathione-dependent reactions.[4,6] Both proteins operate by thiol-disulfide exchange mechanisms via their disulfide/dithiol active sites.

Thioredoxin and glutaredoxin are both linked to the pyridine nucleotide NADPH via the FAD-containing enzymes thioredoxin reductase and glutathione reductase.[3,4] These enzymes contain active site redox-active disulphide bonds and are widely distributed. This article summarizes some current knowledge about the structure, function and localization of mammalian and plant thioredoxin and glutaredoxin systems. In addition possible relationships to dehydrogenases of the plasma membrane[7] and transplasma-membrane redox systems[8] will be discussed.

THIOREDOXIN SYSTEM

Thioredoxin is a small protein, M_r 12,000, present in all procaryotes and eucaryotes so far investigated.[5] The protein

consists of a single polypeptide chain with about 108 amino residues
(Escherichia coli) with only amino acid residues. No metals or other
groups are involed in the thioredoxin mechanism of action, making it
a simple redox protein with the function linked to the oxidation-
reduction of an active site disulfide bond. The oxidized form of
thioredoxin (Trx-S$_2$) contains an active site disulfide with the
structure: -Trp-Cys-Gly-Pro-Cys-. The reduced form (Trx(SH)$_2$)
contains a dithiol.

Trx-S$_2$ is generally reduced by NADPH and thioredoxin reductase
(TR) (Reaction 1). The thioredoxin system is a name used collective-
ly for NADPH, thioredoxin reductase and thioredoxin.[9]

$$\text{(1)} \quad \text{NADPH} + \text{H}^+ + \text{Trx-S}_2 \overset{\text{TR}}{\rightleftharpoons} \text{NADP}^+ + \text{Trx-(SH)}_2$$
$$\text{(2)} \quad \text{Trx-(SH)}_2 + \text{protein-S}_2 \rightleftharpoons \text{Trx-S}_2 + \text{protein-(SH)}_2$$

Trx-(SH)$_2$ is a powerful protein disulfide reductase (Reaction 2).
Reactions 1 and 2 give a system for protein disulfide reduction by
NADPH. However, both Reactions 1 and 2 are reversible and depending
on the reactants thioredoxin may either catalyze net protein reduction or oxidation.[10] So far, the reduction of disulfides has been
studied in different proteins; a recent example is the production of
subunits of human α_2-macroglobulin by specific reduction of
interchain disulfides.[11]

Table 1 shows the redox potential (E$'_0$-values) of thioredoxin
and related compounds. Obviously, NADPH-dependent disulfide reduction of thioredoxin can effect many protein disulfides. However,
only exposed disulfides will react but the factors gowerning the
reactivity are so far largely unknown.

From Table 1 it is also obvious that thioredoxin is a sufficiently reducing agent to react with ferricyanide, cytochrome C and
oxygen and different reduced oxygen species with the exception of
superoxide anion (O$_2^-$).

Table 1.   Oxidation-reduction potentials (E$'_0$-values) for
           thioredoxin and related systems*

System	E$'_0$, V
Trx-S$_2$/Trx-(SH)$_2$	-0.26
L-Cystine/cysteine	-0.22
GSSG/2GSH	-0.23
Lipoic acid ox/red	-0.29
Dithiothreitol ox/red	-0.33
Protein disulfides ox/red	-0.20-0.40
NAD(P)$^+$/NAD(P)H + H$^+$	-0.32
Ferredoxin ox/red (spinach)	-0.43
Pyruvic acid/acetic acid + CO$_2$	-0.70
Ferricyanide/ferrocyanide	+0.36
Fe^{3+}, cytochrome c/Fe^{2+}	+0.22
O$_2$/O$_2^-$	-0.45
O$_2$/H$_2$O	+0.30
O$_2$/2^{2}H$_2$O	+0.82
OH+H$^+$/H$_2$O	+1.35

*Values taken from Principles of Biochemistry by Smith
Hill, Lehman, Lefkowitz, Handler, White, Seventh Edition,
Mc Graw-Hill International, 1983 and ref. 5. E$'_0$ is midpoint
potentials. ox/red, oxidized form/reduced form.

Thioredoxin has been well-characterized from E. coli; the three-dimensional structure of Trx-S$_2$ has been determined by X-ray crystallography to 2.8 Å and now refined to 1.7 Å (Eklund, H. et al. in preparation). As seen from Fig. 1, the active site of Trx-S$_2$ is located in a protrusion at the end of a β-strand (β2) immediately followed by a long α-helix (α$_2$). The structure of Trx-(SH)$_2$ has not been determined yet since no crystals are available. However, evidence for a localized conformational charge upon reduction of Trx-S$_2$ has been obtained from tryptophan fluorescence, CD-spectroscopy and[2] NMR studies.[5] Recently, the structure of thioredoxin in oxidized and reduced form has been analyzed by two-dimensional ^{1}H NMR spectroscopy.[13] The results show that significant changes in structure occur in the disulfide bridge itself, and in residues forcing a flat hydrophobic surface around the active site.[13] Probably, the position of the sulfurs of the Cys-32 and Cys-35 change their position together with significant but local movements of the peptide chain. Analysis of this will be of great interest in explaining the high reactivity of Trx-(SH)$_2$ in disulfide oxido-reduction reactions.

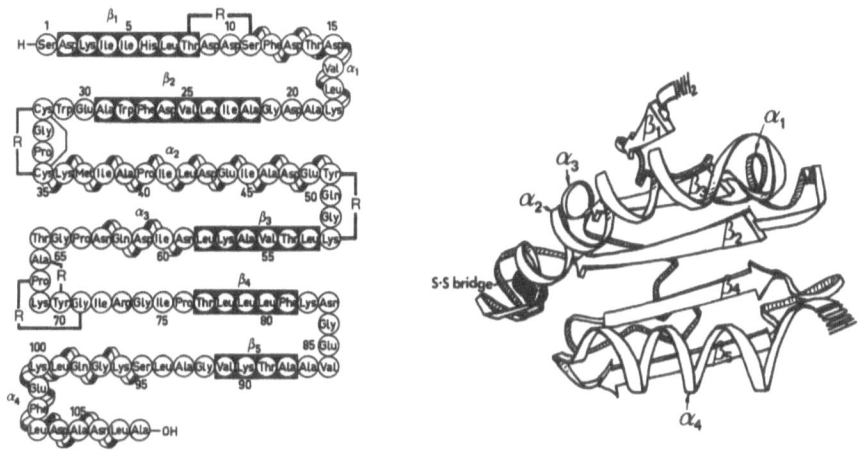

Fig. 1. Structure of E. coli thioredoxin-S$_2$. The amino acid sequence and secondary structure elements (left); R = reverse turn. Schematic drawing of the three-dimensional structure at 2.8 Å resolution by X-ray crystallography[12] (right). α$_1$-α$_4$, α-helices, β$_1$-β$_5$, strands of β-sheet.

MAMMALIAN THIOREDOXIN SYSTEMS

Thioredoxin and thioredoxin reductase have been purified to homogeneity from calf liver,[14] and thymus[15,16] as well as rat liver.[17] These mammalian thioredoxin systems are very similar but show interesting similarities and differences to the corresponding E. coli system (Table 2). So far, only one thioredoxin has been isolated from the mammalian systems using NADPH-dependent insulin disulfide reduction as an assay.[14,15,17]

Thioredoxin

Thioredoxin has the same M$_r$ as the E. coli protein. The protein has 104 amino acid residues.[18] The active center is identical and the structure is 28% homologous to E. coli thioredoxin.[18] The activity of rat or bovine Trx-(SH)$_2$ as a disulfide reductase is the same as E. coli Trx-(SH)$_2$. One important difference

Table 2. Properties of thioredoxin systems

	E. coli	calf thymus
Thioredoxin	M_r:12.000 108 aa 1 S-S bridge	M_r: 12.000 104 aa 1 S-S bridge + 2 SH groups inactivated by oxidation
Thioredoxin reductase	M_r: 66.000 2 subunits high specificity stable	M_r: 116.000 2 subunits broad specificity inactivated by oxidation of SH-groups

between Trx from E. coli and calf thymus is the presence of two
additional half-cystine residues in the C-terminal part of the
structure (Cys-61 and Cys-72)[18] These extra half-cystine residues
easily are oxidized to an intramolecular disulfide resulting in
inactivation of the thioredoxin.[14, 15, 18] This is an important
control mechanism of the activity of thioredoxin in vitro and
probably in vivo.[18]

Thioredoxin reductase

The main difference between the thioredoxin system of E. coli
and mammalian cells concerns thioredoxin reductase. The E. coli
enzyme is a dimer with 33 K subunits each containing an FAD and a
redox-active disulfide. The enzyme is specific for Trx-S$_2$ from E.
coli and will not show activity with Trx-S$_2$ from yeast or rat
liver.[15] In contrast, the mammalian thioredoxin reductase is a dimer
with 58 K subunits each with an FAD and a redox-active disulfide.
The enzyme has a broad substrate specificity and reacts also with
Trx-S$_2$ from E. coli and other species.[5] In addition other non-disul-
fide substrates are reduced by the enzyme (Table 3). We recently
showed that $Fe(CN)_6^{3-}$ is also reduced by thioredoxin reductase from
calf thymus (Fig. 2). The significance of this in relation to
plasma-membrane redox reactions of ferricyanide is not known. The
distribution of thioredoxin and thioredoxin reductase in mammalian
cells (see below), which shows enrichment at the plasma membrane
suggests a relationship; obviously further studies are required.

Localization of thioredoxin system

The localization of thioredoxin and thioredoxin reductase in
the rat has been studied by immunohistochemical techniques,[20-22]
using polyclonal rabbit antisera against the pure proteins. Both
proteins are widely distributed in the tissues and organs of adult
rats but varied between cell types. Hepatocytes, different
epithelial and secretory cells, plasma cells and neurons contained
particularly prominent immunoreactivity of both thioredoxin and
thioredoxin reductase. Mesenchymal cells of connective tissue, bone,
muscle and fibrous tissue showed lower content. However, proliferating cells generally contained stronger immunoreactivity of thioredoxin. In general, thioredoxin and thioredoxin reductase were
mainly localized to the cytoplasm of cells. However, the distribution in the cytoplasm varied and enrichments at secretory granules
of both endocrine and exocrine cells were seen.[20, 21] Of particular
interest for plasma membrane redox activity, a high immunoreactivity
was localized at the plasma membrane and in the subplasma membrane

Table 3. Substrates reduced by NADPH and mammalian thioredoxin reductase

Substrate	Reference
Trx-S$_2$ (E. coli, rat liver, etc.)	15
Vitamin K$_3$ (Menadione)	17
Cu^{2+}	17
Alloxan	19
DTNB	15,17
Selenite	S. Kumar & A. Holmgren in preparation
Diamide	A. Holmgren, unpublished
H$_2$O$_2$	A. Holmgren, unpublished

zone in secretory cells. Probably thioredoxin and thioredoxin reductase may be located at the inner surface plasma of the plasma membrane as extrinsic or peripheral proteins. Whether they can be transmembrane or located exposed to the surface of cells is unknown at present.

The ultrastructural localization of thioredoxin and thio-redoxin reductase in rat hepatocytes has been analyzed with immuno-gold techniques and electron microscopy.[22] Both proteins were recognized all over the cells both in the cytosol and the chromatin of the nucleus. In particular an association with the granular endoplasmatic reticulum and the cisternae of the Golgi complex was apparent. This distribution of thioredoxin and thioredoxin reductase suggests a role in formation of protein disulfides and secretion.

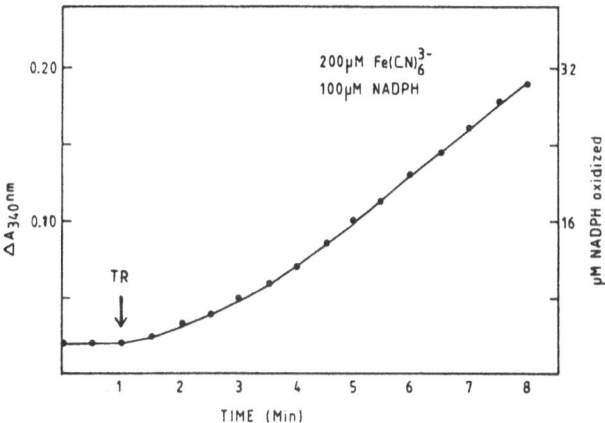

Fig. 2. Reduction of K$_3$Fe(CN)$_6$ by NADPH and calf thymus thioredoxin reductase. Two cuvettes contained 1.0 ml of 50 mM potassium phosphate - 1 mM EDTA, pH 7.0 with 100 μM NADPH and 200 μM Fe(CN)$_6^{3-}$. The absorbance at 340 nm was followed with a Zeiss PMQIII spectrophotometer. Calf thymus TR, 7 x 10^{-8} M was added (arrow).

THIOREDOXIN AND GLUTAREDOXIN IN DNA SYNTHESIS

Deoxyribonucleotides for DNA synthesis are formed de novo by the essential enzyme ribonucleotide reductase.[1,2,23,24]

As seen from Fig. 4 one enzyme in E. coli or mammalian cells will reduce all four ribonucleotides to the corresponding deoxyribo-nucleotides. In E. coli and mammalian cells,[1,2,24] ribonucleotide reductase consists of two subunits. The B1 (E. coli) or M1 (mamma-lian cells) contain substrate binding sites, allosteric regulatory

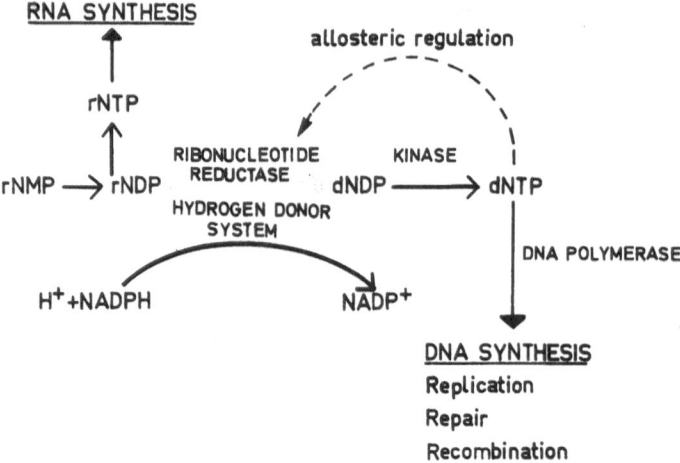

Fig. 3. Reduction of ribonucleotides to deoxyribonucleotides.

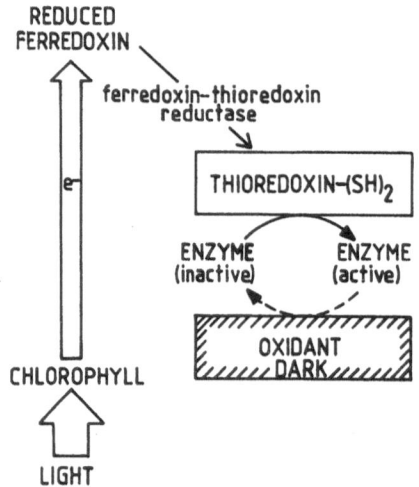

Fig. 4. Regulation of photosynthetic enzymes in the chloroplast by light through the ferredoxin-thioredoxin system.[26]

sites and redox-active dithiols. The other subunit B2 (E. coli) or M2 (mammalian cells) has iron in the form of Fe^{3+} in a μ-oxo-bridge and a unique tyrosyl free radical.[1,23] The iron in M2 is required for activity. The formation of the free radical in M2 requires iron, a dithiol reducing agent and O_2. In contrast to bacterial cells it may be formed spontaneously. In E. coli an enzyme system has been found.[24] During isolation of ribonucleotide reductase it is separated from its natural hydrogen donor system.[1] The isolated enzyme requires a dithiol like dihydrolipoic acid or dithiothreitol. Two hydrogen donor systems in vitro have been isolated.[2] The first hydrogen donor for ribonucleotide reductase (RDRase) was thioredoxin (Reaction 3). Subsequently, GSH and glutaredoxin was isolated from E. coli mutant cells that lacked thioredoxin (Reaction 4). Trx-S$_2$ and GSSG are reduced by NADPH via thioredoxin reductase and gluta-

(3) $\quad$ Trx-(SH)$_2$ + rNDP $\xrightarrow{\text{RDRase}}$ Trx-S$_2$ + dNDP + H$_2$O

(4) $\quad$ 2 GSH + rNDP $\xrightarrow[\text{Glutaredoxin}]{\text{RDRase}}$ GSSG + dNDP + H$_2$O

thione reductase. Both the thioredoxin and the glutaredoxin systems exist in mammalian cells[2,4]; their relative role is not known at present. The recent isolation of viable E. coli mutants lacking both thioredoxin and glutaredoxin[25] strongly suggests the existence of a third system, so far unknown.

THIOL REDOX CONTROL

Regulation of protein functions via reversible oxidation of sulfhydryl groups to disulfides has been called thiol redox control.[9] The 2-electron oxidation of two sulfhydryl groups to a disulfide is a reversible covalent modification that may activate or inactivate the biological activity of a protein.[26,27] Intracellular enzymes are often sensitive to oxidation in vitro and require the presence of thiols to stay active. This is a reflection of the reducing environment in the cell, where high concentrations of GSH (1-10 mM) are present.[27]

Thioredoxin has a powerful activity as a protein disulfide reductase[2,29,30] and will catalyze reactions of the type:

$$R\text{-(SH)}_2 + \text{protein-S}_2 \underset{}{\overset{\text{Trx}}{\rightleftharpoons}} R\text{-S}_2 + \text{protein-(SH)}_2$$

This is due to very rapid reactions of both Trx-S$_2$ and Trx-(SH)$_2$ with dithiols and disulfides, respectively.[29,30] Reduction of Trx-S$_2$ with dithiothreitol is about 10^3 times faster than that of insulin disulfides; Trx-(SH)$_2$ reacts with insulin about 10^5 faster than dithiothreitol reacts with insulin disulfides.[29,30]

A problem in evaluating the biological importance of thiol redox control mechanisms, has been the lability of the SH-groups and disulfides and the presence of fast thiol-disulfide interchange mechanisms.[28] Thus, so far thiol redox control has not been proven in many systems although the mechanism may potentially be as important as phosphorylation.[9]

A system with a well-proven thioredoxin-dependent regulation of enzyme activity is the chloroplast.[26] Light regulates the activity of a number of photosynthetic enzymes in the Calvin cycle (Fig. 4). The chloroplast contains two types of thioredoxin, Trx$_f$ and Trx$_m$ which regulate a number of enzymes including fructose bis-phospha-tase (Trx$_f$) and malate dehydrogenase (Trx$_m$). For a review see ref. 26. In the dark disulfides are formed and the enzymes become inactive.

Glutaredoxin catalyzes the reactions of GSH by which enzymes may be covalently modified in one-electron reduction:

$$\text{GSH} + \text{protein-S}_2 \rightleftharpoons \text{protein} \begin{cases} \text{S-SG} \\ \text{SH} \end{cases}$$

Proteins containing mixed disulfides with GSH are known[28] and the role of glutaredoxin in regulation of such proteins is presently unknown.

Acknowledgements. This investigation was supported by grants from the Swedish Medical Research Council, the Swedish Cancer Society and the Lundberg Foundation.

REFERENCES

1. L. Thelander and P. Reichard, Annu. Rev. Biochem. 48:133 (1979).
2. A. Holmgren, Curr. Top. Cell. Reg. 19:47 (1980).
3. A. Holmgren, Annu. Rev. Biochem. 54:237 (1985).
4. A. Holmgren, Methods Enzymol. 113:525 (1985).
5. A. Holmgren, C.-I. Brändén, H. Jörnvall, and B.-M. Sjöberg, Eds. "Thioredoxin and Glutaredoxin Systems: Structure and Function", Proceedings of the Ninth Karolinska Institute Nobel Conference, May 27-31, 1985. Raven Press, New York, pp. 1-411 (1986).
6. H. Eklund, C. Cambillau, B.-M. Sjöberg, A. Holmgren, H. Jörnvall, J.-O. Höög, C.-I. Brändén, EMBO J. 3:1443 (1984).
7. F.L. Crane, H. Goldenberg, D.J. Morré, and H. Löw, in "Subcellular Biochemistry" Vol. 6, D.B. Roodyn, ed., Plenum Publishing Corporation, New York pp. 345-399 (1979).
8. F.L. Crane, I.L. Sun, M.G. Clark, C. Grebing, and H. Löw, Biochim. Biophys. Acta 811:233 (1985).
9. A. Holmgren, TIBS 6:26 (1981).
10. A. Holmgren, Methods Enzymol. 107:295 (1984).
11. L.-J. Larsson, A. Holmgren, B. Smedsröd, T. Lindblom, and I. Björk, Biochemistry 27:983 (1988).
12. A. Holmgren, B.-O. Söderberg, H. Eklund, and C.-I. Brändén, Proc, Natl. Acad. Sci. USA 72:2305 (1975).
13. J.H. Dyson, A. Holmgren, and P.E. Wright FEBS Letters 228:254 (1988).
14. N.E. Engström, A. Holmgren, A. Larsson, and S. Söderhäll, J. Biol. Chem. 249:205 (1974).
15. A. Holmgren, J. Biol. Chem. 252:4600 (1977).
16. A. Holmgren and M. Luthman, Biochemistry 17:4071 (1978).
17. M. Luthman and A. Holmgren Biochemistry 21:6628 (1982).
18. A. Holmgren, C. Palmberg, and H. Jörnvall, J. Biol. Chem. submitted (1988).
19. A. Holmgren and C. Lyckeborg, Proc. Natl. Acad. Sci. USA 77:5149 (1980).
20. B. Rozell, H.A. Hansson, M. Luthman, and A. Holmgren, Eur. J. Cell Biol. 38:79 (1985).
21. H.A. Hansson, A. Holmgren, B. Rozell, and I.-B. Täljedal, Cell Tissue Res. 245:189 (1986).
22. B. Rozell, A. Holmgren, and H.A. Hansson, Eur. J. Cell Biology in press (1988).
23. P. Reichard and A. Ehrenberg, Science 221:514 (1983).
24. P. Reichard, Biochemistry 26:3245 (1987).
25. M. Russel and A. Holmgren, Proc. Natl. Acad. Sci. USA 85:990 (1988).
26. B.B. Buchanan, in "Glutaredoxin Systems: Structure and Function" A. Holmgren, C.-I. Brändén, H. Jörnvall, and B.-M. Sjöberg, eds. Raven Press, New York, pp. 233-242 (1986).
27. D.M. Ziegler, Annu. Rev. Biochem. 54:305 (1985).
28. H.F. Gilbert, Methods Enzymol. 107:330 (1984).
29. A. Holmgren, J. Biol. Chem. 254:9113 (1979).
30. A. Holmgren, J. Biol. Chem. 254:9627 (1979)

CYTOCHROME b$_{561}$ AND THE MAINTENANCE OF REDOX POISE IN SECRETORY VESICLES [1]

David Njus, Norma J. Kusnetz, Yvonne Vivar Pacquing and Patrick M. Kelley

Department of Biological Sciences
Wayne State University
Detroit MI 48202

The membranes of a variety of secretory vesicles contain an electron transfer system which supports redox reactions catalyzed by enzymes contained within the vesicles. These enzymes include two important monooxygenases, dopamine ß-hydroxylase and peptide amidating monooxygenase. Dopamine ß-hydroxylase, responsible for the biosynthesis of norepinephrine and epinephrine, is found in catecholamine-storing vesicles such as the chromaffin vesicles of the adrenal medulla. Peptide amidating monooxygenase, responsible for the α-amidation of peptide hormones, is found in peptide-storing vesicles such as those in the pituitary and gut. Both monooxygenases use ascorbic acid to reduce the second atom of molecular oxygen to water. While oxygen will diffuse across the secretory vesicle membrane, ascorbate will not. It is to maintain intravesicular ascorbate that the secretory-vesicle membrane has an electron transfer system.

The mechanism of ascorbate regeneration in secretory vesicles is now reasonably well understood, particularly in the chromaffin vesicles of the adrenal medulla. Dopamine ß-hydroxylase uses ascorbic acid as a one-electron donor and forms the free radical, semidehydroascorbate. This semidehydroascorbate is quickly reduced back to ascorbate by cytochrome b$_{561}$, a transmembrane protein in the chromaffin-vesicle membrane. The cytochrome is reduced, in turn, by taking an electron from cytosolic (extravesicular) ascorbic acid. Cytochrome b$_{561}$, therefore, functions as a transmembrane electron carrier. Electron flow from cytosolic ascorbate to intravesicular ascorbate is driven by the pH gradient (inside acidic) and membrane potential (inside positive) established by the H$^+$-translocating ATPase[2] in the chromaffin-vesicle membrane. In this system, a proton gradient drives electron flow in an

---

[1] This work was supported by NIH grant no. GM-30500. D. Njus is an Established Investigator of the American Heart Association.

[2] Abbreviations are: ATPase, adenosine triphosphatase; AH$^-$, ascorbate; A$\dot{}$, semidehydroascorbate; B, cytochrome b$_{561}$; FeCy, ferri/ferrocyanide; ox, oxidized; red, reduced.

interesting reversal of the mechanism originally proposed by Mitchell to account for proton gradients created by electron flow.

The evidence for cytochrome b_{561}-mediated electron transfer is now quite good and has been reviewed elsewhere (Njus et al., 1986; Njus et al., 1987). The issues that remain focus on the protein chemistry of cytochrome b_{561} (Flatmark and Gronberg, 1981; Apps et al., 1984; Fleming and Kent, 1987) and on the physical chemistry of the electron transfer process (Kelley and Njus, 1986; 1988; Wakefield et al., 1986a; 1986b). We are particularly interested in the latter because the apparent simplicity of cytochrome b_{561}-mediated electron transfer makes it reasonable to hope that the phenomenon may be understood on a theoretical level.

Reversal of Electron Transfer

A number of experiments have shown that the ascorbate regenerating system works: Turnover of dopamine ß-hydroxylase consumes intravesicular ascorbate in the absence but not in the presence of external electron donors (Levine et al., 1985; Beers et al., 1986; Menniti et al., 1986; Herman et al., 1988). This direct approach has not yielded much insight into the mechanism of electron transfer, however, because the process is slow and involves electron donors and acceptors that are inconvenient to monitor and, in the case of semidehydroascorbate, unstable. For quantitative studies of electron transfer, we have found it advantageous to reverse the direction of electron flow (Figure 1). Then, internal ascorbate serves as the electron donor and external semidehydroascorbate is the electron acceptor. The equilibrium between ferri/ferrocytochrome b_{561} and semidehydroascorbate/ascorbate is defined by

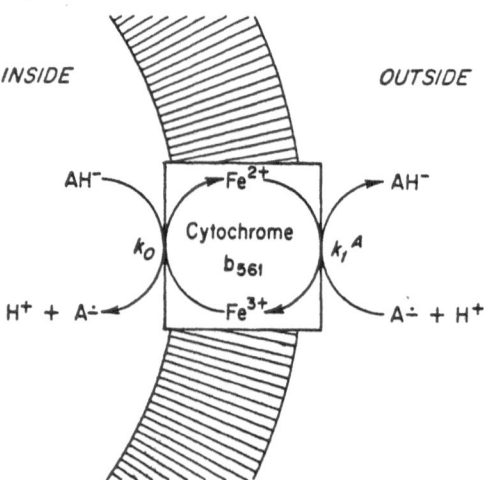

Figure 1. Reverse Electron Flow. Ascorbate (AH-) trapped within ascorbate-loaded chromaffin-vesicle ghosts reduces cytochrome b_{561} in the vesicle membrane. The cytochrome, in turn, reduces semidehydro ascorbate (A$\dot{-}$) present externally. These electron transfer steps are described by the rate constants k_0 and $k_1{}^A$ respectively.

known midpoint reduction potentials, so the rate constants for electron transfer in one direction can be calculated from the rate constants for electron transfer in the opposite direction. Therefore, we can measure rate constants for electron transfer from internal ascorbate to cytochrome b_{561} (k_0) and from cytochrome b_{561} to external semidehydroascorbate ($k_1{}^A$) and then determine the rate constants for the reactions in the physiological direction.

To study reverse electron flow, we use chromaffin vesicles that have been lysed and resealed in a medium containing 100 mM ascorbate (ascorbate-loaded ghosts). This preparation has a defined internal medium containing only the desired electron donor. By contrast, intact chromaffin vesicles are a poor subject for quantitative studies because the endogenous catecholamines are also potential electron donors.

Electron Flow to External Ferricyanide

Before addressing electron flow from internal ascorbate to external semidehydroascorbate (Figure 1), it is instructive to consider electron flow from internal ascorbate to external ferricyanide. Ferricyanide is an experimentally more convenient electron acceptor because it is stable and its reduction may be followed spectrophotometrically. This allows us to measure the rate of electron transfer (Harnadek et al., 1985b). We have recently determined rate constants for electron transfer from internal ascorbate to external ferricyanide. While the details are given elsewhere (Kelley and Njus, 1988), the analysis will be reviewed here because the same logic will be applied to electron transfer from internal ascorbate to external semidehydroascorbate.

In ascorbate-loaded ghosts, cytochrome b_{561} is nearly completely reduced. Upon addition, ferricyanide becomes reduced, and the cytochrome undergoes a concomitant spectral change indicative of oxidation (Kelley and Njus, 1986). Moreover, a membrane potential (inside positive) arises presumably caused by electron flow across the chromaffin-vesicle membrane (Njus et al., 1983; Harndadek et al., 1985a).

When the rate of ferricyanide reduction is examined as a function of the external ferricyanide concentration, a maximum rate is obtained, and this maximum rate depends upon the internal ascorbate concentration. We have interpreted these findings by assuming that electron transfer across the chromaffin-vesicle membrane can be analyzed as two simple steps: the reduction of oxidized cytochrome b_{561} (B_{ox}) by internal ascorbate (AH^-) and the oxidation of reduced cytochrome b_{561} (B_{red}) by external ferricyanide ($FeCy_{ox}$). Each of these steps may be expressed as a simple bimolecular reaction for which a simple rate constant may be determined.

I) $AH^- + B_{ox} \xrightarrow{\ k_0\ } A^{\underline{\cdot}} + B_{red} + H^+$

II) $FeCy_{ox} + B_{red} \xrightarrow{\ k_1{}^F\ } FeCy_{red} + B_{ox}$

Assuming steady-state electron transfer through cytochrome b_{561} and ignoring the reverse reactions, the rate of electron transfer (V) will be

1) $1/V = 1/k_0[AH^-]_{in} + 1/k_1{}^F[FeCy_{ox}]_{out}$

A plot of the reciprocal of the electron transfer rate against the reciprocal of the ferricyanide concentration is linear. The rate constants $k_1{}^F$ and k_0 may be calculated from the slope and intercept of this double-reciprocal plot.

According to this analysis, external ferricyanide oxidizes cytochrome b_{561} with a rate constant $k_1{}^F$ that is ~5 x 10^4 $M^{-1}s^{-1}$ at pH 7.0 (Kelley and Njus, 1988). Internal ascorbate reduces cytochrome b_{561} with a rate constant k_0 that is ~50 $M^{-1}s^{-1}$. As discussed above, the rate constants for the reverse reactions may be inferred from these. The equilibrium constant K_{eq} is equal to the ratio of the rate constants and is related to the midpoint reduction potentials of the two redox pairs. For electron transfer between semidehydroascorbate/ascorbate and ferri/ferrocytochrome b_{561},

2) $K_{eq} = k_{-0}/k_0 = exp\{(E°_{asc} - E°_{b561})(F/RT)\}$

Using values of +0.33 V and +0.14 V for the midpoint reduction potentials of ascorbate and cytochrome b_{561} respectively (Iyanagi et al., 1985; Flatmark and Terland, 1971), the rate constant k_{-0} is about 9 x 10^4 $M^{-1}s^{-1}$ at pH 7.0. The midpoint reduction potential of ascorbate is pH-dependent (+0.42 V at pH 5.5) and, for reasons discussed elsewhere (Kelley and Njus, 1988), we believe this pH-dependence affects k_{-0} rather than k_0. Consequently, k_{-0} should be about 3 x 10^6 $M^{-1}s^{-1}$ at pH 5.5.

Electron Flow to External Semidehydroascorbate

Having characterized electron flow using ferricyanide as the external electron acceptor, let us return to the case in which semidehydroascorbate is the acceptor (Figure 1). We can experimentally create this electron flow by suspending ascorbate-loaded ghosts in a medium containing ascorbic acid and ascorbate oxidase. Ascorbate oxidase oxidizes the external ascorbate to semidehydroascorbate. This, in turn, oxidizes cytochrome b_{561} and initiates electron transfer from internal ascorbate. This electron transfer may be observed by monitoring the oxidation of cytochrome b_{561} and by following the generation of an electron-transfer-dependent membrane potential (Figure 2).

To determine the external semidehydroascorbate concentration present in this experiment, we mixed ascorbate and ascorbate oxidase and followed the absorbance of semidehydroascorbate at 360 nm. (Because semidehydro-ascorbate has a relatively small extinction coefficient (3300 M^{-1} cm^{-1}) and a low concentration, its absorbance was measured in the absence of chromaffin-vesicle ghosts to eliminate interference from light scattering.) This indicates that the concentration of semidehydroascorbate present in the experiment above was about 2.7 μM immediately after ascorbate oxidase addition and declined to zero after three minutes when the medium became anoxic. Since 2.7 μM semidehydroascorbate creates a membrane potential and cytochrome b_{561} oxidation comparable to that caused by 10-fold higher concentrations of ferricyanide, we imagine that the rate constant for electron transfer from cytochrome b_{561} to semidehydro ascorbate ($k_1{}^A$) must be larger than the rate constant for electron transfer to ferricyanide ($k_1{}^F \sim 5 \times 10^4 \, M^{-1} \, s^{-1}$).

To examine this more precisely, let us extend our kinetic analysis to this electron transfer experiment. Reactions I and II above translate into reactions III and IV below:

III) $AH^-{}_{in} + B_{ox} \xrightarrow{\quad k_0 \quad} A^{\bullet}{}_{in} + B_{red} + H^+{}_{in}$

IV) $A^{\bullet}{}_{out} + B_{red} + H^+{}_{out} \xrightarrow{\quad k_1{}^A \quad} AH^-{}_{out} + B_{ox}$

In this case, we will include the reverse reactions, but we again assume steady-state electron transfer through cytochrome b_{561}. Then, the steady-state level of cytochrome b_{561} reduction is

3) $$\frac{B_{red}}{B_{total}} = \frac{k_{-1}{}^A[AH^-]_{out} + k_0[AH^-]_{in}}{k_1{}^A[A^{\bullet}]_{out} + k_{-0}[A^{\bullet}]_{in} + k_{-1}{}^A[AH^-]_{out} + k_0[AH^-]_{in}}$$

and the rate of electron transfer V is

4) $$V = \frac{k_0[AH^-]_{in}k_1{}^A[A^{\bullet}]_{out} - k_{-1}{}^A[AH^-]_{out}k_{-0}[A^{\bullet}]_{in}}{k_1{}^A[A^{\bullet}]_{out} + k_{-0}[A^{\bullet}]_{in} + k_{-1}{}^A[AH^-]_{out} + k_0[AH^-]_{in}}$$

Equation 3 gives us the oxidation state of cytochrome b_{561} observed in Figure 2A. Equation 4 gives us the rate of electron transfer, which cannot be measured directly, but is related to the membrane potential observed in Figure 2B. The generation of the membrane potential ($\Delta\psi$) may be estimated by calculating the charge transfer associated with electron flow. Electron efflux will create an inward current (I_{in}) equal to VF. There will also be an outward current attributable to leakage, and we will assume that this obeys Ohm's Law ($I_{out} = \Delta\psi \, G$) where G is the conductance of the membrane. Then,

5) $d\psi/dt = (I_{in} - I_{out})/C = (VF - \Delta\psi G)/C$

There is admittedly considerable uncertainty about the electrical properties of the chromaffin-vesicle membrane, but we estimate the capacitance (C) to be 10^{-6} farad/cm^2 and the conductance (G) to be 10^{-6} S/cm^2. Then, we can numerically integrate equation 5 to simulate the experiment shown in Figure 2. We assume that $[A^{\bullet}]_{out}$ is initially 2.7 µM and that it drops linearly to zero after 3 minutes. We assume that $[AH^-]_{out}$ is initially 600 µM and that it drops linearly to 90 µM after 3 minutes. We assume that $[AH^-]_{in}$ is initially 100 mM and that this drops as internal ascorbate is stoichiometrically oxidized by electron transfer. Finally, we assume that the internal $A^{\bullet}$ generated by electron transfer is destroyed by disproportionation and that its concentration is determined by the steady state. The result of this computer simulation is shown in Figure 3 for two different

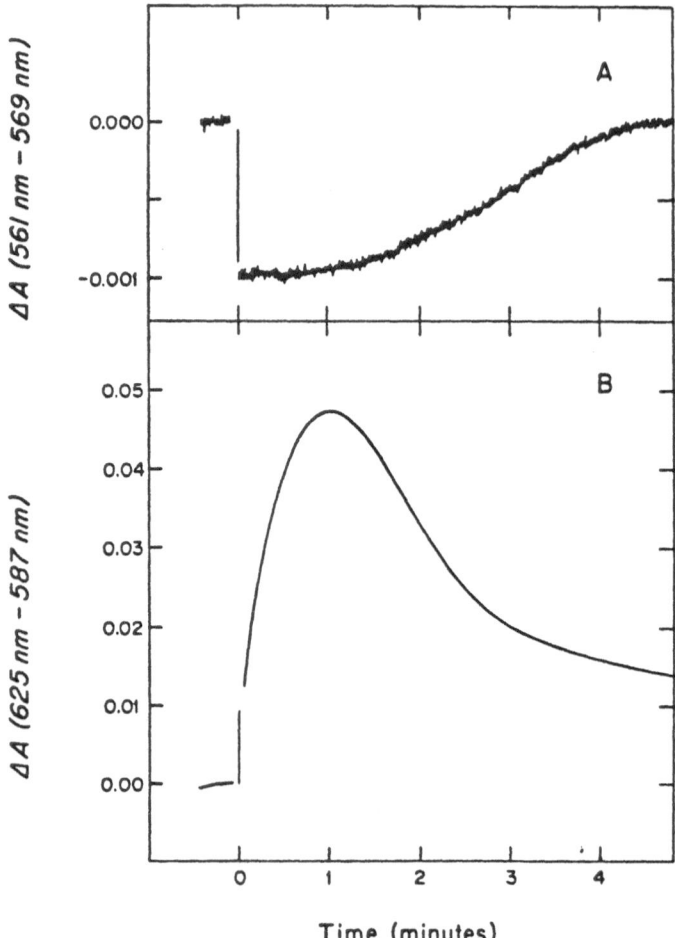

Time (minutes)

Figure 2.   Cytochrome b_{561} oxidation and membrane potential generation concomitant with reverse electron flow.   Chromaffin-vesicle ghosts (350 µg protein) loaded with 0.1 M ascorbate, 0.15 M Tris(PO$_4$), pH 7.0, were suspended in 2.5 ml of 340 mM sucrose, 29 mM (NH$_4$)$_2$SO$_4$, 600 µM ascorbate, 210 µM KCN, 8 mM gluconate, 8 mM Hepes(KOH),   pH 7.0. Absorbance was monitored at 22°C in an Aminco DW-2 spectrophotometer operated in the dual wavelength mode.   In A, the absorbance of reduced cytochrome b_{561} was monitored at 561 nm using the isosbestic point at 569 nm as a reference.   In B, the membrane potential was followed using the optical probe oxonol VI at a concentration of 5.6 µM.   In both cases, ascorbate oxidase (4 units) was added at t=0.

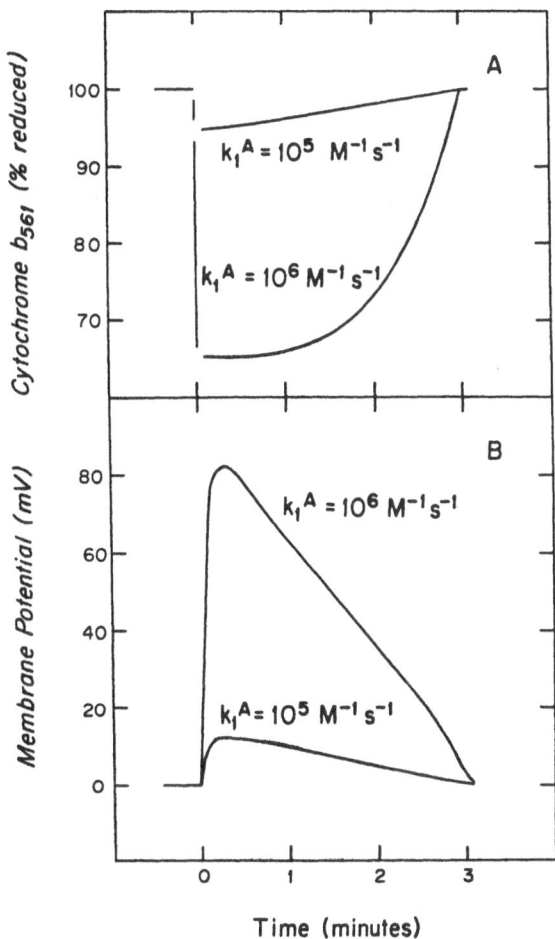

Figure 3. Computer simulation of the experiment shown in Figure 2 calculated as discussed in the text.

values of $k_1{}^A$. This indicates that $k_1{}^A$ is probably larger than 10^5 $M^{-1}s^{-1}$ but less than 10^6 $M^{-1}s^{-1}$.

An alternative approach to determining $k_1{}^A$ is to apply the classical theory of electron-transfer reactions in solution. If we assume that electron transfer between reduced cytochrome b_{561} and an external electron acceptor occurs by an outer-sphere mechanism, then the rate constant for electron transfer from cytochrome b_{561} to acceptor 1 ($k_1{}^1$) is related to the self exchange rates for each species (k_{BB} and k_{11}), to the equilibrium constant for the reaction (K_{B1}), and to the collision factor f_{B1} which we will assume is approximately equal to unity (Marcus and Sutin, 1985):

(6) $\quad k_1{}^1 = (k_{BB}k_{11}K_{B1}f_{B1})^{1/2}$

The rate constant for electron transfer from the cytochrome to a second acceptor may then be calculated from the relation

(7) $\quad k_1{}^2/k_1{}^1 = (k_{22}K_{B2}/k_{11}K_{B1})^{1/2}$

We know $k_1{}^F$, the cross exchange rate constant for ferricyanide/cytochrome b_{561}. The self-exchange rate constants for ferricyanide (k_{11}) and ascorbate (k_{22}) are 5 x 10^3 M^{-1} s^{-1} and 10^6 $M^{-1}s^{-1}$ respectively (Butler et al., 1981; Willams and Yandell, 1983). The equilibrium constants may be calculated from the midpoint reduction potentials of cytochrome b_{561}, ferricyanide/ferrocyanide and ascorbate/semidehydroascorbate using equation 2. Therefore, equation 7 predicts that $k_1{}^A$ is 1.2 x 10^5 $M^{-1}s^{-1}$ which falls within the range estimated above. This value is on the low end, but a larger value could be rationalized on the basis of unequal work terms. For example, the negatively charged membrane surface should repel ferricyanide with its greater negative charge more strongly than it repels semidehydroascorbate. Thus, the true rate constant for electron transfer from cytochrome b_{561} to external semidehydroascorbate ($k_1{}^A$) may well be larger than the value calculated from the rate constant obtained with ferricyanide ($k_1{}^F$).

Implications of the Rate Constants

We have now estimated the rate constants k_0 and $k_1{}^A$ for the electron transfer reactions illustrated in Figure 1. We can also calculate the rate constants for the reverse reactions. This describes quite well the reactions mediated by cytochrome b_{561} in vivo. Three implications in particular are worth noting.

First, we may ask whether cytochrome b_{561} reduces semidehydro ascorbate fast enough to prevent the free radical from disappearing via other pathways. Semidehydroascorbate is a fairly nonreactive free radical and prefers to react with other molecules of semidehydroascorbate. These two molecules disproportionate yielding one molecule of ascorbate and one of dehydroascorbate. At pH 5.5, the second order rate constant for disproportionation is 5 x 10^7 M^{-1} s^{-1} (Bielski et al., 1981). At the intravesicular semidehydroascorbate concentration likely to prevail in vivo ($\sim 2 \times 10^{-7}$ M, Diliberto and Allen, 1981), semidehydroascorbate will disproportionate at a rate of about $\sim 10^{-6}$ M/s. Since k_{-0} is 3×10^6 M^{-1} s^{-1} at pH 5.5 and the cytochrome b$_{561}$ concentration is about 700 µM in terms of internal volume, cytochrome b$_{561}$ should reduce intravesicular semi dehydroascorbate at a rate of $\sim 10^{-4}$ M/s. Thus, reduction of intravesicular semidehydroascorbate should occur two orders of magnitude faster than decay of semidehydroascorbate by disproportionation. This supports the hypothesis that cytochrome b$_{561}$ functions in vivo to reduce intravesicular semidehydroascorbate and regenerate intravesicular ascorbic acid.

Second, we may ask whether the heme of cytochrome b$_{561}$ is positioned near the inner surface of the membrane so as to facilitate reduction of internal semidehydroascorbate. We calculated the rate constant for the transfer of electrons from cytochrome b$_{561}$ to semidehydroascorbate on the inner surface of the membrane (k_{-0}) to be 9×10^4 M^{-1} s^{-1} at pH 7.0. This is comparable to the rate constant for electron transfer from cytochrome b$_{561}$ to semidehydro ascorbate on the outer surface of the membrane ($k_1{}^A$). Thus, the cytochrome appears to be comparably accessible to substrates at either surface of the membrane.

Finally, we may ask whether electron transfer across the chromaffin-vesicle membrane can be mediated by a single heme or whether additional electron-carrying groups are needed. In fact, electrons can be transferred through proteins across fairly long distances (Isied, 1983). Since the rate constants measured here are not particularly fast, it would not be unreasonable for cytochrome b$_{561}$ to mediate transmembrane electron transfer by itself.

References

Apps, D.K., Boisclair, M.D., Gavine, F.S. and Pettigrew, G.W., 1984, Unusual redox behaviour of cytochrome b-561 from bovine chromaffin granule membranes, Biochim. Biophys. Acta 764:8.
Bielski, B.H.J., Allen, A.O. and Schwarz, H.A., 1981, Mechanism of disproportionation of ascorbate radicals, J. Am. Chem. Soc. 103:3516.
Beers, M.F., Johnson, R.G. and Scarpa, A., 1986, Evidence for an ascorbate shuttle for the transfer of reducing equivalents across chromaffin granule membranes, J. Biol. Chem. 261:2529.
Butler, J., Davies, D.M. and Sykes, A.G., 1981, Kinetic data for redox reactions of cytochrome c with Fe(CN)$_5$X complexes and the question of association prior to electron transfer, J. Inorg. Biochem. 15:41.
Diliberto, E.J., Jr. and Allen, P.L., 1981, Mechanism of dopamine-ß-hydroxylation, J.Biol. Chem. 256:3385.
Flatmark, T. and Gronberg, M., 1981, Cytochrome b-561 of the bovine adrenal chromaffin granules. Molecular weight and hydrodynamic properties in micellar solutions of Triton X-100, Biochem. Biophys. Res. Commun. 99:292.
Flatmark, T. and Terland, O., 1971, Cytochrome b$_{561}$ of the bovine adrenal chromaffin granules. A high potential b-type cytochrome, Biochim. Biophys. Acta 253:487.

Fleming, P.J. and Kent, U.M., 1987, Secretory vesicle cytochrome b_{561}: A transmembrane electron transporter, Ann. N.Y. Acad. Sci. 493:101.

Harnadek, G.J., Callahan, R.E., Barone, A.R. and Njus, D., 1985a, An electron transfer dependent membrane potential in chromaffin-vesicle ghosts, Biochemistry 24:384.

Harnadek, G.J., Ries, E.A. and Njus, D., 1985b, Rate of transmembrane electron transfer in chromaffin-vesicle ghosts, Biochemistry 24:2640.

Herman, H.H., Wimalasena, K., Fowler, L.C., Beard, C.A. and May, S.W., 1988, Demonstration of the ascorbate dependence of membrane-bound dopamine ß-monooxygenase in adrenal chromaffin granule ghosts, J. Biol. Chem. 263:666.

Isied, S.S., 1984, Long-range electron transfer in peptides and proteins, Prog. Inorg. Chem. 32:443.

Iyanagi, T., Yamazaki, I. and Anan, K.F., 1985, One-electron oxidation-reduction properties of ascorbic acid, Biochim. Biophys. Acta 806: 255.

Kelley, P.M. and Njus, D., 1986, Cytochrome b_{561} spectral changes associated with electron transfer in chromaffin-vesicle ghosts, J. Biol. Chem. 261:6429.

Kelley, P.M. and Njus, D., 1988, A kinetic analysis of electron transport across chromaffin vesicle membranes, J. Biol. Chem. 263:3799.

Levine, M., Morita, K., Heldman, E. and Pollard, H.B., 1985, Ascorbic acid regulation of norepinephrine biosynthesis in isolated chromaffin granules from bovine adrenal medulla, J. Biol. Chem. 260:15598.

Marcus, R.A. and Sutin, N., 1985, Electron transfers in chemistry and biology, Biochim. Biophys. Acta 811:265.

Menniti, F.S., Knoth, J. and Diliberto, E.J., Jr., 1986, Role of ascorbic acid in dopamine-ß-hydroxylation: The endogenous enzyme cofactor and putative electron donor for cofactor regeneration, J. Biol. Chem. 261:16901.

Njus, D., Kelley, P.M. and Harnadek, G.J., 1986, Bioenergetics of secretory vesicles, Biochim. Biophys. Acta 853:237.

Njus, D., Kelley, P.M., Harnadek, G.J. and Pacquing, Y.V., 1987, Mechanism of ascorbic acid regeneration mediated by cytochrome b_{561}, Ann. N.Y. Acad.Sci. 493:108.

Njus, D., Knoth, J., Cook, C. and Kelley, P.M., 1983, Electron transfer across the chromaffin granule membrane, J. Biol. Chem. 258:27.

Wakefield, L.M., Cass, A.E.G. and Radda, G.K., 1986a, Functional coupling between enzymes of the chromaffin granule membrane, J. Biol. Chem. 261:9739.

Wakefield, L.M., Cass, A.E.G. and Radda, G.K., 1986b, Electron transfer across the chromaffin granule membrane, J. Biol. Chem. 261:9746.

Williams, N.H. and Yandell, J.K., 1983, Ruthenium and iron complexes of dipicolinic acid: Synthesis, solution properties and kinetics of electron transfer reactions with ascorbate ions, Aust. J. Chem. 36:2377.

THE RETINOIC ACID-INDUCED CASCADE OF EVENTS LEADING TO GRANULOCYTE
DIFFERENTIATION IN HL60 THROUGH A POSTULATED GTP-REGULATED INDUCING
MOLECULE

Edward S. Golub*, Teresita Diaz de Pagan, Iris Sun
and F. L. Crane

Purdue University
West Lafayette, IN   47907

INTRODUCTION

Cells which are self renewing and whose progeny have the potential
of establishing more than one lineage are called stem cells. When stem
cells are induced to differentiate they lose their multipotency and
establish lineages leading to the development of functional end cells. Of
the many differentiating systems which use the strategy of the stem cell
(1,2), the hemopoietic system is one of the more intensely studied
because the process occurs continuously in the adult animal (see ref 3).
The cells of the blood arise from multipotent hemopoietic stem cells
which give rise to progenitor cells of committed lineage. The
granulocyte-macrophage progenitor cell has at least one more lineage
establishment decision to make in the progression toward functional end
cells and can be used as a model for decision making in lineage
establishment.

The hemopoietic system appears to be a good system for studying the
events in commitment and lineage establishment but the frequency of the
multipotent cells is so low in normal mammalian bone marrow that direct
study has been impossible (3). The use of cloned tumor lines which are
"frozen" at a progenitor stage but can be induced to continue their
differentiation process in vitro overcomes this problem and offers a
workable system to investigate the events of lineage establishment.
Erythroid differentiation has been studied extensively using cells
transformed with Friend leukemia virus (4) which can be induced by a
variety of agents in vitro. This allows the events associated with the
expression of globin message during late erythrocyte differentiation to
be studied but not the early events in the process (5).   Furthermore,
because Friend cells can only be induced to differentiate along the
erythroid lineage it is reasonable to assume that lineage is already
established and they are not a good model for studying the decision
process.

The human promyelomonocytic cell HL60 (6), on the other hand, is a
tumor cell apparently "frozen" at the granulocyte-macrophage progenitor
stage. These cells, originally isolated from a patient with acute
promyelocytic leukemia, can be induced to differentiate along either the
granulocyte or the macrophage lineage depending upon the nature of the

---

*To whom correspondence should be addressed. Present Address; Ortho
Pharmaceuticals, Biotech Division, Raritan, NJ 08869.

inducer. Dimethyl sulfoxide DMSO (6,7) and retinoic acid (RA) (8) induce granulocyte differentiation and phorbol esters such as TPA induce the macrophage lineage (9). Because one can direct the developmental decision that the cells will make by the selective use of inducers, the nature of the events involved in lineage establishment can be studied. Commitment to a lineage by HL60 has been shown to be an all-or-none event (10) and we have shown that the process can be analyzed as two discrete events (11).

The study of the process of differentiation is complex because it has three phases; 1) the transduction of the inducing signal at the membrane leading to the activation of lineage-specific genes, 2) the activation of these genes and 3) the expression of the gene products which leads to characteristic cellular changes. In this paper we examine the first phase and propose a cascade of events leading to the activation of granulocyte-specific genes in HL60 by RA.

RA is known to be able to induce differentiation in many systems (12) but the mechanism(s) by which it acts are not known. We have taken advantage of a RA$^-$ variant of HL60 (termed HL60.P) (11) to show that RA is a potent inhibitor of membrane electron transport and postulate that this is the the initial step in the differentiation cascade which results in gene activation of the granulocyte but not the macrophage-specific genes. We postulate that this cascade consists of the following events.

1. RA interacts with a membrane receptor causing inhibiton of plasmamembrane electron transport.
2. The inhibition of electron transport results in a decline in the concentration of NAD$^+$.
3. The decline in NAD$^+$ results in down-regulation of inosine monophosphate dehydrogenase (IMPD).
4. The down-regulation of IMPD results in lowered concentration of GTP.
5. The lowered level of GTP causes derepression of a hypothetical GTP-regulated inducer molecule.

The hypothetical GTP-regulated molecule has been postulated to be involved in differentiation of bacteria and yeast by Freese (13) who suggest that it could also be a molecule able to bind DNA and activate genes which are ready to be transcribed. In this paper we present the evidence for the first four steps of the cascade.

RESULTS

We have shown that HL60 can be induced along the granulocyte lineage by surrogate inducers which alter the ionic mileu of the cells (11,14). The ion modifying agents induce differentiation at a different rate than RA. With RA as the inducing agent the first differentiated cells appear after two days of culture when the RA is present throughout the culture period. In contrast, induction with ouabain, A23187 or EGTA occurs very rapidly. In these cases there is a great deal of cell death, but differentiated cells are seen by 24 hr (11,14). With both RA and the ion modulating agents the events leading to differentiation are initiated at the time of the addition of the inducer. The RA-induced events continue gradually for 4 days but the ion modifying-induced events have reached maximum after a few hours of contact. This is consistent with the observation that RA induction of HL60 is a continuously driven process (15). From this data it is clear that the events leading to lineage-specific gene activation can be studied as two discrete phases involving early and late events in the cascade. In this paper we focus on the early RA-induced events which begin at the plasmamembrane and will present the data for the cascade of events from the membrane to the hypothetical induced molecule.

1. RA Inhibits Plasmamembrane Electron Transport

Trans-plasmamembrane redox systems have been implicated in the growth and development of cells (16). The effect of RA on electron transport in both HL60 and HL60.P (the RA$^-$ variant) was determined by measuring the rate of reduction of diferric transferrin. Both HL60 and HL60.P reduce diferric transferrin at the same rate in the absence of RA but as seen in Figure 1, RA inhibits diferric transferrin reduction in HL60 but not in HL60.P.

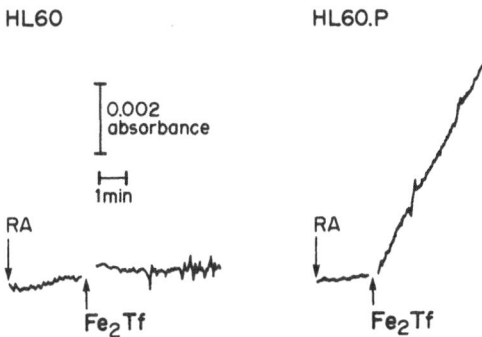

Figure 1. The effect of RA on electron transport in intact cells.

For the experiment HL60 and HL60.P were washed and resuspended in T.D. buffer (140mM NaCl, 25mM KCl, 0.6 mM Na$_2$HPO$_4$, 25mM Trizmabase) pH7.4 without serum. Transplasma membrane diferric transferrin reduction was measured in an Aminco-DW 2a spectrophotometer in the dual beam mode measuring formation of ferrous bathophenanthroline sulfonate according to the method of Avron amd Shavit (27) by subtracting the absorbance at 600 nm from the absorbance at 535 nm. With this wavelength pair the extinction coefficient for the Fe(BPS)$_3$ is 17.6 mM^{-1} cm^{-1}. The reaction mixture of 2.8 ml contained TD buffer, 10 µM bathophenanthroline sulfonate, 10 µM diferric transferrin and 0.01 - 0.03 gww cells. The reaction was started by the addition of cells followed by 10^{-6}M RA and then diferric transferrin.

This result is in contrast to TPA which has no effect on the reaction in either HL60 or HL60.P (Table 1) even though it is a potent inhibitor of electron transport in an SV40 transformed liver cell line (data not shown). Since TPA induces macrophage differentiation in both HL60 and HL60.P we conclude that the establishment of this lineage (probably via protein kinase C, the phorbol ester receptor) does not involve electron transport changes at the initial steps. DMSO induces granulocyte differentiation in both HL60 and HL60.P but does not inhibit electron transport in either (Table 1), suggesting that it functions via a different mechanism than RA.

Table 1. Effect of DMSO and TPA on electron transport.

Diferric transferrin reduction
nmole/min/gww

	HL60	HL60.P
Control	5.0	5.4
RA 10^{-6}	0.4	5.3
TPA 10^{-8}M	4.0	5.0
TPA 10^{-9}M	5.3	5.6
DMSO 1%	4.8	4.9

See caption to Figure 1 for experimental details.

It was essential to determine if RA exerts its inhibition of electron transport at the plasmamembrane. The data in Figure 2 shows that RA inhibits electron transport on isolated plasmamembrane preparations of HL60 but not HL60.P. Examination of the membrane preparations in the electron microscope showed that they were virtually free of mitochondria (not shown).

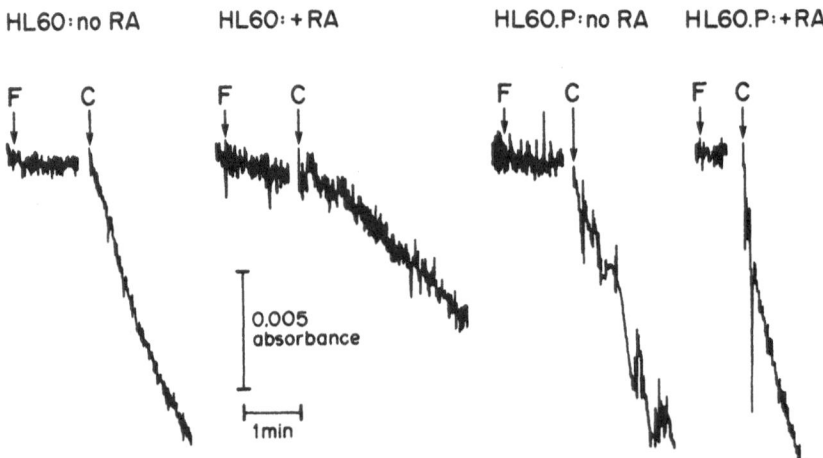

Figure 2. Effect of RA on electron transport in isolated plasma membranes.

Membranes were isolated by two-phase partition using a modification of the general method of Albertson et al (28). Electron transport was measured by following the oxidation of NADH (50uM) in the presence of diferric transferrin (10 µM) and 1 µM potassium cyanide in 2.8 ml TD buffer (see caption to Figure 1). Absorbance at 430 nm is subtracted from 340 nm with the dual-wavelength machine. The extinction coefficient is 6.2 $mM^{-1}cm^{-1}$ with this wavelength pair.

Retinoids are bound to other molecules for solubility and transport and in many systems cellular retinoid binding proteins (CRBP) have been identified (17). While cellular retinoic acid binding proteins (CRABP) have been identified in some cell types there is general agreement that they have not been demonstrated in HL60 (18). Similarly, there is at least one report of a membrane-associated CRABP but their general existence has not been firmly established (19). The data in Figure 2 shows that RA acts to inhibit electron transport at the membrane which is consistent with data which shows that RA can initiate differentiation of HL60 even when immobilized to spheres or plastic petri dishes (15).

Electron transport can be accompanied by proton release which is based on the activation of the Na^+/H^+ antiport. RA has recently been shown to stimulate the Na^+/H^+ antiport (20). Because amilioride inhibits this antiport system (21) its effect on the release of protons from HL60 and HL60.P was tested. Table 2 shows that RA stimulates the release of protons from HL60 but not from HL60.P. The proton release is sensitive to amiloride but, as seen in the Table, amiloride has no effect on the induction of differentiation. Sodium nitroprusside also causes amiloride sensitive proton release from HL60 but it does not induce differentiation (data not shown). From this we conclude that proton release is not a factor in the induction of the granulocyte lineage in HL60.

Table 2. Effect of amiloride on proton release and differentiation in HL60.

	Amiloride	Proton Release	Differentiation
HL60:	-	387	89
	+	96	81
HL60P:	-	96	<10
	+	53	<10

Proton release was measured in a thermoregular cuvette with a Corning combination electrode. Cells (0.01 gww) are suspended in 28 ml sucrose salts buffer (0.1M surcrose, 10mM NaCl, 10mM KCl, 10mM $CaCl_2$, CsCl with 2.5 mM Tris chloride pH7.4.) The suspension was constantly bubbled with air to remove CO_2. The pH change was measured in the range from 7.4 to 7.2 and titrated by additions of 50nmoles standard HCl.

2. Inhibition of Electron Transport Results in a Decline in NAD^+

One consequence of the inhibition of electron transport should be a reduction in NAD^+. Therefore the concentrations of NAD^+ in HL60 and HL60.P were determined at various times after treatment with RA. These results are seen in Figure 3. It can be seen that the NAD^+ levels in HL60 fall by close to 50% but there is a rise in the levels in HL60.P over 24 hours. Part of the NAD^+ level comes from mitochondria and there is a 97% drop in the cytosolic levels of HL60 at 24 hr with a drop of 60% in HL60.P in the same time period. (data not shown).

3. Decline in NAD^+ Results in an Inhibition of Inosine Monophosphate Dehydrogenase

There could be many consequences of a decline in NAD^+ but we have chosen to focus on the effect on GTP. Inosine monophosphate dehydrogenase (IMPD) is a crucial enzyme in GTP biosynthesis and is extremely NAD^+ dependent. Wright and his coworkers (22) have shown that inhibition of inosine monophosphate dehydrogenase with mycophenolic acid (MA) or 2-β-D-ribofuranosylthiazole-4-carboxamide (tiazofurin) cause drastic reduction in GTP concentration and induce granulocyte differentiation of HL60. As seen in Table 3, we have repeated this observation and extended it to show that HL60.P is not induced by these agents.

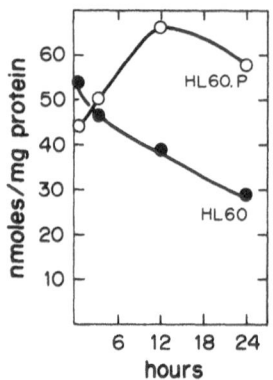

Figure 3. Concentration of NAD^+ and NADH after treatment with RA.

HL60 and HL60.P (2.5×10^5/ml) were treated with RA (10^{-5}M)
for the indicated times after which NAD^+ and NADH were
determined by the method described by Matsumaura and Miyachi (29).

Table 3. Effect of IMPD inhibitors

Percent NBT^+ at day 4.

Reagent		Interval[a]	HL60	HL60.P
Mycophenolic acid 10^{-6}M		4 days	24.5	22.4
	3×10^{-6}M	4 days	94.4	32.7
Retinoic acid	10^{-5}M	4 days	94.5	9.4
Tiazafurin	10^{-4}M	0-20hr[b]	8.7	3.9
	10^{-4}M	4 days	91.9	33.4
Retinoic acid	10^{-5}M	4 days	93.6	25.9

a. Cells incubated in serum-containing medium. MA and RA present for
   all 4 days.
b. Cells incubated in serum free medium (29) from 0-20hr., washed and
   incubated in serum-containing medium until assayed at day 4. One
   group was in the presence of tiazafurin for only 0-20hr, the other
   group was in the presence of tiazafurin during both phases of the
   experiment. RA was present in both phases.

4. Effect of Guanosine on A23187-Induced Differentiation

The inhibition of IMPD seen above results in a down-regulation of GTP biosynthesis. If a decline in GTP is a stop in the cascade leading to granulocyte-specific gene activation, then maintaining GTP levels should prevent induction by maintaining GTP levels via the salvage pathway. We have previously shown that the calcium ionophore A23187 induces granulocyte differentiation in H260(11). In Table 3 we whow that added quanosine inhibits the induction.

Table 4. Effect of Guanosine on Induction of Differentiation by A23187.

				$ZNBT^{+}$ at 24 hr.
A23187	(hr. 0-6)			63
A23187	"	+ guanosine	(hr. 0-6)	16
A23187	"	+ guanosine	(hr. 6-24)	15
A23187	"	+ guanosine	(hr. 0-24)	2

A23187 (4 μM) was added to cultures of HL60 at time 0 and washed out at 6 hr. The cells were grown in complete medium until 24 hours when they were assayed. Guanosine (10^{-4}M) was present at the times indicated.

DISCUSSION

In this paper we have presented the evidence for a RA-induced cascade of events leading to granulocyte differentiation in HL60. The events begin at the plasma membrane when 1) RA causes an inhibition of electon transport. This inhibition 2) causes a fall in the rate at which NAD^{+} is generated. The decline in NAD^{+} concentration 3) causes a down-regulating the enzyme inosine monophosphate dehydrogenase (IMPD). It has been shown by others that RA also causes an increase in NADase (23) and that IMPD is NAD^{+}-dependent and extremely sensitive to changes in concentration of the cofactor (24). We now address the question of how a down-regulation of IMPD could result in granulocyte-specific gene activation.

The precedence for this model comes from work on both HL60 and microbial systems. In HL60 Wright and his coworkers have carried out a remarkable series of experiments (22) in which they showed first that GTP levels fall as a consequence of RA induction. Reasoning that this fall may be due to an alteration in biosynthetic rate they found that IMPD activity was reduced. They then showed that inhibition of IMPD results in induction of granulocyte differentiation which can be overcome by maintaining the levels of GTP through the salvage pathway. Freese and his coworkers (13) in an equally elegant series of experiments have shown that inhibition of IMPD induces differentiation (sporulation) in both B. subtilis and yeast as well as in cultured glial cells (E.Freese and R. Lipsky, personal communication). They have postulated that the fall in GTP derepresses an inducer molecule which then binds to methylated DNA resulting in new gene transcription.

We propose that the granulocyte, but not the macrophage, lineage is established in HL60 by this mechanism. Diferric transferrin is an essential growth factor for mammalian cells and in the absence of RA, HL60 like most cells, reduces it as one means of maintaining homeostatic concentrations of NAD^{+}. This NAD^{+} concentration is sufficient for the functioning of IMPD and is thus necessary for the maintenace of the homeostatic level of GTP. The postulated inducer molecule is repressed when bound to GTP and the cell remains in the uninduced state. RA induces a distrubance of the homeostatic balance by the inhibition of the membrane electron transport system resulting in a fall in NAD^{+}

concentration which results in a down regulation of IMPD and a consequent fall in GTP. We postulate that the fall in GTP results in a dissociation of GTP from the inducer molecule, causing it to be derepressed and perhaps bind to DNA resulting in the activation of granulocyte lineage-specific genes. Because HL60 is arrested at the promyelomonocyte stage, the granulocyte and macrophage lineage-specific genes are in a state in which they are ready to be transcribed. The hypothetical inducer acts only on the granulocyte-specific genes.

Given the fact that RA can induce differentiation even when immobilized outside the cell (15) it is likely that RA reacts with a receptor molecule which then activates a cytoplasmic molecule which may have no RA binding capacity. We have previously shown that HL60, but not HL60.P, can be induced with rapid kinetics to differentiate along the granulocyte line by altering the ionic mileu of the cells(11). We now propose that altering the ionic mileu causes a disocciation of GTP from the inducer molecule and show that the induction is prevented by maintaining a high level of GTP. Rather than postulate two separate mutational events in HL60.P the parsimonius explanation is that the same molecule is involved in both phenomena and is absent in the RA$^-$ variant. If the molecule is a cellular retionic acid binding protein (CRABP), there is a precedence for its translocation to the nucleus where it could also be a DNA-binding protein (25). The fact that there are differentiating microbial systems which use this mechanism will allow us to search for the putative inducing molecule in a simplified system.

The induction to monocyte differentiation in HL60 by phorbol esters would appear to be controlled by a different mechanism. A variant of HL60 which differentiates along the macrophage lineage in response to RA has recently been reported (26) and it will be interesting to see if this variant uses the proposed GTP-dependent induction mechansim to induce the other pathway into which HL60 is able to be induced.

ACKNOWLEDGEMENTS

This work was supported by NSF grant to Edward S. Golub and USPHS grant to Fred L. Crane.

REFERENCES

1. R. Levenson, and D. Housman, Cell 25,5 (1981).
2. A.K. Hall, Cell 33,11 (1983).
3. E.S.Golub, Cell 28,687 (1982).
4. C.W. Friend, J.G. Holland and T. Sato, Proc. Nat. Acad. Sci. USA 68,378 (1971); Marks and Rifkind Ann. Rev. Biochem.
5. N. Kimura and T.W. Mak, J. Cell. Physiol. 128,41 (1978).
6. S.J. Collins, R.C. Gallo and R.E. Gallagher, Nature 270,347 (1977).
7. S.J. Collins, F.W. Ruscetti, R.E. Gallagher and R.C. Gallo, Proc. Nat. Acad. Sci. USA 75;2485, (1978).
8. T.R. Breitman, S.E. Selonick and S.J. Collins, Proc. Nat. Acad. Sci. USA 77,2936 (1980).
9. G.D. Rovera, D. Sanoli, and D. Damsky, Proc. Nat. Acad. Sci. USA 76,2779 (1979); J. Lotem and L. Sachs, Proc. Nat. Acad. Sci. USA 76;5158 (1979);
10. A.W. Boyd and D. Metcalf, Leuk. Res. 8,27 (1984); H. von Melchner and K. Hoffken, J. Cell. Physiol. 125,573 (1985).
11. Golub, E.S. and T. Pagan, Prog Clin. and Biol. Res. 226,235 (1986).
12. R. Lotan and G.L. Nicolson, J. Nat. Cancer. Inst. 59,1717 (1977); A.B. Roberts and M.B. Sporn, in M. B. Sporn, A.B. Roberts and D.S. Goodman (eds), The Retinbids vol II, Academic Press, Orlando, FL, pp209; T.T. Amatruda III and H.P. Koeffler, in M.I. Sherman (ed) Retinoids and Cell Differentiation, CRC Press, Boca Raton,FL, pp79; M.I. Sherman, ibid pp162.

13. E. Freese, E.B. Freese, E.R. Allen, Z. Olempska-Beer, C. Orrego, A. Varma, and H. Wabiko, Molecular Biology of Microbial Differentiation (1985), Am. Soc. Microbiology pp 194.
14. T. Diaz de Pagan and E.S. Golub in preparation.
15. A. Yen, S.L. Reece nd K.L. Albright, J. Cell. Physiol. 118,277 (1984).
16. F.L. Crane, I.L. Sun, M.G. Clark, C. Grebing and H. Low, Biochim. Biophys Acta 811,233 (1985); I.L. Sun, W. Liu, F.L. Crane, D.J. Morre and H. Low, submitted.
17. F. Chytil and D. E. Ong, in Sporn et al op cit p89.
18. J. Nugent and S. Clark (eds),Retinoids, differentiation and disease CIBA Foundation Symposium vol 113 (1985), discussion p90.
19. B. Sani, Biochem. Biophys. Res. Comm. 91,502 (1979); F.E. Cope and R.K. Boutwell, Fed. Proc. 42,1317 (1983) abstract.
20. A. Ladoux, E.J. Cragoe, Jr., B. Gery, J.P. Arbita and c. Frelin, J. Biol. Chem. 26,811 (1987).
21. G.L'Allemain, A. Francki, E. Cragoe, Jr., and J. Pouyssegur, J. Biol. Chem. 259,4313 (1984).
22. D.L. Lucas, H.K. Webster and D.G. Wright, J. Clin. Invest. 72,1889 (1983); D.L. Lucas, R.K. Robins, R.D. Knight, and D.G. Wright, Biochem. Biophys. Res. Comm. 115,971 (1983); D.G. Wright, Blood 69,334 (1987); R.D. Knight, J. Mangum, D.L. Lucas, D.A. Cooney, E.C. Kahn, and D.G. Wright, Blood 69,634 (1987).
23. H. Hemmi and T.R. Breitman, Biochem. Biophys. Rec. Comm. 109,669 (1982).
24. J. H. Anderson and A.C. Sartorelli, J. Biol. Chem. 243,4762 (1968).
25. B. Sani, Biochem. Biophys. Res. Comm. 75,7 (1977); B. Sani and M.K. Donovan, Cancer Res. 39,2492 (1979).
26. Imaizumi, M, J. Uozumi and T.R. Breitman, Cancer Res. 47,1434 (1987).
27. Avron, M. and N. Shavit, Anal. Biochem. 6,549 (1963).
28. Albertson,P.A., B. Anderson, C. Larson, and H.E. Akerland, Methods Biochem. Anal. 28,115 (1982).
29. Matsumara, H. and S. Miyachi, Methods in Enzymol. 69,465 (19).

CHANGES IN PLASMALEMMA FUNCTIONS INDUCED BY PHYTOPATHOGENIC BACTERIA

C. I. Ullrich-Eberius[1], J. Pavlovkin[2], J. Schindel[1],

K. Fischer[1], and A. Novacky[3]

Institut für Botanik, Technische Hochschule[1]
Schnittspahnstr. 3, D-6100 Darmstadt, FRG

Institute of Experimental Phytopathology and Entomology[2]
Slovak Academy of Sciences, 90028-Ivanka pri Dunaji, CSSR

Department of Plant Pathology, University of Missouri[3]
108 Waters Hall, Columbia, Missouri 65211, USA

INTRODUCTION

Plants are attacked by pathogens such as viruses, bacteria, and fungi, not only at the roots by soil-borne pathogens, but also quite severely at the leaves. In our studies we are investigating the interaction of phytopathogenic bacteria with leaves. Bacteria colonize the intercellular space of plant tissues. In the case of disease development, bacteria multiply and induce the plant cells to provide nutrients and water. Some phytopathogenic bacteria are known to produce cellulase and/or pectinases, as e.g. Erwinia pathovars; others are known to produce toxins which drastically interfere with the host cell metabolism, such as tabtoxin produced by Pseudomonas syringae pv. tabaci that specifically interferes with glutamine synthetase (Turner, 1981).

However, for most phytopathogenic bacteria the primary determinants for disease induction are still unknown. Also, no protective agents or reliable methods for resistance induction are available, apart from some doubtful antibiotics. On the other hand, epidemics of increasing severity still occur, e.g. those of bacteria-induced cotton blight in Nicaragua (Lehmann-Danzinger, 1987). There are several highly resistant or tolerant Gossypium cultivars with well defined genes for major or minor resistance (Naumann, 1986) or defined plant defense genes (Ecker and Davis, 1987), but the actual physiological and biochemical basis for resistance mechanisms is not known.

When leaves of these resistant plants are invaded by virulent bacteria or susceptible plants are invaded by incompatible bacteria, resistance is expressed in the hypersensitive reaction (HR) between bacteria and plant cells (Sequeira, 1976; Klement, 1982; Goodman et al., 1986). This is a reaction in which both bacteria and the adjacent individual plant cell are destroyed, leading to the well known necrotic spots.

Evidence is accumulating that the critical early events of recognition and defense elicitation are triggered at the cell wall and the plasmalemma (Novacky, 1980 and 1983). Hours before macroscopic or microscopic symptoms develop and before electrolyte leakage can be measured changes of electrochemical, biophysical, and biochemical properties of the plasmalemma can be detected.

Here, we follow backwards the cascade of rapid events triggered by some phytopathogenic bacteria as revealed in a combination of various kinds of experiments. The synopsis of our recent investigations, considered together with the characteristics of plant senescence (Leshem et al., 1986; Thompson et al., 1987), infers that the bacteria-induced hypersensitive reaction may be regarded as an extremely rapid process of cell senescence.

MEMBRANE POTENTIAL CHANGES LINKED TO DISEASE DEVELOPMENT

Cotyledons of susceptible cotton (Gossypium hirsutum, Acala 44, 14 d old) were inoculated with virulent Xanthomonas campestris pv. malvacearum (race 10, 10^7 cells/ml). Measurements of the electrical membrane potential difference (E_m) at the plasmalemma carried out with glass microelectrodes (Novacky and Ullrich-Eberius, 1982) revealed that in the light the membrane potential was maintained in diseased tissues at the same level as in healthy tissue (Table 1). Thus membranes appear still intact. In the dark, however, E_m rapidly decreased to the level of the energy-independent diffusion potential (E_D) in the diseased but not in the healthy tissue (Table 1).

Table 1. Membrane potential (E_m) in light and dark in healthy (control) and in Xanthomonas c. pv. malvacearum (X.m.) inoculated (48 h) cotton cotyledons. Sections were treated or not with 15 µM fusicoccin during the electrophysiological measurements.

		E_m (mV)	
	FC	control	X.m.
Light	-	- 179	- 188
	+	- 210	- 216
Dark	-	- 173	- 100
	+	- 225	- 95

Energy supply as determined by measurements of photosynthetic O_2 evolution, respiratory O_2 uptake, and the ATP levels in light and dark, was not yet affected by the bacterial infection, so this cannot be the reason for different E_m levels (Novacky and Ullrich-Eberius, 1982). Since in the light, DCMU reduced E_m in the diseased, but not in the healthy tissue, though the ATP levels in the dark in both tissues were the same, some other central cellular process that depends on photosystem II, must be involved. During plant pathogenesis toxic amounts of ammonia are produced (Lovrekovich et al., 1970; Ullrich et al., 1984). Initially ammonia is produced by bacteria from apoplastic amino acids, later from cell wall protein degradation and finally by cellular proteolysis (Kunz, 1986). Ammonia accumulates to high levels in the dark, but much less and later in the light. It was shown to rapidly depolarize cotton cell plasmalemma to the diffusion potential (Ullrich et al., 1984) and to alkalinize the cytosol by up to 0.5 pH units even at much lower concentrations (Bertl et al., 1984).

From these data we conclude that during development of cotton blight, in the light ammonia produced by bacteria is still assimilated by the photosystem II-dependent GS-GOGAT cycle. In the dark, however, NH_4^+ accumulates in the host cell, inducing an alkalinization that leads to inactivation of the H^+-extruding ATPases at the plasmalemma and tonoplast and hence to membrane depolarization, without membrane disintegration. This leads to a shift in plant cell metabolism towards a slow catabolism of storage products and to nutrient efflux.

EARLY MEMRBANE-RELATED EVENTS DURING THE HYPERSENSITIVE REACTION (HR)

Electrolyte Efflux

A strong K^+ efflux is typical for HR and is stoichiometric to an extracellular alkalinization. According to the fact that the pH is determined by the strong ion difference of a solution (Stewart, 1981; Guern and Ullrich-Eberius, 1988), the alkalinization usually observed during bacteria-induced HR development in a plant tissue or in a cell suspension does not originate from an actual H^+ absorption, as interpreted by Atkinson et al. (1985a and 1985b). The correspondent anion is most probably HCO_3^-. A strong and rapid CO_2 evolution was concomitantly recorded (Fig. 3). An additional alkalinization is caused by the remarkable ammonia efflux (Ullrich et al., 1984). In cotton cotyledons inoculated with incompatible Pseudomonas syringae pv. tabaci, 10^8 cells/ml, electrolyte efflux which consisted of 80% potassium ions was detected only 4 h after infection (Pavlovkin et al., 1986).

Electrical Membrane Potential

Pavlovkin et al. (1986) showed that within the first 2 h of inoculation in the same incompatible plant/pathogen combination both the energy-dependent

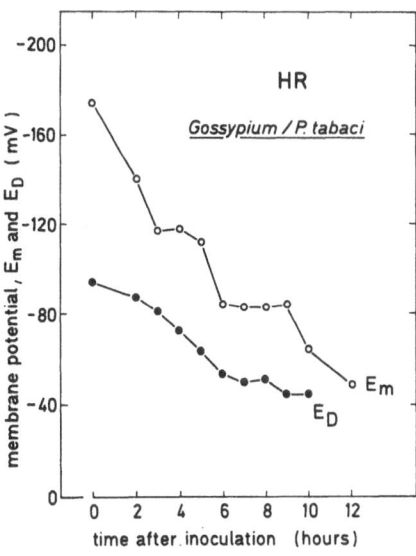

Fig. 1. Membrane potential decay of cotton cotyledons induced by Pseudomonas syringae pv. tabaci in dark. E_m, energy-dependent component plus energy-independent component; E_D, energy-independent, diffusion component, measured in nitrogen-flushed experimental solution.

(E_p) and simultaneously the energy-independent (E_D) component of E_m decreased (Fig. 1) in contrast to the compatible combination. These changes were not reversible in light, in contrast to the compatible host/pathogen combination indicating a very rapid and deleterious damage of the membrane structure. Therefore K^+ efflux associated with HR development is most probably induced by non-specific membrane matrix perturbation and not by the activation of specific exchange transport systems. Since both electrical components are affected simultaneously a voltage-dependent opening of specific K^+ channels is unlikely.

Water Transport Parameters: P, ϵ, L_p

By means of the miniaturized pressure probe, developed by Hüsken et al. (1978) water transport parameters were determined in single cells within the healthy or diseased plant tissue (Fig.2). Very rapid changes could be detected during the HR development. Within the first hour of inoculation of Nautilocalyx forgetii (Gesneriaceae) with the incompatible P.s. pisi (10^8 cells per ml) the hydraulic conductivity (L_p) increased from 19 to $42 \cdot 10^{-8}$ m $sec^{-1} \cdot MP\bar{a}^1$ and within 8 h to $465 \cdot 10^{-8}$ m $sec^{-1} \cdot MP\bar{a}^1$. The half time of water exchange (T 1/2) after endosmosis and exosmosis decreased significantly to 25% of the control. The volumetric elastic modulus of the cell wall (ϵ), a parameter characterizing the rigidity of the cell wall, decreased to 12% of the control. The turgor pressure (P) decreased from 0.30 MPa to 0.085 MPa. The change of this parameter, P, therefore reflects perturbed cell wall properties and the increase in L_p indicates an increased water permeability of the plasmalemma.

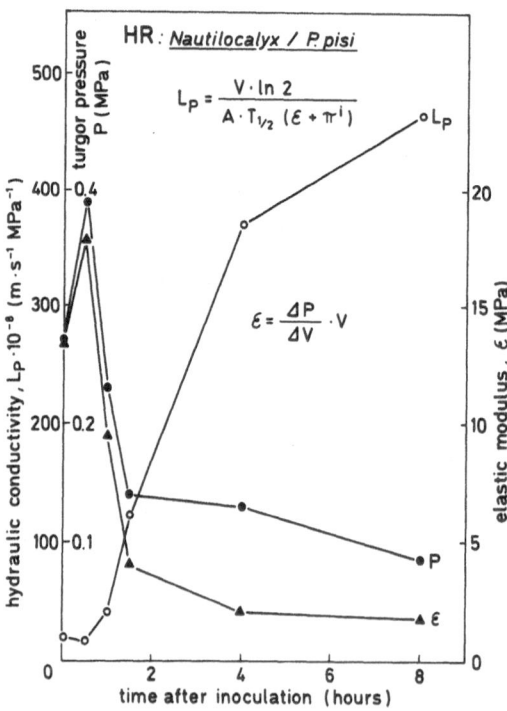

Fig. 2. Time course of changes of the hydraulic conductivity (L_p), the volumetric elasticity modulus of the cell wall (ϵ), and the turgor pressure (P) in Nautilocalyx forgetii inoculated with Pseudomonas syringae pv. pisi.

Apoplastic Peroxidase

There are several reports concerning an increase of peroxidase activity in the intercellular leaf space upon inoculation of resistant plant tissue with HR-inducing bacteria (Rathmell and Sequeira, 1974; Sequeira, 1976; Venere, 1980). Also in our studies with tobacco leaves inoculated with 10^8 cells/ml of P.s. pisi, the extracellular peroxidase activity determined in the intercellular washing fluid (obtained by the method described by Klement, 1965) was twice that in control plants. It increased from 8 to 16 µl H_2O_2/ml fluid/min and in the tissue from 38 to 42 µmol H_2O_2/g fresh wt/min within 3 h, in light and in dark (measured by H_2O_2-induced production of octadehydroguaiacol (Kunz, 1986).

Apoplastic Protease

Another enzyme which might be involved in changes of cell wall properties was followed simultaneously. The protease activity in the intercellular space fluid in tobacco leaves increased during HR development to 3.56 µmol/ml fluid within 6 h, whereas in the water-infiltrated control tobacco leaf apoplast no protease activity could be detected. This enzyme most likely is produced by the bacteria (measured by casein degradation and ninhydrin reaction; Kunz, 1986).

Lipid Peroxidation and Lipoxygenase

Lipid peroxidation is one of the fundamental mechanisms involved in alteration of the membrane lipid phase during plant senescence (Leshem et al., 1986). An increase of lipid peroxidation could account for the observed voltage difference changes and ion efflux during HR development. In cucumber cotyledons inoculated with the incompatible P. pisi the lipid peroxidation activity was measured by following the conversion of lipids to malondialdehyde by the thiobarbituric acid reaction. The lipid peroxidation increased by 80% within 4.5 h. No differences could be detected in tissue inoculated with heat-killed P. pisi or with the saprophyte P. fluorescens and also none with the compatible pathogen P. s. pv. lachrymans (Keppler and Novacky, 1986).

Lipid peroxidation may be initiated by free radicals or lipoxygenase. Lipoxygenase activity was found to increase by 25% in cucumber tissue within 4.5 h, but only when the cotyledons were inoculated with P. pisi, not in the compatible combination nor with the saprophyte. Catalase and superoxide dismutase activity did not change during HR development (Keppler and Novacky, 1987). These results were recently confirmed with tobacco suspension cells inoculated simultaneously with P. s. pv. syringae. Lipid peroxidation transiently increased together with O_2^- production. Both preceded initial increase in K^+ efflux. In contrast a Tn 5 insertion mutant of P. s. syringae, which does not induce HR, showed no lipid peroxidation, no O_2^- production, and no K^+ efflux. Concomitant tests for recovery of P. syringae after separation from the suspension cells undergoing HR, showed that wild type bacteria growth was slower than that of the mutants which did not induce these effects (Keppler et al., 1988).

In a few other examples of hypersensitive plant/pathogen interaction similar results are obtained. When membrane fractions from potato tuber tissue were inoculated with HR-inducing incompatible races of Phytophthora infestans, production of superoxide radicals could be detected. The same was observed during the HR development in potato leaves (Doke, 1983 and 1985; Chai and Doke, 1987).

As a test for free radical initiation by lipid peroxidation cucumber cotyledons were infiltrated with P. pisi plus the free radical scavengers, superoxide dismutase (25 μg/ml), catalase (500 μg/ml), α-tocopherol-acetate
(0.1 mM), and dipyridamol (0.1 mM). A significantly smaller electrolyte leakage was induced by P. pisi or the herbicide paraquat by the addition of SOD:
the relative rate of electrolyte leakage from H_2O-infiltrated cotyledon disks
was 1.2, that from disks inoculated with P. pisi was 2.1, with P. pisi plus
SOD 1.7, with paraquat 1.9, and with paraquat plus SOD 1.7. Catalase and α-
tocopherol-acetate had only a very small scavenging effect. Though these compounds did not completely prevent HR development, the results could indicate
that superoxide radicals are involved in the increase in lipoxygenase activity. It may be a response to the release of polyunsaturated fatty acids
(PUFA) by lipoxygenase or phospholipase activation (Keppler and Novacky, 1987).

Also in tobacco cell suspensions inoculated together with P. pisi, the
addition of free radical scavengers, i.e. SOD and Tiron (1 mM, 1,2-dihydroxy-
benzene-3,5-disulphonic acid) suppressed increased O_2^- production and peroxidation, and increased the recovery of bacteria (Keppler et al., 1988).

Cytokinins, known to function as juvenile hormones in plants, were demonstrated to prevent HR development in mature tobacco leaves. Leaves were
pretreated for 6 to 96 h with kinetin (6-furfurylaminopurine), 2ip (6-(3-me-
thyl-2-butenylamino)purine), BA (6-benzylaminopurine), and zeatin (6-(4-hy-
droxy-3-methyl-trans-2-butenylamino)purine). Depending on the time of pre-
tratment, the subsequent injection of P. pisi did not induce necrotic le-
sions (Novacky, 1972). Similarly, cytokinins retard plant senescence. This
is now interpreted as an ability of cytokinins either to catalyze the disproportionation of the basic superoxide or even more to prevent the formation of free radicals (Leshem et al., 1986).

Following the chain of events backwards, it is important to know the
early events during wounding of plant cells and exposing them to an increased
partial O_2 pressure which is different from that in an intact tissue. The
main question is what might be tne triggering mechanism for rhe O_2^- radical
generation? One of the likely candidates is the peroxidase in the cell walls
and apoplast. Peroxidase could act as catalyst for the membrane-bound NADH
oxidase, yielding O_2^- and oxyperoxidase. Activities of both enzymes could
be demonstrated to be bound at the plasmalemma in purified membrane fractions from cauliflower inflorescences (Askerlund et al., 1987). Recently
Ishii (1987) discovered that during preparation of plant protoplasts from
oat leaves by the combined action of highly purified preparations of pectin
lyase, xylanase, and cellulase C_1, O_2^- radicals and H_2O_2 are generated from
peroxidase-containing cell wall fractions and by the protoplast plasmalemma.
If this applies also to our plant/pathogen hypersensitive reaction, the next
question is, which agent is triggering plant pectinase activity? The phytopathogenic bacteria, used in our studies do not produce pectinases themselves. A strong argument for early cell wall alterations is the rapidly
decreasing volumetric elastic modulus (ε) during HR development, together
with early and drastic cell wall thickening, demonstrated in micrographs
by Goodman et al. (1986). It has been known for a long time that ethylene
rapidly induces alterations in plant pectinase activity.

Ethylene Production

During gas exchange measurements of cotton cotyledons inoculated with
P. tabaci we also analyzed the gas stream for some ethylene production in-
duced by bacteria. Measurements were carried out with a gas chromatograph
(Siemens process gas chromatograph U 180) connected to a tight gas exchange

chamber, flushed with synthetic air. Indeed, there was a strong and sudden but transient ethylene burst already within the first hour of inoculation (Fig. 3). In cotyledons only infiltrated with water no increase in ethylene concentration could be detected. When the cotyledon cells started to collapse due to the bacteria-induced HR, a second outburst of ethylene was recorded. P. pisi and P. tabaci are known to be unable to produce ethylene by themselves (Goto et al., 1985). A similar pattern of an initial and a second late ethylene generation was detected in Citrus leaves infected with Xanthomonas citri (Goto et al., 1980).

CONCLUSIONS

A precondition of HR induction is the attachement of bacteria to plant cell walls. With few exceptions all phytopathogenic bacteria are Gram negative bacteria. This group is characterized by an additional outer membrane rich in proteins and long lipopolysaccharide chains. These polysaccharide residues may have an ethylene triggering effect. At least there is an indication that D-galactose, sucrose, and lactose stimulate ethylene production in tobacco leaf disks (Meir et al., 1985). However, heat-killed bacteria do not function as HR inducers. Thus active bacterial metabolism is the other precondition for successful HR induction. Bacteria are known to produce plant hormones such as auxins. In the presence of auxins the carbohydrate-induced ethylene production could be elicited already within less than 2 h (Meir et al., 1985). Auxin promotes the activation of 1-aminocyclopropane-1-carboxylic acid synthase (Yang et al., 1980). Like during plant senescence, Ca^{2+} ions seem also to be strongly involved in the HR expression (Cook and Stall, 1971) and in intracellular ethylene release (Leshem et al., 1986). Taking all these factors into consideration, probably the HR induced by phytopathogenic bacteria is initiated by synergistic effects of the outer bacterial surface together with some metabolically produced, still unknown substance that perhaps induces a rapid Ca^{2+} influx, which then initiates the cascade of events catalyzed by the plasmalemma redox complex, resulting in a deleterious effect for the bacteria and the locally surrounding plant tissue.

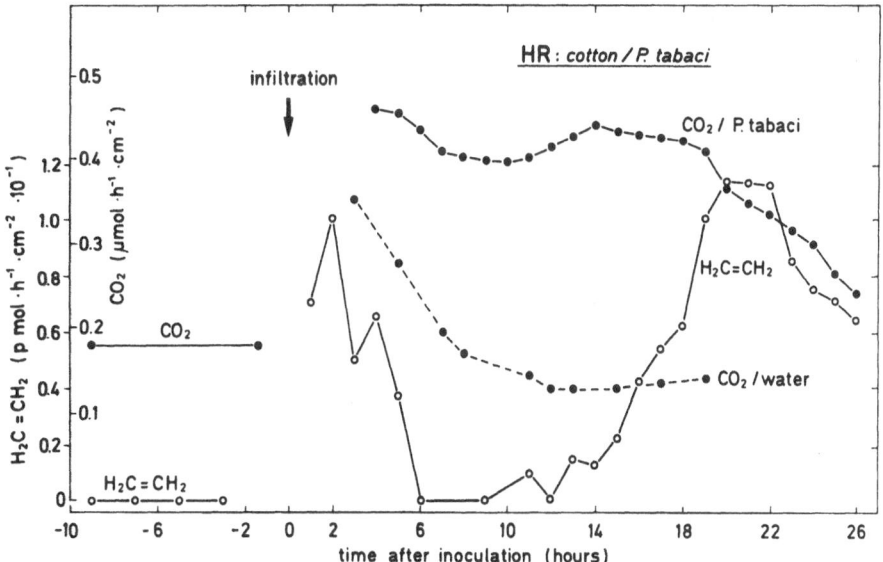

Fig. 3. Ethylene production and CO_2 release by cotton cotyledons (cultivar Acala SJ 2) induced by Pseudomonas syringae pv. tabaci, 10^8 cells/ml.

Inspite of the accumulated indications of primary events proceeding at the cell wall and plasmalemma of plant cells inoculated with incompatible bacteria, the critical question of the very first triggering mechanism remains still to be answered. All plant reactions can very well be compared with reactions observed during senescence of plants (Leshem et al., 1986; Leshem, 1987). Thus the hypersensitive reaction could be defined as an extremely rapid cell senescence, resulting in plant protection against pathogens.

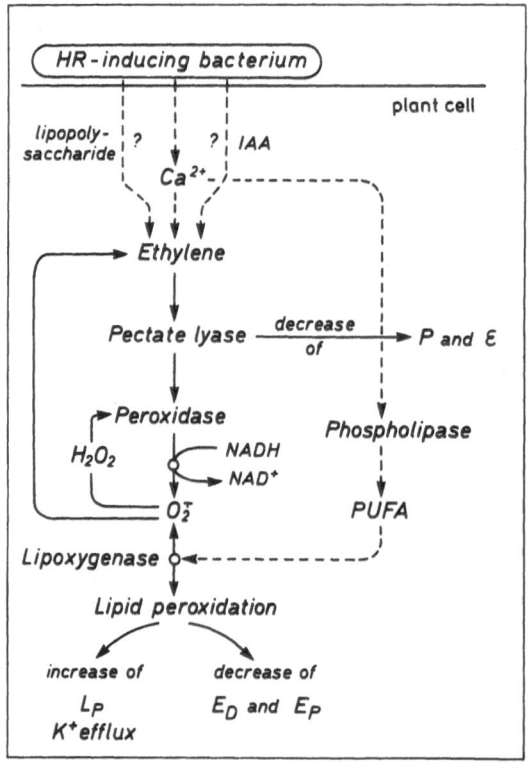

Fig. 4. Cascade of primary events during the development of the hypersensitive reaction induced by pathogenic bacteria in plant tissue. Lp, hydraulic conductivity; P, turgor pressure; $\mathcal{E}$, volumetric elastic modulus; PUFA, polyunsaturated fatty acids; E_D, electrical membrane diffusion potential; E_P, electrical membrane pump potential.

ACKNOWLEDGEMENTS

We wish to acknowledge funding support from the Deutsche Forschungsgemeinschaft to C.U.-E. and from the National Science Foundation to A.N. (DMB-8516038). Special thanks are due to the Alexander von Humboldt-Stiftung for a generous stipend to J.P. and to Dr. Enno Brinckmann (University of Bayreuth) for his undefatigable and invaluable help with the pressure probe measurements.

330

REFERENCES

Askerlund, P., Larsson, C., Widell, S., and Møller, I.M., 1987, NAD(P)H oxidase and peroxidase activities in purified plasma membranes from cauliflower inflorescences, Physiol. Plant., 71:9-19

Atkinson, M. M., Huang, J., and Knopp, J. A., 1985, Hypersensitivity of suspension-cultured tobacco cells to pathogenic bacteria, Phytopathology, 75:1270-1274.

Atkinson, M. M., Huang, J., and Knopp, J. A., 1985, The hypersensitive reaction of tobacco to Pseudomonas syringae pv. pisi:Activation of a plasmalemma K^+/H^+ exchange mechanism, Plant Physiol., 79:843-847.

Bertl, A., Felle, H., Bentrup, F.-W., 1984, Amine transport in Riccia fluitans. Cytoplasmic and vacuolar pH recorded by a pH-sensitive microelectrode, Plant Physiol.,76:75-78.

Chai, H.B., and Doke, N., 1987, Superoxide anion generation:a response of potato leaves to infection with Phytophthora infestans, Phytopathology, 77:645-649.

Cook, A. A., and Stall, R. E., 1971, Calcium suppression of electrolyte loss from pepper leaves inoculated with Xanthomonas vesicatoria, Phytopathology, 61:484-487.

Doke, N., 1983, Involvement of superoxide anion generation in the hypersensitive response of potato tuber tissues to infection with an incompatible race of Phytophthora infestans and to the hyphal wall components, Physiol. Plant Pathol., 23:345-357.

Doke, N., 1985, NADPH-dependent O_2^- generation in membrane fractions isolated from wounded potato tubers inoculated with Phytophthora infestans, Physiol. Plant Pathol., 27:311-322.

Ecker, J. R., and Davis, R. W., 1987, Plant defense genes are regulated by ethylene, Proc. Nat. Acad. Sci. USA, 84:5202-5206.

Goodman,R. N., Király, Z., and Wood, K. R., 1986, The biochemistry and physiology of plant disease, University of Missouri Press, Columbia.

Goto, M., Yaguchi, Y., and Hyodo, H., 1980, Ethylene production in Citrus leaves infected with Xanthomonas citri and its relation to defoliation, Physiol. Plant Pathol.16:343-350.

Goto, M., Ishida, Y., Takikawa, Y., and Hyodo, H., 1985, Ethylene production in the Kudzu strains of Pseudomonas syringae pv. phaseolicola causing halo blight in Pueraria lobata (Willd) Ohwi, Plant Cell Physiol., 26:,141-150.

Guern, J., and Ullrich-Eberius, C. I., 1988, The ferricyanide-driven redox system at the plasmalemma of plant cells: origin of the proton production and reappraisal of the stoichiometry e^-/H^+, in:"Plasma Membrane Oxidoreductase in Control of Animal and Plant Growth," F.L. Crane et al., eds., Plenum Press, New York.

Hüsken, D., Steudle, E., and Zimmermann, U.,1978, Pressure probe technique for measuring water relations of cells in higher plants. Plant Physiol., 61:158-163.

Ishii, S., 1987, Generation of active oxygen species during enzymatic isolation of protoplasts from oat leaves, In Vitro Cellular and Developmental Biology, 23:653-658.

Keppler, L. D., and Novacky, A., 1986, Involvement of membrane lipid peroxidation in the development of a bacterially induced hypersensitive reaction, Phytopathology, 76:104-108.

Keppler, L. D., and Novacky, A., 1987, The initiation of membrane lipid peroxidation during bacteria-induced hypersensitive reaction, Physiol. Mol. Plant Pathol., 30,233-245.

Keppler, L. D., Atkinson, M. M., and Baker, C.J., 1988, O_2^- -initiated lipid peroxidation in the initiation of bacteria-induced hypersensitive reaction in tobacco cell suspensions, Phytopathology, in press.

Klement, Z., 1965, Method of obtaining fluid from intercellular spaces of foliage and the fluid's merit as substrate for phytobacterial pathogens, Phytopathology, 55:1033-1034.

Klement, Z., 1982, Hypersensitivity, in:"Phytopathogenic Procaryotes," M. S. Mount, and G. H., Lacy, eds., Academic Press, New York, 149-177.

Kunz, G., 1986, Die Rolle des Ammoniums bei Bakterien-induzierten Krankheits- und Überempfindlichkeitsreaktionen in Blättern von Tabak und Baumwolle, PhD Thesis, Technische Hochschule Darmstadt.

Lehmann-Danzinger, H.,1987,Epidemie der Eckigen Blattfleckenkrankheit(Xantho- monas campestris pv.malvacearum) in Nicaragua, Phytomedizin, 1:22.

Leshem, Y. Y., Halevy, A., H., and Frenkel, C., 1986 ,"Processes and control of plant senescence," Elsevier, Amsterdam.

Leshem, Y. Y., 1987, Membrane phospholipid catabolism and Ca^{2+} activity in control of senescence, Physiol. Plant.69:551-559.

Lovrekovich, L., Lovrekovich, H., and Goodman, R. N., 1970, Ammonia as a ne- crotoxin in the hypersensitive reaction by bacteria in tobacco leaves, Can. J. Bot., 48:167-171.

Meir, S., Philosoph-Hadas, S., Epstein, E., and Aharoni, N., 1985, Carbohy- drates stimulate ethylene production in tobacco leaf discs, Plant Physiol., 78:131-138.

Naumann, K., 1986, Stand der Erkenntnisse über die Genetik der Resistenz ge- genüber pflanzenpathogenen Bakterien,Arch.Phytopathol.Pfl.schutz,22:167.

Novacky, A., 1972, Suppression of the bacterially induced hypersensitive re- action by cytokinins, Physiol. Plant Pathol.2:101-104.

Novacky, A., 1980,Disease-related alteration in membrane function, in:"Plant Membrane Transport: Current Conceptual Issues," R. M.,Spanswick,W.J. Lucas, and J. Dainty,eds.,Elsevier/North-Holland Biomed.Press,369-378.

Novacky, A., and Ullrich-Eberius, C.I., 1982, Relationship between membrane potential and ATP level in Xanthomonas campestris malvacearum infected cotton cotyledons, Physiol. Plant Pathol., 21:237-249.

Novacky, A., 1983, The effect of disease on the structure and activity of mem- branes, in: "Biochemical Plant Pathology," J.A. Callow, ed., Wiley, Chichest, 347.

Pavlovkin, J., Novacky, A., and Ullrich-Eberius, C. I., 1986, Membrane po- tential changes during bacteria-induced hypersensitive reaction, Phy- siol. Mol., Plant Pathol.,28:125-135.

Rathmell, W. G.,and Sequeira, L., 1974, Soluble peroxidase in fluid from the intercellular spaces of tobacco leaves, Plant Physiol., 53:317-318.

Sequeira, L., 1976, Induction and suppression of the hypersensitive reaction caused by phytopathogenic bacteria: specific and non-specific compo- nents, in:"Specificity in Plant Disease", R. K. S. Wood, and A. Graniti, eds., Plenum Press, New York, 289-306.

Stewart, P. A., 1981, "How to understand acid-base:A quantitative acid-base primer for biology and medicine," E. Arnold, London.

Turner, J. G., 1981, Tabtoxin, produced by Pseudomonas tabaci, decreases Ni- cotiana tabacum glutamine synthetase in vivo and causes accumulation of ammonia, Physiol. Plant Pathol., 19:57-67.

Thompson, J. E., Legge, R. L., and Barber, R. F., 1987, The role of free ra- dicals in senescence and wounding. New Phytol., 105:317-344.

Ullrich, W. R., Novacky,A., Kunz, G., 1984, The role of ammonium in bacteri- al plant disease: accumulation and the effect on the electrical mem- brane potential in cotton and tobacco, in: "Membrane Transport in Plants," W.J. Cram, K. Janácek, R. Rybová, and K. Sigler, eds. Academia Praha, 514.

Venere, R. J., 1980, Role of peroxidase in cotton resistant to bacterial blight, Plant Sci. Lett., 20:47-56.

Yang, S. F., Hoffmann, N. E., McKeon, T., Riov, J., Kao, C. H., and Yung, K. H., 1982, Mechanism and regulation of ethylene biosynthesis, in: "Plant Growth Substances," P.F. Wareing, ed., Academic Press, New York, 239-248.

ON THE OCCURRENCE OF OXIDOREDUCTASES IN THE APOPLAST

OF LEGUMINOSAE AND GRAMINEAE AND THEIR SIGNIFICANCE IN THE STUDY

OF PLASMAMEMBRANE-BOUND REDOX ACTIVITIES

Rodolfo Federico, Chiara Alisi and Riccardo Angelini

Dipartimento di Biologia Vegetale
Università "La Sapienza"
p.le Aldo Moro 5, 00185-Rome, Italy

INTRODUCTION

Growing evidence is currently being gathered on the occurrence and physiological importance of plasmamembrane-bound redox systems in plants[1,2]. These activities are able to accept electrons from NADH and/or NADPH and to reduce both natural (O_2,Fe^{3+}) and artificial (ferricyanide, cytochrome c) substrates. Several physiological roles have been envisaged for NAD(P)H oxidases including: proton translocation[3] , extracellular Fe^{3+} reduction [4] and blue light response[5] .

Studies with intact organs and isolated vescicles, have suggested that these systems occur on both faces of the plasma membrane [2] . As concerns the redox components facing on the outer apoplastic surface, it must be pointed out that their activity in vivo can be influenced by other systems present in the extracellular compartment. In this context it is worth considering that several proteins are associated to the cell wall[6] . Some of them are likely to be structural proteins and these are either insoluble (extensin) or wall-bound through ionic interactions (arabinogalactan proteins). Most of apoplastic proteins are hydrolytic enzymes such as glycosidases, pectinases, xylanases, these being ionic-bound at a variable degree to the cell wall network. A significant fraction of wall proteins are oxidoreductases, namely amine oxidases, peroxidases (POD), IAA oxidases (IAAox)[7,8,9] . These enzymes are likely to be involved in wall stiffening, in the final stages of lignin synthesis and in the regulation of apoplastic IAA and polyamine (PA) levels[9,10,11,12] . It is important to note that in some monocot and dicot families, very high levels of di- and polyamine oxidases (DAO,PAO) are present in the apoplast. These enzymes, aside from being involved in apoplastic PA level regulation, have been considered functionally integrated with peroxidases: DAO and PAO produce, through oxidative deamination of extracellular PA, hydrogen peroxide required for peroxidase activity[12] . It is clear that the assessment of oxidoreductases occorrence in the apoplast is of primary importance in the study of redox systems facing out of the cell. This being particulary true when an in vivo strategy, i.e. ferricyanide reduction by intact plant segments, is utilized. In this case, the possible production of H_2O_2 by apoplastic amine oxidases may interfere with measurements of ferricyanide reductase activity, as hydrogen peroxide would oxidize the reduced form of the artificial substrate used. The situation is further complicated by the fact that amine oxidase level in the apoplast is highly dependent on the physiological status of the plant[13] ,hormones[14] and phytochrome[15] .

CELL WALL PROTEINS: A CRITICAL ASSESSMENT

Cell walls contain both free and bound proteins, the latter being "ionic bound", and thus extractable with high ionic strength buffers from detergent-washed cell wall preparations, or "covalently linked". This later fraction can be obtained with an enzymatic digestion of the cell wall residue after ionic extraction (most used are commercially available combinations of hydrolytic enzymes such as Driselase, Fluka, Bucks, Switzerland[8]). It must be pointed out that apoplastic soluble enzymes cannot be determined by techniques of subcellular fractionation inasmuch as they are lost from the cell wall network during homogenization and are found in the cytoplasmic fraction. To avoid this problem we have utilized an in vacuum infiltration technique that allowed us to obtain the so-called "extracellular fluids" [16]. This consist of the infiltration, under vacuum, of stem or leaf sections with a suitable buffer, followed by centrifugation (~1000xg) of the tissues packed in a syringe. The buffer which has permeated intercellular spaces and cell walls will solubilize apoplastic proteins depending on the ionic strength of the solution and the interaction between the proteins and the cell wall network. Increasing ionic strength or changing ionic composition of the infiltration solution will result in a better yield of those proteins more tightly bound to the cell wall.

OXIDOREDUCTASES IN THE APOPLAST OF LEGUMINOSAE AND GRAMINEAE

In the study of apoplastic oxidoreductases, our attention was drawn to the occurrence of high levels of redox activities in the extracellular fluids obtained from Cicer aretinum L. and Zea mais L. (Table 1). In these fluids no detectable amount of glucose-6-phosphate dehydrogenase was detected, thus showing the absence of any contamination by intracellular contents. From a comparison of the enzyme activity and protein content in the extracellular fluids and in the crude homogenate, it follows that the specific activity of the enzymes assayed can be as much as 100 fold higher in the apoplast than inside the cell.

Table 1. Oxidoreductase activities in crude homogenate and extracellular fluids obtained from 7 day-old light grown Zea mais and Cicer arietinum seedlings. Homogenates were obtained in potassium phosphate buffer 0.5M pH 6.5 (Zea mais) or 0.1M pH 6.5 (Cicer arietinum); enzyme activities were tested as previously described [7,9,18]. Extracellular fluids were obtained as described by Terry and Bonner [16], using a NaCl solution (pH 5.5) 0.5M (Zea mais) or 0.1M (Cicer arietinum) for in vacuum infiltration. Similar oxidoreductase specific activity increases in extracellular fluids were found for oat and pea.

	DAO	PAO	IAAox	POD
		U/mg protein		
Zea mais				
homogenate	–	0.015	3.13	0.32
extracellular fluid	–	0.97	221	27.7
Cicer arietinum				
homogenate	0.059	–	1.53	0.039
extracellular fluid	4.75	–	2.60	0.119

SDS-PAGE of crude homogenate and extracellular fluids evidence the marked difference in protein composition of the intra- and extracellular environment (Fig.1). DAO and PAO recognition after the electrophoretic procedure was performed by Western immunoblotting using rabbit affinity-purified policlonal anti-DAO or anti-PAO antibodies (Fig.1).

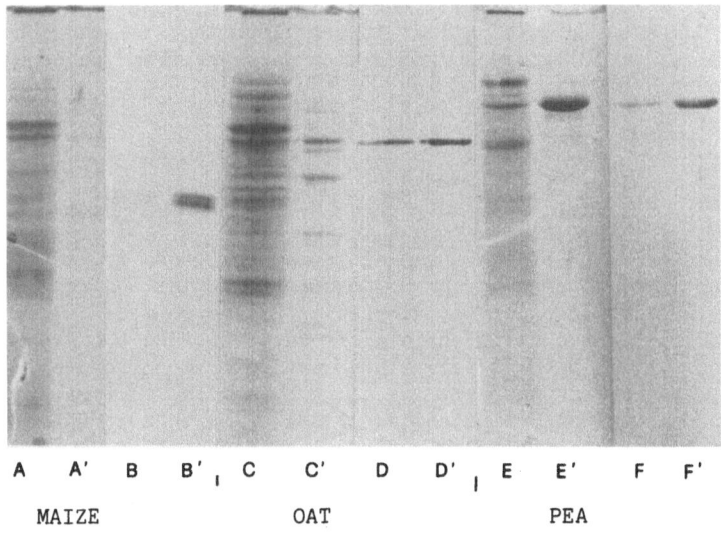

A   A'   B   B' , C   C'   D   D' , E   E'   F   F'

MAIZE   OAT   PEA

Fig.1. SDS-PAGE (A-A', C-C',E-E') and Western immunoblot (B-B', D-D', F-F') of crude homogenate (A, B, C, D, E, F) and extracellular fluid (A', B', C', D', E', F') from 7 day-old light grown maize, oat and pea seedlings. Western immunoblot was developed using anti-rabbit HRP-coupled IgG.

A native PAGE (pH 8.8 for DAO and peroxidase;pH 4.5 for PAO) was performed and protein bands with enzyme activity were revealed by specific colouration. These procedures allow us to asses that DAO, PAO, POD and IAAox are the most plentiful proteins in the apoplast of the species examined.

In order to have better insight into the cellular and tissue distribution of these enzyme activities,we performed both a histochemical study and an investigation on their distribution pattern along the stem. In the in vivo histochemical experiments, we utilized fresh, 200 m thick sections from basal zone of pea and cick-pea (Cicer arietinum) epicotyls or maize mesocotyls obtained from light-grown seedlings. Upon addition of putrescine (pea and cick-pea) or spermine (maize) to sections incubated in syringaldazine, a strong colouration appeared in the cell walls of xylem, xylem parenchyma, schlerenchyma and epidermis. Control sections treated with aminoguanidine (strong inhibitor of DAO and PAO) or catalase were colourless. The explanation of such phenomenon is shown by this scheme:

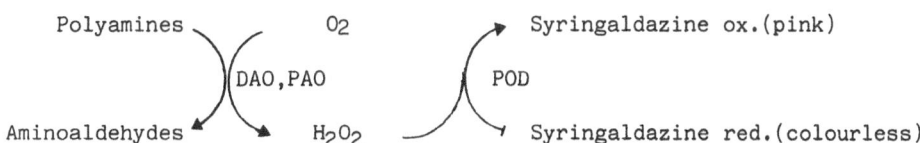

Polyamines   O₂   →   Syringaldazine ox.(pink)

              DAO,PAO   POD

Aminoaldehydes   H₂O₂   ⟶   Syringaldazine red.(colourless)

335

Distribution of DAO and POD along the pea stem, has been determined by enzyme activity measurement in crude homogenates of adjacent sections [13] . Both enzyme activities expressed as UE/mg protein decrease from the base to the top of the epicotyls in agreement with the parallel distribution pattern of lignifying or lignified tissue as determined by the pholoroglucine/HCl test [17] . Cell walls of epidermal tissue, which do not contain lignin, are however a crucial site of activity of the DAO/POD system as judged from the strongly positive syringaldazine test. This observation is in agreement with the primary role of epidermis in the control of growth. In this context is very important to also consider that epidermal activity of DAO/POD system decrease from the basal zone to the apex of the seedlings.

CONCLUSION

In consideration of the above mentioned evidence, in the study of redox activity of plasmamembrane components in the apoplast, a very careful evaluation of experimental results has to be made especially in in vivo experiments using whole plant sections. The occurrence of interfering systems such as polyamine oxidases or peroxidases could produce artifacts or mask the activity of redox systems facing out the cell. Moreover, an additional complication arises owing to the differential distribution of the interfering system along a certain organ or to the temporal display of the activity.

Acknowledgements- This research was supported by the Italian Research Council (C.N.R.), special grant IPRA, subproject 1, paper N.0000.

REFERENCES

1. F. L. Crane, H. Löw and M. G. Clark, Plasma membrane redox system in: "The Enzymes of Biological Membranes", 4, A. N. Martonosi,ed. Plenum Press, New York (1985).
2. I. M. Moller and W. Lin, Membrane-bound NAD(P)H dehydrogenases in higher plant cells, Ann.Rev.Plant Physiol.37:309-334 (1986).
3. R. Federico and C. E. Giartosio, A transplasmamembrane electron transport system in maize roots, Plant Physiol.73:182-184 (1983).
4. H. F. Bienfait, Regulated redox processes at the plasmalemma of plant root cells and their function in iron uptake, J.Bioenerg.Biomemr. 17:73-83 (1985).
5. P. Kjellbom, C. Larsson, P. Askerlund, C. Schelin and S. Widell, Cytochrome P-450/420 in plant plasma membranes: a possible component of the blue-light-reducible flavoprotein-cytochrome complex, Photochem.Photobiol.42:779-783 (1985).
6. D. T. A. Lamport, Structure and function of plant glycoproteins, in "The Biochemistry of Plants", 3, J. Preiss, ed., Academic Press, New York (1980).
7. R. Federico and R. Angelini, Occurrence of diamine oxidase in the apoplast of pea epicotyls, Planta 167:300-302 (1985).
8. R. Grison and P. E. Pilet, Cytoplasmic and wall isoperoxidases in growing maize roots, J.Plant Physiol.118:189-199 (1985).
9. J. D. Waldrum, E. Davies, Subcellular localization of IAA oxidase in peas, Plant Physiol.68:1303-1307 (1981).
10. J. M. Harkin and J. R. Obst, Lignification in trees: indication of exclusive peroxidase partecipation, Science 180:296-298(1973).

11. R. Kaur-Sawhney, H. E. Flores and A. W. Galston, Polyamine oxidase in oat leaves: a cell wall localized enzyme, Plant Physiol.68:494-498 (1981).

12. R. Angelini and R. Federico, Occurrence of diamine oxidase in the apoplast of Leguminosae seedlings, in: "Biomedical studies of natural polyamines", C. M. Caldarera, C. Clò, C. Guarnieri, eds., CLUEB, Bologna (1985).

13. R. Federico and R. Angelini, Distribution of polyamines and their related catabolic enzymes in etiolated and light-grown Leguminosae seedlings, Planta 173:317-321 (1988)

14. A. Cona, R. Angelini and R. Federico, A preliminary study of hormonal regulation on diamine oxidase activity in lentil seedlings, Ann.Bot.45:135-141 (1987).

15. R. Angelini, R. Federico and A. Mancinelli, Phytochrome-mediated control of diamine oxidase level in etiolated lentil seedlings, Plant Physiol.in press (1988).

16. H. E. Terry and B. A. Bonner, An examination of centrifugation as a method of extracting an extracellular solution from peas, and its use for the study of indolacetic acid-induced growht, Plant Physiol.66:321-325 (1980).

17. P. B. Gahan, "Plant histochemistry and cytochemistry: an introduction", Academic Press, London (1984).

18. U. H. Bergmeyer, "Methods of enzymatic analysis",2, Academic Press, New York and London (1974).

RELATIONSHIP BETWEEN NAD[+]/NADH LEVELS AND ANIMAL CELL GROWTH

P. Navas[1], I.L. Sun[2], D.J. Morré[3] and F.L. Crane[2]

[1] Departamento de Biología Celular, Universidad de Córdoba, 14071 Córdoba, Spain and Departments of [2] Biological Sciences and [3] Medicinal Chemistry and Pharmacognosy, Purdue University, West Lafayette, IN 47907, USA

INTRODUCTION

Studies have shown a close relationship between pyridine nucleotide levels and the proliferating stage of some tissues[1]. Also NAD(H) and NADP(H) pools have been studied as a function of growth in both normal and SV40 transformed 3T3 cells[2,3]. Decrease of NAD(H) levels in rat liver follows carcinogen (2-acetylaminofluorene) administration[4]. Therefore, it is reasonable to postulate that pyridine nucleotide pools are important in the control of cell division[2,3,5]. The regulation of the biosynthesis of pyridine nucleotides is complex since several alternative pathways have been described in mammalian cells[3]. Here, we show evidence that levels of pyridine nucleotides and relative redox states are altered in cells growing under the stimulation of growth factors, especially as a result of both short- and long-term treatments with external electron acceptors which react with the transplasma membrane redox system[5].

TRANSPLASMA MEMBRANE REDOX SYSTEM AND CELL GROWTH

All eukaryotic cells have a transplasma membrane redox enzyme system[6-9]. The electron flow through the plasma membrane is accompanied by proton release from the cell[7,10] and in some cells the redox enzyme activity is sensitive to hormones[11-13]. Diferric transferrin is essential for growth of many cultured cells in serum free media[14,15] (Fig. 1), and acts as an electron acceptor for the transplasma membrane redox system[16]. Activation of transplasma membrane electron transport with ferricyanide stimulated the growth of melanoma cells in serumdeficient media[17] and promotes HeLa cell proliferation in the absence of fetal calf serum and other growth factors[18] (Fig. 2). Ferricyanide also induces proton extrusion[19]. Therefore, the transmembrane redox system might increase the pH of the cytoplasm, which is often associated with mitosis[20].

A series of impermeable oxidants with redox potentials down to -125 mV were shown to have similar growth promoting effects on HeLa cells[19]. All of these oxidants are reduced by the transplasma membrane electron transport system. Oxidants which are not reduced by the transmembrane electron transport do not stimulate growth[19].

Insulin, a growth-promoting hormone, stimulates growth of HeLa cells in the absence of serum (Fig. 3) and also stimulates transmembrane redox activities[18]. An additive effect on the growth of HeLa cells by insulin and ferricyanide has also been demonstrated[18]. Fetal calf serum has been the most common supplement used to stimulate growth (Fig. 4).

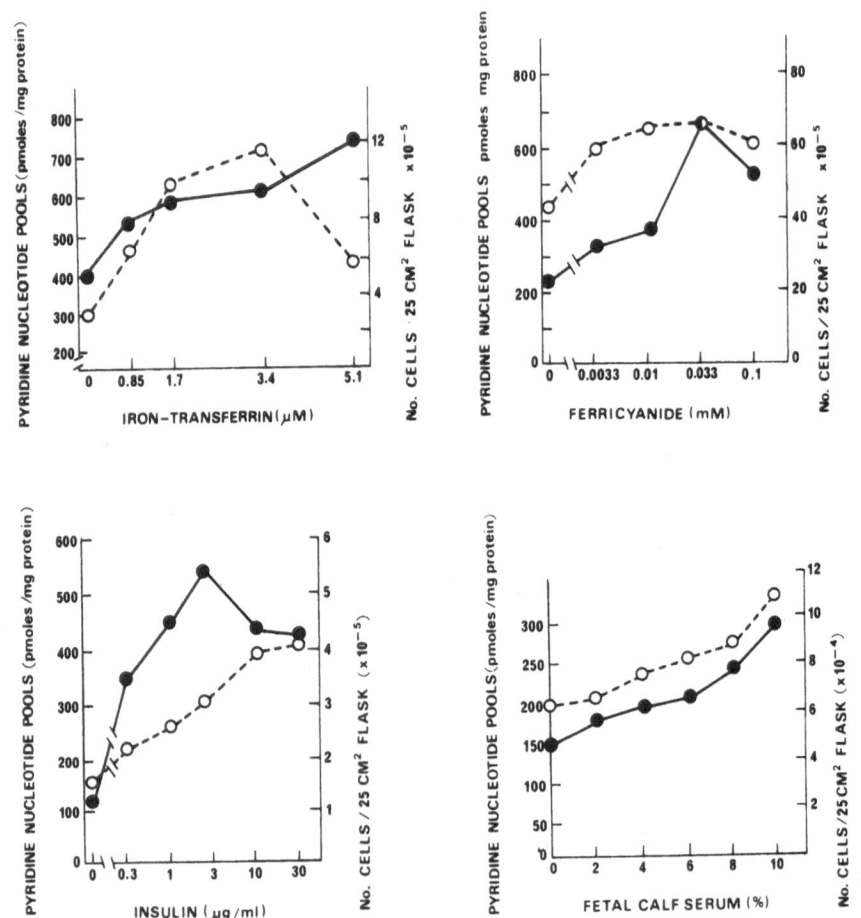

Figs. 1-4. Effect of different growth promoting factors on the grown (O-----O) and on pyridine nucleotides (●——●) of HeLa cells after 48 h growth in a serum-free medium.

The response of pyridine nucleotide concentrations to these exogenous stimuli, which greatly promote cell proliferation in a serum-depleted media, has therefore been tested.

NADH AS ELECTRON DONOR FOR THE TRANSMEMBRANE REDOX SYSTEM

All plasma membrane which have been prepared show NADH dehydrogenase activities[9,13,16], but a major problem has been to identify the cytoplasmic electron donor[5]. From indirect evidence, NADH appears to be the endogenous electron donor in erythrocytes[21], HeLa cells[22], and plants[7]. NADPH[23] also may act as an electron donor for iron reduction by some plants[23].

Therefore, NAD^+/NADH were determined in HeLa cells during a short-term incubation with diferric transferrin or ferricyanide. Both oxidants induced a decrease of NADH, which was proportional to the increase of the oxidized form NAD^+ (Fig. 5). Insulin increased the effect of both diferric transferrin and ferricyanide on NADH levels, but insulin itself did not produce any effect on NAD(H) levels after 15 min incubation. NADP/NADPH did not change in diferric transferrin- or ferricyanide-treated HeLa cells[24]. Apotransferrin and ferrocyanide, which do not stimulate cell growth[5], did not cause NADH oxidation in HeLa cells[24]

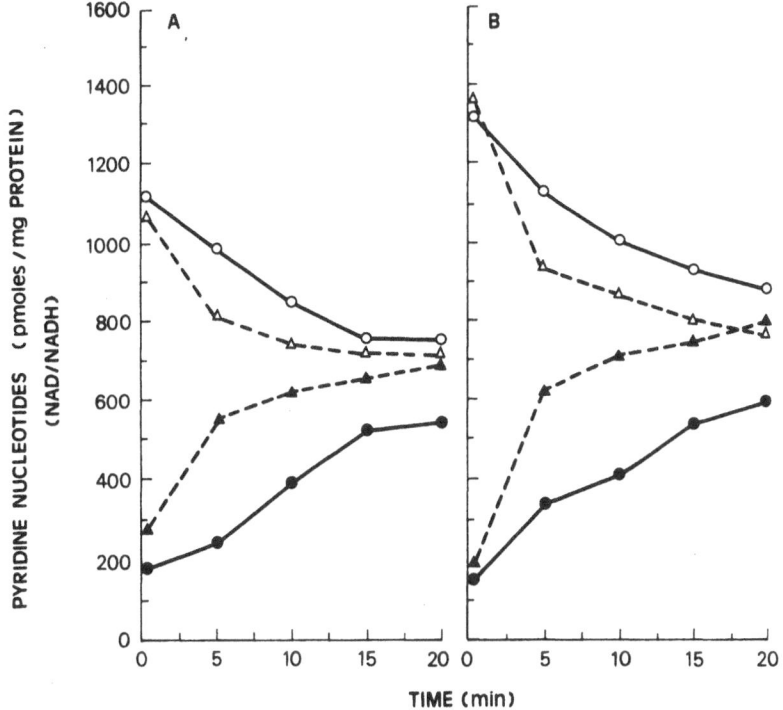

Fig. 5. Short-term effect of diferric transferrin (A) and ferricyanide (B) on pyridine nucleotide levels in HeLa cells. Open circles: NADH and black circles: NAD^+ in the presence of 3.4 uM diferric transferrin or 0.033 mM potassium ferricyanide. Open triangles: NADH and black triangles: NAD^+ in the presence of 3.4 uM diferric transferrin or 0.033 mM potassium ferricyanide plus 30 ug/ml insulin.

Changes in NAD^+/NADH were proportional to diferric transferrin and ferricyanide concentrations (Table I). Both NADPH and total pyridine nucleotides were not affected significantly.

It has been proposed that adriamycin and related antitumor drugs can inhibit cell growth by action at the plasma membrane without penetration into the cell[25], and these drugs act as inhibitors of the transmembrane electron transport at concentration which inhibit growth[5,22].

When HeLa cells are treated with adriamycin three minutes before the incubation with diferric transferrin, there is less oxidation of internal NADH (Table II). This is consistent with adriamycin inhibition of NADH diferric transferrin reductase of isolated plasma membranes[26].Inhibition of diferric transferrin induced oxidation of internal NADH shows that the

TABLE I. Short—term effects of growth stimulation on the distribution of
pyridine nucleotides in HeLa cells.

Reductant	Concentration	Pyridine nucleotide (pmoles/mg protein)				
		NAD^+	NADH	NADP	NADPH	Total
None		224	1,367	12.5	4.4	1,608
Diferric						
Transferrin	1.70 uM	333	1,247	12.3	4.4	1,597
	3.40 uM	553	1,028	12.6	3.7	1,597
Ferricyanide	0.01 mM	285	1,316	13.7	4.9	1,620
	0.10 mM	580	1,021	15.3	5.1	1,621

adriamycin sensitive NADH dehydrogenase could be the basis for the
inhibited electron flow.

The relative interchange of the reduced and oxidized forms of NADH
induced by external electron acceptors transferrin and ferricyanide does
not involved formation or loss of pyridine nucleotides. Insulin stimulates
the decrease of internal NADH and reduction of the external electron
acceptors at concentratrion of insulin which stimulates cell growth.
Therefore, NADH appears to be a primary endogenous electron donor for the
transplasma membrane redox system of HeLa cells and the stimulation of
cell growth by external impermeable oxidants may be correlated to the
oxidation of the cytolosic NADH.

TABLE II. Effect of adriamycin on pyridine nucleotide concentration of
HeLa cells in presence of 3.4 uM diferric transferrin. Pyridine
nucleotides are expressed as pmoles/mg protein.

Addition	Time (min)	NAD^+	NADH
None	0	108	285
	5	130	260
	10	165	228
Adriamycin (10^{-7}M)	5	96	290
	10	127	278

NAD(H) AND CELL TRANSFORMATION

Transformed cells are characterized by less of growth control compared to non transformed cells. The comparison of pyridine nucleotide
levels of cells showing non—transformed and transformed phenotypes will be
useful for studying the relation between NAD^+/NADH and cell growth.

Changes in pyridine nucleotide pools have been associated to the
growth of normal and transformed 3T3 cells[3]. NAD(H) or NADP(H) may
increase[2,27] or decrease[1] as a result of transformation. Therefore, NAD(H)
levels and redox state have been measured as a function of growth in fetal
rat livers cells (RLA 209—15) infected with a temperature sensitive strain
of SV40 which exhibit density—dependent inhibition of growth at 40ºC and
have lost this property at 33ºC.

With RLA 209-15 cells growing at 40ºC, showing the non-transformed phenotype, the NAD$^+$ concentration decreased and NADH concentration increased proportionally with time (Fig. 6A). The opposite pattern was observed in cell grown at 33ºC (Fig. 6B). In order to establish if the difference between the two phenotypes was a consequence of the interconversion between reduced and oxidized forms of pyridine nucleotides, NAD(H) pool was measured in RLA 209-15 cells for 48 h at 40ºC and then shifted down to 33ºC for and additional 48 h. The change of pyridine nucleotide levels was reversible (Fig. 6C) with the phenotype. The high NADH concentration reached after 48 h growth at 40ºC decreased and a proportional increase of NAD$^+$ was observed after the shift to 33ºC.

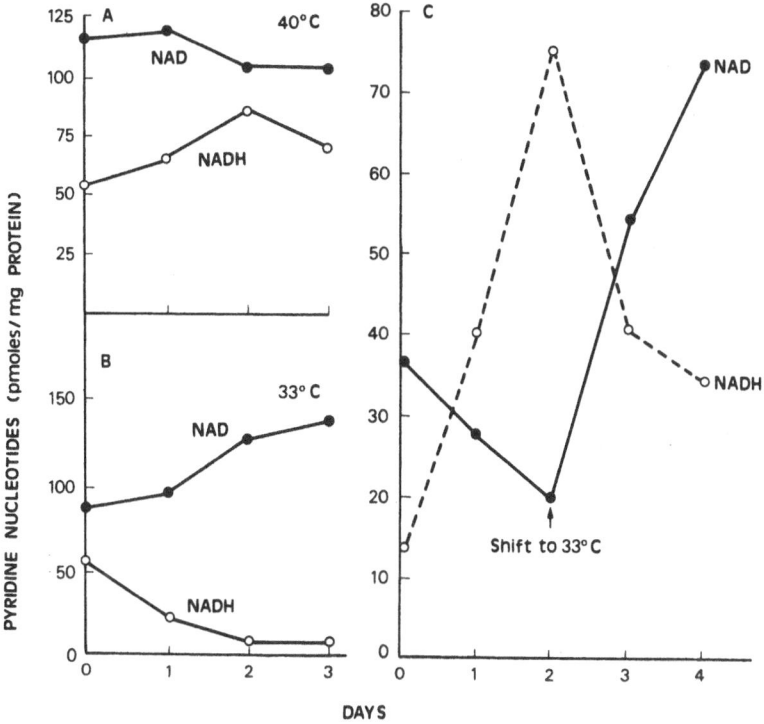

Figure 6. Changes in pyridine nucleotide redox state in transformed (33ºC) and non-transformed (40ºC) RLA 209-15 cells. Two sets of cultures were plated at 33ºC and grown for 2 days, maintained at 33ºC (B). Another set of culture was grown for 2 days at 44ºC and then shifted to 33ºC (C).

Decrease of both NAD(H) levels[1,4,28,29] and NADH ferricyanide reductase activity[4] have been demonstrated after transformation. NADP(H) pools run in a similar pattern and in a similar range in RLA 209-15 grown at 33 and 40ºC[30]. Similar results were observed for 3T3 cells[3].

Levels of NAD(H) appear to be significantly higher in nontransformed cells than in transformed cells. The lowering of the NADH pool and the increase of the oxidized form is a unique pattern found in the transformed phenotype.

The regulation of NAD(H) biosynthesis is potentially complex, since several pathways have been described in mammalian cells[3]. Regulation may affect the activity of enzymes, and the activity of the transplasma membrane dehydrogenase may contribute to control of the redox state of the pyridine nucleotide pool. The decrease in transmembrane dehydrogenase activity which is observed in transformed cells may reflect the decline in NADH within the cell so that the NADH concentration becomes rate limiting for the dehydrogenase.

PYRIDINE NUCLEOTIDE POOLS AND GROWTH PROMOTING FACTORS

Pyridine nucleotide pools have been determined in HeLa cells grown 48 h in a serum free medium supplemented with different growth factors.

Diferric transferrin induced a two-fold increase of total pyridine nucleotides as well as cell replication (Fig. 1). The effect of diferric transferrin during culture on the distribution of oxidized and reduced forms of pyridine and phosphopyridine nucleotides was expressed principally by an increase in NADH (Table III).

TABLE III. Effect of diferric transferrin on the distribution of NAD(H) and NADP(H) in HeLa cells after 48 h culture. Pyridine nucleotides are expressed as pmoles/mg protein.

Addition	Concentration	NAD$^+$	NADH	NADP	NADPH	Total
Control		133	265	19	12	429
Diferric						
transferrin	0.85 uM	170	375	25	12	582
	1.70 uM	190	443	39	17	689

Ferricyanide stimulated total nucleotides as well as cell growth reaching an optimal concentration of ferricyanide at 0.033 mM (Fig. 2). This correlation between growth and pyridine nucleotides was observed for both NAD(H) and NADP(H) in reduced and oxidized forms (Fig. 7).

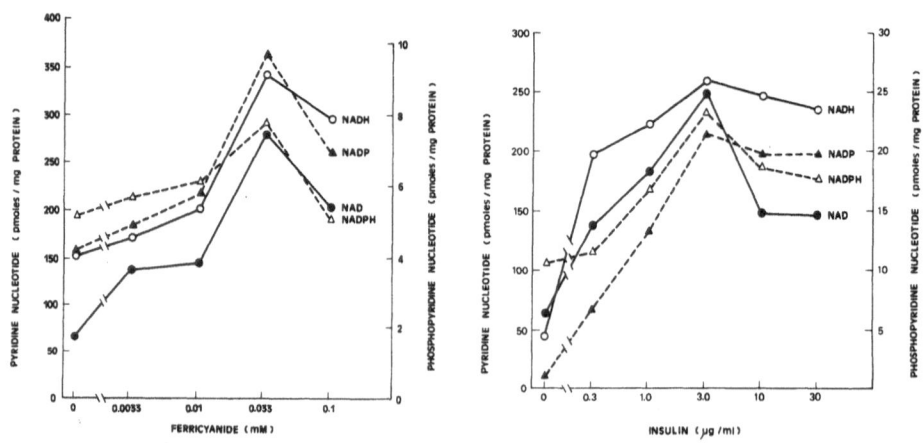

Fig. 7 and 8. Effect of long-term incubation with ferricyanide and insulin on pyridine nucleotides of HeLa cells.

When HeLa cells were grown in serum depleted but insulin supplemented medium, total pyridine nucleotides increased with growth (Fig. 3). Levels of NADP and NADPH were maintained a constant increase in a similar range whereas the level of NADH increased more and was higher than that of NAD$^+$ (Fig. 8).

Fetal calf serum induced a progressive increase in pyridine nucleotide levels corresponding to cell density (Fig. 4). Each individual pyridyne nucleotide concentration increased markedly specially with 10% serum concentration (Fig. 9).

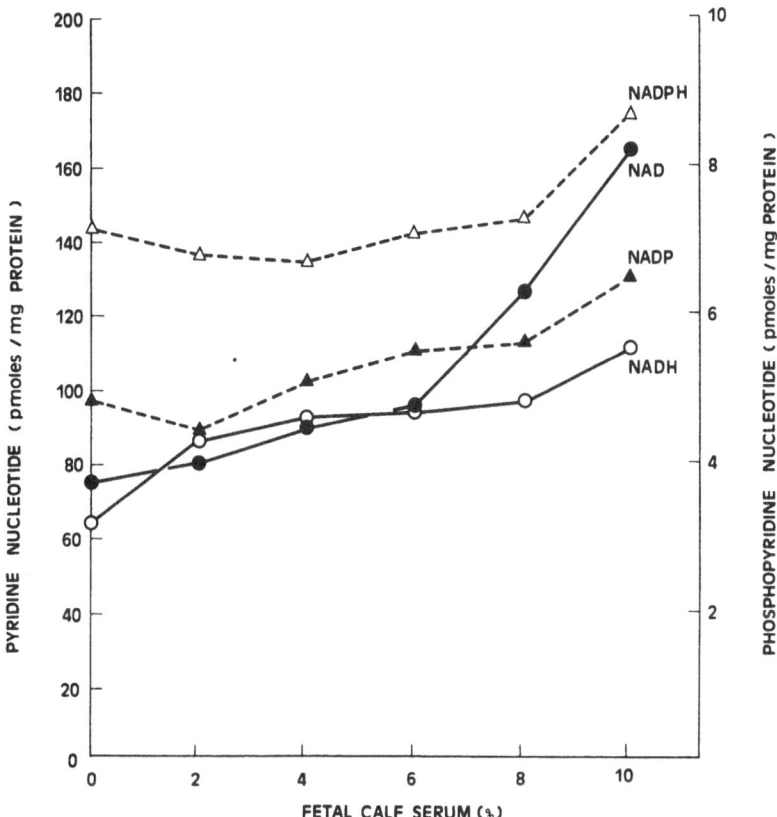

Fig. 9. Long-term effect of fetal calf serum on NAD(H) and NADP(H) of HeLa cells.

The results described above show that following growth promotion by various agents, HeLa cells acquired increased pools of pyridine nucleotides due largely to elevated NADH. At the same time, less difference was observed in phosphopyridine nucleotides.

Most dehydrogenase reactions and an important part of metabolism which changes during growth are mediated by NAD(H) and NADP(H) as coenzymes. There are several mechanisms by which the levels of pyridine nucleotide are controlled. For example, transdehydrogenases regulate the ratio of oxidized and reduced forms, NADH kinase affects the synthesis of NADP from NAD$^+$ and NADPase will regenerate NAD$^+$ from NADP[31].

On the other hand, NADH can serve as a substrate in the formation of poly (ADP-ribose) by mammalian cell nuclei[32]. Some evidence suggests that the accumulation of poly (ADP-ribose) may be involved in termination of DNA synthesis, and especially for DNA repair[33]. A wide variety of DNA damaging agents cause a marked lowering of cellular NAD^+ levels[34].

REFERENCES

1. E.L. Jedeikin, and S. Weinhouse, Metabolism of neoplastic tissue. VI. Assay of oxidized and reduced diphosphopyridine nucleotide in normal and neoplastic tissues, J. Biol. Chem. 213:271 (1955).
2. J.P. Schwarth, J.V. Passoneau, G.S. Johnson and I. Pastan, The effect of growth conditions on NAD^+ and NADH concentrations and the NAD^+ :NADH ratio in normal and transformed fibroblasts, J. Biol. Chem. 249:4138 (1974).
3. E.L. Jacobson, and M.K. Jacobson, Pyridine nucleotide levels as a function of growth in normal and transformed 3T3 cells, Arch. Biochem. Biophys. 175:627 (1976).
4. I.L. Sun, W. MacKellar, F.L. Crane, R. Barr, W.L. Elliot, R.L. Varnold, P.F. Heinstein and D.J. Morré, Decreased NADH-oxidoreduct-ase activities as an early response in rat liver to the carcinogen 2-acetylaminofluorene, Cancer Res. 45:157 (1985).
5. F.L. Crane, I.L. Sun, M.G. Clark, C. Grebing and H. Löw, Transplasma-membrane redox systems in growth and development, Biochin. Biophys. Acta 811:233 (1985).
6. M.G. Clark, E.J. Patrick, G.S. Patten, F.L. Crane, H. Löw and C. Grebing, Evidence for the extracellular reduction of ferricyanide by rat liver, Biochem. J. 200:565 (1981).
7. F.L. Crane, H. Roberts, A.W. Linnane, and H. Löw, Transmembrane ferricyanide reduction by cells of the yeast Saccharomyces cerevisiae, J. Bionerg. Biomemb. 14:191 (1982).
8. T.A. Craig and F.L. Crane, Evidence for a transplasma membrane electron transport system in plant cells, Proc. Indiana Acad. Sci. 90:150 (1981).
9. F.L. Crane, H. Löw and M.G. Clark, Plasma membrane redox enzymes, in: "Enzymes of Biological Membranes" A. Martinosi ed., Vol. 4, Plenum, New York (1985).
10. T.L. Dormandy and Z. Zarday, The mechanism of insulin action: The immediate electrochemical effects of insulin on red cell systems, J. Physiol. 180:684 (1965).
11. M.G. Clark, E.J. Patrick and F.L. Crane, Properties and regulation of a transplasma membrane redox system in rat liver, Biochem. J. 204:795 (1982).
12. F.L. Crane, H.E. Crane, I.L. Sun, W.C. MacKellar, C. Grebing and H. Löw, Insulin control of a transplasma membrane NADH dehydrogenase in erythrocyte membranes, J. Bioenerg. Biomemb. 14:425 (1982).
13. H. Goldenberg, Plasma membrane redox activities, Biochim. Biophys. Acta 694:203 (1982).
14. D.Barnes and G. Sato, Serum-free cell culture: a unifying approach, Cell 22:649 (1980).
15. M-S. Tsao, G.H.S. Sanders and J.W. Grisham, Regulation of growth of cultured hepatic epithelial cells by transferrin, Exp. Cell Res. 171:52 (1987).
16. I.L. Sun, P. Navas, F.L. Crane, D.J. Morré and H. Löw, NADH diferric transferrin reductase in liver plasma membrane, J. Biol. Chem. 262:15915 (1987).
17. K.A.O. Ellem and G.F. Kay, Ferricyanide can replace pyruvate to stimulate growth and attachment of serum restricted human melanoma cells, Biochem. Biophys. Res. Commun. 112:183 (1983).
18. I.L. Sun, F.L. Crane, C. Grebing and H. Löw, Transmembrane redox in

control of cell growth. Stimulation of HeLa cell growth by ferricyanide and insulin, Exp. Cell Res. 156:528 (1985).

19. I.L. Sun, F.L. Crane, H. Löw and C. Grebing, Transplasma membrane redox stimulates HeLa cell growth, Biochem. Biophys. Res. Commun. 125:649 (1984).

20. W.H. Moolenaar, R.Y. Tsien, P.T. van der Saag and S.W. DeLeat, Na^+/H^+ exchange and cytoplasmic pH in the action of growth factors in human fibroblasts, Nature 304:645 (1983).

21. R.K. Mishra and H. Passow, Induction of intracellular ATP synthesis by extracellular ferricyanide in human red blood cells, J. Membr. Biol. 1:214 (1969).

22. I.L. Sun, F.L. Crane, H. Löw and C. Grebing, Inhibition of plasma membrane NADH dehydrogenase by adriamycin and related anthracycline antibiotics, J. Bioenerg. Biomemb. 16:209 (1984).

23. P.C. Sijmons, W van der Briel and H Bienfait, Cytosolic NADPH is the electron donor for extracellular ferric reduction in iron-deficient bean roots. Plant Physiol. 75:219 (1984).

24. P. Navas, I.L. Sun, D.J. Morré and F.L. Crane, Decrease of NADH in HeLa cells in the presence of transferrin or ferricyanide, Biochem. Biophys. Res. Commun. 135:110 (1986).

25. T.R. Tritton and G. Yee, The anticancer drug adriamycin can be cytotoxic without entering cells, Science 217:248 (1982).

26. I.L. Sun, P. Navas, F.L. Crane, D.J. Morré and H. Löw, Diferric transferrin reductase in the plasma membrane is inhibited by adriamycin, Biochem. Int. 14:119 (1987).

27. G. Warburg, Enzymology of cancer cells, N. Engl. J. Med. 296:486 (1977).

28. M.K. Jacobson, V. Levi, H. Juarez-Salinas, R.A. Barton and E.L. Jacobson, Effect of carcinogenic N-alkyl-N-nitroso compounds on nicotinamide adenine dinucleotide metabolism, Cancer Res. 40:1797 (1980).

29. P.W. Rankin, M.K. Jacobson, J.R. Mitchell and D.B. Busbee, Reduction of nicotinamide adenine dinucleotide levels by ultimate carcinogens in human lymphocytes, Cancer Res. 40:1803 (1980).

30. I.L. Sun, P. Navas, F.L. Crane, J.Y. Chou and H. Löw, Transplasma membrane electron transport is changed in Simian Virus 40 transformed liver cells, J. Bioenerg. Biomemb. 18:471 (1986).

31. N.O. Kaplan, Current Topics in Cellular Regulation, R.L. Levine and A. Ginsburg eds., Vol. 26, Academic Press, Hands (1985).

32. O. Hayaishi and K Ueda, Poly(ADP-ribose) and ADP-ribosylation of proteins, Ann. Rev. Biochem. 46:95 (1977).

33. M.R. Parnell, P.R. Stone and W.J.D. Whish, ADP-ribosylation of nuclear proteins, Biochem. Soc. Trans. 8:215 (1980).

34. C.J. Skidmore, M.I. Davies, P.M. Goodwin, H. Halldorsson, P. Lewis, S. Shall and A-A. Zia'ee, The involvement of poly(ADP-ribose) polymerase in the degradation of NAD caused by -radiation and N-methyl-N-nitrosourea, Eur. J. Biochem. 101:135 (1979).

REDOX CONTROL OF GLUTATHIONE AND THIOREDOXIN REDUCTASES

Juan López-Barea and Jose Antonio Bárcena

Depto. de Bioquímica y Biología Molecular (F. Veterinaria)
Universidad de Córdoba, Avda. de Medina Azahara s/n
14071 Córdoba, ESPAÑA

INTRODUCTION

Redox interconversion was discovered in the seventies, when activation of several photosynthetic enzymes by reduced thioredoxin (1), and redox control of NO_3^--, NO_2^--, and $NADP^+$-reductases were first described (2). Regulation of many enzymes by -SH/-SS- exchange has been recently reported, either by direct reaction with low Mw thiols or disulfides, or in a process catalyzed by thiol-transferases (3). Glutathione and thioredoxin reductases are two FAD-containing $NADP^+$-disulfide oxidoreductases, using NADPH to reduce the -SS- groups of glutathione and thioredoxin (1, 4, 5). Both enzymes and their reduced products connect the $NADP^+$/NADPH couple with the pools of low and high Mw thiols and disulfides; then, a general regulatory mechanism could be envisaged if the activities of such enzymes were redox controlled.

Redox inactivation of mouse glutathione reductase by NADPH was described in 1979 (6). A detailed study of the yeast enzyme interconversion led to our current idea assigning such a process to formation and disappearance of an erroneous disulfide at the GSSG-binding site of the reduced enzyme (7, 8). Interconversion under physiological conditions has been demonstrated in E. coli and S. cerevisiae (9-11). Recently, redox interconversion of thioredoxin reductase has been shown also (12, 13). In the present article we will summarize our knowledge of redox interconversion of both enzymes, their physiological significance, and the possible implications of their self-regulatory behaviour in the control of different membrane proteins by the intracellular redox state.

GLUTATHIONE REDUCTASE INTERCONVERSION

As shown in Table 1, the activity of mouse, yeast, E. coli, and horse glutathione reductases decrease rapidly in the presence of the reduced but not the oxidized forms of several redox couples, indicating that the process is due to reduction rather than to conformational changes which should also be induced by the oxidized compounds (6-9, 14). Dithionite and dithiothreitol (DTT) also promote significant activity decrease. The loss of activity under reducing conditions is assigned to a covalent modification of the reduced enzyme (6-10): the mouse and yeast enzymes remain inactive even after removal of the excess NADPH (6, 7), and inactivation depends on time and temperature of incubation (7, 9, 14). Inhibition by NADPH-X, a product of

349

Table 1. Redox Inactivation Of Glutathione Reductase

Compound Added	Glutathione Reductase Activity From			
	Mouse Pure Enzyme[a]	Yeast Pure Enzyme[b]	E. coli Pure Enzyme[c]	E. coli Perm. Cells[d]
None	100	100	100	100
NADPH	5 |0.3|[e]	4 |0.1|	0 |0.02|	18 |0.02|
NADP+	102 |0.3|	121 |0.1|	107 |0.02|	102 |0.02|
NADH	6 |0.3|	2 |0.1|	6 |0.02|	90 |0.02|
NAD+	72 |0.3|	95 |0.1|	98 |0.02|	102 |0.02|
GSH	–	48 |1.0|	22 |2.00|	50 |2.00|
GSSG	–	98 |0.5|	90 |1.00|	86 |1.00|
Ferricyanide	–	69 |1.0|	98 |1.00|	–
Dithionite	–	77 |18|	67 |25.0|	12 |25.0|
DTT	–	–	37 |1.00|	54 |1.00|

[a] Activities after 3 h at 37ºC; 100% was 1.2 U/ml (taken from reference 6).
[b] After 1 h at 0ºC; 100% was 9 U/ml (taken from reference 7).
[c] After 2 h at 20ºC: 100% was 0.24 U/ml (taken from reference 9).
[d] After 2 h at 37ºC; 100% was 0.39 U/ml (taken from reference 10).
[e] Final mM concentration are shown between bars.

NADPH degradation, has been ruled out (6). Redox inactivation of the E. coli enzyme has a E_{act} = 13.4 Kcal/mol (9); the inactivation is fast: the pure E. coli enzyme has 2 min half-life at 37 ºC with 20 µM NADPH (9).

Glutathione reductase inactivation depends on reductant and enzyme concentration: a 50% inactivation of the pure mouse, yeast, and E. coli enzymes is promoted by 1.6, 0.5, and 0.3 µM NADPH, respectively (6, 7, 9), while 45 µM NADH is needed for 50% inactivation of the E. coli enzyme (9), reflecting its lower affinity for the later compound. The NADPH concentrations required for 50% inactivation are similar to those of subunits and well below Km for NADPH, suggesting that one NADPH molecule is enough to inactivate each subunit (7, 9). Inactivation decreases as the enzyme concentration increases, excluding aggregation of the reduced enzyme as a cause; actually, the NADPH-inactivated mouse and E. coli enzymes show the same Mw as their active forms (6, 9). The pure yeast and E. coli enzymes are fully inactivated by NADPH at pH below 5.5 or over 7.5, while only partial inactivation is observed between such pH values; two pKa, around 5 and 7, have been determined for inactivation of such enzymes, suggesting that two ionizable groups are involved in the process (7, 9).

Protection and Reactivation

Since the intracellular NADPH concentrations are higher than those needed for GSSG-reductase inactivation, the enzyme should be continuously inactive in vivo, unless it were protected and reactivated. Table 2 shows the protection from NADPH-inactivation by different compounds: GSSG (physiological substrate of the enzyme) and FeCy (ferricyanide, substrate of its diaphorase activity, 15), are highly effective, probably by reoxidizing the enzyme before the inactivation takes place (6-9); GSH and DTT protect the enzyme also very efficiently. NADP+ and NAD+ at high concentrations slightly protect the pure enzymes (14), probably by interaction with the NADPH-binding site preventing the enzyme from reduction (6-9), as does 2´,5´-ADP (6).

Table 2. Protection Of Glutathione Reductase From Redox Inactivation

Compound Added	Glutathione Reductase Activity From			
	Mouse Pure Enzyme[a]	Yeast Pure Enzyme[b]	E. coli Pure Enzyme[c]	E. coli Perm. Cells[d]
None	100	100	100	100
NADPH	9 |0.3|[e]	7 |0.01|	0 |0.02|	20 |0.02|
NADPH, NADP$^+$	15 |0.3, 1|	9 |0.01, 1|	7 |0.02, 1|	39 |0.02, 1|
NADPH, NAD$^+$	-	10 |0.01, 1|	0 |0.02, 1|	52 |0.02, 1|
NADPH, FeCy	-	67 |0.01, 1|	75 |0.02, 1|	-
NADPH, GSSG	81 |0.3, 1|	88 |0.01, 1|	106 |0.02, 1|	91 |0.02, 1|
NADPH, GSH	19 |0.3, 1|	60 |0.01, 1|	42 |0.02, 2|	52 |0.02, 2|
NADPH, DTT	-	74 |0.01, 1|	38 |0.02, 1|	85 |0.02, 1|

[a] Activities after 7 h at 0ºC; 100% was 2.1 U/ml (taken from reference 6).
[b] After 1 h at 0ºC; 100% was 7 U/ml (taken from reference 7).
[c] After 2 h at 20ºC; 100% was 0.33 U/ml (taken from reference 9).
[d] After 1 h at 37ºC; 100% was 0.47 U/ml (taken from reference 10).
[e] Final mM concentration are shown between bars.

Table 3 shows the reactivation of inactive GSSG-reductase: the oxidized compounds, GSSG, FeCy, and DPIP (dichloroindophenol), are highly effective in reactivation, as the thiols GSH and DTT; no reactivation is promoted by NADP$^+$ or NAD$^+$ (6-9). At very short times DTT promotes full reactivation of the yeast enzyme, that recovers more than 100% activity of the initial control (7). The dependence on time and temperature indicate that reactivation requires a chemical reaction taking place in the inactive enzyme (6-9, 14). Reactivation by GSSG is maximal at 30-37 ºC, favoured at acidic pH values (7, 9, 14), and occurs even after NADPH removal, indicating that GSSG effect is not mediated by GSH (7). The effect of several thiols has been studied: 0.2 mM 2,3-dimercaptopropanol, 0.6 mM DTT, 1.9 mM GSH and 7 mM 2-mercaptoethanol promote 50% reactivation of the yeast enzyme (8).

Figure 1-A shows the protection by different glutathione concentrations from NADPH-inactivation of yeast and E. coli glutathione reductase. Oxidized glutathione fully protects at concentrations equimolecular with NADPH, probably by oxidizing this compound, but still protects at lower concentrations (7, 9, 14). The reduced form of glutathione only protects at concentrations much higher than GSSG, although it does not reach such a high effect; in fact, GSH protection decreases with time (9) and very low GSH concentrations potentiate NADPH-inactivation (14). Figure 1-B shows the reactivation promoted by glutathione on NADPH-inactivated E. coli glutathione reductase. Oxidized glutathione is again much more effective and specific than GSH: a significant reactivation is observed at 10 μM GSSG, and maximal activity is promoted by 50 μM GSSG, its effect diminishing at higher concentrations. On the contrary, GSH requires much higher concentrations to produce its maximum effect, and only a very limited effect is observed in the millimolar range (9). The reported intracellular GSSG and GSH concentrations are in the order of 10-40 μM and 1-10 mM respectively (16), making it clear that GSSG, rather than GSH, plays a physiologically significant role in protection and reactivation of glutathione reductase. As Figure 1 shows, small changes in GSSG concentration, but not in GSH concentration, could provide a signal for modulating this enzymatic activity (9).

Table 3. Reactivation of Pure Glutathione Reductase

Compound Added	Glutathione Reductase Activity[a] From					
	Mouse[b]		Yeast[c]		E. coli[d]	
None	3		2		0.4	
NADP+	9	|2|[e]	-		0	|1|
NAD+	-		-		0	|1|
DPIP	34		-		-	
FeCy	48	|2|	71	|5|	36	|1|
GSSG	65	|2|	77	|5|	41	|1|
GSH	-		80	|5|	53	|1|
DTT	-		72	|5|	3	|1|

[a] After inactivation with 20–300 µM NADPH, the enzymes were incubated with the compounds shown, the activities assayed and expressed as percentage.
[b] After 8 h at 37ºC; 100% activity was 2.1 U/ml (taken from reference 6).
[c] After 3 h at 37ºC; 100% activity was 10.2 U/ml (taken from reference 7).
[d] After 9.5 h at 37ºC; 100% activity was 0.32 U/ml (taken from reference 9).
[e] Final mM concentration are shown between bars.

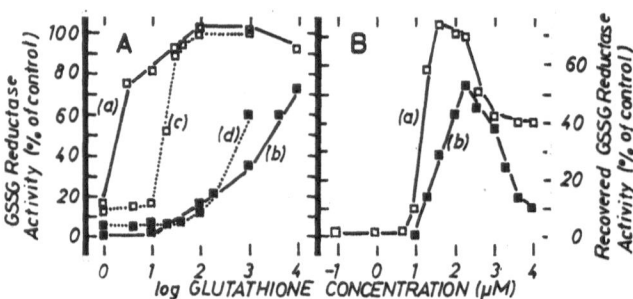

Fig. 1.   Protection and reactivation of pure enzymes by glutathione. A, Protection: (a, b) yeast enzyme plus GSSG or GSH, respectively; (c, d) E. coli enzyme plus GSSG or GSH, respectively. B, Reactivation of the E. coli enzyme: (a) plus GSSG; (b) plus GSH. (refs. 5 and 7)

Mechanism Of Redox Interconversion

The structure and catallytic mechanism of human erythrocyte glutathione reductase have been established by X-ray crystallography (17, 18). Each subunit of the dimeric enzyme has a binding site for NADPH and another for GSSG, separated by the isoalloxazine ring of FAD. A redox-active disulfide group, present at the GSSG-binding site near FAD, is transiently reduced to dithiol during electron transfer from NADPH to GSSG via FAD (17, 18). The reduced enzyme shows a characteristic spectrum with higher absorbance than its oxidized form between 500 and 700 nm, and maximum absorbance increase at 530 nm, attributed to the formation of a charge-transfer complex between the closest active-site thiolate and the oxidized FAD (18, 19). Besides its main catalytic activity, glutathione reductase shows a diaphorase activity which catalyses reduction of ferricyanide by NADPH (15), and probably indicates the functional state of the NADPH-binding site (8).

The maintenance of full diaphorase activity during glutathione reductase inactivation, and the inactivation of GSSG-reductase by reduced methyl viologen (donor for a new enzymatic assay), indicate that inactivation does not take place at the NADPH-binding site, but at the GSSG-binding site (8, 20). Figure 2-A shows the evolution of the visible absorption spectrum during inactivation: immediately after NADPH addition the enzyme displays the reduced spectrum, followed by a very marked decrease of the absorbance region due to the charge-transfer complex (8). Figure 2-B shows the parallelism between the changes in enzymatic activity along interconversion and 530 nm absorbance: NADPH-inactivation parallels a decrease of 530 nm absorbance, while GSH-reactivation is associated with full recovery of the 530 nm absorbance; the subsequent decline in 530 nm absorbance could be attributed to formation of mixed disulfides with GSH (8). Such results indicate that the charge-transfer complex between proximal thiolate and FAD disappears gradually during inactivation (8).

NADPH-inactivation of the yeast enzyme requires metal cations: it is fully protected by 10 μM EDTA or DETAPAC (11), and highly stimulated by Zn^{2+} and Cd^{2+}. Both the redox inactivation and the reactivation promoted by GSSG are similar under aerobic or anaerobic conditions, * thus excluding participation of oxygen or oxygen radicals in the inactivation process. Protection and reactivation by GSSG, GSH, thiols and $NADP^+$, the constant diaphorase activity during interconversion of glutathione reductase, and the spectral changes previously described, have led us to attribute redox interconversion to formation and disappearance of an erroneous disulfide bridge between the proximal thiolate formed at the GSSG-binding site of the reduced enzyme and another close cysteine (7-9). Binding to a different type of residue, or blocking of the thiolate by metal cations can be alternatively proposed.

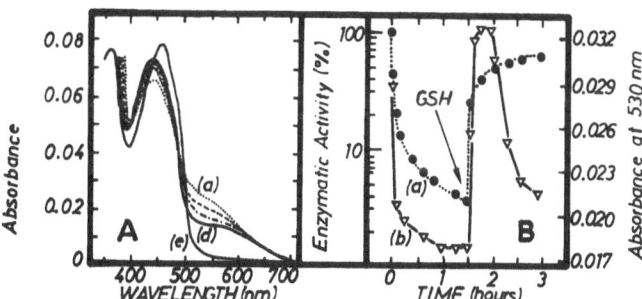

Fig. 2. Spectral changes during interconversion of yeast GSSG-reductase. A, Visible spectra along NADPH-inactivation: (a-d) spectra after 2, 6, 10, and 40 min after NADPH addition; (e) spectrum of the oxidized enzyme. B, Parallelism between enzymatic activity and 530 nm absorbance: (a) enzymatic activity during inactivation by NADPH and reactivation by GSH; (b) changes in 530 nm absorbance. (ref. 8)

* C. García-Alfonso, E. Martínez Galisteo, J. A. Bárcena, and J. López-Barea, unpublished results.

Physiological Significance Of Glutathione Reductase Interconversion

Table 1 shows the in situ inactivation of GSSG-reductase in permeabilized E. coli cells: NADPH, dithionite, GSH, and less efficiently NADH, inactivate the enzyme; half-life for inactivation with 20 µM NADPH is only 30
seconds at 37 ºC, although after 10 min the enzyme progressively reactivates
due to oxidation of NADPH by cellular oxidases (10). Inactivation depends on
temperature and reductant concentration, the NADH concentrations required
for 50% effect are 350-fold higher than those of NADPH (10). Table 2 shows
the protection by several compounds against NADPH-inactivation in permeabilized E. coli cells: GSSG and DTT fully protect , while GSH protect partially; at difference with the pure enzyme, NADP$^+$ and NAD$^+$ also protect significantly (10). The enzyme from permeabilized E. coli cells is reactivated by
the same compounds positive in Table 3 for the pure enzyme (10).

Figure 3-A shows the in vivo inactivation of the E. coli GSSG-reductase
in intact cells incubated with permeable substrates which upon intracellular
oxidation increase the NAD(P)H concentration: malate produces the maximum
inactivation, followed by isocitrate and, at much higher concentrations,
lactate (10). Figure 3-B shows the time-course of inactivation by malate,
and the full reactivation promoted by diamide, which chemically oxidizes GSH
(21) and also protects the enzyme from malate inactivation (10). Tert-butyl
or cumene hydroperoxides, which enzymatically oxidize GSH (21), yield a 100-
200% increase in the GSSG-reductase activity of intact E. coli cells (10).

Interconversion of GSSG-reductase has been demostrated also in cell-free
extracts, and permeabilized, or intact S. cerevisiae cells. Figure 3-C shows
the fast and complete inactivation of the yeast enzyme in cell-free extracts
treated with 0.2 mM NADPH at 30 ºC, and the full reactivation produced by 1
mM GSSG (11). Several oxidizable substrates inactivate the yeast enzyme when
added to cell-free extracts primed with NAD(P)$^+$: gluconate and glucose-6-P
are the most efficient, followed by isocitrate and malate, while citrate or
glucose do not inactivate; the effect promoted by each substrate correlates
with the specific activity of the corrresponding dehydrogenase (11); similar
conclusions have been reached in situ with permeabilized yeast cells. Glutathione reductase from intact yeast cells is inactivated after 2 hours growth

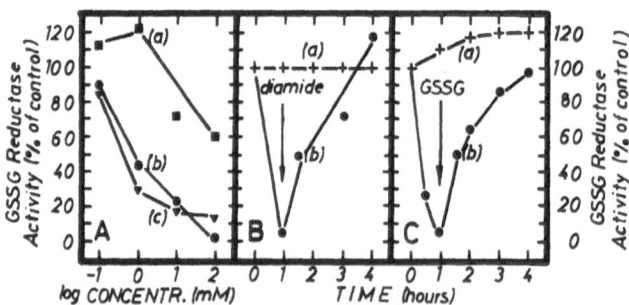

Fig. 3.   Physiological significance of GSSG-reductase interconversion. A.
          Activity of intact E. coli cells enzyme after 2 h with: (a) lac-
          tate; (b) malate; (c) isocitrate. B, Interconversion of intact E.
          coli cells enzyme: (a) control; (b) plus malate, then diamide. C,
          Interconversion of the enzyme of yeast cell-free extracts: (a)
          control; (b) plus NADPH, then GSSG. (references. 10 and 11)

with gluconolactone as carbon source, [**] with a parallel increase in the $|NADPH|/|NADP^+|$ ratio, thus confirming that the inactivation is due to a more reducing intracellular redox state.

Glutathione reductase and the NADPH-regenerating systems are central in antioxidant defense and tighly coordinated (23). In normal conditions (with high NADPH and low GSSG concentrations) GSSG-reductase would be partially inactive: several cellular functions are activated by GSSG, making necessary the maintenance of higher than threshold GSSG concentrations, incompatible with a fully active glutathione reductase (3, 4, 22). In such circumstances, the activity of the pentose-phosphate cycle would be low, since reoxidation of NADPH to $NADP^+$ by GSSG-reductase controls glucose-6-P dehydrogenase (23). In oxidative stress, however, the direct quenching by GSH of reactive oxygen species and the glutathione peroxidase activity would rapidly increase the GSSG concentration, reactivating GSSG-reductase, what would subsequently increase the $NADP^+$ concentration and deinhibit the pentose-phosphate pathway, thus producing more NADPH to counteract the oxidative stress (22).

THIOREDOXIN REDUCTASE INTERCONVERSION

Table 4 shows that the activity of pure E. coli or calf-liver thioredoxin reductases decrease drastically upon incubation with NADPH, or less markedly NADH, while $NADP^+$, NAD^+ and E. coli thioredoxin promote a certain stimulatory effect, thus suggesting that the enzymes are modulated by redox interconversion (12, 13). The inactivation is very fast, the half-life for inactivation of the mammalian enzyme is about 2 minutes at 30ºC in the presence of 0.15 mM NADPH (13). The redox inactivation of thioredoxin reductase depends also on reductant concentration and pH (13).

Table 4 shows also the complete protection from NADPH-inactivation induced by E. coli thioredoxin (which in fact yields higher activity than that of untreated control); a partial protection is afforded also by $NADP^+$ and NAD^+, and by GSSG and CDE (cystine dimethyl ester); cystine or cystamine do not produce any significant effect, while lipoate potentiates the redox inactivation (12, 13). Finally, the NADPH-inactivated enzymes are fully reactivated after 30 minutes incubation with 0.1 mM E. coli thioredoxin, while partial reactivation is promoted by $NADP^+$ or GSSG; no significant effect is produced by NAD^+, CDE, or cystamine, while again lipoate further inactivates the enzyme beyond the level due to NADPH (12, 13).

The redox inactivation is fully protected by EDTA, thus implicating some metal cations in the process (12, 13). NADPH-inactivation of thioredoxin reductase does not require oxygen, as shown by the full inactivation promoted in anaerobiosis, thus excluding again the participation of oxygen or oxygen radicals in the inactivation process (12).

The structural and functional similarities between glutathione and thioredoxin reductases (5, 24) allow us to propose analogies between their redox interconversion mechanisms. The complete protection against NADPH-inactivation of thioredoxin reductase and the full reactivation promoted by E. coli thioredoxin suggest the involvement of thiol-disulfide exchange reactions; by analogy with GSSG-reductase (8), an erroneous disulfide near the active site could be suggested as cause of interconversion (12). The location of the erroneous disulfide at (or close to) the active site is supported by the full protection and reactivation afforded by thioredoxin (physiological substrate) as compared to other disulfides. A unspecific effect of thioredoxin on another disulfide bridge can not be excluded at the moment (12, 13).

[**] J. Florindo, J. Peinado, and J. López-Barea, unpublished results.

Table 4. Redox Interconversion Of Thioredoxin Reductase

Compound Added	Concentration (mM)	Thioredoxin Reductase Activity	
		Calf-Liver[a]	E. coli[b]
Inactivation And Protection:			
None	-	100	100
NADPH	0.24	45	25
NADP$^+$	1	113	110
E. coli OxTx	0.1	115	138
NADPH, NADP$^+$	0.24, 1	51	94
NADPH, E. coli OxTx	0.24, 0.1	105	125
Reactivation:			
None[c]	-	45	23
NADP$^+$	1	46	43
E. coli OxTx	0.1	100	101
GSSG	2.5	36	42

[a] After 45 min at 30ºC: 100% activity was 1.1 U/mg (taken from ref. 12).
[b] After 30 min at 30ºC; 100% activity was 7 U/mg (taken from ref. 12).
[c] After inactivation for 30 min at 30 ºC with 0.24 mM NADPH, compounds were added, the activities assayed 30 min later, and expressed as percentage.

POSSIBLE IMPLICATIONS OF GLUTATHIONE AND THIOREDOXIN REDUCTASE INTERCONVERSION IN REDOX CONTROL OF MEMBRANE PROTEINS

In addition of its role as universal reductant in biosynthetic processes, NADPH is the main source of electrons counteracting oxidative stres,
which affect the -SH/-SS- status of different molecules (22). Both glutathione and thioredoxin reductases connect the NADP$^+$/NADPH couple with the
pools of low and high Mw thiols and disulfides (1, 4, 5), interconnected
themselves by different proteins displaying thioltransferase activity:
protein-disulfide isomerase, low Mw thiol-transferases, glutaredoxin or even
thioredoxin; some of them are glycoproteins (25, 26) located at or near
membranes (25, 27).

The activity of several membrane proteins is regulated by thiol/disul-
fide exchange. Thus, the fat cell hexose transport system is activated by
low-Mw disulfides and other oxidizing compounds (28), while adenylate and
guanylate cyclases are activated by low Mw thiols and inactivated by disulfides (3). The beta-adrenergic receptors are activated by low-Mw thiols
reducing an intramolecular disulfide bridge; a similar mechanism is postulated for activation of receptors for glucagon, opiate, interleukin-2 and -3,
and LH/hCG (29). In neuroblastoma cells, it has been suggested that cluster-
ing of opiate receptor involves a thiol-disulfide exchange, independent of
binding (30). GSH and other small thiols modulate ADP-induced platelet aggregation, probably by formation of mixed disulfides with essential -SH
groups of the ADP receptor (31). Reduction of a disulfide at the binding
site of the acetylcholine receptor of Torpedo californica electric tissue
alters both the binding selectivity and the transduction of binding into
cation specific channnel opening (32). The thiol redox status of IGF-I
receptor is an important determinant of its conformation, and therefore of
the biological responses to IGF-I (33). Likewise, the disulfides of insulin
receptor/kinase class I play multiple roles in the function of the receptor-
kinase activity (34).

The opening of anion (Cl⁻) conductance in tubulovesicles and apical membrane of gastric parietal cells, necessary for operation of the gastric H,K-pump responsible for acid secretion, is triggered by -SS- crosslinking between H,K-ATPase subunits (35). Oxidizing perturbations of the -SH/-SS-status, by diamide treatment or GSSG-reductase inhibition, stimulate acid secretion in isolated canine parietal cells, even in the absence of stimulants (36). The inner mitochondrial membrane NAD(P)-transhydrogenase contains a dithiol group essential for catalysis and for proton translocation (37).

Redox interconversion of glutathione and thioredoxin reductases in response to changes in the NADP⁺/NADPH could alter the redox status of low or high molecular weight thiols and disulfides. In consecuence, the autoregulation of such two cytosolic enzymes could affect not only the activity of many soluble enzymes, but also that of several enzymes, receptors, channels, and transport systems located at the plasma membrane regulated by thiol-disulfide exchange reactions.

ACKNOWLEDGEMENTS

This work was carried out with grants number 4079-79, 2965-83, and PB86-0146 from the CAICYT and CICYT, España.

REFERENCES

1. B. B. Buchanan, Role of light in the regulation of chloroplast enzymes, Annu Rev Plant Physiol. 31:341 (1980).
2. M. G. Guerrero, J. M. Vega, and M. Losada, The assimilatory NO₃-reducing system and its regulation, Annu Rev Plant Physiol. 32:169 (1981).
3. D. M. Ziegler, Role of reversible oxidation-reduction of enzyme thiols-disulfides in metabolic regulation, Annu Rev Biochem. 54:305 (1985).
4. A. Meister, and M. E. Anderson, Glutathione, Annu Rev Biochem. 52:711 (1983).
5. A. Holmgren, Thioredoxin, Annu Rev Biochem. 54:237 (1985).
6. J. López-Barea, and C. Y. Lee, Mouse-liver glutathione reductase. Purification, kinetics, and regulation, Eur J Biochem. 98:487 (1979).
7. M. C. Pinto, A. M. Mata, and J. López-Barea, Reversible inactivation of S. cerevisiae glutathione reductase under reducing conditions, Arch Biochem Biophys. 228:1 (1984).
8. M. C. Pinto, A. M. Mata, and J. López-Barea, The redox interconversion mechanism of S. cerevisiae glutathione reductase, Eur J Biochem. 151:275 (1985).
9. A. M. Mata, M. C. Pinto, and J. López-Barea, Redox interconversion of glutathione reductase from E. coli. A study with pure enzyme and cell-free extracts, Mol Cell Biochem. 67:65 (1985).
10. A. M. Mata, M. C. Pinto, and J. López-Barea, Redox interconversion of E. coli glutathione reductase. A study with permeabilized and intact cells, Mol Cell Biochem. 68:121 (1985).
11. J. Peinado, A. Llobell, and J. López-Barea, Effect of EDTA and oxidizable substrates on the redox interconversion of glutathione reductase in S. cerevisiae cell-free extracts, submit. to Biochim Biophys Acta.
12. J. A. Barcena, E. Martínez-Galisteo, C. Gª-Alfonso, and J. López-Barea, NADPH and oxidized thioredoxin mediate redox interconversion of calf-liver and E. coli thioredoxin reductase, submit. to Biochem J.
13. E. Martínez-Galisteo, "Ph. D. Dissertation," Universidad de Córdoba, Spain (1987).
14. C. García-Alfonso, "Ph. D. Dissertation," Universidad de Córdoba, Spain (1987).
15. C. S. Tsai, and J. R. P. Godin, Multifunctional activities of yeast glutathione reductase, Int J Biochem. 19:337 (1987).

16. T. P. M. Akerboom, and H. Sies, Assay of glutathione, glutathione disul-
    fide, and glutathione mixed disulfides in biological samples, Methods
    Enzymol. 77:373 (1981).
17. P. A. Karplus, and G. E. Schulz, Refined structure of glutathione reduc-
    tase at 1.54 A resolution, J Mol Biol. 195:701 (1987).
18. E. F. Pai, and G. E. Schulz, The catalytic mechanism of glutathione re-
    ductase as derived from X-ray diffraction analyses of reaction inter-
    mediates, J Biol Chem. 258:1752 (1983).
19. V. Massey, and S. Ghisla, Role of charge-transfer interactions in flavo-
    protein catalysis, Ann N Y Acad Sci. 227:446 (1974).
20. A. Llobell, V. M. Fernández, and J. López-Barea, Electron transfer bet-
    ween reduced methyl viologen and GSSG: a new assay of S. cerevisiae
    glutathione reductase, Arch Biochem Biophys. 250:373 (1986).
21. J. L. Plummer, B. R. Smith, H. Sies, and J. R. Bend, Chemical depletion
    of glutathione in vivo, Methods Enzymol 77:50 (1981).
22. H. Sies, Biochemistry of oxidative stress, Angew Chem. 25:1058 (1986).
23. A. Llobell, A. López-Ruiz, J. Peinado, and J. López-Barea, Glutathione
    reductase directly mediates the stimulation of yeast glucose-6-phos-
    phate dehydrogenase by GSSG, Biochem J. 249:293 (1988).
24. M. E. O'Donnell, and C. H. Williams Jr., Proton stoichiometry in the re-
    duction of FAD and disulfide in Escherichia coli thioredoxin reduct-
    ase. Evidence for a base at the active site, J Biol Chem. 258:13795
    (1983).
25. N. Lambert, and R. B. Freedman, The latency of rat liver microsomal pro-
    tein disulphide isomerase, Biochem J. 228:635 (1985).
26. K. Axelsson, S. Eriksson, and B. Mannervik, Purification and characteri-
    zation of cytoplasmic thioltransferase (glutathione:disulfide oxido-
    reductase) from rat liver, Biochemistry 17:2978 (1978).
27. B. Rozell, H. A. Hansson, M. Luthman, and A. Holmgren, Immunohistochemi-
    cal localization of thioredoxin and thioredoxin reductase in adult
    rats, Eur J Cell Biol. 38:79 (1985).
28. M. P. Czech, Differential effects of sulfhydryl reagents on activation
    and deactivation of the fat cell hexose transport system, J Biol
    Chem. 251:1164 (1976).
29. C. C. Malbon, S. T. George, and C. P. Moxham, Intramolecular disulfide
    bridges: avenues to receptor activation?, Trends Biochem Sci. 12:172
    (1987).
30. E. Hazum, K.- J. Chang, and P. Cuatrecasas, Role of disulfide and
    sulphydryl groups in clustering of enkephalin receptors in neuroblas-
    toma cells, Nature. 282:626 (1979).
31. G. Thomas, V. A. Skrinska, and F. V. Lukas, The influence of glutathione
    and other thiols on human platelet aggregation, Thromb Res. 44:859
    (1986).
32. P. N. Kao, and A. Karlin, Acetylcholine receptor binding site contains a
    disulfide cross-link between adjacent half cysteinyl residues, J Biol
    Chem. 261:8085 (1986).
33. H. A. Jonas, and L. E. Harrison, Disulfide reduction alters the immuno-
    reactivity and increases affinity of insulin-like growth-factor-I
    receptors in human placenta, Biochem J. 236:417 (1986).
34. L. J. Pyke, A. T. Eakes, and E. G. Krebs, Characterization of affinity-
    purified insulin receptor/kinase. Effect of dithiothreitol on recep-
    tor/kinase function, J Biol Chem. 261:3782 (1986).
35. N. Takeguchi, and Y. Yamazaki, Disulfide cross-linking of H,K-ATPase
    opens Cl⁻ conductance, triggering proton uptake in gastric vesicles.
    Studies with specific inhibitors, J Biol Chem. 261:2560 (1986).
36. C. E. Olson, A. H. Soll, and N. Kaplowitz, Modulating effect of thiol-
    disulfide status on (^{14}C)aminopyrine accumulation in the isolated
    parietal cell, J Biol Chem. 260:8020 (1985).
37. B. Persson, and J. Rydström, Evidence for a role of a vicinal thiol in
    catalysis and proton pumping in mitochondrial nicotinamide nucleotide
    transhydrogenase, Biochem Biophys Res Commun. 142:573 (1987).

CARDIAC Ca^{2+} CHANNELS AND SARCOLEMMA REDOX

Michael G. Clark, Stephen Rattigan, Perry J.F. Cleland, Stephen J. Edwards and Aidan G.M. Davison

Department of Biochemistry, University of Tasmania, Hobart, Australia

INTRODUCTION

Alpha-1-adrenergic receptors have been identified in both animal and human cardiac tissue (Schumann, 1978; Clark and Patten, 1984a; Bruckner et al., 1985) and several alpha adrenergic mediated events have been reported. Alpha agonists cause increases in inotropy of the heart (Bruckner et al., 1985) although this is not exclusive to alpha agonists as beta agonists also cause increases in inotropy and chronotropy. Alpha agonists also cause changes in metabolism such as activation of phosphofructokinase (Clark and Patten, 1984a) and increased glucose transport and uptake (Clark and Patten, 1984b; Rattigan et al., 1986; Abel et al., 1987), leading to increased glycolysis in the heart. Alpha stimulation has also been postulated to control hypertrophy of the heart (Simpson et al., 1986) and alpha agonists have been shown to cause the expression of c-myc and c-fos genes (Starksen et al., 1986; Barka et al., 1987).

The mechanism by which alpha adrenergic agonists mediate these events is not precisely known, but increases in myoplasmic Ca^{2+} concentration may be involved (Clark and Patten, 1984a; Rattigan et al., 1986). The increased Ca^{2+} appears to be due to the involvement of Ca^{2+} channels in the sarcolemma that are sensitive to Ca^{2+} channel blockers such as nifedipine (Bruckner et al., 1985; Clark and Rattigan, 1986). Beta agonists also regulate the same Ca^{2+} channels via phosphorylation of membrane components (e.g. calciductin) by the cAMP-dependent protein kinase (Rinaldi et al., 1982). However it is apparent that alpha adrenergic stimulation does not involve cAMP or cAMP-dependent kinases (Clark and Patten, 1984a; Bruckner et al., 1985).

In other tissues it has become apparent that phosphoinositide (PI) hydrolysis occurs as a result of alpha-adrenergic stimulation (Berridge, 1982; Exton, 1985). Hydrolysis of PI leads to the formation of two second messengers, inositol 1,4,5 trisphosphate (IP$_3$) and 1,2 diacyl-glycerol (DAG). IP$_3$ has been shown to cause release of Ca^{2+} from intra-cellular stores and DAG causes the activation of protein kinase C (Berridge, 1982; Exton, 1985). As in the adenylate cyclase system there is evidence that a G-protein couples the alpha receptor to phospholipase C, the enzyme which hydrolyses PI. In some cells this G-protein appears to resemble the G$_i$ subunit of the adenylate cyclase system in terms of

359

its sensitivity to pertussis toxin but is otherwise unique compared with G_i (Berridge, 1987).

In cardiac tissue PI hydrolysis and the formation of inositol phosphates in response to alpha stimulation has been shown in isolated duck, guinea pig and rat ventricular myocytes (Brown and Jones, 1987; Leung et al., 1986; Brown et al., 1985) and perfused rat heart (Woodcock et al., 1987). However production of DAG and activation of protein kinase C (PKC) have yet to be shown. It is unknown whether a G-protein is involved in the PI hydrolysis in the heart, but PI hydrolysis is not sensitive to pertussis toxin (Schmitz et al., 1987). A question also remains as to whether IP_3 can directly activate the sarcolemmal Ca^{2+} channels. Thus the mechanism of Ca^{2+} channel activation remains to be elucidated.

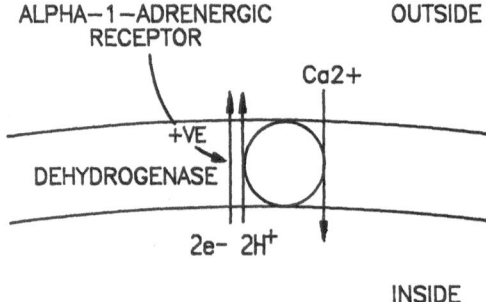

Figure 1. Proposed relationship between alpha-adrenergic effects on trans-sarcolemmal electron efflux and uptake of calcium through calcium channels in heart cells.

Transmembrane redox systems are present in all cells. Although no efinitive role for this system has yet been established it has been shown that such systems are related to cell growth, facilitation of iron uptake and defense against bacteria, each of which involve specific ion movements (Crane et al., 1985). In the heart trans-sarcolemmal electron efflux, measured as extracellular ferricyanide reduction in perfused rat heart, is increased in response to alpha adrenergic stimulation (Low et al., 1984, 1985). Time and dose relationships indicate a close association between ferricyanide reduction and changes in contractility in the rat heart (Low, 1985; Clark and Rattigan, 1986). As changes in contractility reflect levels of myoplasmic Ca^{2+} concentration it has been postulated (Fig. 1) that the increase in reduction of extracellular ferricyanide results from an increase in electron efflux that is closely coupled to Ca^{2+} influx through sarcolemmal Ca^{2+} channels (Crane et al., 1985; Clark and Rattigan, 1986). Since NADH has now been identified as one of the probable intracellular donors for the trans-plasma membrane redox system (Navas et al., 1986) it appears likely that electron efflux is accompanied by proton efflux (Fig. 1).

Therefore in the present study the relationship of the PI signal transduction system and trans-sarcolemmal electron efflux in activating the sarcolemmal Ca^{2+} channels was investigated. Three possibilities were considered:

(1) alpha-adrenergic stimulation that causes PI hydrolysis, DAG formation and PKC activation causes the phosphorylation of proteins that increase trans-sarcolemmal electron efflux and opening of Ca^{2+} channels.

(2) alpha-adrenergic stimulation that causes PI hydrolysis and IP_3 formation causes opening of Ca^{2+} channels and stimulation of trans-sarcolemmal electron efflux mediated by IP_3.

(3) alpha-adrenergic stimulation leads directly to increased trans-sarcolemmal electron efflux that opens Ca^{2+} channels without the involvement of PI hydrolysis.

These possibilities were investigated by using synthetic diacylglycerol and the phorbol ester, PMA, activators of protein kinase C (Berridge, 1982); compound C48/80 and neomycin, inhibitors of phospholipase C (Bronner et al., 1987; Downes and Michell, 1981; Whipps et al., 1987; Slivka and Insel, 1987); aluminium tetrafluoride, an activator of G-proteins involved in PI hydrolysis (Blackmore and Exton, 1986).

MATERIALS AND METHODS

Materials. L-phenylephrine, L-isoproterenol, DL-propranolol, phorbol-12-myristate-13-acetate (PMA), 4α-phorbol 12,13-didecanoate (PDD), A23187, bathophenanthroline disulphonate, compound 48/80 and neomycin were obtained from Sigma. Diacylglycerol (1-oleoyl-2-acetyl glycerol) was obtained from either Serdary Research Laboratories or Sigma.

Heart Perfusion. Male Hooded Wistar rats (180-200 g) that had been fed ad libitum were used. Hearts were removed from the anaesthetized animals (pentobarbitone) and perfused in the Langendorff manner at 37°C with Krebs-Henseleit bicarbonate buffer pH 7.4 equilibrated against O_2 + CO_2 (95:5) and containing 1.25 mM $CaCl_2$, 5 mM glucose and 0.5 mM sodium ferricyanide. After an equilibration period of 10 min experiments were commenced with perfusion being carried out in a non-recirculating manner at a constant perfusion pressure of 80 cm water. Agents were infused directly above the heart at a rate of 1/100th the perfusate flow rate.

Contractility was determined by the tension developed between two hooks placed 8 mm apart on the ventricle wall. Tension was measured by a isometric force transducer and initial (unstimulated) tension was set at 2.0 g peak to peak.

During experiments perfusate from the heart was collected into a fraction collector set on time mode (26 sec/fraction). Normally 10 fractions were collected before the agent was added (control rate) and then fractions were collected for a further 5 min after addition of the agent being investigated.

Assays. Samples of perfusate were taken and assayed for ferrocyanide using bathophenanthroline disulphonate (Avron and Shavit 1963). The rate of ferricyanide reduction (nmol/min/g wet wt. of heart) was calculated according to the following equation:

$$\text{Rate} = \frac{[\text{ferrocyanide}](\mu M) \times \text{perfusate flow rate (ml/min}}{\text{wt. of heart (g)}}$$

RESULTS

To investigate whether PKC was involved with increases in trans-sarcolemmal electron efflux and activation of Ca^{2+} channels, activators of PKC were added to the perfused rat heart. The effect of these activators on ferricyanide reduction and tension development are shown in Table 1.

TABLE 1. Effect of PKC activators on ferricyanide reduction and tension development in the perfused rat heart. Mean ± S.E.M. (n).

Addition	Ferricyanide Reduction (% of preaddition rate)$^\Delta$		Increase in Tension (%)
Saline	92 ± 6	(5)	0
Phenylephrine (20 μM) + Propranolol (10 μM)	153 ± 13	(4)*	82
PMA (200 nM)	80 ± 11	(3)	0
OAG (30 μg/ml)	102 ± 3	(3)	0
A23187 (0.4 μM)	97 ± 7	(3)	ND
OAG (30 μg/ml) + A23187 (0.4 μM)	111 ± 9	(3)	ND

* significantly different from saline addition P<0.01
ND – not determined
$^\Delta$ preaddition rate of ferricyanide reduction was 60.3 ± 3.8 nmol/min/g wet wt. (n = 21)

Neither phorbol-12-myristate-13-acetate (PMA) or 1-oleoyl-2-acetyl glycerol (OAG) caused a stimulation of ferricyanide reduction. In some systems to elicit maximum responses the calcium ionophore A23187 is required with DAG (Kaibuchi et al., 1983) but this combination was also ineffective in the perfused heart (Table 1).

No sustained effect of PMA or OAG on tension development in the heart was observed (Table 1) but transient increases between 10 and 30 sec after the addition occurred. The magnitude of these increases were 22% for PMA and 27% for OAG.

Both PMA and OAG caused a decrease in perfusate flow rate and the time course for the PMA effects is shown in Figure 2. The decrease in flow rate caused by PMA was not reversible when PMA was removed.

The effects of the PKC activators on the alpha and beta adrenergic stimulation of the perfused heart are shown in Table 2. PMA but not OAG inhibited the alpha agonist (phenylephrine + propranolol) stimulation of ferricyanide reduction and contractility. PMA did not however block the increased contractility caused by the beta agonist isoproterenol. The inhibition of ferricyanide reduction caused by PMA was dose-dependent and appeared to be more sensitive than the reduction in perfusate flow caused by PMA (Figure 3).

The effect of agents that either stimulate G-proteins (NaF + AlCl$_3$) or inhibit phospholipase C (neomycin and C48/80) are shown in Table 3. NaF + AlCl$_3$ addition did not stimulate ferricyanide reduction and C48/80 treatment had no effect on ferricyanide reduction and contractility or on the alpha stimulation of these processes.

Neomycin was found to directly react with ferricyanide in the perfusate and thus its effect on ferricyanide reduction are unknown. Neomycin also had dramatic effects on contractility of the perfused heart leading to a virtual cessation of contraction. This effect was rapid and readily reversible with the removal of neomycin. Phenylephrine stimulation of the heart was able to overcome the inhibition of contraction caused by neomycin (results not shown).

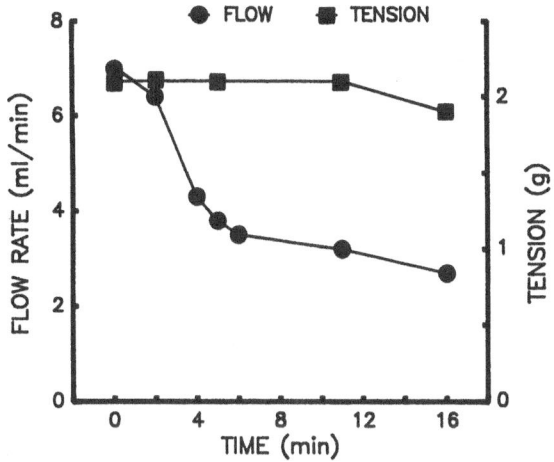

Figure 2. Time course for the effect of PMA (200nM) on flow rate and tension development in the perfused rat heart.

DISCUSSION

Alpha adrenergic stimulation of the perfused rat heart causes increased trans-sarcolemmal electron efflux that is closely associated with opening of sarcolemmal Ca^{2+} channels. This leads to increased myoplasmic Ca^{2+} concentration and contractility (Low et al., 1985; Clark and Rattigan, 1986). Alpha adrenergic stimulation in many tissues and in the heart causes the activation of the PI signal transduction system (Berridge, 1982; Exton, 1985; Woodcock et al., 1987). The PI signal transduction system has two branches, which involve the second messengers DAG and IP$_3$ (Berridge, 1982).

TABLE 2. Effect of PKC activators on alpha and beta adrenergic stimulated perfused rat heart. Mean ± S.E.M. (n).

Pretreatment	Treatment	Ferricyanide Reduction (% of Pretreatment Rate)$^\Delta$	Increase in Tension Development (%)
None	Saline	91.6 ± 5.9 (5)	0
"	Phenylephrine (20 μM) + Propranolol (10 μM)	153.3 ± 12.9 (4)*	82
20' PMA (200 nM)	Phenylephrine (20 μM) + Propranolol (10 μM)	85.6 ± 2.0 (3)	0
20' PDD (200 nM)	Phenylephrine (20 μM) + Propranolol (10 μM)	127.2 ± 10.1 (3)*	ND
None	Isoproterenol (20 μM)	NPD	73
PMA (200 nM)	Isoproterenol (10 μM)	NPD	150
OAG (6 μg/ml)	Phenylephrine (20 μM) + Propranolol (10 μM)	ND	82

ND — not determined
NPD — not possible to determine since agent directly reacts with ferricyanide
* significantly different from saline addition P<0.01
Δ pretreatment rate — 60.3 ± 3.8 nmol/min/g wet wt. (n = 21)

The production of DAG in other tissues causes the activation of PKC (Berridge, 1982; Exton, 1985). Although it has not been shown that alpha adrenergic stimulation in the heart leads to production of DAG capable of activating PKC it has been shown that PMA, a direct activator of PKC, does cause the translocation of PKC from the cytosol to the particulate fraction in perfused rat heart (Davison et al., unpublished). Translocation is often considered an indication of activation of PKC (Nishizuka, 1984). In the present study addition of PMA did not cause any change to the trans-sarcolemmal electron efflux (Table 1). There was a transient rise in contractility, which might reflect Ca^{2+} release from intracellular stores (Exton, 1985), but no sustained increase in contractility occurred reflecting that sarcolemmal Ca^{2+} channels were not activated. The synthetic diacylglycerol, OAG, another activator of PKC in other systems (Evans and Farrar, 1987), was also ineffective, whether used alone or in combination with the calcium ionophore A23187 (Table 1).

Both PMA and OAG did decrease the perfusate flow rate (Figure 2) indicating they were capable of causing a physiological response in the perfused heart system. Also PMA pretreatment caused a loss of the alpha adrenergic stimulation of trans-sarcolemmal electron efflux and contractility (Table 2) in a dose-dependent manner (Figure 3). In other tissues PKC activation has been shown to cause phosphorylation and down-regulation of the alpha adrenergic response (Cotecehia et al., 1985; Carvera et al., 1986; McMillan et al., 1986; Woods et al., 1987; Kunos and Ishac, 1987). This may occur at the receptor level or at the

G-protein that couples the receptor to the response (Kunos and Ishac, 1987; Orellana et al., 1987).

The results of the present study indicate that PKC activation is not involved in the stimulation of trans-sarcolemmal electron efflux or opening of the sarcolemmal Ca^{2+} channels. However PKC may regulate the alpha adrenergic stimulation of these responses by down-regulation at the receptor level.

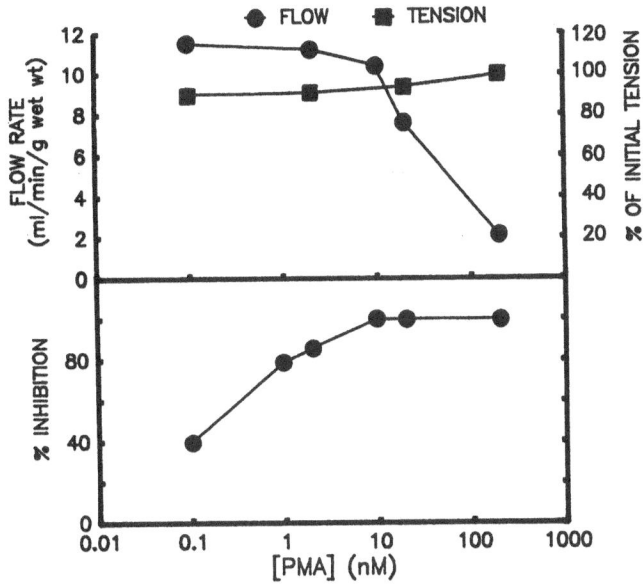

Figure 3.   Dose curve for the effect of PMA on flow rate, tension development and the inhibition of the phenylephrine stimulation of ferricyanide reduction in the perfused rat heart.

The other second messenger of the PI signal transduction system is IP_3. IP_3 production can be blocked in other tissues by inhibition of phospholipase C with neomycin (Downes and Michell, 1981; Whipps et al., 1987; Slivka and Ingel, 1987) or with C48/80 (Bronner et al., 1987). The effect of neomycin addition to the perfused heart was complex. Contractility was rapidly decreased indicating that neomycin was acting as more than just an inhibitor of phospholipase C. As neomycin reacted directly with ferricyanide in the perfusate it was not possible to assess its effects on trans-sarcolemmal electron efflux. Alpha adrenergic stimulation was able to overcome the effects of neomycin on the contractility of the heart suggesting that phospholipase C activation and thus IP_3 production was not necessary for sarcolemmal Ca^{2+} channel activation. This was further supported by the observation that C48/80 was also ineffective in blocking the alpha adrenergic stimulation of contractility and trans-sarcolemmal electron efflux (Table 3). However

TABLE 3. Effect of NaF + AlCl₃, neomycin and C48/80 on ferricyanide reduction and tension development in the perfused heart.

Pretreatment	Treatment	Ferricyanide Reduction (% of Control)$^\phi$	Tension Development (% of Control)
None	NaF (5 mM) + AlCl₃ (10 µM)	104.7	ND
"	Neomycin (2 mM)	677$^\Delta$	<10
"	C48/80 (5 µg/ml)	89.4	75
C48/80 (5 µg/ml)	Phenylephrine (20 mM) + Propranolol (10 µM)	125 ± 7 (3)*	140 ± 10 (3)*

$^\Delta$ neomycin reacted directly with ferricyanide
* significant increase P<0.01
$^\phi$ control rate 60.3 ± 3.8 nmol/min/g wet wt. (n = 21)

conclusive evidence that IP₃ is not involved awaits studies showing that IP₃ production in the heart is blocked by these agents.

In conclusion these results indicate that the PI signal transduction system is not likely to be involved in the stimulation of either trans-sarcolemmal electron efflux or activation of sarcolemmal Ca^{2+} channels.

Recently Yatani et al. (1987) have shown that a G-protein can directly regulate mammalian cardiac Ca^{2+} channels. Aluminium tetra-fluoride causes activation of the G-protein coupled to the alpha adrenergic receptor in liver tissue (Blackmore and Exton, 1986). Thus NaF + AlCl₃ were added to the perfused heart, but no effect on trans-sarcolemmal electron efflux was observed (Table 3). This may be due to the cardiac G-protein being different to the liver G-protein and thus not being sensitive to aluminium tetrafluoride. It is already known that pertussis toxin does not inhibit the alpha adrenergic mediated effect on IP production in the heart unlike the inhibition that occurs in liver (Schmitz, 1987).

The mechanism by which alpha adrenergic stimulation of the rat heart causes the activation of trans-sarcolemmal electron efflux and the sarcolemmal Ca^{2+} channels as well as the relationship between each await further research.

REFERENCES

Abel, K.C., Rattigan, S. and Clark, M.G. (1988). Comparison of adrenergic agonist and insulin effects on 3-0-methyl D-glucose efflux and sarcolemmal cytochalasin B binding by perfused rat heart. Int. J. Biochem. 20, 291-295.

Avron, M. and Shavit, N. (1963). A sensitive and simple method for determination of ferrocyanide. Anal. Biochem. 6, 549-554.

Barka, T., van de Noen, H. and Shaw, P.A. (1987). Proto-oncogene fos (c-fos) expression in the heart. Oncogene 1, 439-443.

Berridge, M.J. (1982). A novel cellular signalling system based on the integration of phospholipid and calcium metabolism. In: Calcium and Cell Function Vol. III, pp.1-36. Academic Press, New York.

Berridge, M.J. (1987). Inositol lipids and cell proliferation. Biochem. Biophys. Acta 920, 301-305.

Blackmore, P.F. and Exton, J.H. (1986). Studies on the hepatic calcium-mobilizing activity of aluminium fluoride and glucagon. J. Biol. Chem. 261, 11056-11063.

Bronner, C., Wiggins, C., Monte, D., Marki, F., Capron, A., Landry, Y. and Franson, R.C. (1987). Compound 48/80 is a potent inhibitor of phospholipase C and a dual modulator of phospholipase A_2 from human platelet. Biochem. Biophys. Acta 920, 301-305.

Brown, J.H., Bruxton, I.L. and Brunton, L.L. (1985). α_1-Adrenergic and muscarinic cholinergic stimulation of phosphoinositide hydrolysis in adult rat cardiomyocytes. Circ. Res. 57, 532-537.

Brown, J.H. and Jones, L.G. (1987). Phosphatidyl inositol turnover in the heart. In: Phosphoinositides and receptor mechanisms (Putney, J.W. ed.), pp.245-270. Alan R. Liss Inc., New York.

Bruckner, R., Mugge, A. and Scholz, H. (1985). Existence and functional role of α_1-adrenoceptors in the mammalian heart. J. Mol. Cell. Cardiol. 17, 639-645.

Clark, M.G. and Patten, G.S. (1984a). Adrenergic control of phospho-fructokinase and glycolysis in rat heart. Curr. Top. Cell. Reg. 23, 127-176.

Clark, M.G. and Patten, G.S. (1984b). Adrenergic regulation of glucose metabolism in rat heart. A Ca^{2+}-dependent mechanism mediated by both α- and β-adrenergic receptors. J. Biol. Chem. 259, 15204-15211.

Clark, M.G. and Rattigan, S. (1986). Alpha adrenergic receptor mechanism: biochemical events. J. Mol. Cell. Cardiol. 18 Suppl. 5, 69-77.

Corvera, S., Schwartz, K.R., Graham, R.M. and Garcia-Sainz, J.A. (1986). Phorbol esters inhibit α_1-adrenergic effects and decrease the affinity of liver cell α_1-adrenergic receptors for (-) epinephrine. J. Biol. Chem. 261, 520-526.

Cotecchia, S., Leeb-Lunberg, L.M.F., Hagen, P-O., Lefkowitz, R.J. and Caron, M.C. (1985). Phorbol ester effects on α_1-adrenoceptor binding and phosphatidylinositol metabolism in cultured vascular smooth muscle cells. Life Sci. 37, 2389-2398.

Crane, F.L., Sun, I.L., Clark, M.G., Grebing, C. and Low, H. (1985). Transplasma-membrane redox systems in growth and development. Biochem. Biophys. Acta 811, 233-264.

Davison, A.G.M., Cleland, P.J.F., Rattigan, S. and Clark, M.G. (1988). Unpublished observations.

Downes, C.P. and Michell, R.H. (1981). The polyphosphoinositide phospho-diesterase of erythrocyte membranes. Biochem. J. 198, 133-140.

Evans, S.W. and Farrar, W.L. (1987). Interleukin 2 and diacylglycerol stimulate phosphorylation of 40S ribosomal S6 protein. J. Biol. Chem. 262, 4624-4630.

Exton, J.H. (1985). Mechanisms involved in α-adrenergic phenomena. Am. J. Physiol. 248, E633-E647.

Kaibuchi, K., Takai, Y., Sawamura, M., Hoshijima, M., Fujikiwa, T. and Nishizuka, Y. (1983). Synergistic functions of protein phosphoryl-ation and calcium mobilization in platelet activation. J. Biol. Chem. 258, 6701-6704.

Kunos, G. and Ishac, E.J.N. (1987). Mechanism of inverse regulation of $alpha_1$- and beta-adrenergic receptors. Biochem. Pharmacol. 36, 1185-1191.

Leung, E., Johnson, C.I. and Woodcock, E.A. (1986). Stimulation of phosphatidylinositol metabolism in the heart. Clin. Exp. Pharm. Physiol. 13, 359-363.

Low, H., Crane, F.L., Partick, E.J., Patten, G.S. and Clark, M.G. (1984). Properties and regulation of a trans-plasma membrane redox system of perfused rat heart. Biochem. Biophys. Acta **804**, 253-260.

Low, H., Crane, F.L., Partick, E.J. and Clark, M.G. (1985). α-Adrenergic stimulation of trans-sarcolemma electron efflux in perfused rat heart. Possible regulation of Ca^{2+}-channels by a sarcolemma redox system. Biochem. Biophys. Acta **844**, 142-148.

McMillan, M., Chernow, B. and Roth, B.L. (1986). Phorbol esters inhibit alpha$_1$-adrenergic receptor-stimulated phosphoinositide hydrolysis and contraction in rat aorta: evidence for a link between vascular contraction and phosphoinositide turnover. Biochem. Biophys. Res. Commun. **134**, 970-974.

Navas, P., Sun, I.L., Morre, D.J. and Crane, F.L. (1986). Decrease of NADH concentration in HeLa cells in the presence of transferrin or ferricyanide. Biochem. Biophys. Res. Commun. **135**, 110-115.

Nishizuka, Y. (1984). The role of protein kinase C in cell surface signal transduction and tumour promotion. Nature **308**, 693-698.

Orellana, S., Solski, P.A. and Brown, J.H. (1987). Guanosine 5'-0-(thiotriphosphate)-dependent inositol trisphosphate formation in membranes is inhibited by phorbol ester and protein kinase C. J. Biol. Chem. **262**, 1638-1643.

Rattigan, S., Edwards, S.J., Hettiarachchi, M. and Clark, M.G. (1986). The effects of α- and β-adrenergic agents, Ca^{2+} and insulin on 2-deoxyglucose uptake and phosphorylation in perfused rat heart. Biochem. Biophys. Acta **889**, 225-235.

Rinaldi, M.L., Capony, J-P. and Demaille, J.G. (1982). The cyclic AMP-dependent modulation of cardiac sarcolemmal slow calcium channels. J. Mol. Cell. Cardiol. **14**, 279-289.

Schmitz, W., Scholz, H., Scholz, J., Steinfath, M., Lohse, M., Puurunen, J. and Schwabe, U. (1987). Pertussis toxin does not inhibit the α_1-adrenoceptor-mediated effect on inositol phosphate production in the heart. Eur. J. Pharm. **134**, 377-378.

Schumann, H.J., Wagner, J., Knorr, A., Reidemeister, J.C., Sadony, V. and Schramm, G. (1978). Demonstration in human atrial preparations of α-adrenoceptors mediating positive inotropic effects. Naun-Schm. Arch. Pharmacol. **302**, 333-336.

Slivka, S.R. and Insel, P.A. (1987). α_1-Adrenergic receptor-mediated phosphoinositide hydrolysis and prostaglandin E_2 formation in Madin-Darby canine kidney cells. J. Biol. Chem. **262**, 4200-4207.

Simpson, P., Bishopric, N., Coughlin, S., Karliner, J., Ordahl, C., Starksen, N., Taso, T., White, N. and Williams, L. (1986). Dual trophic effects of the alpha$_1$-adrenergic receptor in cultured neonatal rat heart muscle cells. J. Mol. Cell. Cardiol. **18** Suppl. 5, 45-58.

Starksen, N.F., Simpson, P.C., Bishopric, N., Coughlin, S.R., Lee, W.M.F., Escobedo, J.A. and Williams, L.T. (1986). Cardiac myocyte hypertrophy is associated with c-myc protooncogene expression. Proc. Natl. Acad. Sci. **83**, 8348-8350.

Whipps, D.E., Armston, A.E., Pryor, H.J. and Halestrap, A.P. (1987). Effects of glucagon and Ca^{2+} on the metabolism of phosphatidyl-inositol 4-phosphate in isolated rat hepatocytes and plasma membranes. Biochem. J. **241**, 835-845.

Woodcock, E.A., White, B.S., Smith, I. and McLeod, J.K. (1987). Stimulation of phosphatidylinositol metabolism in the isolated, perfused rat heart. Circ. Res. **61**, 625-631.

Woods, N.M., Cutherbertson, K.S.R. and Cobbold, P.H. (1987). Phorbol-ester-induced alterations of free calcium ion transients in single rat hepatocytes. Biochem. J. **246**, 619-623.

Yatani, A., Codina, J., Imoto, Y., Reeves, J.P., Birnbaumer, L. and Brown, A.M. (1987). A G protein directly regulates mammalian cardiac calcium channels. Science **238**, 1288-1292.

ROLE OF Na^+/H^+-ANTIPORTER IN GROWTH STIMULATION BY HA-RAS

Hans Grunicke, Karl Maly, Hermann Oberhuber,
Wolfgang Doppler, Johann Hoflacher, Boris W.
Hochleitner, Rolf Jaggi* and Bernd Groner*

Institute of Medical Chemistry and Biochemistry
University of Innsbruck, Austria and
*Ludwig Institute for Cancer Research
Inselspital, Bern, Switzerland

ABSTRACT

Expression of the transforming Ha-ras oncogene in MMTV-LTR-Ha-ras transfected 3T3 fibroblasts causes a growth factor independent intracellular alkalinization by a dimethylamiloride sensitive mechanism. The phenomenon is accompanied by a desensitization of the intracellular calcium mobilizing system to serum growth factors and an increase in the generation of inositol phosphates. Data are presented which support the assumption that the mobilization of intracellular Ca^{2+} occurs by a Na^+-dependent mechanism. The results are discussed as indicating a constitutive activation of growth factor signal transduction by transforming Ha-ras. The exact sequences of events initiated by the oncogene, however, remains to be elucidated.

INTRODUCTION

Alterations in the cytosolic Ca^{2+} and H^+-concentrations are common events following mitogenic stimulation of eukaryotic cells[1-3]. It is still unclear, however, by which biochemical mechanisms these changes in H^+- or Ca^{2+}-

concentrations stimulate cellular replication. Especially subject to discussion is whether the increase in pH_i is a prerequisite or merely a facilitating condition for processes involved in cell proliferation.

We decided to employ Ha-ras transfected cells in order to investigate the biological significance of the mitogen induced cytosolic alkalinization. In order to permit an expression of the ras oncogene under controlled conditions, NIH 3T3 fibroblasts were transfected with an MMTV-LTR Ha-ras construct. Recombination with the MMTV-LTR sequence subjects the transcription of the oncogene to control by glucocorticoids. Addition of dexamethasone to MMTV-LTR-ras transfected cells leads to a rapid induction of LTR-ras-transcription. The gene product $p21^{ras}$ is detectable within 1 hr after addition of the hormone and reaches a maximum at 24 hr as described by Jaggi et al.[4]. Expression of the transforming Ha-ras oncogene which differs from the proto-oncogene by a nucleotide substitution at codon 12 specifying a glycine in the proto-oncogene and a valine in the transforming oncogene leads to a "rounded" transformed morphological phenotype. The cells grow in soft agar and cause tumors in nude mice[4]. Addition of dexamethasone to G1 arrested cells causes growth factor independent triggering into S-phase[5]. None of these effects are observed if cells carrying the corresponding MMTV-LTR Ha-ras proto-oncogene construct are treated with dexamethasone. Thus, in this system, cell proliferation and tumorigenesis can be controlled by the expression of a single gene and it seemed interesting to investigate whether alterations in pH_i are correlated with these biological parameters.

METHODS

NIH 3T3 fibroblasts transfected with MMTV-LTR-Ha-ras constructs were grown in DMEM supplemented with 10% fetal calf serum in the presence of 5% CO_2 as described[5,6]. The LTR-oncogene constructs contain the coding regions of either transforming Ha-ras from a human bladder carcinoma cell line or the normal human placenta Ha-ras proto-oncogene homologue. Plasmids were prepared and cells transfected as described by Jaggi et al.[4]. Intracellular pH was determined by measuring the intra- and extracellular distribution of ^{14}C-benzoic acid as described[5]. Ca^{2+} was measured by fluorescence spectrometry employing fura-2 according to Grynkiewicz et al.[7]. Isolation of inositol phosphates and analysis of phosphatidylinositols were performed as described[6].

RESULTS

I. Expression of transforming Ha-ras causes an intracellular alkalinization.

Expression of the transforming Ha-ras in serum starved, growth-arrested cells leads to a gradual increase in cytosolic pH (fig. 1) which reaches a maximum after approx. 20 hr. Addition of serum growth factors does not lead to a further enhancement of the maximum value which is obtained by dexaméthasone alone. This is consistent with the assumption that Ha-ras and serum growth factors affect intracellular pH by the same mechanism and that the transforming Ha-ras alone is sufficient for a maximal activation of this system. The rise in cytosolic pH can be blocked by dimethylamiloride, a known inhibitor of the Na^+/H^+-antiporter[8], indicating that the decrease in cytosolic H^+-concentration is the result of an activation of the Na^+/H^+-antiporter. Expression of the Ha-ras proto-oncogene does not affect the pH_i values (fig. 1).

II. Effect of Ha-ras on phosphatidylinositol metabolism.

By what mechanism does Ha-ras activate the Na^+/H^+-antiporter? In a variety of systems, protein kinase C seems to be involved in the stimulation of the Na^+/H^+-antiporter[2,9], although, this is not the only pathway[10]. Protein kinase C in turn has been shown to be activated by diacylglycerol[11]. Diacylglycerol is a product of the hydrolytic cleavage of phosphatidylinositol-4,5-bisphosphate

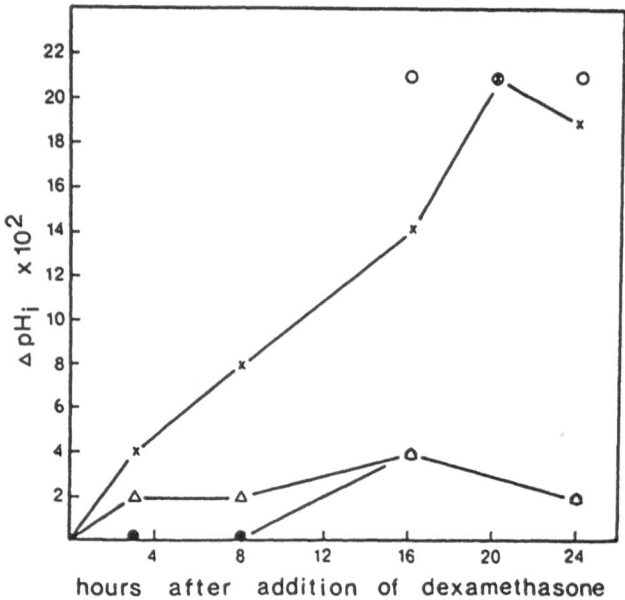

Fig. 1. Effect of Ha-ras on Intracellular pH

Cells were growth arrested by a 24 hr incubation in a 1:1
mixture of Ham's F12/DMEM containing 1.5% fetal calf serum.
Oncogene expression was induced by the addition 1 μM
dexamethasone to the growth medium at time 0. Intracellular
pH was measured at the time points indicated as described
under Methods. Where indicated fetal calf serum was added
10 min prior to the pH-determinations. The ordinate shows
the increase in intracellular pH relative to the value at 0
time. (Absolute values at time 0 (means ± SE) were 7.32 ±
0.02 for cells transfected with transforming Ha-ras and 7.34
± 0.02) for cells transfected with Ha-ras proto-oncogene).
(x) transfected with transforming Ha-ras; (o) transfected
with transforming Ha-ras plus additional 50 μM
dimethylamiloride; ( ) transfected with Ha-ras proto-
oncogene; (o) transfected with transforming Ha-ras plus
additional fetal calf serum (5%).

From: Maly et al., Adv.Enzyme Regul. 27 (1988),121-131.
With permission of Pergamon Press.

and other phospholipids by C-type phospholipases. The degradation of phosphatidylinositol-4,5-bisphosphate (PIP_2) by PIP_2-specific phospholipase C has been shown to be an early event in signal transduction after stimulation by a variety of growth factors[12]. This PIP_2-specific phospholipase C is known to be under control of a GTP-binding protein (G-protein) which may act as a signal transducer between the growth factor receptor and the enzyme[12]. The remarkable homology between p21[ras] and the α-subunit of various G-proteins has led to the suggestion that p21[ras] may act in a similar way. Indeed, normal p21[ras] has been shown to couple growth factor receptors to inositol phosphate production and the growth stimulatory effect of serum could be blocked by anti-ras-antibody (review by Masters and Bourne[13]). Evidence for an increase in phosphatidylinositol-4,5-bisphosphate turnover in ras transformed cells has been presented[15]. However, alterations in ras-transformed cells may not be directly related to the oncogene product but could represent secondary phenomena which occur as a result of the transformation process. It seemed interesting, therefore, to study the effect of a controlled ras-expression in LTR-ras transfected cells on PIP_2 metabolism. The data of table 1 demonstrating an increased generation of inositol phosphates after expression of the transforming Ha-ras, are in accordance with a ras-induced activation of the PIP_2 specific phospholipase C. As can be seen induction of the Ha-ras proto-oncogene is not sufficient to cause a serum growth factor independent stimulation in inositol phosphate production.

III. Effect of Ha-ras on intracellular Ca^{2+}.

The data from table 1 suggest a Ha-ras induced stimulation of phosphatidylinositolphosphate degradation, probably catalyzed by phospholipase C. The products of this reaction are diacylglycerol which presumably acts as a stimulator of protein kinase C and inositol-1,4,5-tris-phosphate (IP_3). IP_3 is known to mediate the release of Ca^{2+} from intracellular stores[13]. The continuous accumulation of IP_3 which is observed when the ras oncogene is expressed,

Table 1. Effects of Ha-ras and Serum Growth Factors on Phosphatidylinositol Turnover and Inositol Phosphate Accumulation.

	InsP$_x$ accumulation[+] dpm x 10^3 per 10^5 cells	PtIns-turnover[*] % increase (control = 100)
Cells transfected with activated Ha-ras		
controls	5.8 ± 1.2[x]	
+ Dex	13.8 ± 0.8	186
+ serum	25.6 ± 2.7	442
+ serum + Dex	31.2 ± 3.6	414
Cells transfected with Ha-ras proto-oncogene		
controls	17.6 ± 1.7	
+ Dex	17.1 ± 1.4	100
+ serum	41.9 ± 1.7	270
+ serum + Dex	28.3 ± 2.1	160

[*] Phosphatidylinositol turnover is defined as percentage of label in total phosphatidylinositol pool appearing as inositolphosphates during the 45 min in presence of 25 mM LiCl. Absolute values of controls transfected with activated Ha-ras are 10.7 ± 0.8.

[+] InsP$_x$ = sum of total inositolphosphates (distribution in controls transfected with isolated Ha-ras was IP$_1$ = 5.3 x 10^3 dpm; IP$_2$ = 3.1 x 10^2 dpm; IP$_3$ = 1.8 x 10^2 dpm.

[x] means ± SE

Cells were growth arrested by incubation at 1% serum for 48 hr. During this time phosphatidylinositols and inositolphosphates were labeled by incubation with 7 μCi per ml 3H-inositol as described under Methods. Oncogene expression was induced by the addition of 1 μM dexamethasone (DEX) 18 hr before harvesting the cells. Where indicated, fetal calf serum was added to a final concentration of 10% 20 min before terminating the incubation. LiCl was added to all samples to a final concentration of 25 mM and remained present during the last 45 min of the incubation period. Phosphatidylinositols and inositolphosphates were analyzed as described under Methods.

should lead to a depletion of the corresponding intracellular Ca^{2+}-stores. Thus, exposure of cells to an agent which causes a release of Ca^{2+} from intracellular stores should mobilize less Ca^{2+} in cells in which Ha-ras is continuously expressed than in corresponding controls which are not under the influence of an activated transforming Ha-ras oncogene. Fig. 2 demonstrates that this is indeed the case.

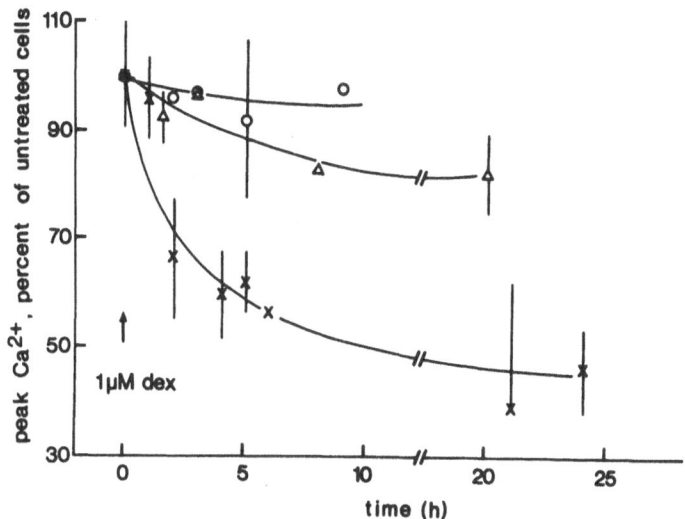

Fig. 2. Effect of Ha-ras on Ca^{2+} Mobilization by Exogenous Growth Factors.

Cells were growth arrested by incubation in presence of 1% fetal calf serum for 24 hr. Expression of the transforming Ha-ras-oncogene and the Ha-ras proto-oncogene was induced by addition of 1 μM dexamethasone at time 0. Where indicated, fetal calf serum was added to a final concentration of 5% and the resulting increase in cytosolic Ca^{2+} was measured as described under Methods. For Ca^{2+} determinations, cells were kept in a calcium free medium containing 10 mM EGTA. Values are expressed as % increase in cytosolic Ca^{2+} relative to Ca^{2+} levels in non-serum treated cells. (o) nontransfected NIH 3T3 control cells; ( ) cells transfected with Ha-ras proto-oncogene; (x) cells transfected with transforming Ha-ras. Bars indicate SEM.

The level of cytosolic free Ca^{2+} was not found to be permanently increased in cells in which the transforming Ha-ras is expressed compared to corresponding controls[6]. Although this might be the result of the experimental conditions in which cells were kept in calcium free medium in the presence of EGTA, one has to assume that the cytosol is rapidly cleared from excess Ca^{2+} by either a Ca^{2+} ATPase or Na^+/Ca^{2+}-exchanger. However, in view of the mechanism outlined before, one should postulate that a transitory increase in cytosolic Ca^{2+}-level did occur which was sufficient to trigger the replication machinery, but the duration of this increase may have been too short to be detectable under the experimental conditions employed here. On the other hand, it cannot be excluded that the observed desensitization of the Ca^{2+}-releasing system under the influence of Ha-ras is caused by some kind of feed-back inhibition leading to a "down regulation" of the releasing system and is not the result of the depletion of Ca^{2+} stores. Indeed, a protein kinase C mediated feed-back inhibition of phospholipase C has been described[15,16]. This kind of "down regulation", however, does not seem to operate in our cells as cells in which the transforming Ha-ras has been induced by dexamethasone continue to generate inositol phosphates at an increased rate for more than 36 hr which is beyond the time intervals studied here (data not shown).

Table 2. Effects of Dimethylamiloride (DMA) and Extracellular $(Na^+)_o$ on Intracellular Ca^{2+} Mobilization by Bombesin.

	$(Na^+)_o$	increase in cytosolic Ca^{2+} (nmol)
Bombesin	+	260 ± 54
	-	121 ± 20
Bombesin + DMA	+	111 ± 14
	-	106 ± 26

Cells were growth arrested as described in the legend to fig. 2. Bombesin was added to a final concentration of 3 μM and the increase in cytosolic free calcium determined as described under Methods. Ca^{2+}-determination was performed in a Ca^{2+}-free medium containing 10 mM EGTA. Where indicated, 130 mM extracellular Na^+ $(Na^+)_o$ was substituted by an equimolar concentration of choline. Dimethylamiloride (DMA) was added were indicated at a final concentration of 50 μM. Data indicate means ± SE.

IV.   Is there an interdependence between the Na^+/H^+-antiporter and the intracellular Ca^{2+}-releasing system?

Table 2 demonstrates that the bombesin induced increase in cytosolic Ca^{2+} can be blocked by dimethylamiloride (DMA) a well known inhibitor of the Na^+/H^+-antiporter[8]. These results seem to suggest that the mobilization of Ca^{2+} from the endoplasmic reticulum requires a drop in cytosolic H^+-concentration.   Such a conclusion is further supported by the observation that the Ca^{2+}-mobilization by bombesin depends on extracellular Na^+.

Substitution of the extracellular Na^+ by choline also eliminates the bombesin induced Ca^{2+}-mobilization (table 2). Similar findings have been obtained with thrombin stimulated platelets[17,18]. In platelets, however, the calcium release does not seem to require a preceding alkalinization[18,19]. In the later system, N-ethyl-N-isopropyl amiloride (EIPA), a highly specific inhibitor of the Na^+/H^+-antiporter, blocks thrombin induced Ca^{2+} mobilization at concentrations which are 100 fold above those required to depress the Na^+/H^+-antiporter[18]. Furthermore, an experimentally imposed change in pH_i did not produce significant alterations in Ca^{2+}_i[20]. The conclusion that the data shown in table 2 indicate a pH dependence of the intracellular Ca^{2+}-releasing system cannot be justified.   However, our findings as well as data obtained with human platelets are consistent with a hitherto unknown Na^+-dependence of the intracellular Ca^{2+} releasing system.   The exact role of Na^+ in this system, however, remains to be elucidated.   As an alternative one may speculate that cytosolic Ca^{2+} stimulates the Na^+/H^+-antiporter.   Evidence in favour of this hypothesis has been presented[20].   Recently, it has been demonstrated, however, that in lymphocytes, concanavalin A induced alkalinization is independent of changes in cytosolic calcium[11].

CONCLUSION

Expression of the transforming Ha-ras, but not of the Ha-ras proto-oncogene leads to an increase of cytosolic pH by an amiloride sensitive mechanism.   In addition to the changes of the pH_1 the transforming Ha-ras causes a growth factor independent activation of the phosphatidylinositol metabolism and a desensitization of the intracellular calcium releasing system to serum growth factors. In view of these data it is concluded that the transforming Ha-ras causes a growth factor independent activation of PIP_2 specific phospholipase C which generates diacylglycerol and IP_3. Diacylglycerol activates protein kinase C which in turn stimulates the Na^+/H^+-antiporter. IP_3 mediates a release of Ca^{2+} from endoplasmic reticulum stores.

Although, there are some conflicting reports[21,22], findings similar to ours which support the proposed mechanism have been described by other authors. A microinjection of p21[ras] into cells has been shown to elevate cytosolic pH[23]. Evidence for an increase of phosphatidylinositol metabolism in ras-transformed cells has also been published[15]. It should be emphasized, however, that definite proof for the proposed sequence of events following expression of Ha-ras in LTR-ras transfected cells is still lacking. It has been demonstrated that the Na^+/H^+-antiporter can be activated by a protein kinase C independent mechanism[11]. Thus, the change in phosphatidylinositol metabolism may represent an accompanying, independent phenomenon. Furthermore, ras-transformed cells have been shown to produce other growth factors, such as EGF and TGF-α[24], which could cause additional subsequent metabolic alterations. The cells employed here, however, do not respond to EGF which makes a susceptibility to TGF-α also unlikely considering the homology of these two growth factors.

REFERENCES

1. T. R. Hesketh, J. P. Moore, J. D. H. Morris, M. V. Taylor, J. Rogers, T. A. Smith, and J. C. Metcalfe, A common sequence of calcium and pH signals in the mytogenic stimulation of eukaryotic cells, Nature 313:481 (1985).

2. W. H. Moolenaar, Effects of growth factors on cytoplasmic pH regulation, Ann.Rev.Physiol. 48:363 (1986).

3. R. A. Lagarde, and J. M. Pouyssegur, The Na$^+$/H$^+$-antiport in cancer, Cancer Biochem.Biophys. 9:1 (1986).

4. R. Jaggi, B. Salmons, D. Muellener, and B. Groner, The v-mos and H-ras oncogene expression represses glucocorticoid hormone dependent transcription from the mouse mammary tumor virus LTR, EMBO J. 5:2609 (1986).

5. W. Doppler, R. Jaggi, and B. Groner, Induction of v-mos and activated Ha-ras oncogene expression in quiescent NIH 3T3 cells causes intracellular alkalinization and cell-cycle progression, Gene 54:145 (1987)

6   K. Maly, H. Oberhuber, W. Doppler, J. Hoflacher, R. Jaggi, B. Groner, and H. Grunicke, Effect of Ha-ras on phosphatidylinositol metabolism Na$^+$/H$^+$-antiporter and mobilization of intracellular calcium, Adv.Enzyme.Regul. 27:121 (1988)

7. G. Grynkiewicz, N. Poenie, and R. Y. Tsien, A new generation of Ca^{2+} indicators with greatly improved fluorescence properties, J.Biol.Chem. 260:3440 (1985).

8. G. L'Allemain, A. Franchi, E. J. Cragoe jr, and J. Pouyssegur, Blockade of the Na$^+$/H$^+$-antiporter abolishes growth factor induced DNA-synthesis in fibroblasts, J.Biol.Chem. 259:4313 (1984).

9. S. Grinstein, and A. Rothstein, Mechanism of regulation of the Na$^+$/H$^+$-exchanger, J.Membrane Biol., 90:1 (1986).

10. S. Grinstein, J. D. Smith, C. Rowatt, and S. J. Dixon, Mechanism of activation of lymphocyte Na$^+$/H$^+$-exchange by concanavalin A, J.Biol.Chem. 262:15277 (1987).

11. Y. Nishizuka, Turnover of inositol phospholipids and signal transduction, Science 225:1365 (1984).

12. M. J. Berridge, and R. F. Irvine, Inositol tris-phosphate, a novel second messenger in cellular signal transduction, Nature 312:315 (1984).

13. S. B. Masters, and H. R. Bourne, Role of G-proteins in transmembrane signaling: Possible functional homology with the ras proteins, in: "Oncogenes and growth control", P. Kahn, and Th. Graf, eds., Springer Verlag, Berlin (1986).

14. L. F. Fleischman, S. B. Chahwala, and L. Cantley, Ras-transformed cells: Altered levels of phosphatidyl-inositol 4,5-bisphosphate and catabolites, Science 231:407 (1986).

15. K. D. Brown, D. M. Blakeley, M. H. Hamon, M. S. Laurie, and A. N. Corps, Protein kinase C mediated negative-feedback inhibition of unstimulated and bombesin-stimulated polyphosphoinositide hydrolysis in Swiss mouse 3T3 cells, Biochem.J. 245:631 (1987).

16. A. Pandiella, L. M. Vicentini, and J. Meldolesi, Protein kinase C mediated feedback inhibition of the Ca^{2+}-response at the EGF-receptor, Biochem.Biophys.Res.Comm. 149:145 (1987).

17. W. Siffert, G. Siffert, P. Scheid, T. Riemens, G. Gorter, and J. W. Akkerman, Inhibition of Na^+/H^+-exchange reduces Ca^{2+}-mobilization without affecting the inositol cleavage of phosphatidylinositol-4,5-bisphosphate in thrombin stimulated platelets, FEBS-Letters 212:123 (1987).

18. L. Hunyady, B. Sarkadi, J. R. Cragoe, jr, A. Spät, and G. Gardos, Activation of sodium-proton exchange is not a prerequisite for Ca^{2+}-mobilization and aggregation in human platelets, FEBS-Letters 225:72 (1987).

19. A. W. M. Simpson, and T. J. Rink, Elevation of pH_i is not an essential step in calcium mobilization in fura-2-loaded human platelets, FEBS-Letters 222:144 (1987).

20. N. E. Owen, and M. L. Villereal, Evidence for a role of calmodulin in serum stimulation of Na^+-influx in human fibroblasts, Proc.Natl.Acad.Sci. 79:3537 (1982).

21. G. Parries, R. Hoebel, and E. Racker, Opposing effects of ras-oncogene on growth factor stimulated phosphoinositide hydrolysis: Desensitization of platelet-derived growth factor and enhanced sensitivity to bradykinin, Proc.Natl.Acad.Sci. 84:2648 (1987).

22. J. C. Lacal, J. Moscat, and S. A. Aaronson, Novel source of 1,2-diacylglycerol elevated in cells transformed by Ha-ras oncogene, Nature 330:269 (1987).

23. N. Hagag, J. C. Lacal, M. Graber, S. Aaronson, and M. V. Viola, Microinjection of ras p21 induces a rapid rise in intracellular pH, Mol.Cell Biol. 7:1984 (1987).

24. D. Ozanne, R. J. Fulton, and P. L. Kaplan, Kirsten murine sarcoma virus transformed cell lines and a sponanteously transformed rat cell line produce transforming factors, <u>J.Cell Physiol.</u> 105:163 (1980)

PROTEIN KINASE C : PLASMA MEMBRANE TO NUCLEUS

Anant N. Malviya, Ahmed Masmoudi, Gérard Labourdette,
Marcel Mersel, Patrick Rogue and Guy Vincendon

Centre de Neurochimie du CNRS and INSERM U44
5 rue Blaise Pascal, 67084 Strasbourg Cedex, France

INTRODUCTION

It seems well founded that one of the major pathways of transmembrane
signalling is mediated via activation of calcium-phospholipid dependent
protein kinase known as protein kinase C (1). Ever since its discovery by
Nishizuka in 1979, the enzyme has been implicated in a range of cellular
response including that of growth and proliferation. The mechanism of action
of protein kinase C seems to require association of the enzyme with plasma
membrane to prime the activity. In this scenario diacylglycerol acts as the
endogenous kinase C activator. Diacylglycerol is one of the breakdown
products of phosphatidylinositol (2) catalyzed by specific phospholipase C
activated by the binding of growth modulators to their respective receptors
located on the plasma membrane. The other hydrolytic product of the inositol
cycle is 1,4,5 inositol trisphosphate which releases calcium from endoplasmic reticulum. Physiological calcium levels may be modulated ranging
from 0.1-1.0 μM, by a variety of stimuli. In fact, the elevated calcium
levels do not lead to protein kinase C activation but elicit rapid and reversible association of the enzyme with the membrane. These calcium levels
increase the apparent affinity of phorbol esters (TPA) to whole cells.

Alternative route of diacylglycerol formation, other than phosphoinositides, is also documented. In transformed cells containing Ha-ras
oncogenes increase in diacylglycerols levels, in the absence of any
detectable increase in inositol phosphatides, is recently reported (3).

Another signalling pathway is represented by cyclic-AMP and cyclic-AMP
dependent (protein) kinase. There seems to be a "cross-talk" (4) between
protein kinase C mediated pathway and cyclic-AMP dependent protein kinase
pathway, although the mechanism by which such a dialogue operates is
currently debated (5).

A cytosolic rise in cyclic-AMP level has been known to cause
translocation of cyclic-AMP dependent protein kinase from the cytosol to the
nucleus. In B lymphocytes increased cyclic-AMP production causes protein
kinase C (6) translocation to the nucleus.

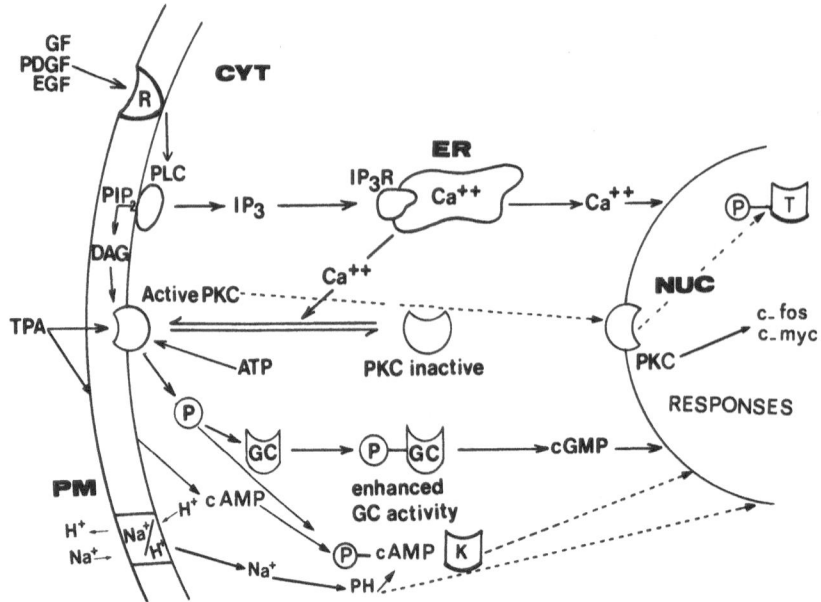

Fig. 1. An overview of signal transduction and mechanism of protein kinase C action.
Growth modulators (GF, PDGF, EGF) receptor (R). 4,5-bisphosphate (PIP$_2$), 1,4,5-trisphosphate (IP$_3$), diacylglycerol (DAG), 12-O-tetradecanoyl phorbol-13-acetate (TPA), guanylate cyclase (GC), cyclic AMP-protein kinase (c-AMP-K), protein kinase C (PKC), phosphoinositide specific phospholipase C (PLC), topoisomerase II or transcription factor (T) phosphorylated (P).

The mechanism of signal transduction as mediated via protein kinase C activation at the site of plasma membrane has received considerable attention. But how the nuclear events, activated during signal transduction, are regulated by protein kinase C remains less clear. There is considerable controversy regarding a nuclear location of protein kinase C. Earlier immunocytochemical studies of Nishizuka (1) and later observations of Misra and Sahyoun (7) showing calcium dependent binding of purified protein kinase C to isolated rat liver nuclei disfavour endogenous nuclear location of the enzyme. Immunocytochemical data (8) in neurons and immunochemical studies in rat liver nuclear preparations (9) provide evidence in favour of a nuclear location of protein kinase C. Although nuclear location of a number of serine-threonine protein kinases is well known (10) the role played by phosphorylation-dephosphorylation at the site of nuclei has not received deserved attention.

With this preamble new evidence is advanced here in favour of endogenous location of protein kinase C in rat liver nuclei. Subcellular distribution of protein kinase C activity in rat liver is presented with due comparison of cofactors requirement for cytosolic versus nuclear enzymatic activity.

PROTEIN KINASE C ACTIVITY IN LIVER HOMOGENATE

When the rat liver homogenate was examined for protein kinase C activity as a function of calcium concentrations (Fig. 2), two calcium optima were observed for maximum kinase activity.

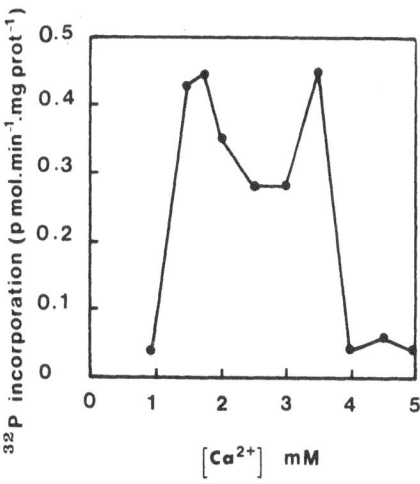

Fig. 2. Protein kinase C activity in rat liver homogenate. The enzymatic activity was determined by substracting the activity measured in the absence of phosphatidylserine from the total activity determined in the presence of calcium plus phosphatidylserine.

This rather intriguing pattern of protein kinase C activity was further verified in an assay condition containing Triton X-100 (11) at two calcium optima i.e. 1.75 mM and 3.5 mM.

Protein kinase C activity, at the two calcium concentrations monitored, is similar with respect to inhibition by Triton X-100 at concentrations above 0.02 % and prevention of the Triton X100 inhibition by diacylglycerol. This confirms that the maximum activity of protein kinase C in rat liver homogenate is really manifest at two calcium concentrations.

Table 1. Protein kinase C activity in rat liver homogenate monitored in the presence of Triton X-100 and diacylglycerol.

Additions	Protein kinase C activity	
	pmol ^{32}P/min/mg protein	
	Ca^{++} [1.75 mM]	Ca^{++} [3.5 mM]
Ca^{++}	0.20	0.40
Ca^{++} + PS	0.69	0.89
Ca^{++} + PS + DAG	0.82	0.95
Ca^{++} + PS + Triton X-100 (0.05%)	0.52	0.75
Ca^{++} + PS + Triton X-100 (0.05%) + DAG	0.71	0.92
Ca^{++} + PS + Triton X-100 (0.02%)	0.69	0.95
Ca^{++} + PS + Triton X-100 (0.02%) + DAG	0.95	1.03

The assay of protein kinase C activity was essentially the one described (12) and diacylglycerol was prepared as reported (13). Diacylglycerol (100 μM) and Triton X-100 were added directly to the assay medium.

Table 2

Subcellular fractions	Protein mg	5'-Nucleotidase		Na⁺-K⁺ ATPase		Protein kinase C			
						[Ca^{++} 1.75mM]		[Ca^{++} 3.5 mM]	
		Specific activity (a)	Activity (b)	Specific activity (a)	Activity (b)	Specific activity (c)	Activity (d)	Specific Activity (c)	Activity (d)
Homogenate (H)	751(100)	4 100	4 920(100)	760	912(100)	0.46	414(100)	0.45	405(100)
Particulate (P)	536(71)	16 000	5 300(107)	2 900	966(106)	2.1	380(92)	0.02	20.0(5)
Cytosolic (C)	333(44)	500	526(10)	N.D.	–	0.21	82(19)	N.D.	–
Nuclear (N)	43 (6)	300	14(0.29)	50	2(0.24)	N.D.	–	3.4	78 (19)

The values in parentheses are the percentage of the activity. (a) Specific activity is expressed as pmol inorganic phosphate released/min/mg protein ; (b) same as (a) expressed per liver ; (c) pmol ^{32}p/min/mg protein, (d) same as (c) expressed per liver. N.D. : Not Detected.

SUBCELLULAR DISTRIBUTION

In the absence of any systematic study on subcellular distribution of protein kinase C activity in rat liver, we have undertaken such a venture. Cytosol, particulate and nuclear fractions were isolated from the rat liver. A typical distribution of protein kinase C activity (at two calcium optima) along with the distribution of Na^+/K^+ ATPase and 5'-nucleotidase activities is shown in Table 2.

It is obvious that the isolated nuclei are not contaminated with plasma membrane. The activity of plasma membrane marker enzymes is associated with the particulate fraction. Secondly, the protein kinase C activity assayed at 3.5 mM calcium concentration is mainly observed in the nuclear fraction. In the light of these data it may be suggested that 19% of the total protein kinase C activity observed at 3.5 mM calcium belongs to the rat liver nuclei. A large portion of the protein kinase C activity (92%) requiring 1.75 mM calcium is present in the particulate fraction and is masked. The particulate kinase C activity could only be seen after Triton X-100 treatment.

PARTIAL PURIFICATION OF NUCLEAR KINASE C

The next step was purification of rat liver nuclear protein kinase C activity. A summary of the partial purification is listed in Table 3.

Table 3.

Purification step	Total protein (mg)	Specific activity pmol ^{32}P/min/mg protein	Purification fold	% recovery
Nuclear extract	239.36	1.93	–	–
DEAE	7.92	354.7	1	100
Phenyl–Superose	0.243	1863.1	5.3	16
Mono Q	0.008	17416.6	49	5

Since contaminating inhibitors or activators of protein kinase C may mar the assessment of true activity in the nuclear extract, the enzyme activity observed after DEAE-cellulose chromatography is taken as the first step of purification.

The partially purified nuclear protein kinase C after Mono Q chromatographic step reveals two protein bands at 80 kDa and 66 kDa respectively (Fig. 3, lane d).

When protein kinase C derived either from cytosol, or particulate or nuclear fractions was immunoreacted with goat anti-rat brain protein kinase C antibody a single immunoreactive protein corresponding to 80 kDa molecular weight was revealed in all the three subcellular compartments (details to be published elsewhere). This differs from a report (14) giving a molecular weight for rat liver protein kinase C as 64 kDa.

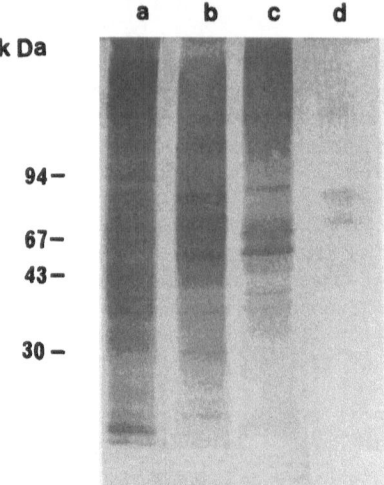

Fig. 3. SDS—polyacrylamide gel electrophoresis of protein kinase C derived from various purification steps. Protein bands are revealed by silver nitrate staining. Lane a (nuclear extract, 2.0 µg protein), lane b (DEAE-pooled fraction, 1.0 µg protein), lane C (Phenyl-Superose pooled fraction 0.5 µg protein), and lane d (Mono Q pooled fraction 0.01 µg protein).

PHORBOL DIBUTYRATE (PDBu) BINDING

Table 4 lists phorbol dibutyrate binding to rat liver nuclei protein kinase C during various purification steps. It may be worth noting that isolated rat liver nuclei do not reveal any PDBu binding, or protein kinase C activity. After isolated nuclei are treated with Triton X-100 and sonicated appropriately the kinase C activity is seen as well as the binding with phorbol dibutyrate. The lack of phorbol dibutyrate binding in isolated nuclei preparations, prior to Triton X-100 extraction, may also support the argument that the nuclei preparation is devoid of cytoplasmic or plasma membrane contamination. In the nuclei extracted with Triton X-100 there is higher degree of PDBu binding as compared to protein kinase C enzymatic activity. This underestimation of kinase C activity could be attributed to the presence of enzyme inhibitors which may not affect PDBu binding.

A second important observation consistant with this result is a progressive increase in phorbol ester binding with the purification steps. However, a perusal of enzymatic activity over phorbol ester bound reveals an almost identical value after DEAE-cellulose chromatography and Phenyl-Superose chromatography. This puts forward a strong argument in favour of nuclear protein kinase C serving as a phorbol ester receptor as does its cytosolic counterpart.

COFACTOR REQUIREMENT FOR NUCLEAR VERSUS CYTOSOLIC PROTEIN KINASE C ACTIVITY

For studying various cofactor requirements for protein kinase C activity of nuclei and cytosol, each of these subcellular fractions were subjected to a single DEAE-cellulose chromatography and active fractions were pooled and analyzed enzymatically.

Table 4. Phorbol dibutyrate [^{3}H]-PDBu binding to rat liver nuclear protein kinase C

Purification steps	Proteins mg/ml	Specific activity pmol ^{32}P/min/ mg protein	[^{3}H]-PDBu bound pmol/mg protein	Enzyme activity/PDBu bound
Isolated nuclei	12.7	Not Detected	Not Detected	–
Nuclei extracted with Triton X-100	8.93	2.07	0.44	4.7
DEAE	0.264	394.0	5.7	69.0
Phenyl-Superose	0.021	3593.0	58.3	62.0

Concentration of [^{3}H]-PDBu was 40 nM and TPA was 32 μM. Isolated nuclei were whole unbroken nuclei. The isolated nuclei were suspended in requisite volume of a medium containing 2 mM EDTA, 0.05% Triton X-100, 20 mM Tris-HCl pH 7.5 and sonicated in aliquots for 10 seconds six times with one minute interval in between two sonications, incubated in ice for 30 min, centrifuged at 100,000 x g for 30 min. The resulting supernatant served as nuclear extract and the source of nuclear protein kinase C. Protocol for PDBu binding was followed as recommended (17).

Two most salient distinctions appear between the nuclear versus cytosolic kinase C activity. Firstly, a higher calcium optimum for nuclear protein kinase C activity. Secondly, in the absence of added calcium, the nuclear enzyme activity monitored with PS + TPA (or DAG) did not reach the magnitude of the activity attained when optimum calcium plus PS were present. While the cytosolic enzymatic activity was similar in both assay conditions. That is when the assay was done with Ca^{++} + PS or TPA (or DAG) + PS. In this regard the rat liver cytosolic protein kinase C activity behaves like the rat brain cytosolic enzyme. It is well established that in the presence of phorbol ester (TPA) the K_a for calcium is decreased and protein kinase activity is fully manifested in the presence of PS + TPA (15).

The only other similar observation reported is the protein kinase C activity of COS cell requiring 3.0 mM calcium concentration in the presence of saturating concentrations of diacylglycerol (16).

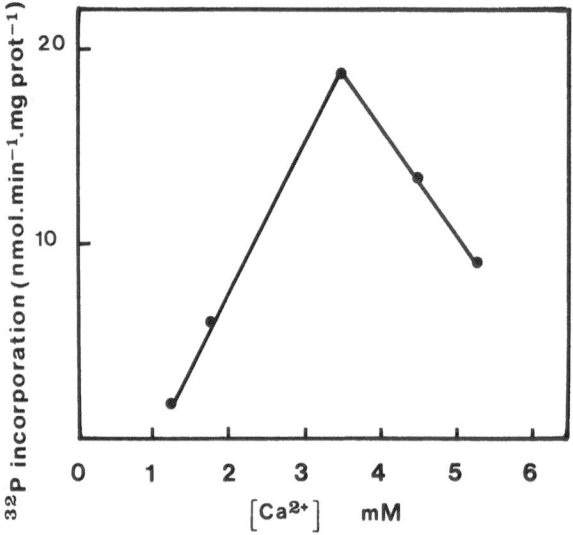

Fig. 4. A (nuclei) and B (cytosolic) protein kinase C activity. Dotted bars represent the enzymatic activity in the presence of calcium plus phosphatidylserine (PS).

CHARACTERISTICS OF PARTIALLY PURIFIED NUCLEAR ENZYME

Active fractions obtained from the Mono Q chromatography (containing two protein bands on SDS-polyacrylamide gel) were pooled and the calcium (Fig. 5) and cofactors requirements (Table 5) were examined. Maximum enzymatic activity was manifested at 3.5 mM calcium concentration. In fact starting right from the liver homogenate up to partially purified nuclear enzyme, a peak of kinase C activity has been consistently manifested at a calcium level of 3.5 mM (Fig. 2).

Table 5.　Partially purified rat liver nuclei protein kinase C activity determined with various cofactors

Cofactors added	Enzymatic activity nmol ^{32}P/min/mg protein
Ca^{++}	0.08
PS	0.10
Ca^{++} + PS	18.75
Ca^{++} + PS + TPA	22.50
Ca^{++} + PS + DAG	24.33
PS + TPA	7.2
PS + DAG	7.4
Ca^{++} + TPA	1.8
Ca^{++} + DAG	1.9

Tris-HCl 20 mM, pH 7.5, EGTA 1.2 mM, Ca^{++} 3.5 mM, PS 16.6 µg, histone H1 20 µg, phenylmethylsulphonyl fluoride 2.0 mM, TPA 81 nM, DAG 100 µM in a total volume of 100 µl constituted the standard assay medium.

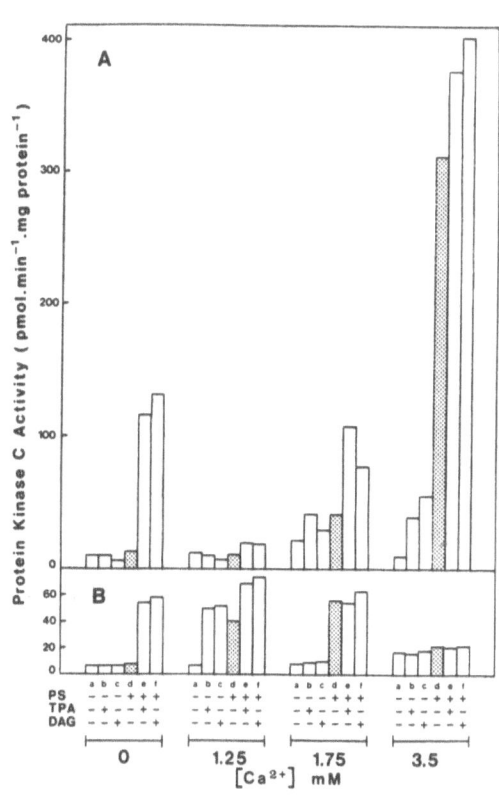

Fig. 5.　Calcium optimum for partially purified rat liver nuclei protein kinase C activity. The enzymatic activity was determined in the presence of phosphatidylserine minus the negligible activity measured in the absence of PS.

A close perusal of these results reveals that the characteristics of protein kinase C purified from rat liver nuclei remain identical to those observed after single step DEAE-cellulose chromatography (fig. 4 A). The enzyme is striclty dependent on high calcium and phospholipid. These results strongly support the location of protein kinase C in rat liver nuclei. This information sets the stage for seeking a pertinent role for protein kinase C at the site of cell nuclei during cellular growth and gene expression.

The nuclear location and characterization of protein kinase C is important and useful in view of its three main sites of action : the plasma membrane, the cytoplasm and the nucleus.

It is hoped that phosphorylation-dephosphorylation mechanism at the site of nuclei (16) will be better understood and will in turn enhance the knowledge of role played by protein kinase C in cell growth and gene expression.

Acknowledgements

We thank Dr. J.P. Zanetta for helpful discussions and Mrs S. Ott for secretarial assistance. Financial support from the Association pour la Recherche sur le Cancer (A.N.M. and G.L., grant n° 6451) is acknowledged.

REFERENCES

1. Y. Nishizuka, Studies and perspectives of protein kinase C, Science 233:305 (1986)
2. M.J. Berridge and R.F. Irvine, Inositol triphosphate, a novel second messenger in cellular signal transduction, Nature 312:315 (1984).
3. J.C. Lacal, J. Moscat and S.A. Aaronson, Novel source of 1,2-diacylglycerol elevated in cells transformed by Ha-ras oncogene, Nature 330:269 (1987).
4. T. Yoshimasa, D.R. Sibley, M. Bouvier, R.J. Lefkowitz and M.G. Caron, Cross-talk between cellular signalling pathways suggested by phorbol ester-induced adenylate cyclase phosphorylation, Nature 327:67 (1987).
5. J. Miller and E. Rozengurt, Protein kinase C and cyclic AMP pathways cross-talk, Nature 331:492 (1988).
6. J.C. Cambier, M.K. Newell, L.B. Justement, J.C. McGuire, K.L. Leach and Z.Z. Chen, LA binding ligands and cAMP stimulate nuclear translocation of PKC in B lymphocytes, Nature 327:629 (1987).
7. U.K. Misra and N. Sahyoun, Protein kinase C binding to isolated nuclei and its activation by a calcium-phospholipid-independent mechanism, Biochem. Biophys. Res. Commun. 145:760 (1987).
8. J.G. Wood, P.R. Girard, G.J. Mazzei and J.F. Kuo, Immunocytochemical localization of protein kinase C in identified compartments of rat brain, J. Neurosci. 6:2571 (1986).
9. S. Capitani, P.R. Girard, G.J. Mazzei, J.F. Kuo, R. Berezney and F.A. Manzoli, Immunochemical characterization of protein kinase C in rat liver nuclei and subnuclear fractions, Biochem. Biophys. Res. Commun. 142:367 (1987).
10. A. M. Edelman, D.K. Blumenthal and E.G. Krebs, Protein serine/threonine kinases, Ann. Rev. Biochem. 56:567 (1987).
11. Y.A. Hannun, C.R. Loomis and R.M. Bell, Activation of protein kinase C by Triton X-100 mixed micelles containing diacylglycerol and phosphatidylserine, J. Biol. Chem. 260:10039 (1985).
12. J. Zwiller, M.O. Revel and A.N. Malviya, Protein kinase C catalyzes the phosphorylation of guanylate cyclase in vitro, J. Biol. Chem. 260:1350 (1985).
13. A.N. Malviya, J.C. Louis and J. Zwiller, Separation from protein kinase C – a calcium independent TPA-activated phosphorylating system, FEBS Lett. 199:213 (1986).

14. S. Azhar, J. Butte and E. Reaven, Calcium-activated phospholipid-dependent protein kinases from rat liver : subcellular distribution, purification, and characterization of mutliple forms, Biochemistry 26, 7047 (1987).
15. J.P. Arcoleo and I.B. Weinstein, Activation of portein kinase C by tumor promoting phorbol esters, teleocidin, aplysiatoxin in the absence of added calcium. Carcinogenesis 6:213 (1985).
16. T. Hunter, A thousand and one protein kinases, Cell 50:823 (1987).
17. N.A. Sharkey and P.M. Blumberg, Specific binding of [20-^{3}H] 12-deoxy-phorbol 13-isobutyrate to phorbol ester receptor subclasses in mouse skin particulate preparations. Cancer Res. 45:19 (1985).

PLASMA MEMBRANE CYTOCHROME COMPONENTS: MIDPOINT REDOXPOTENTIALS, LIGHT- AND

NADH MEDIATED REDUCTIONS

Han Asard, Mireille Venken, Roland Caubergs and Jan A. De Greef

University of Antwerpen, R.U.C.A.
Department of Biology
Groenenborgerlaan, 171, B-2020 Antwerpen, Belgium

The thorough characterization of PM redox components is highly depend-ent on the method for isolation and purity of the membranes. The purity and cytochrome content of PM fractions from several plant species, obtained by aqueous two phase partitioning, was investigated. Preparations are generally enriched in vanadate-sensitive ATPase activity (PM marker) and show strongly decreased (6 to 50 fold) levels of NADH-CCR activity (presumptive ER marker). The remaining activity might originate from specific PM redox constituents. Variable amounts of latent IDPase activity (Golgi marker) were detected. PM preparations oxidize NADH (in presence of duroquinone and KCN) at rates between 9 and 40 nmol NADH.min^{-1}.mg^{-1} protein. Triton X-100 stimulation indicates the presence of tightly sealed right-side out PM vesicles.

In previous work the presence of a specific high potential b-type cyto-chrome in the PM of cauliflower was demonstrated (Caubergs et al., this volume). Ascorbate reduces 30-70% of the total cytochromes, indicating E'o values above +80 mV. Blue light sensitive components (LIAC) varied from 5-20%, but independent from the level of ascorbate reduction. In addition redox titrations were performed on cauliflower, beans and oat leaves. The major component in each case has a potential between +120 and +160 mV. Statistical analysis further resolved the presence of varying amounts of other cytochromes.

In order to further characterize the PM cytochrome components we attempted to detect P-450 like spectra. Difference spectra after CO binding show a dominant maximum at 450 nm both in microsomes and intracellular mem-branes. Calculated amounts were 26 $pmol.mg^{-1}$ (40% of total) and 20 $pmol.mg^{-1}$ (20%) respectively. On the other hand purified PM's show small amounts (11.2 $pmol.mg^{-1}$, 10%). Subcellular localization was further investigated after density gradient fractionation of cauliflower microsomes. The distribution of P-450/420 clearly correlates with CCR activity at 20 to 25% sucrose, whereas no absorption was found in the PM region of the gradient. In con-trast some NADH-CCR activity is located at the maximum of vanadate-sensitive ATPase (35-40%), consistent with the presence of PM specific NADH oxido-reductases.

Finally, in recent experiments the ability of NADH to reduce PM b-type cytochromes was tested. No reductions were obtained in aerobic conditions. However after addition of O_2-consuming systems, significant reductions are measurable in both cauliflower (up to 40% of total cytochromes) and zucchini (up to 20%) PM's. Addition of duroquinone increases the degree of reduction. It was also demonstrated that in addition to NADH unaltered LIAC's were obtained. The question raises whether these amounts can be explained on the basis of contaminating ER cytochromes in two phase purified PM fractions. These data possibly suggest a role for NADH as electron donor in cytochrome containing electron transfer reactions at the PM.

LOCALISATION OF DONOR AND ACCEPTOR SITES OF NAD(P)H DEHYDROGENASE(S)

USING "INSIDE-OUT" AND "RIGHT-SIDE-OUT" PLANT PLASMA MEMBRANE VESICLES,

AND CHARACTERIZATION OF PLASMA MEMBRANE-BOUND b-CYTOCHROMES

Per Askerlund, Christer Larsson and Susanne Widell

Department of Plant Physiology
Box 7007
S-220 07 Lund, Sweden

"Inside-out" and "right-side-out" plasma membranes (PM) were prepared from cauliflower as described for sugar beet leaf PM (1). Based on the latency of the K,Mg^{2+} ATPase, the "inside-out" fraction contained 51 % "inside-out" vesicles whereas the "right-side-out" fraction contained 10 % "inside-out" vesicles. Observe that with sugar beet leaf PM a much more extreme separation was obtained (1). The "inside-out" vesicles from cauliflower showed ATP-dependent proton uptake and were thus sealed.
Cytochrome c (cyt c), ferricyanide (FeCN) and phenyl-p-quinone could all accept electrons from the PM-bound NADH dehydrogenase in the presence of antimycin A and KCN. The latencies for these different activities with "inside-out" and "right-side-out" PM showed patterns similar to the ATPase. Since neither NADH, nor cyt c, nor FeCN can penetrate the membrane, these results indicate that the binding sites for NADH, cyt c and FeCN are on the cytoplasmic side of the PM. However, we can not exclude that FeCN and cyt c also accept electrons on the apoplastic side.
With all electron acceptors NADH gave much higher activities than NADPH. No NAD(P)H could be oxidized in the absence of added electron acceptors under the same conditions. The NADH: FeCN and NADH: cyt c reductase activities were differently affected by concentrations of Triton X-100 above 0.015 % (w/v); the activity with cyt c decreased dramatically in contrast to that with FeCN, probably due to the disruption of a redox chain between a dehydrogenase and a b-type cyt.

PM-bound cytochromes were characterized with difference spectrophotometry at 77 K. In PM from spinach leaf and cauliflower, there was a main component reducible by ascorbate, with a symmetric α-band at 556-557 nm. In addition a component with a split α-band (552 + 558 nm), probably identical with cyt b_5, was seen when the dithionite minus ascorbate spectrum was recorded. Reduction with NADH gave a symmetrical peak similar to that with ascorbate, but smaller. The absence of a split α-band with NADH as reductant suggests that there is no cyt b_5 reductase in the PM. Similar results were obtained with PM from barley leaf and root. PM from cauliflower (2) and spinach leaf contained significant amounts of cytochrome P-420 and P-450.

REFERENCES

1. Larsson, C., Widell, S. & Sommarin, M. (1988) FEBS Lett. 229: 289-292

2. Kjellbom, P., Larsson, C., Askerlund, P., Schelin, C., & Widell, S. (1985)
   Photochem. Photobiol. 42: 779-783

NADH-FERRICYANIDE OXIDOREDUCTASE ACTIVITY IN MAIZE ROOT CELL

PLASMA MEMBRANE

Jean-Pierre Blein, Isabelle Bourdil and René Scalla

Laboratoire des Herbicides
INRA, BV 1540
21034 Dijon, France

Plasma membranes from maize roots were prepared by two-phase partition (mb.A) (1 g of microsomal pellet/15 g of 6.4 % (w/w) Dextran T500, 6,4 % (w/w) polyethylene glycol 4000), or by differential centrifugation, sedimentation through a 30 % sucrose cushion, and washing with Triton X-100 plus KBr (mb. B). Properties of NADH-ferricyanide oxidoreductase activity of these membrane preparations were compared.

Both preparations were enriched in plasma membrane ATPase. That activity was Mg^{2+}-dependent, inhibited with vanadate, and unaffected by molybdate, oligomycin and nitrate. The preparations were almost devoid of NADH-cytochrome c reductase and IDPase activities. Cytochrome oxidase was not detectable.

Specific activities of NADH-ferricyanide oxidoreductase were 930 and 450 nmol ferricyanide reduced per min and mg protein, in preparations A and B respectively. These activities increased in the presence of detergents. Maximal stimulation by Triton X-100 was reached at a concentration of 0.01 %, (mb. A 40 %, mb. B 57 %). Lysolecithin, at a concentration of 0.05 mg/ml stimulated activity of mb. A by 100 % and mb. B by 90 %. The rate of NADPH oxidation amounted to about 70 % of that of NADH oxidation. This activity reached a maximum at 0.5 mg/ml lysolecithin with mb. A and was independent of detergent concentration (between 0.1 to 1 mg/ml) with mb. B.

In both preparations, NADH-ferricyanide oxidoreductase activity was maximal at pH 7. In the presence of lysolecithin (0.1 mg/ml), another peak of activity appeared at pH 5.

In membranes A and B, ferricyanide reduction showed biphasic kinetics. Double reciprocal plots indicated the existence of two apparent Km and Vmax. These atypical kinetics were found independently of the presence of detergent during membrane preparations or activity measurements. NADH oxidation also followed biphasic kinetics in membrane A preparations, and was inhibited by substrate excess in membrane B. The latter property was found to be induced by Triton X-100 washing of membranes.

In conclusion, pH effects and reaction kinetics suggest that plasma membrane preparations contain more than one NADH-ferricyanide oxidoreductase. Alternatively, the system could possess several affinity sites for its substrates.

THE EFFECT OF SETHOXYDIM ON THE NAD(P)H OXIDATION

ASSOCIATED WITH ISOLATED PLASMALEMMA VESICLES

Elke Fischer, Ulrich Lüttge, and Zeno Varanini

Institut für Botanik
Technische Hochschule Darmstadt
Darmstadt, FRG

Plasmalemma enriched vesicles of two day old maize roots were purified using differential and discontinuous sucrose gradient centrifugation according to the technique described by De Michelis and Spanswick, (1986). The 30 - 40 % interface was enriched in K^+-stimulated, vanadate-sensitive, nitrate-insensitive and azide-insensitive ATPase activity.

Plasmalemma vesicles exhibited NAD(P)H-dependent ferricyanide reduction activity which was insensitive to inhibitors of mitochondrial electron transport such as KCN, and antimycin A. The activity of the plasmalemma electron transfer from NAD(P)H to ferricyanide was higher in the presence of NADH (K_m = 67 µM; V_{max} = 5.3 µmol (mg protein)$^{-1}$ min^{-1}) than NADPH (K_m = 65 µM; V_{max} = 2.0 µmol (mg protein)$^{-1}$ min^{-1}).

The selective post-emergence herbicide sethoxydim, a cyclohexanedione derivative, specifically inhibited the NAD(P)H-dependent ferricyanide reduction of plasmalemma vesicles at pH values of the test-medium below pH 6.5. The degree of inhibition increased with decreasing pH. The reduction of NAD(P)H-dependent ferricyanide reductase by 500 µM sethoxydim at pH 5 was approximately 50 % of the control. The ATPase activity and the ATP-dependent H^+ transport of the plasmalemma vesicles were not affected by sethoxydim.

This work was supported by BASF AG, D-6703 Limburgerhof, FRG.

REFERENCES

De Michelis, M.I., and Spanswick, R.M., 1986, H^+ pumping driven by vanadate-sensitive ATPase in membrane vesicles from corn roots, Plant Physiol., 81:542.

THE RESOLUTION OF TWO PHASES OF FERRICYANIDE REDUCTION IN CULTURED CARROT

CELLS

J.D.C. Chalmers[1] and J.O.D. Coleman

Department of Plant Sciences, University of Oxford, U.K.
[1]Present address: School of Biological Sciences, University
of East Anglia, Norwich, U.K.

Ferricyanide reductase activity is popularly determined over a
relatively short period frequently as short as 10 mins. Over this time
period the rate of ferricyanide reduction versus time is generally
interpreted as being linear. By employing an amperometric method of
ferricyanide measurement (a ferricyanide recycling method) (Chalmers et al
1984) rapid rates of electron transport may be followed over extended
periods of time with no change in the redox potential of the bulk solution.
A second advantage of this electrochemical method is that the output from
the electrode (current) is directly proportional to the turnover of
ferricyanide at any point in time rather than a measure of the remaining
bulk solution ferricyanide concentration. The current-time profile is thus
a direct representation of rate versus time. With an assay system showing
these two features it becomes clear that in cultured carrot cells the rate
of ferricyanide reduction with incubation time is not constant even over the
initial relatively short time period, the first 10-15 mins, indeed the
greatest rate of change in rate is observed at this stage.

We have shown that the ferricyanide current-time profile is sigmoidal
in shape and is suggestive of two phases of ferricyanide reduction described
here as phase A and phase B. These two phases can be further resolved by
the use of inhibitors. A flexible correlation exists between these changing
rates of electron flux and H^+ extrusion.

The most simple type of explanation for this phenomenon would be one invoking
a release of substrate from non cytoplasmic pools as the incubation
progresses and available substrate is consumed. Alternatively the faster
rate of ferricyanide reduction may involve a mobilization of new proteins in
the membrane in response to a diminished arrival of electrons at the
physiological electron acceptor due to interception by ferricyanide. Lastly
these phases may in fact be a feature peculiar to cultured cells, either
involving the consumption of stored carbohydrate, cultured cells contain
large amyloplasts, or purely an equilibration phenomenon or a response to
harvesting shock.

REFERENCES

Chalmers, J.D.C., Coleman, J.O.D. and Walton, N.J. (1984) Use of an
Electrochemical Technique to Study Plasmamembrane Redox Reactions in
Cultured Cells of Daucus carota L. Plant Cell Reports 3: 243-246.

MODIFICATION OF PLASMA MEMBRANE DT-DIAPHORASE ACTIVITY UPON TRANSFORMATION: A COMPARISON BETWEEN ASTROCYTES IN PRIMARY CULTURE AND C6 GLIOBLASTOMA CELLS

M. Mersel, L. Vitkovic[a], G. Vincendon and
A.N. Malviya

Centre de Neurochimie du CNRS et de l'Unité 44
de l'INSERM, 5 rue Blaise Pascal, 67084 Strasbourg Cedex, France and [a]Laboratory of Molecular Biology, NINCDS, National Institutes of
Health, Bethesda, MD, USA

Acceptor oxidoreductase activity using NADH or NADPH as electron donnors, dichloroindophenol (DCIP) and potassium ferricyanide [$K_3 Fe(CN)_6$] as external electron acceptors and rotenone, dicoumarol and antimycin as inhibitors, was investigated in plasma membrane isolated from neonatal rat astrocyte primary culture and the plasma membrane of glioblastoma cell lines (C6). Astrocytic plasma membranes were enriched approximately 10 fold in both NADH and NADPH-DCIP oxidoreductase specific activities (300 nmol min/mg protein), whereas plasma membranes derived from C6 cells possessed NADH and NADPH-ferricyanide oxidoreductase activity (9 nmol ming/mg protein) - typical of DT-diaphorase system, but no enrichment of those activities was observed in the isolated C6 plasma membrane. NADPH-DCIP reductase activity of astrocyte plasma membranes was sensitive towards dicoumarol and antimycin (45% and 71% inhibition, respectively). The NADH or NADPH-ferricyanide reductase activity of C6 plasma membranes was inhibited by rotenone, dicoumarol and antimycin (40%, 60% and 10%, respectively). Both cell types displayed NADH-DCIP cell surface reductase activity. In addition, NADPH-DCIP reductase activity was found at the surface of C6 cells. Astrocyte cell surface activity was sensitive towards dicoumarol exclusively (70% inhibition) whereas in the case of C6 cells, rotenone, dicoumarol or antimycin inhibited the enzymatic activity.

These results raise a possibility that the transformation phenomenon may alter the topological orientation of DT-diaphorase activity located at the plasma membrane. Can cellular division or maturation be attributed to the observed variation in DT-diaphorase activity?

TRANSPLASMA - MEMBRANE REDOX SYSTEM IN ERYTHROCYTE GHOSTS

F. Waerenborgh**, M.P. Bicho* and C. Manso*

*Instituto Quimica Fisiológica
**LNETI ICEN Dep. Radioisótopos
Inst. Quimica Fisiológica, Fac. Medicina
1600 Lisboa, Portugal

NADH diaphorase is present in the membrane of ery-throcytes. Its most potent electron acceptor is ferricya-nide, which was used to measure enzymatic activity, (Mishra and Passow, J. Membrane Biol. 1, 214-224, 1969; Zamudio et al, Arch. Biochem. Bioph. 129, 336-345, 1969).

A pH profile curve was made which is compatible with the presence of cysteine SH groups in the active center.
Sulfhydryl reagents like PCNBS (10^{-5}) and NEM (10^{-3}M) reduced respectively the enzymatic activity by 100% and 80%.

Atebrin (10^{-3}) inhibits the enzyme by 43% and azide (10^{-1}M) also inhibits by 53%. In order to evaluate the possible effect on SH groups of oxygen related radicals we studied the effect of superoxide dismutase, catalase and ethanol.

Catalase (3 000 U, pH 6.1) activated the enzyme acti-vity by 50% and ethanol (10^{-3}M) by 35%. Superoxide dismu-tase (3 000 U, pH 6.1) had no effect.

These results suggest that hydrogen peroxide is gene-rated in the system and can act by itself as an inhibitor.

This inhibitory effect is probably not related to oxidation of sensitive SH groups since glutathione gave no protection and NEM did not agravate the inhibition.

STIMULATION OF NADH-CYTOCHROME C OXIDOREDUCTASE OF ENDOPLASMIC RETICULUM

TRANSITIONAL ELEMENTS BY RETINOL

Dorothy M. Morré, S. Rathinagounder, Jingan Zhao, D. David
Nowack and D. James Morré

Purdue University, West Lafayette, IN 47907

INTRODUCTION

Vesicles, termed transition vesicles, that bleb from part-rough,
part-smooth regions of endoplasmic reticulum (ER) and are thought to be
involved in membrane trafficking between the ER and Golgi apparatus, are
increased in numbers in livers of rats gavaged with excess retinol (Morré
and Morré, 1987). Here, we report studies to elucidate possible
biochemical effects on this special vesicle-forming portion of the ER.

MATERIALS AND METHODS

Microsomes and ER fractions were isolated as described (Nowack et
al., 1987) from livers of male rats. NADH-cytochrome c was assayed as
described (Miyake et al., 1968). Vitamin A compounds were added in
ethanol with an equivalent amount of ethanol alone in controls. The
retinol/retinoic acid-membrane mixtures were preincubated for 3 min at
37°C prior to initiation of the reaction by addition of NADH.

RESULTS AND DISCUSSION

Retinol (2 and 200 μM) both stimulated (2-20 μM) and inhibited (20-
200 μM) the NADH-cytochrome c reductase activity in a dose-dependent
manner related to the protein/retinol ratio. The retinol concentrations
were well below those normally associated with unspecific detergent
effect of retinol on membranes. When subfractionated on sucrose density
gradients, the retinol-responsive fraction was concentrated in the
lightest ER fractions, those consisting of part-rough, part-smooth
transitional ER elements. Retinoic acid and Triton X-100, a detergent,
at high concentration inhibited the activity but did not stimulate in the
same range of concentrations as retinol. Retinol in the concentration
range of 5 to 50 μM decreased membrane microviscosity of the ER. The
findings suggest an interaction of retinol with the ER portions of
transition vesicle formation and ER-Golgi apparatus associations.

REFERENCES

Miyake, Y., Gaylor, J. L., and Mason, H. S. (1968) J. Biol. Chem. 243, 5788-5797

Morré, D. J., and Morré, D. M. (1987) Int. Cell Biol. Repts. 11, 89-93.

Nowack, D. D., Morré, D. M., Paulik, M., Keenan, T. W., and Morré, D. J. (1987) Proc. Natl. Acad. Sci. U.S.A. 84, 6098-6102.

NADH-DEHYDROGENASES IN SYNAPTIC PLASMA MEMBRANES

J.-L. Dreyer and T. Treichler

Dept. Biochemistry, Univ. of Fribourg
rue du Musée 5, CH-1700 Fribourg

Transplasma-membrane redox systems have received increasing attention in the last years as they have been found in all cells examined and appear to be a general characteristic of cells. NADH-dehydrogenase activities, using several electron acceptors, are observed in highly purified synaptic membrane preparations. These activities cannot be accounted for by microsomal or mitochondrial contamination. The NADH-dehydrogenases of synaptic membranes show a selective response to several agents: they are insensitive to rotenone, antimycin, chelating agents such as EDTA, o-phenanthroline, alpha-picolinate or nitrilotri-acetate and N-ethylmaleimide; inhibition is observed with atebrin, azide, p-chloro-mercuribenzoate, and the activity vanishes upon limited proteolysis; the neuroexcitatory aminoacids Glu and Asp and their agonists markedly stimulate whereas dopamine and adrenaline strongly inhibit the NADH-acceptor oxido-reductase activities. The electron transport system is further markedly activated by antidepressants, but is unaffected by alpha- and beta-agonists or antagonists, neither by peptide hormones, opiates, adenosine, GABA and benzodiazepines, nor by nicotinic and muscarinic acids or acetylcholine.

Chemical analysis revealed the presence of Fe and acid labile sulphide, flavin, and ubiquinone. The intact preparations have been investigated by EPR spectroscopy at 13° K, but exhibited no EPR-detectable signal whether untreated, oxidized with ferricyanide or reduced with dithionite. However treatment of the synaptic membranes with sodium cholate unmasks an isotropic signal around $g = 2.01$ with features typical for a (3Fe-xS) center as established from power saturation and temperature studies. The possibility of an association of this redox component with receptors for neurotransmitters was investigated by means of selective co-solubilization studies.

The effects of NADH-dehydrogenases on adenylate cyclase in highly purified synaptic membranes has also been investigated. Oxidizing agents such as ferricyanide or dichlorophenol-indophenol in the presence of reduced pyridine nicotinamides produce a significant activation of adenylate cyclase which was stimulated 2.4 to 3.6 fold. A number of neuro-active agents which affect the rate of electron transfer in NADH-dehydrogenases were tested for their effects on adenylate cyclase mediated through the activation of the NADH-dehydrogenases. A marked synergistic activation of adenylate cyclase activity was observed under these conditions with dopaminergic agents (pindolol, bromo-ergocryptine, apomorphine), whereas naloxone and clonidine (which both do not directly affect NADH-dehydrogenase activities) and atebrin inhibit the activation of adenylate cyclase mediated through the NADH-dehydrogenases. Furthermore, inhibition of adenylate cyclase activity by a variety of drugs (e.g. antidepressants, excitatory aminoacids, spiperone) could be prevented by the NADH-dehydrogenases. Finally, NADH-dehydrogenase markedly potentiates the activation of adenylate cyclase by forskolin. These data establish that NADH-dehydrogenases in synaptic membranes are involved in post-receptor signal transduction and that the redox environment in the cell modulates in some manner neurotransmitter-signaling by influencing the production of the second messenger cAMP.

ARE PLASMALEMMA REDOX PROTONS INVOLVED

IN GROWTH CONTROL BY PLANT CELLS?

R. Barr and F. L. Crane

Department of Biological Sciences, Purdue University
West Lafayette, Indiana, 47907, U.S.A.

It has been established (Barr 1987,1988) that in cultured carrot cells protons are excreted from two different sources: by the action of the plasma membrane H^+-ATPase, located on the cytoplasmic side of the membrane, and through plasmalemma NADH-dehydrogenase (Brightman et al., 1987,1988), a transmembrane enzyme, which transports electrons in association with protons across the plasmalemma from the cytoplasmic side toward the outside of the cell. Evidence for this comes from studies with inhibitors of the H^+-ATPase and plasma membrane redox. In this study, two new inhibitors of plasma membrane redox reactions, acridine orange and neutral red, were tested on H^+ excretion and growth of carrot cells. H^+ excretion, potassium ferricyanide reduction, NADH oxidase and peroxidase activities were also assayed on carrot cells grown for 8 days in presence of high NaCl concentrations (100-300 mM). It was found that acridine orange in concentrations of 3-15 μM inhibited H^+ excretion from 50-85% and ferricyanide reduction 100% upon direct addition of acridine orange to carrot cells, but growing carrot cells with acridine orange in similar concentrations stimulated the H^+-ATPase-generated H^+ but inhibited ferricyanide reduction and associated redox H^+ 100%. Carrot cell growth, NADH oxidase and peroxidase were inhibited from 80-100%. Neutral red inhibited the action of the plasma membrane H^+-ATPase > 50%, and growth and associated plasma membrane redox activities from 50-100%. Growth of carrot cells on high NaCl concentrations also inhibited the H^+-ATPase-excreted H^+ 38% and growth, measured as dry weight of carrot cells, 66%, while plasma membrane redox functions - ferricyanide reduction, NADH oxidation and peroxidase activity gave from 90-100% inhibition. From these data it appears that H^+ excretion by plasma membrane redox reactions controls H^+ in some special membrane domain that is associated with carrot cell proliferation.

References

Barr, R., 1987, in "Electron Transport in Plasma Membranes," J. Ramirez, ed., Publications Office CSIC, Madrid, pp. 27-40.
Barr, R., 1988, Physiologia Plantarum (in press).
Barr, R., and Crane, F. L., 1987, Plant Physiol. 83 (supplement):53 (No. 317).
Brightman, A. O., Barr, R., Crane, F. L., and Morré, D. J., 1987, Plant Physiol. (supplement):93 (No. 562).
Brightman, A. O., Barr, R., Crane, F. L., and Morré, D. J., 1988, Plant Physiol. (in press).

IS THE PROTON EXTRUSION COUPLED TO FERRICYANIDE REDUCTION MEDIATED BY

THE H $^+$ -ATPase ?

Jean-Pierre Blein and René Scalla

Laboratoire des Herbicides
INRA, BV 1540
21034 Dijon (France)

In sycamore cell cultures, medium acidification can be induced by addition of fusicoccin (known to stimulate ATPase activity), or by addition of ferricyanide.

It has been reported that fusicoccin and ferricyanide exert additive effects on H $^+$ -extrusion, which suggests the existence of two separate proton transport systems. We have observed such an additivity, but we think that, before the existence of two systems can be postulated, it must established that ATP-dependent H $^+$ - e x t r u s i o n proceeds at its maximal rate when ferricyanide is added to the cells.

We have noticed that H $^+$ -extrusion resulting from addition of fusicoccin does not follow a linear time course, but slows down after a few minutes. We hypothesize that the initial velocity is limited by the available pool of protons, and that the second phase is limited by the cytosolic ATP concentration. In order to test this possibility, we have pretreated cells with malate. This treatment resulted in an increase of the initial rate of fusicoccin-dependent H $^+$ -extrusion. This increase, which was proportional to the malate concentration, suggested that fusicoccin-dependent proton extrusion does not necessarily works at its maximal rate before addition of ferricyanide. In these conditions, when ferricyanide was added during the second phase of fusicoccin-stimulated proton extrusion, no stimulation of proton extrusion was noticed, i. e. there was no longer additivity of rates of proton extrusion.

On another hand, we have found that all the inhibitors of ferricyanide reduction that we have tested (carbanilate derivatives, oligomycin) also inhibit fusicoccin-stimulated (ATP-dependent) proton extrusion. Moreover, we have been able to establish a correlation between the inhibition by carbanilate derivatives of proton extrusion resulting from ferricyanide reduction or from treatment by fusicoccin.

We conclude that the activity of the redox system and that of ATPase could be coupled, the latter being responsible for all proton extrusion, resulting from ferricyanide reduction and from stimulation by fusicoccin.

RELATIONSHIP OF PLASMA MEMBRANE REDOX ACTIVITY TO PROTON TRANSPORT IN

ROOTS OF IRON-DEFICIENT CUCUMBER

E. Alcántara and M.D. de la Guardia

Dpto. de Agronomia, Escuela de Ingenieros Agrónomos
Universidad de Córdoba
Córdoba, Spain

Under iron deficiency, cucumber (<u>Cucumis</u> <u>sativus</u> L. cv. Pickler) plants respond with a great increase in the capacity to reduce ferric chelates at their root surface and with an acidification of the root medium. The transmembrane e^- transfer to extracellular e^- acceptors is associated with a H^+ efflux which should be differentiated from the one induced by iron deficiency without the need of an extracellular electron acceptor. In the last years diverse plasma membrane redox systems have been studied. In some cases the ratio e^-/H^+ efflux during the reduction process has been established. For instance, $1e^-/1H^+$ in isolated mesophyll cells of <u>Asparagus</u> <u>sprengeri</u> Regel (1) or $2e^-/1H^+$ in roots of iron deficient <u>Phaseolus</u> <u>vulgaris</u> L. (3). However, Rubinstein and Stern (2) have shown in roots of <u>Zea</u> <u>mays</u> L. a redox system which involves only the transport of e^- across the plasmalemma. In this case H^+ efflux is an indirect effect of the depolarization of the membrane potential, causing the activation of the membrane H^+-ATPase.

In this work we have estudied the iron deficiency induced plasma membrane redox system of roots of cucumber plants. We have used plants less than 12 days old that had been deprived of Fe in the nutrient solution for 2 or 3 days before the redox activity determination. This was achieved with Fe III-EDTA as e^- acceptor and BPDS to measure the Fe II produced. The pH and in some cases K^+ efflux was also measured. The effect of the H^+-ATPase inhibitor DCCD (50 μM) was compared with a control. Results show that in control the ratio of e^-/H^+ efflux was higher than 10 at the begining and decreased with time. DCCD totally inhibited H^+ efflux and only nearly 50% the e^- efflux. During the reduction process K^+ efflux increased with time and was higher in the DCCD treatment than in control and in control higher than in the treatment without Fe as e^- acceptor. These results are in agreement with those obtained by Rubinstein and Stern (2) and suggest that the redox system induced under iron deficiency in cucumber only transports e^- , and the associated H^+ efflux is the result of an indirect effect on the membrane H^+-ATPase.

Supported by CAICYT, Project 2011/83.

REFERENCES

1. Neufeld,E. and A.W.Bown. 1987. Plant Physiol. 83, 895–899.
2. Rubinstein,B. and A.I.Stern. 1986. Plant Physiol. 80, 805–811.
3. Sijmons,P.C. , F.C.Lanfermeijer, A.H. de Boer, M.B.A. Prins, and H.F.Bienfait. 1984. Plant Physiol. 76, 943–946.

INTRACELLULAR pH MODIFICATIONS LINKED TO THE ACTIVITY OF THE FERRICYANIDE

DRIVEN ACTIVITY OF THE PLASMALEMMA REDOX SYSTEM IN *Elodea densa* LEAVES,

Acer pseudoplatanus AND *Catharanthus roseus* CELLS

Jean Guern[1], Yves Mathieu[1] , Geneviève Ephritikhine[1],
Cornelia I. Ullrich-Eberius[2], Ulrich Lüttge[2],
Maria-Térésa Marré[3] and Erasmo Marré[3]

Laboratory of Plant Cell Physiology, CNRS, 91198 Gif sur
Yvette, France[1] - Botanisches Institut, Technische
Hochschule, D-6100 Darmstadt, RFA[2] - Dipartimento di
Biologia, Universita di Milano, 20133 Milano, Italia[3]

The activity of the ferricyanide-driven redox system at the plasmalemma of plant cells is potentially able to modify the concentration of protons in the cytoplasm. This problem was studied by using the ^{31}P NMR technique to measure the cytoplasmic and vacuolar pH in *Elodea densa* leaves, *Acer pseudoplatanus* and *Catharanthus roseus* cells. *Elodea densa* leaves (about 2 g FW) or *Acer pseudoplatanus* and *Catharanthus roseus* cells (about 7 g FW) were packed in a 20 mm diameter NMR tube and perfused at a rate of 15-30 ml. min^{-1} with a buffered calcium sulfate solution. Spectra were recorded at 161.93 MHz with a WM 400 Brucker spectrometer by accumulating either 512 scans 1.2 sec each (*Acer pseudoplatanus* and *Catharanthus roseus* cells) or 1024 scans (*Elodea densa* leaves).

The cytoplasm of *Acer pseudoplatanus* cells which display a moderate redox activity, is not acidified by ferricyanide (1 mM) alone while a subsequent addition of Erythrosin B (EB, 500 µM), a putative inhibitor of the plasmalemma ATPase, drastically acidifies the cytoplasmic and vacuolar compartments. In *Elodea densa* leaves the redox activity is higher and ferricyanide (1 mM) induces a significant (0.15-0.3 pH unit) acidification of the cytoplasm. Fusicoccin (FC, 100 µM) is able to reverse this acidification with a marked increase of the vacuolar pH and EB (200 µM) inhibits the action of FC. Clear-cut results have been obtained with *Catharanthus roseus* cells which display the highest redox activity. Ferricyanide reduction, associated with K$^+$ release, induces a marked acidification of the cytoplasm (about 0.3 pH unit). The cytoplasmic acidification lasts as long as external ferricyanide is available but is totally reversed when all ferricyanide is used up.

The results obtained give support to the idea that the redox system at the plasmalemma transfers only electrons, depolarizes the cells and induces K$^+$ release. The consequences in terms of cytoplasmic pH depend on the relative activities of the redox system and the proton pump ATPase.

PRELIMINARY EVIDENCE FOR A REGULATION OF THE PLASMA MEMBRANE REDOX CHAIN BY THE ACTIVITY OF THE ATP-DRIVEN H^+ PUMP IN ELODEA DENSA LEAVES

M.T. Marrè , F. Albergoni, A. Moroni and E. Marrè

Department of Biology
University of Milano
Via Celoria 26, 20133 Milano, Italy

The reduction of extracellular electron acceptors such as $Fe(CN)_6^{3-}$ by a plasma membrane redox system is associated with the release of H^+ into the medium. Recent evidence indicates that the $Fe(CN)_6^{3-}$-induced net efflux of H^+ depends on an increase in activity of the ATP-driven H^+ pump, interpreted as a secundary consequence of the operation of the redox pump. According to this view, the H^+ pump would be activated by the depolarization of the transmembrane potential and also by the cytoplasm acidification accompanying $Fe(CN)_6^{3-}$ reduction by the redox system. On the other hand, fusicoccin, a toxin thought to stimulate the activity of the ATP-driven H^+ pump, has been repeatedly reported to increase the rate of $Fe(CN)_6^{3-}$ reduction.

As an interpretation of this effect of FC, we suggested that the increase in cytosolic pH and the potential hyperpolarization consequent to the FC-induced stimulation of the ATP-driven H^+ pump might stimulate the redox chain inasmuch as the activity of the latter leads to opposite effects (i.e. cytosol acidification and potential depolarization). Thus FC would counteract the feedback effect of these two factors on the redox system.

The results presented here bring a preliminary support to this hypothesis. In fact they show that in Elodea densa leaves the reduction of $Fe(CN)_6^{3-}$, in conditions in which it is stimulated by FC, and is associated with a depolarization of the potential and an increase in acidity of the cytosol and of the cell sap:

1) is inhibited by erythrosin B and by diethylstilbestrol, two inhibitors of the H^+ transporting ATPase presumably mediating the effects of FC on H^+ extrusion;

2) is inhibited by the cytoplasm acidifying treatment with isobutyric acid, under conditions in which the transmembrane potential difference is not influenced;

3) is stimulated by white light (30 Watt m^{-2}), which induces cell sap alkalinization by about 0.1 and 0.3 pH units respectively in the absence and in the presence of $Fe(CN)_6^{3-}$ and strongly stimulates ATP-driven H^+ extrusion (Albergoni et al., 1987, in: B.P. Marin, ed., "Plant Vacuoles. Their Importance in Plant Cell Compartimentation", NATO ASI, Series, Plenum Pubb. Corp., New York).

These results suggest that changes in cytoplamic pH may modulate the activity of the plasma membrane redox chain independently of changes of the transmembrane electrical potential.

RELATIONSHIP BETWEEN AUXIN-DEPENDENT PROTON EXTRUSION AND PLASMALEMMA REDOX SYSTEM IN MAIZE COLEOPTILE SEGMENTS

S. del Valle-Tascón, J. Salguero and
F. González-Darós

Departamento de Biología
Universitat de València
Vegetal, E-46100 Burjasot, Valencia, Spain

Abraded coleoptile segments reduce ferricyanide utilizing endogenous substrates. The kinetics of ferricyanide reduction shows an initial rapid rate followed by a slow rate of reduction. The duration of the initial phase is about 5 minutes. The slow phase is observed for more than 30 minutes. A Lineweaver-Burk plot of the slow phase was linear and revealed maximum velocity of the reaction of 42 nmols ferricyanide reduced/hour*coleoptile segment and an apparent Km (ferri) of 2.25mM.

The apparent rate of proton extrusion was stimulated by ferricyanide reduction. Ferricyanide-dependent acidification occurs immediately after the addition of ferricyanide. This stimulation decays continuously. Ferrocyanide does not affect the rate of proton extrusion.

The stoichiometry of ferricyanide reduced to proton extruded changes markedly over the course of the experiment. A steady state ratio usually between ferricyanide reduced and proton extruded is attained after 10 minutes. Because different ratios of ferricyanide reduced to proton extruded can be obtained it is possible that the trans-plasmamembrane redox system involves transport only of electron across the plasmalemma.

Exogenous NADH was immediately oxidized by corn coleoptiles. NADH oxidation could be fitted to normal Michaelis-Menten kinetics yielding an apparent Km of .8 mM in the Lineweaver-Burk plot. NADH oxidation strongly inhibits the rate of proton extrusion.

These experimental results suggest that trans-plasmamembrane redox activity regulates the proton extrusion in corn coleoptiles.

This work was supported by grant n 3469/83 to SVT from CAIYT (Spain).

POSSIBLE EXPLOITATION OF AN H+ RELATED MECHANISM IN THE TREATMENT OF BREAST AND COLON TUMORS: A NEW APPROACH TO THE ADJUVANT TREATMENT OF CANCER

S. Harguindey, J. Catalán, and L.M. Antón Aparicio*

Clínica La Esperanza, c/Esperanza 2, Vitoria, Spain
and Residencia Juan Canalejo, La Coruña, Spain *

SUMMARY

There is much of evidence, from basic research (Table I) and
intermediary metabolism (Table II) and in the different situations where certain tumors arise (Table III) to consider
that chronic abnormalities in microenvironmental pH may have
both a direct etiological role in epithelial human cancer as
well as an indirect one secondary to its mediating effects in
the "in vivo" activation of N-nitroso compounds and other mu-
cosal carcinogens (1). At the intracellular level, a pH rise
has recently been thought to represent a universal and/or
terminal, however non-specific mechanism of action of v-mos
and ras oncogens (2,3), other growth factors such as phorbol
esters, and external oxidants (4).

RESULTS

Tables I to III will show the accumulated data on the cause-
effect relationships among pH abnormalities and cancer, ranging from basic to clinical levels, in an attempt to integrate
the understanding of the many effects of a low microenvironmental hydrogen ion concentration ($\downarrow$[H+] or $\uparrow$pH) on a wide
array of biological systems.

DISCUSSION

Amiloride (Modamide, France), an inhibitor od the Na+/H+ antiport system, by inducing a dose-related intracellular acidification, has been reported to suppress oncogen ras p21 activity(2).
Growth and dissemination of breast and colon cancer have been
reported to be related to the expression and activity of this
oncogen (5). It is concluded that the clinical use of Amiloride
and/or related compounds which act as tumor cell acidifiers
(bioflavonoids- quercetin?) may represent a most rational and

Table I. Direct biological consequences of pH elevation
on basic cell behavior (ref.1).

. Increase in cell detachment and disaggregation
. Increase in cell migration
. Decrease in contact inhibition
. Stimulation of DNA synthesis and replication
. Mitogenic effect on resting cells
. Stimulation of cell division and multiplication
. Detachment of membrane protein
. Increase in trophoblastic invasion
. Increase in cell motility
. Increase in cell permeability
. Abnormal mitochondrial swelling
. Normal mithocondria behaves as tumoral mitochondria
. Normal liver cells behaves as malignant cells in alkaline
  conditions regarding their dependence and correlation of
  pH and aerobic glycolysis
. Stimulation of lymphocyte mitogenesis
. Change in differentiation pathways
. Common final pathway mediating the activity of oncogen ras
  p21 and other growth factors
. Low pH stimulates cell proliferation and tumor cell
  formation (?)

Table II. Effects of high pH on cell intermediary metabolism and
parallelism of effects among hypoxia and alkalosis.

A) EFFECTS OF DECREASED $[H+]$ (ALKALOSIS) ON CELL METABOLISM

      . Acceleration of glycolyisis
      . Activation of phosphofructokinase
      . Increase in pyruvate production
      . Increase in lactate production (aerobic glycolysis)
      . Decrease in pyruvate oxidation
      . Increase in Na+/K+ ATPase and ATP hydrolysis

B) EFFECTS OF O2 DEFICIENCY (HYPOXIA) ON CELL METABOLISM
      . Acceleration of glycolysis
      . Activation of phosphofructokinase
      . Increase in pyruvate and lactate production (anaerobic
        glycolysis)
      . Decrease in mitochondrial oxidation
      . Decrease in ATP production

Table III. Direct high pH-mediated carcinogenesis:

clinical situations (ref. 1)

CANCER LOCATION	BASIS OF pH ROLE
OROPHARYNGEAL	High pH betel quid tobacco
ESOPHAGEAL	HCL helps to prevent cancer in Plummer-Vinson syndrome.
STOMACH	Pernicious anemia Alkaline duodenal reflux Gastric atrophy with aclorhydria Biliary salt reflux
PANCREAS	Increased bicarbonate elimination in the duodenum
COLON	Elevated pH from diet Postureterosigmoidostomy Villous adenoma
BLADDER	High pH of urine plus carcinogens

promising avenue in the adjuvant treatment of certain human tumors. A preliminary protocol in this approach has recently been implemented in a few institutions. Furthermore, these concepts make more rational the attempts to use nitrosamine scavengers, such as vitamin E and ascorbic acid, in epithelial cancer prevention, at least in the high pH-high risk situations of the gastrointestinal tract (1).   These considerations and this approach also emphasize important aspects of malignant growth as a process of a dynamic and intimate acid-base nature.

REFERENCES

1) Harguindey S., LM Antón Aparicio. Integrated etiopathogenesis of cancer of epithelial surfaces with emphasis in the digestive tract. Neoplasia (Spain) 4:288-294, 1987.

2) Hagag N, Lacal JC, Graeber M, Aaronson S, Viola MV. Microinjection of ras p21 induces a rapid rise in intracellular pH. Mol. Cell Biol. 7:1984-1988, 1987.

3)Doppler W, Jaggi R, Groner B. Induction of v-mos and activated Ha-ras expression in quiescent NIH 3t3 cells causes intracellular alkalinisation and cell-cycle progression. Gene 54: 147-153, 1987.

4) Sun Il, Garcia-Cañero R, Liu W, Toole-Sims W, Crane FL, Mowe DK, Low H. Diferric transferrin reduction stimulates the Na+/H+ antiport of HeLa cells. Biochem. Biophys. Res. Commun. 145:467-473, 1987.

5) Lundy J, Grimson R, Mishirki Y, Chao S, Oravez S, Fromowitz F, Viola MV. Elevated ras oncogen expression correlates with lymph node metastases in breast cancer patients. J. Clin. Oncol. 4:1321-1325, 1986.

CONTROL OF SODIUM/PROTON EXCHANGE BY PLASMA MEMBRANE ELECTRON TRANSPORT

W. Toole-Simms, I.L. Sun, F.L. Crane,
D.J. Morré and H. Löw

Departments of Biological Sciences and
Medicinal Chemistry Purdue University,
West Lafayette, Indiana, USA and
Endocrinology Department, Karolinska Institute,
Stockholm, Sweden

External electron acceptors activate proton release from many animal cell types. Ferricyanide induced proton release has been described [1,2,3]. Ferric ions loosly associated with diferric transferrin are even more efficient in stimulation of proton release than is ferricyanide, in that they give a higher ratio of protons released to electrons transferred[4]. Release of up to 100 protons per electron transferred has been observed. Inhibitors of plasma membrane electron transport such as adriamycin or bleomycin inhibit proton release[5,6]. Since the ratio of protons released to ferric transferrin reduced greatly exceeds 1 or 2, which would be expected of a redox coupled proton pump, proton transfer through a redox activated channel is most likely. Following studies by R. Garcia-Cañero, J. Diaz-Gil and M. Guerra which presented evidence that the oxidant induced proton release is dependent on the Na^+/H^+ exchange system, we find that ferric ion induced proton release with pineal or HeLa cells is dependent on external Na^+ ions and is inhibited by amiloride and amiloride analogs. Apotransferrin does not induce proton release. Inhibition of ferric ion reduction by apotransferrin or monoclonal antibodies to the transferrin receptor (B3/25, GB16) prevents induction of proton release, which indicates that the transferrin receptor is involved in the maximum redox activation of the Na^+/H^+ exchange. Ferricyanide reduction is less dependent on the transferrin receptor and is less efficient in induction of proton release. We propose that oxidant of cytosolic NADH by the transmembrane dehydrogenase can provide a proton for activation of the Na^+/H^+ antiport[7] and that localized proton movement may be favored by the presence of the transferrin receptor. Activation of proton release by impermeable oxidant has been observed with 3T3, SV40 transformed 3T3, rat liver, BK, HL60 and human erythroleukemia cells in addition to HeLa and rat pineal cells.

REFERENCES

1. T.L. Dormandy and Z. Zarday, J. Physiol. 180, 684 (1965)
2. I.L. Sun, F.L. Crane, C. Grebing and H. Löw,
   J. Bioenerg. Biomemb. 18, 583 (1984)
3. I.L. Sun, P. Navas, F.L. Crane, J.Y. Chou and H. Löw,
   J. Bioenerg. Biomemb. 18, 471 (1986)
4. I.L. Sun, W. Toole-Simms, F.L. Crane, D.J. Morré, H. Löw
   and J.Y. Chou, Biochim. Biophys. Acta 938, 17 (1988)
5. I.L. Sun and F.L. Crane. Proc. Indiana Acad. Sci. 93,
   267 (1984)
6. I.L. Sun and F.L. Crane, Biochem. Pharmacol. 34, 617
   (1985)
7. S. Grinstein and A. Rothstein, J. Memb. Biol. 90, 1
   (1986).

NITRATE UPTAKE BY INTACT SUNFLOWER PLANTS

E. Agüera, P. de la Haba, A.G. Fontes and J.M. Maldonado

Departamento de Biología Vegetal y Ecología
Facultad de Ciencias, Universidad de Córdoba
14004 Córdoba, Spain

Nitrate utilization by higher plants involves nitrate uptake from the medium, its reduction to ammonia, either in the roots or in the leaves, and incorporation of the generated ammonia into carbon skeletons for the synthesis of amino acids and other organic nitrogen compounds.

In this study, we have examined the uptake of nitrate by intact sunflower plants in relation to the activity of nitrate reductase, the enzyme which reduces nitrate to nitrite. Net nitrate uptake has been determined from nitrate depletion of external solution. In nitrate-deprived plants, nitrate uptake shows a lag period of 2 to 4 hours. Nevertheless, treatment of the plants with nitrate for 24 hours prior uptake experiments results in maximal rates of nitrate uptake from the beginning. Nitrate uptake does not take place when oxygen is removed from the solution. Plot of the rate of nitrate uptake _versus_ different external nitrate concentrations shows saturation kinetic without linear (diffusion) component. These results indicate that nitrate uptake is an active-carrier mediated process. From the double-reciprocal plot a Km of 32 μM for nitrate uptake is obtained. The optimum pH lies between 5 and 6. Pretreatment of the plants with cyclo-heximide almost completely abolishes nitrate uptake, presumably because the nitrate-induced synthesis of the carrier is inhibited. N-ethylmaleim-ide (NEM), a sulfydryl-group reagent, drastically inhibits nitrate uptake. Nitrate reductase activity in the roots is also negatively affected both by cycloheximide and NEM pretreatments. Addition of ammonia into the external solution also promotes impairment of nitrate uptake and inactivation of nitrate reductase. Methionine sulfoximine (MSO), an inhibitor of glutamine synthetase activity that in consequence blocks ammonia assimilation, enhances the negative effect of ammonia both on nitrate uptake and root nitrate reductase activity. Addition of ammonia and MSO to the plants causes an increase in the nitrate and ammonia contents of the root tissue.

The above results suggest that, in intact sunflower plants, nitrate uptake is controlled by the rate of nitrate reduction in the roots. When nitrate reduction in the root cells is decreased nitrate concentration in the cytoplasm rises and, as a consequence, net nitrate uptake is depressed. Internal nitrate would therefore exert a negative feed back control on its own uptake system.

Research supported by grant no. 2149-83 from CAICYT (Spain)

METABOLIC CHANGES ASSOCIATED WITH FERRICYANIDE REDUCTION

BY ELODEA LEAVES

V. Trockner and E. Marrè

Department of Biology
University of Milano
Via Celoria 26, 20133 Milano, Italy

It has been shown that ferricyanide reduction by <u>Elodea densa</u> leaves, in the dark, is associated with a decrease of intracellular pH, a marked depolarization of the transmembrane potential and a net efflux of H^+ and K^+. The sum of $(H^+ + K^+)$ extruded is approximately equal to the amount of $Fe(CN)_6^{3-}$ reduced. Fusicoccin (FC) increases both $Fe(CN)_6^{3-}$ reduction and the associated efflux of H^+ and decreases K^+ efflux. This effect is interpreted as a consequence of the hyperpolarizing, cytoplasm alkalinizing action of FC, due to its capacity to activate the ATP-driven H^+ pump. The results of a preliminary analysis of the metabolic changes associated with these responses of transport mechanisms to either FC, or $Fe(CN)_6^{3-}$, or both in combination, can be summarized as follows.

I) <u>Respiration and ATP level</u>. FC as well as $Fe(CN)_6^{3-}$, fed either separately or in combination, significantly increase O_2 uptake and depress the ATP level. The increase in O_2 uptake is satisfactorily accounted by the energy requirement of the H^+ extruding mechanism, if a P/O ratio of 3 and an ATP/H^+ extruded ratio of 1 are assumed. Thus the increase of O_2 uptake is interpreted as due to the decrease of the energy charge of the adenylate system.

II) <u>Malate and glucose-6-phosphate</u>. In the absence of $Fe(CN)_6^{3-}$, FC induces a marked increase in the levels of malate and glucose-6-phosphate, while $Fe(CN)_6^{3-}$ alone scarcely influences these two substrates. In contrast, when FC and $Fe(CN)_6^{3-}$ are fed in combination, $Fe(CN)_6^{3-}$ efficiently counteracts the FC-induced increases in malate and, even more strongly, in glucose-6-phosphate. Parallel experiments in which the fate of C-1 and C-2 of ^{14}C-labelled glucose has been followed show that $Fe(CN)_6^{3-}$ reduction is associated with a marked decrease of the C-1/C-2 incorporation ratio into ribose (RNA ribose). These data suggest that $Fe(CN)_6^{3-}$ reduction utilizes as a source of reducing power the oxidative pentose phosphate pathway, with NADPH as the electron donor to the plasma membrane redox chain.

STIMULATION OF C-FOS AND C-MYC PROTO-ONCOGENE EXPRESSION BY FERRICYANIDE, AN EXTRACELLULAR ELECTRON ACCEPTOR

Anthony F. Cutry, Alan J. Kinniburgh[*] and Charles E. Wenner

Departments of Experimental Biology and [*]Human Genetics
Roswell Park Memorial Institute
Buffalo, New York 14263   USA

One of the earliest consequences of mitogenic stimulation of mammalian cells is the activation of a set of "immediate early genes," two of which are the proto-oncogenes c-fos and c-myc. These two genes are the cellular homologs of the FBR and FBJ murine osteogenic sarcoma viruses and the avian myelocytomatosis virus, respectively. Both have been shown to be activated by a variety of agents which elicit a mitogenic response, such as EGF, PDGF and bombesin. Given their role in cell growth (and that of c-fos in differentiation as well), it is of interest to identify the components of the signal transducing pathways responsible for the activation of these genes.

Previous work has indicated that we can achieve a small mitogenic response in C3H 10T1/2 mouse embryo fibroblasts (as measured by radiolabelled deoxythymidine incorporation into the acid insoluble cell fraction) by simply adding ferricyanide to the media at micromolar concentrations. Subsequently, we determined whether this impermeable electron acceptor is capable of stimulating increases in c-fos and c-myc mRNAs. At 100 uM ferricyanide, we observe small increases in c-fos and c-myc mRNA levels, approximately 3.5 and 2.5-fold respectively. We believe that ferricyanide is eliciting this response via a plasma membrane associated charge separation with subsequent alkalinization of the cytosol. Some reports (Malviya and Anglard, 1986) have indicated that the action of ferricyanide is amiloride sensitive, an indicator that sodium/proton antiport activity may be involved. Given the proposed role of the antiport in growth factor mediated signal generation, ferricyanide, since its effects are independent of receptor binding, may be a useful tool in analyzing the role of the sodium/proton antiport system in mitogenic signalling.

References

Malviya, A.N. and Anglard, P. (1986). FEBS Lett. 200, 265-270.

EFFEECTS OF PMA ON PROTEIN KINASE C IN PERFUSED RAT HEART

A.G.M. Davidson, P.J.F. Cleland, S. Rattigan
and M.G. Clark

Department of Biochemistry, University of
Tasmania, Hobart, Australia

Protein kinase C (PKC), the phospholipid, Ca^{2+}-dependent protein kinase has been implicated in cardiac hypertrophy (Simpson et al. J. Molec. Cell. Cardiol. 18, suppl. 5, 45-58, 1985) and more recently in cardiac glucose transport (Van de Werve et al. Diabetes 36, 310-314, 1987). In addition α_1 adrenergic agonists as well as muscarinic agonists increase phosphatidylinositol (PI) turnover in heart thus implying a role for PKC as part of the signal transduction processes. α-Adrenergic agonists also activate cardiac trans-sarcolemma electron efflux (Löw et al. Biochim. Biophys. Acta 844, 142-148, 1985). Since neither PI-turnover nor electron efflux is affected by β-agonists it appeared possible that PKC activation and electron efflux were uniquely involved in α-adrenergic receptor activation and were causally related. In the present study three questions were addressed: - (i) α_1-adrenergic receptor activation associated with PKC activation?; (ii) what is the effect of treatment of the perfused rat heart with tumour-promoting phorbol ester (PMA), a putative activator of PKC?; and (iii) what is the relationship between PKC and trans-sarcolemma electron efflux?

The system used was the Langendorff perfused rat heart. Perfusions were conducted at $37^{o}C$ and employed Krebs Henseleit bicarbonate buffer containing 5 mM glucose, 1.27 mM Ca^{2+} and 0.05 mM EDTA. Hearts were homogenized in Ca^{2+}/EGTA buffers to set the free [Ca^{2+}] at 230 nM. Protein kinase activity was determined following anion exchange chromatography on DE-52 of cytosol and either 0.2% or 2% Triton X-100 extracts of the 100,000 g particulate material using histone IIIs as substrate.

Initial experiments failed to show α_1 adrenergic-mediated activation of PKC as judged by the failure of 1 μM epinephrine + 10 μM DL propranolol to cause translocation from cytosol to membrane fractions. Activation of PKC was masked by a Ca^{2+}-induced translocation that occurred during homogenization such that particulate PKC changed from 33.5% to 78% over the range of 20 nM to 2μM free Ca^{2+}. Treatment of the perfused heart with PMA and extraction wiht 0.2% Triton X-100, sufficient to fully extract PKC activity from control hearts, resulted in a dose-dependent apparent loss of total PKC ($K_{0.5}$ = 6 nM) that occurred rapidly ($t_{0.5}$ = 1.5 min at 200 nM) and was not mimicked by non-tumour promoting phorbol ester. However extraction of the particulate material with 2% Triton X-100 showed that PMA caused a tighter binding of PKC and not a loss of activity. PMA did not stimulate trans-sarcolemma electron efflux but the 60% apparent loss of total PKC activity coincided with the total loss of α_1 adrenergic stimulation of electron efflux and inotropy. PKC would appear to be involved in α_1 adrenergic control of contractility but not causally related to sarcolemmal electron efflux.

ROLE OF ZINC IN PLASMA MEMBRANE INTEGRITY

AND O_2^{-} -GENERATING NADPH OXIDASE ACTIVITY

I. Cakmak and H. Marschner

Dept. Plant Nutrition Univ. Cukurova Adana,Turkey and
Dept. Plant Nutrition Univ. Hohenheim Stuttgart, FRG

The NADPH- dependent superoxide radical (O_2^{-}) generation and plasma membrane permeability were studied in roots of cotton (<u>Gossypium</u> <u>hirsutum</u> c.) plants grown in nutrient solution with different Zn concentrations. Zinc deficiency increased plasma membrane permeability indicated by efflux of K^+, NO_3^-, amino acids, sugars and phenolics. Resupply of Zn to deficient plants for 12 h substantially decreased this leakage (1).

The effects of Zn on membrane permeability were closely correlated in the microsomal membranes and cytosol with the levels of O measured by electron spin resonance (ESR) spectroscopy. The amplitudes of the O_2^{-} derived Tiron ESR-signale also coincided with an O_2^{-} generating oxidase activity which for maximal activity requires a high p^H optimum, a flavin and NADPH, but not NADH. In addition, Zn also inhibited the NADPH oxidation (2,3).

The results indicate that Zn affects integrity of plasma membranes, at least in part, by interfering with O_2^{-} generation during the oxidation of NADPH by a plasma membrane-bound oxidoreductase. This role of Zn may also have ecological implications with respect of nutrient mobilization and microbial activity in the rhizosphere.

References

1. Cakmak, I. and Marschner, H. (1988). J. Plant Physiol. (in press)
2. Cakmak, I. and Marschner, H. (1988). Physiol. Plant. (in press)
3. Cakmak, I. and Marschner, H. (1988). J. Exp. Bot. (submitted)

ANION SUPEROXIDE AND HYDROGEN PEROXIDE FORMATION ELICITED BY NAD(P)H

OXIDATION IN RADISH PLASMALEMMA VESICLES

Francesco Macrì and Angelo Vianello

Sect. of Plant Physiology and Plant Biochemistry
Institute of Plant Protection, University of Udine
I - 33100 UDINE   (ITALY)

Plant plasma membranes contain some redox activities which appear related to some physiological processes (Bienfait 1985, Crane et al. 1985, Lüttge and Clarkson 1985, Møller and Lin 1986).

The aim of this work was to check if NAD(P)H oxidation in radish plasmalemma-enriched vesicles could be related to the formation of oxygen free radicals.

Microsomal vesicles, obtained from 24 h old radish seedlings, were fractionated by a discontinuous sucrose gradient. The fraction at the 35-45% sucrose interface is enriched in the plasma membrane marker vanadate-sensitive, nitrate-insensitive H^+-ATPase. This fraction shows an NAD(P)H-ferricyanide or NAD(P)H-cytochrome c oxidoreductase activity which is insensitive to pH in the range between 4.5 and 8.0. NADH also elicites an O_2 consumption (NADH oxidation), but only at pH 4.5 - 5.0. The NADH-ferricyanide oxidoreduction is stimulated, at pH 7.0, by salicylhydroxamic acid (SHAM), ferulic acid, p-cumaric acid, p-chloromercuribenzoic acid (PCMB), p-chloromercuribenzenesulfonic acid (PCMBS) and mersalyl, whereas it is insensitive to duroquinone. At pH 5.0 this activity is only stimulated by ferulic acid, PCMB, PCMBS and mersalyl. NAD(P)H oxidation is inhibited by SOD and catalase, added either before or after the supply of NAD(P)H. Ferrous ions strongly stimulate NAD(P)H oxidation. Such Fe^{2+}-stimulated activity is also inhibited by SOD and catalase. Hydrogen peroxide does not stimulate NADH oxidation, while it does stimulate Fe^{2+}-induced NADH oxidation. The oxidation induced by NADH is unaffected by SHAM and Mn^{2+}, is slightly stimulated by ferulic acid and inhibited by KCN. In addition, NADH induces the conversion of epinephrine to adrenochrome, indicating that anion superoxide is formed during its oxidation.

These results suggest that radish plasma membranes contain an NAD(P)H ferricyanide or cytochrome c oxidoreductase and an NAD(P)H oxidase active only at pH 4.5 - 5.0. The latter appears able to induce the formation of anion superoxide that is then converted to hydrogen peroxide. Ferrous ions, sparking a Fenton reaction, would stimulate NAD(P)H oxidation.

References

- Bienfait, H.F. (1985). J. Bioenerg. Biomembr. 17, 73-83.
- Crane, F.L., Sun, I.L., Clark, M.G., Grebing, C. and Löw, H. (1985). Biochim. Biophys. Acta 811, 233-264.
- Lüttge, U. and Clarkson, D.T. (1985). Prog. Bot. 47, 73-86.
- Møller, I.M. and Lin, W. (1986). Annu. Rev. Plant Physiol. 37, 309-334.

ASCORBATE FREE RADICAL INVOLVEMENT IN ELONGATION-RELATED GROWTH PROCESSES

IN ONION ROOTS

A. Hidalgo and P. Navas

Departamento de Biología Celular
Facultad de Ciencias, Universidad de Córdoba
E-14004 Córdoba, Spain

Ascorbate free radical (AFR) induced a 30% increase in Allium cepa root growth. Since neither cell cycle nor mitotic index were affected, the nature of this growth stimulation may not be related to cell proliferation. In fact, the increase in cell size induced by AFR in both meristematic and elongating cells indicated that AFR action could be involved in the activation of the elongation-component of root growth.

AFR has been proposed to be involved in the redox processes at the plasma membrane that stimulate proton extrusion to the cell wall (Morré et al., 1986). AFR produced by ascorbate oxidase from ascorbate secreted to the cell surface would be available to function as an electron acceptor for the cytosolic NADH oxidation. Such oxidation would lead to a local cytoplasm acidification that might therefore activate a plasma membrane $|H|$ATPase (Morré et al., 1986; Rubinstein and Stern, 1986). According to the current literature (Lass et al., 1986; Rubinstein and Stern, 1986; Guern et al., 1988; M.T. Marré et al., 1988), the redox system at the plasma membrane, which transfers the electrons from the NADH oxidation to an external electron acceptor, appears not to transfer protons, in fact, protons resulting from this redox activity would be pumped by a $|H|$ATPase of the plasma membrane as we suggest here for the AFR-reductase.

In this regard, a possible role for AFR in elongation growth is discussed. Both plasma membrane- and tonoplast-associated ATPase activities were increased by AFR, the former could explain its role in the plasma membrane-related redox processes while the latter, together with an increase in the vacuole soluble acid phosphatase activity, could be related to the high vacuolization induced by AFR that may lead to elongation after cell wall loosening occurs. Finally, several physiological parameters such as oxygen, nitrate and sugar uptake by onion roots were also stimulated by AFR.

Partially supported by C.I.C.Y.T. grant no. 666/86.

REFERENCES

Guern J., Mathieu Y., Ullrich-Eberius C.I., Marré M.T. and Marré E. (1988). This Workshop. pp., P 13.
Lass B., Thiel G. and Ullrich-Eberius C.I. (1986). Planta 169,251-259.
Marré M.T., Moroni A., Albergoni F.G. and Marré E. (1988). Plant Physiol. 86,000-000.
Morré D.J., Navas P., Penel C. and Castillo F.J. (1986). Protoplasma 133,195-197.
Rubinstein B. and Stern A.I. (1986). Plant Physiol. 80,805-811.

ROLE OF CELL SURFACE ON NADH-MDHA-OXIDOREDUCTASE: LECTIN INHIBITION

P. Navas, A. Estevez, J.M. Villalba and M.I. Burón

Departamento de Biología Celular
Facultad de Ciencias, Universidad de Córdoba
E-14004 Córdoba, Spain

Ascorbic acid is a bivalent oxidation reduction molecule that acts via a free radical intermediate, monodehydroascorbate (MDHA). MDHA has been described as an electron acceptor for NADH oxidation (Crane et al., 1985) by the NADH-MDHA-oxidoreductase activity in plasma membranes (Goldenberg, 1980) and in the endomembranes system (Sun et al., 1983).

This activity has been localized in isolated plasma membrane from rat liver. These fractions were obtained by aqueous two-phase partition (Löw et al., 1986) and a 90% purity was obtained being 40-fold enriched in K^+-stimulated ouabain-inhibited, p-nitrophenyl phosphatase relative to the total homogenate.

Rat liver plasma membrane is able to oxidize NADH using MDHA as electron acceptor at a rate of 4.2 ± 0.3 nmoles/min.mg protein showing a Km for MDHA of 2.5 mM and of 2.0 mM for NADH. Con A, WGA and LFA lectins inhibit about 81%, 95% and 97% respectively this activity with Ki of 0.2, 0.1 and 0.05 ng/ml.

When plasma membrane fractions were digested with trypsin (1%) or neuraminidase (1 IU/ml), NADH-MDHA-oxidoreductase activity was 48% or 64% inhibited respectively but addition of LFA lectin did not show any additional inhibition. It is apparently due to the absence of LFA binding residues in pretreated membranes suggesting a significative relationship between this enzyme activity and the integrity of cell surface.

NADH-MDHA-oxidoreductase might be carried out by the transmembrane redox system as NADH has been demonstrated to be a natural electron donor for the transplasma membrane dehydrogenase (Navas et al., 1986). Therefore, ascorbate free radical emerges as another possible electron acceptor for a plasma membrane dehydrogenase apparently as part of the transmembrane redox system.

Partially supported by C.I.C.Y.T. grant no. 666/86.

REFERENCES

Crane F.L., Löw H. and Clark M.G. (1985) in: Enzymes of Biological membranes. Ed. Martonosi A.N., Plenum, New York, Vol.4, pp. 465-510.

Goldenberg H. (1980) Biochem. Biophys. Res. Commun. 94, 721-726.

Löw H., Sun I.L., Navas P., Grebing C., Crane F.L. and Morré D.J. (1986) Biochem. Biophys. Res. Commun. 139, 1117-1123.

Navas P., Sun I.L., Morré D.J. and Crane F.L. (1986) Biochem. Biophys. Res. Commun. 135, 110-115.

Sun I.L., Crane F.L. and Morré D.J. (1983) Biochem. Biophys. Res. Commun. 115, 952-957

IRON REDUCTION BY SUNFLOWER ROOTS UNDER IRON STRESS

M.D. de la Guardia, E.Alcántara and M. Fernández

Dpto. de Agronomía. Escuela de Ingenieros Agrónomos
Universidad de Córdoba
Córdoba, Spain

Under iron deficiency stress several dicotyledonous species (sun-flower, bean, cucumber, etc.) develop in their roots morphological and physiological changes in order to increase iron uptake: root tips swelling, increase in the ferric reducing capacity (RC), proton efflux, and release of phenolic compounds. In this work we present the response to iron stress of a Fe-efficient inbred line (RHA 274) and the effect of an inhibitor of polar auxin transport 2,3,5-triiodobenzoic acid (TIBA) on the iron stress response.

The pH of external solution of plants 12 to 17 days old held without iron decreased and the RC increased, until the 5th day, after that the RC decreased sharply. However, if pH of external solution was controlled with MES buffer the RC was held for a longer period. When iron was withheld from bathing solution during the RC determination, the H^+ excretion decreased, that indicate the existence of H^+ efflux related with Fe(III) reduction by the oxidoreductase. Reduction of exogenous ferric ion and an associated apparent net proton efflux have been demostrated with corn and bean roots (2,3). Landsberg (1984) has suggested that auxin coming down from apices and young leaves to roots is involved in the Fe-stress response. To test this hypothesis TIBA (50 ppm) has been used in the nutrient solution to study its effect on the process. TIBA produced the following effects: a) decreased root tips swelling and lateral roots production, b) decreased drastically the reducing capacity, c) changed the pattern of lowering the external solution pH producing an early H^+ efflux, but after 4 days the total H^+ efflux was lower in TIBA than in the Fe stress control plants. These results support the idea that auxin play a role in the Fe stress response.

Supported by CAICYT, Project 2011/83

REFERENCES

1. Landsberg,E.C. (1984) J. Plant Nutr. 7(1-5), 609-621.
2. Rubinstein,B. and A.I.Stern. (1986). Plant Physiol. 80, 805-811.
3. Sijmons,P.C., F.C.Lanfermeijer, A.H.de Boer, M.B.A.Prins, and H.F.Bienfait. (1984). Plant Physiol. 76, 943-946.

IRON-STRESS IN TOMATO (*Lycopersicum esculentum* cv. Rutgers) RESULTS IN INCREASED IRON REDUCTION ON THE ROOT PLASMA MEMBRANE

Douglas G. Luster and Thomas J. Buckhout

Plant Photobiology Laboratory, Beltsville Agricultural
Research Center, Agricultural Research Service, Beltsville, MD
20705, USA

Roots of tomato plants were grown for 21 days in complete hydroponic nutrient solution including Fe^{3+}-chelate EDDHA and subsequently switched to nutrient solution withholding Fe for 4 to 5 days to induce iron stress. The iron-stressed plants reduced chelated Fe (measured as ferrozine complex) at rates 7-fold higher than those of roots of plants grown in Fe-sufficient conditions. The response in Fe-deficient plants was localized to root hairs which developed on secondary roots during the period of Fe stress.

Plasma membranes (PM) isolated by aqueous 2-phase partitioning from secondary roots of tomato plants under Fe stress exhibited a 93% increase in rates of NADH-dependent Fe-citrate reduction (measured as bathophenanthrolinedisulfonate complex) compared to PM isolated from roots of Fe-sufficient plants. Detection of the reductase activity required the presence of detergent indicating structural latency. In contrast, NADPH-dependent Fe-citrate reduction was not significantly different in root PM isolated from Fe-deficient *vs* Fe-sufficient plants and proceeded at substantially lower rates than NADH-dependent Fe-citrate reduction. Mg^{2+}-ATPase activity was increased 22% in PM from roots of Fe-deficient plants compared to PM isolated from roots of Fe-sufficient plants. Two-dimensional electrophoretic analysis revealed that at least two polypeptides had dramatically increased incorporation of ^{35}S-methionine between control plants and those which had been Fe stressed for one day. The results demonstrate that induction of Fe reduction in roots of tomato plants under Fe stress is at least partially localized to the plasma membrane.

Author Index

434

Dicoumarol, 402
Dicyclohexylcarbodiimide, 120, 248, 261, 265, 410
DIDS, 120
Diethylenedithiocarbanate, 24
Diethylene triamine pentaacetate, 20, 25, 142, 147
Diethylstilbesterol, 76, 236, 248, 260, 413
Diferric transferrin, 29, 124
  effect on NADH level, 341-342
  extra iron content, 40-41, 156
  in growth stimulation, 29, 40-41, 339
Diferric transferrin reduction, 38, 127, 154-157, 159, 178, 183, 315
  adriamycin inhibition, 183
  anthracycline inhibition, 183
  antitumor drug inhibition, 186
  effect of pH, 160
  by HL60 cells, 315
  by HL60 mutant, 315
  by K562 cells, 145
  retinoic acid inhibition, 315
  in SV40 transformed cells, 185
Differentiation, 313-317, 423
Dihydroorotate dehydrogenase, 7
Dihydroteleleocidin, 8-11
Dimethyl amiloride, 371, 376
Dimethyl sulfoxide, 314, 316
Dithiothreitol, 120
Divalent cations, 130, 353
DNA synthesis, 7, 9-12, 29, 299
  antimycin A effect on, 11
  with iron compounds, 40
  requirement for oxygen, 7, 9
  rotenone effect, 10
  transferrin receptor control, 167
Dopamine, 24, 406
Dopaminergic agents, 407
Doxorubicin, see adriamycin
DT-Diaphorase, 402
Duroquinone, 62, 278-279, 396, 427

E_0', see oxidoreduction potential
e^-/H^+ efflux, 235, 246-247, 253, 257-261, 410, 419, 422
  in iron deficiency, 410
  with transferrin, 419
Electrochemical assay, 71, 401
Electrophoretic K^+ uptake, 222
Electron transport, 192-193
  carriers, 192
  in mitochondria, 193
Elongation growth, 48-49, 428
EGF, 14-15, 165, 378, 423
EGTA, 200, 314, 376
Endocytosis, 115, 117, 123, 127, 139
Endoplasmic reticulum, 404

Endosomes, 116-117, 119, 135
Energy, 197
  availability, 7, 12, 324
  control of, 283
  coupling, 193, 201
  respiratory, 11
Energy uncouplers, effect on iron uptake, 129
Epinephrine, 424
Erythrocyte membranes, 1-2, 403
  and cytochrome b_5, 2
Erythrosine B, 234-236, 261, 412-413
ESR, 62, 406, 426
Ethanol, 106-108, 403
Ethylene, 328-329

FER gene, 90
Ferric ammonium citrate, 38, 154-157, 159-160
Ferric citrate, 431
Ferricoordination complexes, 20-25, 29
  in growth stimulation, 25, 29, 46
  in peroxide removal, 20-25
Ferric reduction, 1-2, 38, 90, 127, 154-156, 430-431
  FER gene effect, 90
  heme inhibition, 116
  in iron deficiency, 1-2, 90, 430-431
  Km for, 159
  pH effect on, 160
  in tomato mutant, 90
  triiodobenzoate inhibition, 430
Ferricyanide, 2, 13
  activation of adenylate cyclase by, 407
  and ATP levels, 237
  and c-fos and c-myc RNA, 13, 423
  and cytoplasmic acidification, 412
  and cytoplasmic streaming, 268
  and growth, 2, 18, 29, 39
  inhibition of iron uptake, 129, 145
  inhibition of transferrin iron reduction, 145
  and metabolism, 422
  and NADH level, 342
  in peroxide removal, 20
  in proton release, 30, 72-73, 239, 246, 254, 266
  and protooncogenes, 423
  in thiol oxidation, 21
  with TPA, 14, 33
Ferricyanide reductase, 27, 46
  activation by mitogens, 31, 51
  with age of cells, 74
  inhibition, 32
  effect of ions on, 31, 33